D0914559

MTP International Review of Science

Reaction Mechanisms in Inorganic Chemistry

MTP International Review of Science

Publisher's Note

The MTP International Review of Science is an important new venture in scientific publishing, which we present in association with MTP Medical and Technical Publishing Co. Ltd. and University Park Press, Baltimore. The basic concept of the Review is to provide regular authoritative reviews of entire disciplines. We are starting with chemistry because the problems of literature survey are probably more acute in this subject than in any other. As a matter of policy, the authorship of the MTP Review of Chemistry is international and distinguished; the subject coverage is extensive, systematic and critical; and most important of all, new issues of the Review will be published every two years.

In the MTP Review of Chemistry (Series One), Inorganic, Physical and Organic Chemistry are comprehensively reviewed in 33 text volumes and 3 index volumes, details of which are shown opposite. In general, the reviews cover the period 1967 to 1971. In 1974, it is planned to issue the MTP Review of Chemistry (Series Two), consisting of a similar set of volumes covering the period 1971 to 1973. Series Three is planned for 1976, and so on.

The MTP Review of Chemistry has been conceived within a carefully organised editorial framework. The over-all plan was drawn up, and the volume editors were appointed, by three consultant editors. In turn, each volume editor planned the coverage of his field and appointed authors to write on subjects which were within the area of their own research experience. No geographical restriction was imposed. Hence, the 300 or so contributions to the MTP Review of Chemistry come from many countries of the world and provide an authoritative account of progress in chemistry.

To facilitate rapid production, individual volumes do not have an index. Instead, each chapter has been prefaced with a detailed list of contents, and an index to the 10 volumes of the MTP Review of Inorganic Chemistry (Series One) will appear, as a separate volume, after publication of the final volume. Similar arrangements will apply to the MTP Review of Physical Chemistry (Series One) and to subsequent series.

Butterworth & Co. (Publishers) Ltd.

Inorganic Chemistry
Series One
Consultant Editor
H. J. Emeléus, F.R.S.
Department of Chemistry
University of Cambridge

Volume titles and Editors

Physical Chemistry
Series One
Consultant Editor
A. D. Buckingham
Department of Chemistry
University of Cambridge

Volume titles and Editors

1 **THEORETICAL CHEMISTRY**
Professor W. Byers Brown, *University of Manchester*

2 **MOLECULAR STRUCTURE AND PROPERTIES**
Professor G. Allen, *University of Manchester*

3 **SPECTROSCOPY**
Dr. D. A. Ramsay, *National Research Council of Canada*

4 **MAGNETIC RESONANCE**
Professor C. A. McDowell, *University of British Columbia*

5 **MASS SPECTROMETRY**
Professor A. Maccoll, *University College, University of London*

6 **ELECTROCHEMISTRY**
Professor J. O'M Bockris, *University of Pennsylvania,*

7 **SURFACE CHEMISTRY AND COLLOIDS**
Professor M. Kerker, *Clarkson College of Technology, New York*

8 **MACROMOLECULAR SCIENCE**
Professor C. E. H. Bawn, F.R.S., *University of Liverpool*

9 **KINETICS**
Professor J. C. Polanyi, F.R.S., *University of Toronto*

10 **THERMOCHEMISTRY AND THERMODYNAMICS**
Dr. H. A. Skinner, *University of Manchester*

11 **CHEMICAL CRYSTALLOGRAPHY**
Professor J. Monteath Robertson F.R.S., *University of Glasgow*

12 **ANALYTICAL CHEMISTRY–PART 1**
Professor T. S. West, *Imperial College, University of London*

13 **ANALYTICAL CHEMISTRY – PART 2**
Professor T. S. West, *Imperial College, University of London*

INDEX VOLUME

Organic Chemistry
Series One
Consultant Editor
D. H. Hey, F.R.S.
Department of Chemistry
King's College, University of London

Volume titles and Editors

1 **STRUCTURE DETERMINATION IN ORGANIC CHEMISTRY**
Professor W. D. Ollis, *University of Sheffield*

2 **ALIPHATIC COMPOUNDS**
Professor N. B. Chapman, *Duke University, North Carolina*

3 **AROMATIC COMPOUNDS**
Professor H. Zollinger, *Swiss Federal Institute of Technology*

4 **HETEROCYCLIC COMPOUNDS**
Dr. K. Schofield, *University of Exeter*

5 **ALICYCLIC COMPOUNDS**
Professor W. Parker, *University of Stirling*

6 **AMINO ACIDS AND PEPTIDES**
Editor to be announced

7 **CARBOHYDRATES**
Professor G. O. Aspinall, *University of Trent, Ontario*

8 **STEROIDS**
Dr. W. D. Johns, *G. D. Searle & Co., Chicago*

9 **ALKALOIDS**
Professor K. Wiesner, *University of New Brunswick*

10 **FREE RADICAL REACTIONS**
Professor W. A. Waters, F.R.S., *University of Oxford*

INDEX VOLUME

Inorganic Chemistry
Series One

Consultant Editor
H. J. Emeléus, F.R.S.

MTP International Review of Science

Volume 9

Reaction Mechanisms in Inorganic Chemistry

Edited by **M. L. Tobe**
University of London

Butterworths · London
University Park Press · Baltimore

THE BUTTERWORTH GROUP

ENGLAND
Butterworth & Co (Publishers) Ltd
London: 88 Kingsway, WC2B 6AB

AUSTRALIA
Butterworth & Co (Australia) Ltd
Sydney: 586 Pacific Highway 2067
Melbourne: 343 Little Collins Street, 3000
Brisbane: 240 Queen Street, 4000

NEW ZEALAND
Butterworth & Co (New Zealand) Ltd
Wellington: 26–28 Waring Taylor Street, 1

SOUTH AFRICA
Butterworth & Co (South Africa) (Pty) Ltd
Durban: 152–154 Gale Street

ISBN 0 408 70247 8

UNIVERSITY PARK PRESS

U.S.A. and CANADA
University Park Press Inc
Chamber of Commerce Building
Baltimore, Maryland, 21202

Library of Congress Cataloging in Publication Data
Tobe, M L
Reaction mechanisms in inorganic chemistry
(Inorganic chemistry, series one, v. 9) (MTP international review of science)
Includes bibliographies
1. Chemistry, Inorganic. 2. Chemical reaction – Conditions and laws. I. Title
QD151.2.15 vol. 9 [QD475] 541'.39 74-160331
ISBN 0-8391-1012-X

Filmset by Photoprint Plates Ltd., Rayleigh, Essex
Printed in England by Redwood Press Ltd., Trowbridge, Wilts
and bound by R. J. Acford Ltd., Chichester, Sussex

Consultant Editor's Note

The problem of keeping abreast of research literature on as broad a front as possible is one that confronts all chemists. In the past this difficulty has been met, in the main, by literature surveys and by several uncorrelated reviews of progress in certain subject areas. There are obvious inadequacies in this approach, which have become increasingly apparent in recent years. I was, therefore, grateful for the opportunity of helping to plan this new series, which has been designed to provide a comprehensive, critical survey of each of the main branches of chemistry.

This section of the MTP International Review of Science deals with progress in Inorganic Chemistry. The subject is developing at an astonishing rate and in many directions. Fortunately, however, it lends itself to a systematic treatment. Ten volumes have been prepared, three dealing with the main group elements and two with the general chemistry of the transition metals. Organometallic derivatives of the main group elements and lanthanides and actinides are covered separately, as is the subject of reaction mechanisms. The two remaining volumes on radiochemistry and solid state chemistry have been planned to avoid, as far as possible, overlap with those that have gone before.

It is a pleasure to thank the many experts who have collaborated as authors and volume editors in making this publication possible. While working to a pre-arranged over-all plan, they have been able to assess and interpret the literature in terms of their own experience in specialised fields. I believe that in this way they will not only provide a record of what has been done, but will stimulate further exploration in this fascinating branch of chemistry.

Cambridge

H. J. Emeléus

Preface

The enormous expansion of interest in inorganic reaction mechanisms that has occurred in the past decade has paralleled very closely the development of new reactions and the availability of new techniques. In the past five years—the period covered by this set of reviews—the main developments have come from the ready availability (provided the money is to hand) of commercially designed equipment for the studies of fast reactions and for the study of nuclear magnetic resonance. Many people are now building upon the sound foundation laid by Eigen and the very few other pioneers of fast reaction kinetics, and the areas amenable to study have spread from a narrow range of slow substitution and redox reactions to cover virtually the whole of the periodic table. It is likely that this area of chemistry will be thoroughly consolidated in the next few years. Nuclear magnetic resonance has not only provided the means to follow reactions that previously were not amenable to kinetic study, but it has also shown us reactions which were hitherto unrecognised. A great deal more will be heard about stereochemical change, non-rigid molecules and fluxional behaviour in the next few years.

In subdividing this volume I have been very much aware of the fact that the subject has reached a watershed but I have not been brave enough to back my own judgement of what lies on the other side and have kept to a fairly conservative set of titles. The next issue of this volume should most surely have a significantly different emphasis. The major growth areas will be in the field of stereochemical change and reactions involving both metal and ligand. Genuine mechanistic studies of oxidative addition are only just starting to be published. The major interest in reactions of co-ordinated ligands is now starting to pay dividends and our understanding of the complicated mechanisms of catalysis in preparative and biological systems is rapidly advancing. It will soon be recognised that, in the growth areas, division in terms of organic and inorganic mechanism is not only artificial but downright inconvenient since it often causes the wrong questions to be asked and the right answers to remain unrecognised.

London M. L. Tobe

Contents

1
The Stability of Complexes in Solution

M. T. BECK
Kossuth Lajos University, Debrecen

1.1 INTRODUCTION

An enormous number of stability data were determined in the period 1950–1965, a period of 'gold fever' in coordination chemistry. The great result of this period was that it became general to assume stepwise complex formation in systems containing metal ions and (potential) ligands; reliable experimental and calculation methods were elaborated. Because equilibrium 'constants' can in most cases be obtained with readily available instruments, many laboratories were engaged in investigations in this field. It was almost inevitable that besides papers offering reliable data obtained by careful experimental work and using well-chosen methods, a great number of papers were published, the data of which are either not precise enough or even completely wrong. It is probably not an overestimation to say that only a fraction (not bigger than 30%) of the data of the monumental compilation by Sillén and Martell[1] can be regarded as reliable. It is not exceptional that the numerical values of the constants obtained by different methods and authors for one and the same system show deviations even in the order of magnitude. This may lead to the misbelief that the concept of complex equilibria is not a sound approach to describe the behaviour of metal complexes in solution.

In the last 5 years – the period surveyed in this review – complex equilibria have continued to be studied intensely. The strengthening of the critical approach and the development of experimental procedures and calculation methods has led to the result that a greater part of the more recent data can be regarded as fairly reliable.

We believe that the modern studies are characterised by the following features:

(a) The number of *l'art pour l'art* studies motivated mainly by the fact that 'this system has so far not been studied' is rapidly decreasing.

(b) The number of experimental methods making possible the study of different types of complex equilibria has increased: essentially all types of complex equilibria can be adequately treated.

(c) The application of high-speed computers in the evaluation of experimental data has become quite general.

(d) It is generally considered now that one must pay relatively more attention to mixed ligand, outer sphere, protonated and polynuclear species instead of to the parent complexes ML_i.

1.2 SOME CONCEPTUAL PROBLEMS

1.2.1 The concept of chemical species in solution

The basic aim of the study of complex equilibria is to identify the different species in the solutions and to determine the corresponding stability constants. The constants are reliable if all of the properties of the system can be quantitatively and qualitatively described by these constants. This approach is evidently based on the applicability of the mass action law. It is obvious, however, that in solution the concept of a chemical species is much more ambiguous than in the gas phase. This is well exemplified by the Eigen mechanism of the formation of complexes[2]. When a hydrated metal ion and ligand approach each other, the hydration shells of both species attenuate gradually; a new entity is obtained if M and L are separated only by the water molecules bound in the inner spheres of both M and L; an outer sphere complex is formed when only one water molecule is between them; and finally a direct interaction of M and L occurs leading to the formation of an inner sphere complex. In this process the following entities can be distinguished: 'free', solvated ions with long-range electrostatic interaction between them; ion-pair $[MOH_2OH_2L]_{aq}$, outer sphere complex $[MOH_2L]_{aq}$ and inner sphere complex $[ML]_{aq}$. $[ML]_{aq}$ and $[MOH_2L]_{aq}$ are well-defined species, $[MOH_2OH_2L]_{aq}$ is much less well defined, and it would be only fair to say that there is a continuous transition between the outer sphere complex and the completely separated ions. According to researches carried out mainly in the last few years[3–7], the most intelligent solution of this dilemma is to make a compromise. Using the armoury of classical and modern methods one must identify as many distinct chemical species as possible and the long-range electrostatic interactions should be taken into account by the activity coefficients. Unfortunately, even the extended forms of the Debye-Hückel equation cannot treat solutions of even moderately high concentration quantitatively. In the case of weak complexes, however, it is necessary to apply fairly high ligand concentrations to achieve a considerable formation of the complex. To avoid the change of activity coefficients in changing the concentration of the ligand, a high concentration of a swamping salt is generally applied.

1.2.2 The constant ionic medium principle

The long dispute whether equilibrium constants determined at high and constant ionic strength, where the gradual formation of complexes by gradually exchanging a small fraction of the inert swamping salt for the ligand involves a negligible change of the medium can be regarded as thermodynamic constants, seems to be settled. It is generally believed that, e.g. 3 M $NaClO_4$ is as exact a reference state as pure water. However, there are surprisingly few studies of ionic activities in a swamping medium[8–11]. According to Prue and Read[8], the activity coefficient of HCl increases by 1.8% when its concentration changes from 0 to 0.2 mol kg^{-1} and the sum of the molality of HCl and $NaClO_4$ is 2.8545. They found that the changes of activity coefficient can be calculated with the help of the specific interaction theory[12]. Sodium perchlorate is still the most popular swamping salt, although there are now plenty of data to show that perchlorate ion forms weak complexes with a number of metal and complex ions[13]. In most cases it is possible to find a suitable inert salt. Bond[14] found that, in the case of thallium^{I} complexes, potassium fluoride is a more convenient swamping salt than sodium perchlorate. The stability of $TlClO_4$ at $I = 0$ has been found to be 0.32 ± 0.04 mol cm^{-3} at 30 ± 0.1°C, while no complex formation could be detected with fluoride ion even at high concentration of Tl^{I}. The application of a constant ionic medium is difficult in a study of outer sphere complex formation, this reaction being less specific than the inner sphere complexation[15].

It is important to realise that stability constants referring to the same ionic medium and not to the same ionic strength can be compared: in fairly concentrated solutions, only the same ionic medium can be regarded as the common reference state. The stability constants determined at the same ionic strength but in different ionic media may show great differences. This appears clearly from the data on the complex $EuNO_3^{2+}$ which were obtained by solvent extraction[16].

Table 1.1 Stability constants (K_1) for $EuNO_3^{2+}$, at 25 °C

Cation	$I = 0.5$ M	$I = 1.0$ M
H^+	2.45 ± 0.04	1.99 ± 0.12
Li^+	2.19 ± 0.06	1.75 ± 0.03
		1.99 ± 0.05(55 °C)
NH_4^+	2.98 ± 0.12	–
Na^+	2.48 ± 0.06	2.10 ± 0.04

1.2.3 The problem of weak complexes

Stability constants exhibit a very wide range of numerical values. Their values in the case of the most stable complexes may reach or even exceed

10^{50}, while the smallest published constant, referring to a charge transfer complex of I_2 with 2,3-dimethylbutane, is only 0.0034 [17]. Most of the stability data for weak complexes were obtained spectrophotometrically. It appears that stability constants with the dimension M^{-1} which are smaller than 0.2 should be regarded as dubious. Prue[18] has pointed out that even random collisions may cause spectral changes from which an 'equilibrium constant' of 0.2 M^{-1} can be calculated. On the other hand, difficulties may arise in the mathematical treatment of the spectrophotometric data referring to very weak complexes. In these cases the two unknown constants, stability constant and molar absorbancy, cannot be separated, and only their product can be obtained. According to Person's analysis[19], in order to get a reliable stability constant for a 1:1 complex, one must achieve a ligand concentration of $0.1 \times K^{-1}$. A similar analysis on the optimum conditions for the spectrophotometric determination of stability constants of weak complexes of various composition has also been made[20]. Although Person's criterion was contradicted by Heric's statistical analysis[21], it is clear that no quantitative information can be obtained for complexes the stability constants of which are much smaller than one.

Stability constants of fairly weak complexes can be evaluated from careful analysis of potentiometric measurements. An excellent example is provided by Butler and Huston's study[22] on the stability constant of $NaCO_3^-$ ($\log K = 0.96 \pm 0.13$) and of $NaHCO_3$ ($\log K = -0.30 \pm 0.13$) at $I = 0$, employing sodium amalgam and sodium-selective glass electrodes. The same authors have also determined the stability constant of NaF too, where $\log K = -0.79 \pm 0.04$ [23].

Most outer sphere complexes are fairly weak. These complexes will be treated in Section 1.5.2.

1.3 A SURVEY OF THE EXPERIMENTAL METHODS

Although no new principle has been introduced into the study of complex equilibria, the progress in methodology in the last few years was nevertheless extremely important, especially in the application of potentiometry and spectroscopy. The application of ion-selective or even ion-specific membrane electrodes will certainly produce as great an impact on the study of complex equilibria as did the introduction of glass electrodes. The application of n.m.r., infrared and Raman spectroscopy has been also broadened and made it possible to determine the donor sites of multidentate ligands.

1.3.1 Potentiometry

The measurement of electrode potentials may give information on the activities of metal ions and of ligands. In recent e.m.f. studies of complex equilibria the application of different types of ion-selective membrane electrodes[24] played a greater role than that of classical electrodes of the first and second kind. The use of ion-specific electrodes seems especially promising in the study of complicated equilibria involving several different ligands and metal ions.

1.3.1.1 *Measurement of metal ion activity*

Recently a number of activity determinations were performed using amalgam electrodes[25,26]. This type of electrode has also been used to determine stability constants of tin(II)[27], zinc(II)[28] and lead(II)[29] complexes. The application of different ion-selective electrodes seems extremely promising in the study of complex equilibria. An excellent monograph has been published on ion-selective glass electrodes[30]. This type of electrode was used in the study of the reactions of silver(I) and nickel(II) with EDTA[31], and in the investigation of equilibria involving tetraphenylborate[32] and of ion association between sodium ion and tetrametaphosphate and trimetaphosphate[33].

The recently developed liquid and solid membrane electrodes sensitive to different metal ions are of great importance both in analytical chemistry and in studying complex equilibria. There are some commercially available electrodes and many can be easily fabricated[34–39]. It is always very important in applying these electrodes to check carefully the dependence of the potential on the metal ion activity in question and determine the effects on the potential of other cations present in the system to be investigated. Most of the equilibrium studies concern calcium-selective electrodes[40–46].

While sodium-selective glass electrodes which are not interfered with by potassium ions are readily available, potassium-selective glass electrodes show poor selectivity to other common cations. Recently it was found that certain antibiotics form more stable complexes with potassium than with sodium ions[47–48] and this was utilised in constructing potassium-selective membrane electrodes[49–52]. This electrode permitted the direct determination of the stability constant of a complex of potassium ion with adenosine triphosphate. Greater credit may be given to the value determined by Mohan and Rechnitz[53] which is about 20 times the earlier indirectly obtained value.

1.3.1.2 *Measurement of ligand activity*

Ion-selective membrane electrodes are available for some anions, too. These electrodes are of different types, but it appears that the electrodes applying crystals are in general more convenient. By far the most thoroughly applied anion-selective electrode in the study of complex equilibria is the fluoride-specific lanthanum fluoride crystal electrode developed by Frant and Ross[54]. This follows from the fact that no electrode of the second kind is available for fluoride ion; this hindered the investigation of fluoride complexes by e.m.f. measurements. The only anion which interferes with fluoride is hydroxide. This effect and the relatively great stability of hydrogen fluoride result in the pH range 4–10 being the most convenient for equilibrium studies. At lower pH the protonation of fluoride has to be considered, at higher pH corrections must be made. So far the fluoride complexes of the following cations have been studied by means of the fluoride electrode:

H^{+} [55, 56, 60], Mg^{2+} [57, 60, 403], Ca^{2+} [57, 56], Pb^{2+} [58], Be^{2+} [59], Sn^{2+} [60], Cd^{2+} [403], Zn^{2+} [403], In^{3+} [61], Ni^{2+} [403], Ag^{+} [403], Al^{3+} [56], Gd^{3+} [60], Y^{3+} [60], Sc^{3+} [60], Eu^{3+} [60], Fe^{3+} [55, 60], Th^{4+} [62], and H_3BO_3 [63].

An iodide-selective membrane electrode was applied in the determination of the stability constant of BiI^{2+} and I_3^- [64], and a perchlorate-sensitive liquid membrane electrode was used to determine the solubility products of $KClO_4$, $Co(NH_3)_6(ClO_4)_3$, Cu(pyridine)$_4$$(ClO_4)_2$ and Fe(phen)$_3$$(ClO_4)_2$ [65]. The membrane electrode developed by Dobbelstein and Diehl[66] is responsive to univalent anions but not to multivalent anions and cations other than hydrogen. A liquid membrane electrode prepared by Coetzee and Freiser[67, 68] does not distinguish uni- and multivalent anions. The acid dissociation constant of acetic acid was determined by means of this latter electrode. It seems that these electrodes can be applied to study many complex equilibria, and furthermore that they may conveniently be applied as *reference electrodes* without liquid junction.

A most interesting effect was observed by Hirsch and Portock[69] when thiocyanate was titrated with silver(I) or mercury(II) using an Orion perchlorate ion electrode. The potential first decreased, then increased, and a U-shaped titration curve was obtained. This finding was explained by the authors by assuming that the electrode is sensitive to the complex anions, too. Such a possibility must always be borne in mind in the evaluation of results obtained with membrane electrodes.

The availability of membrane electrodes has decreased efforts to find new electrodes of the second kind and the application of redox systems to measure ligand activity. A silver phosphate electrode was suggested for the study of equilibria involving phosphate ions[70]. The system Fe^{3+}—Fe^{2+}—F^-, first applied by Brosset and Orring[71], was used by Nordén in a study of the stability of fluoride complexes of hafnium(IV) [72] and zirconium(IV) [73]. The reversible redox reaction of certain sulphur-containing ligands was exploited[74] in the determination of stability constants of their metal complexes.

1.3.1.3 Measurement of hydrogen ion activity

The measurement of hydrogen ion activity is undoubtedly the most frequent task in the study of complex equilibria. Efforts have been directed to increasing the accuracy and reliability of pH measurements by means of the glass electrode[75–77]. With careful experimental work the same accuracy can be achieved as with the less convenient hydrogen electrode. The glass electrode can be applied in some non-aqueous media, too[78–79].

If pH-metry is employed in the study of complex equilibria, one must take into account that it is incapable of defining pL precisely above a certain value, namely above pK + 2. McBryde's analysis[80] of the error resulting from this limitation was recently generalised[81].

A programmable automatic instrument for pH-metric measurements has been developed by Carpéni *et al.*[82]. Although it is a fairly complicated device it may be very useful in equilibrium studies, especially if combined with a computer.

1.3.1.4 Reference electrodes for non-aqueous solutions

The importance of the study of complex equilibria in non-aqueous solutions is rapidly increasing. The prerequisite of the application of e.m.f. methods is the availability of reference electrodes. Recently, reference electrodes were suggested for dimethyl sulphoxide[83], dimethylformamide[83, 85, 86], propylene carbonate[83, 84] and acetonitrile[87].

1.3.2 Polarography

It seems that polarography has lost much of its former popularity for the study of complex equilibria, and we may limit this survey to a few papers. Bond suggested rapid d.c. and a.c. polarographic techniques (drop time only 0.16 s) instead of the conventional ones, and applied them for determination of equilibrium constants in the $Cu^{2+}-F^-$ [88] and $Bi^{3+}-F^-$ [89] systems. Momoki, Sato and Ogawa[90, 91] made an extremely careful experimental study of the Cd^{2+}—SCN^- system applying different supporting media and treated the obtained data statistically by means of a high-speed computer. The mystery of the apparent non-existence of $Cd(SCN)_3^-$ was solved and all β_i values ($1 \leqslant i \leqslant 4$) and the stability constants of two mixed ligand complexes $Cd(SCN)NO_3$ and $Cd(SCN)_2NO_3^-$ were derived.

Polarography[92] and chronopotentiometry[93] were also used to determine the stability constants of outer-sphere complexes.

1.3.3 Spectrophotometry

Visible and u.v. spectrophotometry is still one of the most frequently applied techniques in the study of complex formation in solution, but only a few papers contribute to the theory and instrumentation of the method. Some matrix algebraic methods have been suggested[94–96] for the determination of the number of absorbing species. The calculation of the spectra and stability constants of these species requires the application of sophisticated computer analysis, but a simple graphical procedure has been suggested to determine only the number of absorbing species[97]. Ramette[98] thoroughly analysed the spectrophotometric determination of stability constants of 1:1 complexes and suggested a program for the evaluation of data by high-speed computers. A new null method suited for titration procedures was developed by Wallin[99]. The frequently used mole ratio method was critically examined by Marcus[100]. It seems that it has practically the same limitations as the method of continuous variation.

1.3.4 Vibrational spectroscopy

Although Raman and infrared spectroscopic studies of complexes in solution are mainly directed towards the classification of structural problems, many stability constants have been obtained using these techniques. An infrared study[101] of the silver(I)–thiocyanate system in pyridine gave equilibrium constants for both mono and binuclear species. In the former case the ligand is bound at the S end, while in the latter case both the S and N atoms are involved in complex formation: thiocyanate acts as a bridging ligand. Although a Raman study[102] of zinc(II), cadmium(II) and mercury(II) with thiocyanate did not permit quantitative evaluation of stability constants, it could be concluded that the N atom is coordinated to zinc(II), the S atom to mercury(II), while cadmium(II) forms both thiocyanate and isothiocyanate complexes.

Raman studies[103] indicated the formation of $TlCl_6^{3-}$ complex at high chloride concentration and an approximate value was calculated for the product K_5K_6. Quantitative information was obtained[104] for TaF_6^- and TaF_7^{2-} species. It was found that the nitrate ion forms weak complexes as a monodentate ligand with cadmium(II) [105], while with cerium(IV) [106] it behaves as a bidentate ligand and the complex formed is fairly stable. Vibrational spectroscopy seems very useful in the study of organometallic ions. The association of methylmercuric ion with different anions was studied[107] and it was found that, independently of the charge of the anion, 1:1 species exist in the solution. A comparative study of solids and solutions indicated that dimethylgallium(III) hydroxide is tetrameric in solution, too.

The coordination of a third ethylenediamine[109] to zinc(II) and mercury(II), and that of a third oxalate[110] to zinc(II) was found using a Raman technique. The Raman study[111] of metal complexes of ethylenediaminetetraacetate seems promising, and indicates the formation of protonated species in acidic medium and the coordination of OH^- at higher pH with certain metal ions.

This technique is a useful tool for the investigation of hydrolytic equilibria. No polymeric species was found in the case of arsenic(III) [112], while a comparative study of solids and solutions indicated that in the case of bismuth(III) [113] $Bi_6(OH)_{12}^{6+}$, and in the case of lead(III) [114] $Pb_4(OH_4^{4+}$ and $Pb_6(OH)_8^{4+}$ are the dominating species. Some information could be obtained on the composition of species in solutions of metals of Groups IVA, VA and VIA in aqueous hydrofluoric, hydrochloric and perchloric acids[115]. Polymeric species containing hydroxo bridges are formed in perchloric acid solution only, indicating that with these ions halides do not tend to act as bridging ligands. The Raman and infrared spectra[116] of aluminate ions in alkaline solutions indicate that in the range pH 8–12 polymeric species are present while above pH 13 linear AlO_2^- occurs. It appears that the elucidation of the structure and distribution of the species in alkaline aluminate solutions requires much further work.

1.3.5 N.M.R. and e.s.r. spectroscopy

1.3.5.1 N.M.R. spectroscopy

Most of the researches concern proton resonance, but other nuclei have also been studied. It has sometimes proved useful to attack problems with both vibrational and resonance spectroscopy. This technique is frequently applied to determine solvation numbers and to study coordination in the outer sphere. P.M.R. experiments[117] showed that, in concentrated aqueous aluminium(III) nitrate solution, nitrate ion is coordinated in the outer sphere. Aluminium(III) also forms outer sphere complexes in dimethylformamide solution with chloride, bromide and iodide ions[118]. Interestingly enough, in acetonitrile solutions of aluminium(III) chloride the metal ion is completely solvated, while in solutions of aluminium(III) perchlorate the coordination sites are partially occupied by perchlorate[119, 120]. Aluminium-27 n.m.r. has also been used to detect complex formation[121]. A number of different aluminium(III) complexes have been investigated in several solvents. From the chemical shift and the change of line width it could be stated that tetrahedral, octahedral or intermediate geometric structures are formed depending on the ligand.

A p.m.r. technique was used to detect the hexa-aquocobalt(II) ion in aqueous solution[122], to determine the stability constants of the silver^{I}–benzene complex[123] and of some charge-transfer complexes[124], and to get qualitative information on association between zinc(II) and perchlorate in methanol[125] and between manganese(II) and several anions in water[126].

The n.m.r. study of paramagnetic manganese(II) showed[127] that in solutions of $MnSO_4$ both outer and inner sphere complexes are formed and that the sulphate ion behaves as a bidentate ion in the inner sphere. Dimerisation and stepwise solvation of chlorophyll in methanol solution could be quantitatively studied by p.m.r.[128].

The metal complexes of aminopolycarboxylic acids are studied most thoroughly by p.m.r. An incomplete list is: EDTA complexes: Zr^{IV} [129], Hf^{IV} [129], Pd^{II} [129–130], Rh^{III} [134], U^{IV} [131], Cu^{II} [136], Ni^{II} [136], Fe^{III} [136]; iminodiacetate complexes: Mo^{VI} [135], Pt^{II} [132]; *N*-methyliminodiacetate complexes: Mo^{VI} [133], Pt^{II} [132]; nitrilotriacetate complexes: Pt^{II} [132]; derivatives of iminodiacetate: Co^{II} [137]; propylenediaminetetracetate(PDTA) complexes[138]: K^{I}, Rb^{I} and Ca^{II}. The great value of this technique is well exemplified by the work of Kula[133] who quantitatively treated the complicated equilibria in the molybdenum(VI)–*N*-methyliminodiacetate system and determine the structural and bonding characteristics of the complex species. A very surprising result of the study of interaction between PDTA and of alkali metal ions is that both ends of the ligand are involved in the coordination although the complex is very weak[138]. A comparative p.m.r. study[404] of different aminopolycarboxylates and molybdenum(V) indicated that except for diaminocyclohexanetetraacetate all chelating ligands form stable complexes.

P.M.R. has been used extensively to study *configurational* and *conformational* equilibria. It was found[139] that substituted ethylenediamine ligands coordinated to nickel(II) in aqueous solution exist in various *gauche* forms. ΔG for the *planar* $\rightleftharpoons$ *octahedral* equilibrium in chloroform solution of nickel(II) complexes of different Schiff bases were determined, and the existence of a 5-coordinated species was pointed out[140, 141]. Thermodynamic data for the *planar* $\rightleftharpoons$ *tetrahedral* equilibrium of bis(pyrrole-2-aldimino) nickel(II) complexes in chloroform solution could also be obtained[142].

N.M.R. was successfully applied in determining bonding sites in metal chelates of hydroxycarboxylates[143], 8-quinolinol[144] and aminoacids[145]. One of the important and unexpected results of these studies is that the hydroxy group of threonine is coordinated to Mn^{II} but not to Cu^{II} [146]. It seems likely that the reason is the smaller tendency of Cu^{II} to form hexacoordinated species.

1.3.5.2 E.S.R. spectra

The line widths of e.s.r. signals of paramagnetic metal ions can also be applied in studying outer and inner sphere coordination. Manganese(II) complexes of halides[147], sulphate[148], dithionite[149], nitrate and perchlorate[150], and chromium(III) complexes of a number of anionic ligands have been studied[151, 152]. Many equilibrium constants have been determined. The experiments with the manganese(II)–dithionite system are especially interesting, because both the central ion and the radical ion SO_2^- (the dimer of which is dithionite ion) are paramagnetic.

E.S.R. studies have given information on the dimerisation of chelates of copper(II) with aminoacids[153] and with hydroxycarboxylic acids[154]; however, no data has been obtained on the dimerisation constants.

1.3.6 Calorimetry

The application of thermistors for temperature measurement led to a very rapid development of calorimetry in many fields. In the study of complex equilibria it has produced important progress and seems a very promising subject. In principle, calorimetry is the most generally applicable technique for the study of complex formation in solution because each of these reactions is accompanied by some change in enthalpy. Although there are commercially available extremely accurate – and very expensive – instruments, most investigators use home-made devices. Papers dealing with the construction of calorimeters are continuously published[155–162]. The principle of an automatic calorimeter was described recently[162].

In most cases calorimetric measurements are used to determine only the formation enthalpies of the different species of the solution, the concentration of which is calculated using independently determined stability constants. The combination of free energies and enthalpies then gives the corresponding entropy values. Halide complexes of In^{II} [61], Cd^{II} [163–165], Zn^{II} [166], Ni^{II} [167], Co^{II} [167], Cu^{II} [167], monosulphato[168] and different acetato[169] complexes of tervalent rare earth ions have been studied using this approach. The formation enthalpies of a number of complexes of the following ligands have also been determined: amines[170, 171], aminoacids[172–174], peptides[157, 175], phenol[176], succinic acid[177], aminopolycarboxylic acids[178–182].

However, a number of papers deal with a more ambitious approach to complex equilibria by calorimetry. These papers point out that not only the enthalpies but even the stability constants can be obtained from a single calorimetric titration. The theory and computational problems of the evaluation of stability constants, enthalpies and entropies from calorimetric data have been thoroughly treated[183–190] and special programs for computer calculation have been elaborated[188, 190]. A simple and promising method has been described[158] which is an application of the principle of corresponding solutions to calorimetry.

A comparison[190] of several acid dissociation and complex stability constants obtained by calorimetry and by conventional methods (Table 1.2) seems to indicate that the accuracy of calorimetry is not much less than the other techniques and one may even expect an improvement of the technique in the near future.

Table 1.2 Comparison of *K* values determined by calorimetric titration with those determined by conventional procedures

Reaction	$-\log K$	*Method**
$HSO_4^- = H^+ + SO_4^{2-}$	1.97 ± 0.03	I
	1.99	II
$HPO_4^{2-} + OH^- = PO_4^{3-} + H_2O$	-1.6 ± 0.03	I
	-1.625 ± 0.01	IV
$Had = H^+ + ad^-$	12.35 ± 0.03	I
	12.5	II
$Hcyt = H^+ + cyt^-$	12.15 ± 0.05	I
	12.16	II
$Hmet = H^+ + met^-$	3.76 ± 0.04	I
	3.74	II
$Hpy^+ = H^+ + py$	5.17 ± 0.02	I
	5.18	III
$Him^+ = H^+ + im$	6.99 ± 0.02	I
	6.99	II
$Hthma^+ = H^+ + thma$	8.03 ± 0.37	I
	8.075	II
$Ag^+ + py = Agpy^+$	-2.04 ± 0.06	I
	-2.00	II
$Ag^+ + 2py = Agpy_2^+$	-4.09 ± 0.05	I
	-4.11	II
$Cu^{2+} + py = Cupy^{2+}$	-2.50 ± 0.02	I
	-2.52	III
$Cu^{2+} + 2py = Cupy_2^{2+}$	-4.30 ± 0.05	I
	-4.38	III
$Cu^{2+} + 3py = Cupy_3^{2+}$	-5.16 ± 0.06	I
	-5.69	III
$Cu^{2+} + 4py = Cupy_4^{2+}$	-6.04 ± 0.1	I
	-6.54	III

*I, calorimetry; II, hydrogen electrode (potentiometric measurements); III, glass electrode (pH measurements); IV, spectrophotometry.
met^-, aniline-*m*-sulphonate ion; py, pyridine; im, imidazole; thma, tris(hydroxymethyl)methylamine; ad^-, adenosinate; cyt^-, cytosinate.

Obviously the system must have favourable ΔH and ΔG values for accurate constants to be obtained, especially in the case of multiple equilibria. Stability constants have been calorimetrically measured in the case of the following systems: hydrolysis of Hg^{II}, Cd^{II}, Cu^{II}, Ni^{II}, Fe^{III}, uranyl(VI), molybdate(VI), tungstate(VI) and chromate(VI), (a review of many of his own original papers is given by Arnek[162]); cadmium(II) iodide[191] and acetate[164]; aquo complexes of cobalt(II) and nickel(II) in butanol[158]; a number of acid dissociation constants of carboxylic acids[192] and protonated amines[193]; stability constants of pyridine complexes of silver(I) and copper(II)[188]. The method has been applied to obtain stability constants in fused salt systems, too[194, 195].

1.3.7 Miscellaneous methods

1.3.7.1 Spectropolarimetry

In spite of the long application of spectropolarimetry in the study of complexes in solution, the collection of quantitative information on the stability began only recently; the availability of commercial spectropolarimeters has contributed to progress in this field. There are two types of system investigated: (a) the ligand is optically active and the rotation changes as a consequence of metal complex formation; (b) the complex itself is asymmetric and therefore optically active, and the uptake of ligand(s) in the outer sphere results in a change in the rotatory power. The following systems have been studied by this technique: inner sphere complex formation: aluminium(III)[196] and boric acid[197] complexes of tartrate; zinc(II), cobalt(II) and nickel(II) complexes of mandelate[198]; arsenite complexes of mannitol[199]; outer sphere complex formation: association of tris(ethylenediamine)cobalt(III) complex with selenite[200], thiosulphate[201], EDTA[15] and polyacrylate[202].

1.3.7.2 Solubility measurements

This technique is one of the simplest, but nevertheless is a very powerful tool in the quantitative study of complex equilibria. A good review[203] was published recently on the method, its applicability and limitations. The potential of the method is well illustrated by the determination of the full sets of successive stability constants in the tin(II)–iodide[204], bismuth(III)–bromide[205] and copper(II)–chloride[206] systems. In the latter case the stability constant of a binuclear species $Cu_2Cl_4^{2-}$ was obtained. The problems of the iteration procedures to calculate the free ligand concentration from solubility data have been thoroughly treated[207].

1.3.7.3 Solvent extraction

Two international symposia in the last few years dealing exclusively with solvent extraction studies of complex formation show the importance of this method. The Proceedings of these symposia[208, 209] give a cross-section of the field. An excellent monograph[210] treats authoritatively all aspects of solvent extraction of complexes. Methodically the most important progress was achieved by designing 'an apparatus for continuous measurement of distribution factors in solvent extraction'[211–213]. This instrument is now commercially available under the name AKUFVE (the abbreviation of the Swedish equivalent of the sentence in inverted commas). This apparatus consists of five parts: mixer, separator (a centrifuge with rotation speeds of up to 18 000 rev/min), external liquid system, detectors and a measuring system. Because a large number of data are collected in a short time, an on-line computer is applied. A special program (CKRUT) has been developed. Experiments with the system copper(II)–acetylacetone–benzene–0.1 M Na ClO_4 at one temperature, applying seven different concentrations of acetylacetone and yielding about 200 data, required 25 h, and the total computing time was 10 min.

1.3.7.4 Reaction kinetics

The importance of kinetic methods has been increased considerably by the fact that different techniques have rendered possible rate studies of practically all complex formation reactions. It seems likely that the number of kinetic studies leading to quantitative data on the stabilities of complexes will rapidly increase in the future.

The study of anation and conjugate aquation reactions of inert complexes is a very convenient possibility of obtaining stability data. This approach was used by Elding[214] who measured the equilibrium constant (K_4) of the reaction

$$PtCl_4^{2-} + H_2O \rightleftharpoons PtCl_3H_2O^- + Cl^-$$

by both rate and independent equilibrium experiments. The excellent agreement justifies the kinetic approach. In the case of moderately fast substitution reactions the stopped-flow technique can be applied. The stability constant of the complex $FeCrO_4^+$ has been determined[215]. Detailed kinetic studies of complicated systems also give valuable information on the composition and stability of different species. The empirical rate equations are of diagnostic value as regards the composition of certain species. Kolski and Margerum[216] were led to postulate the $H_2Ni(CN)_4$ in studying the formation and decomposition of tetracyanonickelate(II). Outer sphere complex formation occurs frequently as a pre-equilibrium, the constant of which can easily be obtained. Complex formation has to be considered frequently in the case of redox reactions and catalytic phenomena, too[217–219]. In favourable cases the stability constants, or at least the compositions of the complexes, can be obtained. It must be considered, however, that sometimes different mechanisms (involving different species) may be equally compatible with the same set of experimental data.

The investigation of the mechanism of anation reactions

$$trans\text{-}[Co(cyclam)ClH_2O]^{2+} + NCS^- \rightarrow trans\text{-}Co(cyclam)ClNCS]^+ + H_2O$$

has furnished an excellent example[220]. The following two sets of reactions can be written:*

(a) dissociative mechanism

$$RH_2O^{2+} \rightarrow R^{2+} + H_2O \quad (1.1)$$

$$R^{2+} + H_2O \rightarrow RH_2O^{2+} \quad \text{(fast)} \quad (1.2)$$

$$R^{2+} + NCS^- \rightarrow RNCS^+ \quad \text{(fast)} \quad (1.3)$$

(b) interchange mechanism

$$RH_2O^{2+} + NCS^- \rightleftharpoons RH_2O, NCS^+ \quad \text{(fast)} \quad (1.4)$$

$$RH_2O, NCS^+ \rightarrow RNCS^+ + H_2O \quad (1.5)$$

* cyclam = 1,4,8,11-tetra-azacyclotetradecane
R = *trans*-[Co cyclam Cl^{2+}]

The validity of the interchange mechanism would require the existence of a fairly stable outer sphere complex. However, direct chemical evidence indicates that such a complex does not form in any appreciable concentration, and therefore the dissociative mechanism has to be accepted although it gives the same rate law as the interchange mechanism.

Relaxation measurements make possible the determination of the rate constants of both the forward and the backward reactions of very fast reactions. Stability constants of 1:1 complexes and of mixed ligand complexes have been obtained, and inner sphere and outer sphere complexations have been distinguished[221–224].

1.3.7.5 Viscometry

In most cases the B-coefficients of a series of complexes are compared[225]. Similar B-coefficients indicate structural similarities and vice versa. On the basis of the theory of regular solutions[226], Irving and Smith[227,228] have derived some equations making possible the quantitative study of association between uncharged species. Stabilities of mixed ligand complexes[227] and of solvates of iodine[228] have been determined.

1.3.7.6. Mössbauer studies

This technique is inherently limited to the study of solids. However, it has been recognised that *frozen solutions* can also be investigated and some complex equilibria, particularly the hydrolysis of iron(III) salts, have been studied[229,230]. Nevertheless, one cannot expect to get quantitative information on the concentration distribution of the complex species, i.e. equilibrium data. On the other hand, if such data are independently obtained, Mössbauer studies of frozen solutions may give an insight into structural and bonding problems, provided that the rapid freezing does not result in a shift of the equilibrium.

1.3.7.7 X-ray studies

No equilibrium data, but structural information on dominant species can be obtained by x-ray studies[231–234]. The application of this technique is limited to well-soluble complexes of elements of higher atomic number. Mainly hydrolytic reactions have been studied[231–234].

1.3.7.8 Coagulation studies

Although this technique necessarily requires experiments with fairly complicated systems, reliable stability data can sometimes be obtained, as in the case of the stability of $AlSO_4^+$ [235].

1.4 COMPUTATIONAL METHODS

The standard graphical and numerical methods for calculation of stability constants were developed in the early stage of research and no significant progress has been made in this respect. One cannot even expect any substantial future development in this field. However, the introduction of high-speed computers has revolutionised the calculation of stability constants, particularly in the case of complicated systems. The application of computers to the evaluation of data obtained by certain experimental techniques has already been mentioned in this review, and several special programs will be treated later, but some general comments should be given here.

The most general treatment was given by Sillén *et al.*[236–240]. However, the LETAGROP programs – which were originally written in FORTRAN, and their ALGOL versions are also available – are fairly complicated and frequently simpler programs can be applied more conveniently. A non-linear least-squares approach was applied by Sayce[242] in elaborating SCOGS (stability constants of generalised species) written in FORTRAN IV. It is an extended version of GAUSS[243, 244] and is capable of calculating association constants for any of the species formed in systems containing up to two metal ions and two ligands, provided that the degree of complex formation depends on pH. The program seems versatile and its wide application is expected, especially in equilibria involving mixed ligand, protonated and polynuclear complexes. In the application of this program a correction[245] has to be taken into account. The philosophy of the use of computers in equilibrium studies was treated wisely by Hume[241].

Although their generality is an advantageous feature of computer programs in equilibrium chemistry, it should also be considered that the more general the program, the bigger and the more expensive the computer required; there is an accompanying increase of computation time, the risk of making errors, and the difficulty of finding them. Therefore many simpler computer programs have been developed[246–251], but it has to be carefully considered whether the use of a simpler program in a given case is permissible. From this point of view the method developed by Nagypál, Gergely and Jékel[251] is particularly useful in studying stepwise formation of mononuclear binary complexes, because the existence of other species is strikingly indicated by the program itself. In the general case the complex formation function is symmetrical only if a maximum two ligands are taken up by the metal ion. However, from any unsymmetrical complex formation function ($N > 2$) symmetrical functions can be derived if the successive formation of mononuclear binary complexes occurs. Therefore the asymmetry of derived functions indicates other equilibria: hydrolysis, polynuclear complex formation, etc.

Another important aspect of the application of electronic computers is the calculation of the concentrations of the different species from total concentrations and stability constants. In addition to a number of papers[252–254], a recent book[255] deals with this problem.

1.5 A SURVEY OF SOME IMPORTANT TYPES OF COMPLEXES

It is probably not an overestimation that about 2000 papers providing stability constants have been published in the last 5 years. As a matter of course it is impossible to give a full account of this work, and this review is limited to the discussion of some particularly important types such as outer sphere, mixed ligand and polynuclear complexes and the equilibrium studies of complexes of biological importance.

1.5.1 Outer sphere complexes

Just as inner sphere complex formation is an ubiquitous phenomenon, one always must consider the formation of outer sphere complexes. This is the reason why this type of complex is receiving increasing attention. A nearly complete list of stability data (up to 1967) was published recently[256], and a monograph[257] is under preparation. In dealing with outer sphere complex formation one must distinguish two cases depending on the inertness of the inner sphere. If the inner sphere is inert, the situation is simple and the outer sphere complex formation can be studied separately. However, if the inner sphere is labile, the stability constant is a composite constant, e.g.:

$$\beta_1 = \frac{[M(H_2O)_5L]+[M(H_2O)_6,L]}{[M(H_2O)_6][L]} = \beta_{01}+\beta_{10}$$

where β_1 is the constant determined by the conventional methods, while β_{10} and β_{01} are the stability constants of the first inner and outer sphere complexes, respectively. The resolution of this composite constant to β_{10} and β_{01} is fairly difficult and is not possible in all cases. If the inner sphere complex is very stable, the contribution of the outer sphere complex to the overall stability is usually not too great.

Sometimes it is difficult to decide whether an outer sphere complex is formed or, by the expansion of the coordination sphere, mixed ligand complex formation occurs.

In general the same experimental methods can be applied as in the study of inner sphere complex formation. However, some of the techniques are particularly important. These are as follows: resonance spectroscopy, relaxation kinetic methods, spectropolarimetry and spectrophotometry. Some results obtained with these were mentioned earlier.

A spectrophotometric method was elaborated to follow the association of an asymmetric inert complex with an optically active ligand[258, 259]. The method requires the comparative study of solutions containing either or both antipodes of an optically active ion and the optically active counter ion.

Ultrasonic absorption of aqueous solutions of sulphates of manganese(II)[260], magnesium(II)[261] and some inert complexes[262] has been studied. In the former cases, the stability constants of both the inner and outer sphere complexes could be obtained; in the latter case the inert inner sphere remained intact and the stability constant of the outer sphere complex could be determined. A good agreement with the values of constants obtained by independent methods gives credit to the reliability of the constant and to the exactness of this approach. Outer sphere complex formation has to be taken into consideration frequently in anation reactions slow enough to be studied by conventional kinetic methods. Data for outer sphere complexes of cobalt(III)[266–268], chromium(III)[269, 270] and rhodium(III)[271] have been obtained. Although ambiguities in connection with the evaluation of a stability constant may sometimes arise, rate studies are extremely useful in tracing weak interactions as the formation of outer sphere complexes.

Optical rotatory dispersion is a particularly useful technique for the investigation of outer sphere complexes because a number of inert complexes can be prepared in optically active form. Some disagreement between the results obtained by this and other techniques are awaiting elucidation[263–265].

Dilatometry is a new approach in the investigation of outer sphere complex formation[272]. One might expect a greater volume change accompanying inner than outer sphere complex formation, but in the case of $Co(NH_3)_5H_2O^{3+}$ the same expansion was observed on forming either the inner or outer sphere sulphate complex. Nevertheless, this simple technique deserves further application in the comparative study of inner and outer sphere complex formation.

In most cases the existence of a 1:1 outer sphere complex is pointed out, but there are examples where ligands are successively taken up in the outer sphere. In these cases there is a possibility for mixed ligand complex formation. An early example was given by Larsson[273] who pointed out that salicylate and cyclohexanone are taken up simultaneously by $Co(NH_3)_6^{3+}$. Recent solubility studies[274] led to the conclusion that $Co(NH_3)_6^{3+}$ can simultaneously take up both ClO_4^- and halide ions. However, this result is in conflict with the data of other solubility studies[15]. It is likely that mixed ligand outer sphere complexes are formed when tris(phenanthroline)iron(II) is extracted into some organic solvents from aqueous solutions containing two different anions.

Coordinatively saturated binuclear complexes can also take up further ligands. Although conductance measurements[275] have given clear cut evidence against the formation of *anionic* species in the reaction of $[(H_3N)_5CoO_2Co(NH_3)_5]^{5+}$ with sulphate, a 1:1 species may exist. Garbett and Gillard[276] have found by polarimetric experiments that the following binuclear complexes

$$(-)\left[(en)_2Co\genfrac{}{}{0pt}{}{\diagup NH_2 \diagdown}{\diagdown O_2 \diagup}Co(en)_2\right]^{4+},\quad (+)\left[(en)_2Co\genfrac{}{}{0pt}{}{\diagup NH_2 \diagdown}{\diagdown O_2 \diagup}Co(en)_2\right]^{3+}$$

$$(+)\left[(en)_2Co\genfrac{}{}{0pt}{}{\diagup NH_2 \diagdown}{\diagdown SO_4 \diagup}Co(en)_2\right]^{3+},\quad (+)\left[(en)_2Co\genfrac{}{}{0pt}{}{\diagup NH_2 \diagdown}{\diagdown NO_2 \diagup}Co(en)_2\right]^{4+}$$

form outer sphere complexes with sulphate ion. A most interesting study[277] seems to prove that a cluster of coordinatively saturated inert complexes may be formed around a polydentate ligand. From the promoting effect of benzene hexacarboxylate ion on the aquation of halogenopenta-ammine-chromium(III) the formation of 3:1 complex was concluded and its overall stability constant could be obtained ($\log \beta_3 = 9.88 \pm 0.13$). A potentiometric titration made possible the calculation of all the three constants ($\log K_1 = 4.3 \pm 0.3$; $\log K_2 = 4.1 \pm 0.2$; $\log K_3 = 3.1 \pm 0.3$). Although the sum of the last three values differs considerably from $\log \beta_3$ obtained kinetically, the data agree qualitatively. The proposed structure involves hydrogen bridges between the carboxylate groups and the ammine hydrogens in the inner sphere:

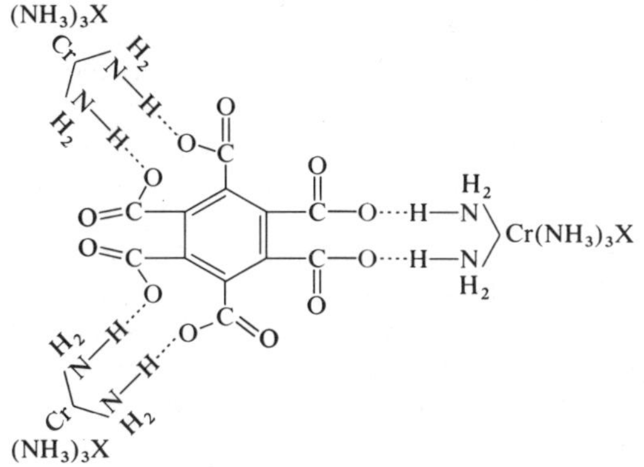

The formation of ion clusters of this type is very unusual and the phenomenon deserves a thorough study.

Most of the stability data can be explained on the basis of Bjerrum's theory of ion pair formation, that is by assuming that only electrostatic interactions are responsible for the formation of outer sphere complexes. However, there are many data which seem to indicate that specific forces also have to be considered. Some of the relevant observations are as follows:

(a) Formation of anionic species by overcompensation of the original positive charge by the uptake of several anions in the outer sphere.

(b) Sometimes there is no evidence for complex formation when it would be expected from a consideration of the charges of the ions, e.g. Larsson[263] found no interaction between $Co(en)_3^{3+}$ and $Fe(CN)_6^{3-}$. It is difficult to understand the extremely small tendency of fluoride ion to be taken up in the outer sphere. (This is probably connected with the firm hydration of fluoride ion.)

(c) In certain cases dipositive complexes form more stable outer sphere complexes with anions than do tripositive complexes of the same type[271]. The protonation of EDTA does not strongly influence the stabilities of its outer sphere complexes[15].

(d) Many data indicate that a considerable charge-transfer type interaction occurs between an inert complex and the ligand in the outer sphere[278]. Using the classical and the more sophisticated methods it is likely that the present inconsistencies will disappear and a general theory of outer sphere complex formation will soon be developed.

1.5.2 Mixed ligand complexes

If a solution contains a metal ion and at least two different ligands there is always a possibility for the formation of mixed ligand complexes. Considering the potential donating ability of the counter anion and of the solvent molecule, there are very few cases indeed when this possibility is out of consideration. Therefore it is rather perplexing to realise that the systematic work in this area began only some 15 years ago. Different types of mixed ligand complexes have been studied, methods to determine their stability constants have been elaborated, their importance in chemical and biochemical processes has been discussed and the factors determining their stabilities have been theoretically treated. Two reviews[279, 280] deal with this field in detail while two others[281, 282] are mainly concerned with the biochemical roles of the complexes.

So far, only mixed ligand complexes containing two different ligands (ternary complexes) have been quantitatively studied, although there is much information on the formation of more complicated species. Three kinds of ternary complexes may conveniently be distinguished: (a) complexes containing two different unidentate ligands; (b) complexes containing one unidentate and one multidentate ligand; (c) complexes containing two different multidentate ligands. It should be considered that mixed ligand complexes may be formed in substitution reaction

$$MA_N + B \rightleftharpoons MA_{N-1}B + A$$

or in addition reaction

$$MA_N + B \rightleftharpoons MA_NB$$

by the expansion of the coordination sphere.

The earlier studies mainly involved type (a) complexes, but type (c) is now at the centre of interest. Only a few papers referring to systems involving unidentate ligands are mentioned here: $HgCl_2-HgI_2$, $HgBr_2-HgI_2$; $HgCl_2-HgBr_2$[283]; $Fe^{3+}-Cl^{-}-SO_4^{2-}$[284]; BiI_3-BiCl_3[285]; $Pd^{2+}-Br^{-}-I^{-}$, $Pd^{2+}-Cl^{-}-I^{-}$ and $Pd^{2+}-Br^{-}-Cl^{-}$[286, 287]. A general method for the spectrophotometric study of MA_3+MB_3 system has been elaborated[285].

Type (b) is only sporadically studied. The equilibrium constants of the following reactions have been determined spectrophotometrically[288]:

$$trans\text{-}Rh(en)_2\,Br_2^+ + I^- \rightleftharpoons trans\text{-}Rh(en)_2\,BrCl^+ + Br^-$$
$$trans\text{-}Rh(en)_2\,Cl_2^+ + I^- \rightleftharpoons trans\text{-}Rh(en)_2\,ICl^+ + Cl^-$$

while a Raman study has indicated the simultaneous coordination of glycine and thiocyanate to zinc(II) and cadmium(II)[289].

In the case of type (c) one must always consider the coordination of ligands the conjugated acids of which are weak: the stability constants can be obtained by pH-metric titration. The situation is relatively simple if the two ligands in question are of very different basicities. Mixed ligand complexes of oxalate and ethylenediamine[290, 291], 2,2′-bipyridyl (and analogous ligands) and various more basic ligands[292–295], 8-quinolinol and arylhydroxycarboxylic acids[296, 297], have been systematically investigated. Even in this simple case the application of computers in the evalution of stability constants is recommended, and the use of high-speed computers is practically inevitable when the basicities of the two ligands are similar. Special methods have been developed for mixed ligand complexes MAB by Näsänen and Koskinen[298, 299] and by Perrin, Sayce and Sharma[300]. More complicated systems have also been studied pH-metrically the results being evaluated by SCOGS[301] and a special version of LETAGROP[302]. Mainly copper(II) complexes of different diamines and different aminoacids have been investigated.

The relationship between the stabilities of a mixed ligand complex and its parent complexes has been long disputed. It was first thought that the mixed ligand complexes are always more stable than could be expected on statistical grounds but there are now many examples of the opposite behaviour. Statistically, mixed ligand complex formation is always favoured. The statistical probability can be obtained by considering the number of ways in which the different complexes can be formed. Table 1.3[303] gives these probabilities for all possible mixed ligand species up to coordination number 6.

Table 1.3 Formation of mixed ligand complexes

	Statistical probabilities relative to the simple complex				
Mixed complex	ML_2	ML_3	ML_4	ML_5	ML_6
MAB	2	—	—	—	—
MABC	—	6	—	—	—
MA_2BC_0	—	3	—	—	—
MAB_2C_0	—	3	—	—	—
MA_0B_2C	—	3	—	—	—
MA_0BC_2	—	3	—	—	—
MABCD	—	—	24	—	—
MABCDE	—	—	—	120	—
MABCDEF	—	—	—	—	720

Elementary electrostatic considerations also indicate that mixed ligand complex formation is favoured. However, a more sophisticated treatment has shown that both stabilisation and destabilisation may occur as a consequence of simultaneous coordination of two different ligands. A formal thermodynamic approach to the problem has been given by Fridman[280]. Sometimes the mixed ligand species are extremely stable. For example in the case of mixed β-diketone complexes of copper(II) no detectable amount of parent complex is present in the solution of the mixed ligand species[304]. Steric effect and back-coordination have also been invoked to account for the preferred formation of mixed ligand complexes[305, 306]. The great increase in stability of the mixed ligand complexes with certain pairs of ligands is thought to be of particular importance from the point of view of the specificity of enzyme reactions. Many model studies have been performed to clarify the role of mixed ligand complex formation in enzyme action[307–310].

An interesting class of mixed ligand complexes involves the simultaneous coordination of two antipodes of an optically active ligand. Mixed ligand complexes of optically active amines[311] and aminoacids[312, 313] have been studied and stereoselectivity, that is the preferred formation of the complex containing both antipodes, has been found only in the case of complexes of histidine[314]. However, the stability of the copper(II)–L-histidine–L-threonine complex is the very same as that of the copper(II)–D-histidine–L-threonine complex[302]. A calorimetric study indicated a difference in stereoselectivity in the formation of copper(II) and nickel(II) bis-histidine complexes[315]. While with nickel(II) the formation of the racemic bis complex is preferred over that of the optically pure species, the opposite behaviour is found with copper^{II}. The biochemical implications of stereoselectivity have been treated by Gillard[316].

1.5.3 Polynuclear complexes

The many different types of polynuclear complexes were recently considered systematically. Most of the recent studies have dealt with the hydrolysis of metal ions and metal complexes, but increased attention is being paid to other types of polynuclear complexes. Just as in the case of mixed ligand complexes, the possible biochemical role is frequently implicated.

1.5.3.1 Hydrolysis of metal ions and metal complexes

Although potentiometry is still the most frequently applied method (e.g. References 317 and 318) in obtaining stability constants of polynuclear complexes of this type, it is of great importance that a number of other methods, such as ultracentrifugation[319, 320, 322], visible[323], vibration[321, 322] and resonance[321] spectroscopy, calorimetry[189], etc., also be used to provide a more reliable approach to the complicated systems. A comprehensive review was published very recently[324]. The newest studies have confirmed the earlier results that a great variety of complexes of different composition, structure and stability is formed; the hydrolyses of even fairly similar metal ions do not follow the same pattern. The concept of 'isohydric law' has been further developed and Carpéni's systematic studies show that in most cases *one* predominant polymeric species is formed[327]. Polynuclear species are formed in the hydrolysis of organometallic cations, too. For example the predominant products of the hydrolysis of both dimethylgold(IV)[325] and dimethylgallium(III)[326] are the dimers $[(CH_3)_2AuOH]_2$ and $[(CH_3)_2GaOH]_2$, respectively.

Polynuclear complexes may be formed in the hydrolyses of certain metal chelates. In these cases, either the multidentate ligand or hydroxo groups, or both, may function as bridges. Investigations by Zompa and Bogucki[328] furnish a particularly interesting example. They studied the copper(II) complex of tetrakis (aminomethyl) methane:

```
H2N-CH2       CH2-NH2
       \     /
          C
       /     \
H2N-CH2       CH2-NH2
```

Only two of the four amine groups are sterically available for a given copper(II) ion of square-planar coordination. At low pH, mono and binuclear chelates are formed but at higher pH the product is an insoluble polymeric species

```
                                                                                               (2n-2)+
HO     NH2-CH2      NH2          [   OH      NH2-CH2       CH2-NH2         ]   OH
  \   /       \    /    \        |  /  \    /       \     /       \        |  /
    Cu          C         Cu     |      Cu            C             Cu     |
  /   \       /    \    /        |  \  /    \       /     \       /        |  \
HO     NH2-CH2      NH2          [   OH      NH2-CH2       CH2-NH2         ]   OH
                                                                            n-1
```

in which the value of n is in the range 10–20. Addition of ethylenediamine leads to the decrease of nuclearity and the formation of soluble complexes.

A binuclear mixed ligand complex is formed when copper(II) is hydrolysed in the presence of equimolar 2,2′-bipyridyl, 1,10-phenanthroline or histamine[329]. The hydrolysis of certain chelates of iron(II) has also been studied. There is a simple 2 monomer $\rightleftharpoons$ dimer equilibrium when the iron(III) complex of *N*-hydroxyethylenediamine-triacetic acid is hydrolysed[330], but in the hydrolysis of an equimolar mixture of iron(III) and citrate a highly polymeric species is formed at about pH 8–9[331]. In the presence of excess citrate no polymerisation occurs[332].

1.5.3.2 *Polynuclear complexes involving monodentate and multidentate ligands*

The formation of polynuclear complexes is due to the fact that a ligand bound to a metal ion still has the ability to donate a further pair of electrons to another metal ion. This is the case when the coordinated donor atom still has an unshared pair(s) of electrons, or the number of donor groups in the ligand exceeds the maximum coordination number of the central ion, or the steric arrangement of the donor atoms of the multidentate ligand prevents the coordination of all of them to the same central ion. A binuclear mixed-valence copper acetate complex is formed almost quantitatively when a solution of $Cu(CH_3CN)_4ClO_4$ is added to a solution of $Cu(ClO_4)_2 \cdot 6H_2O$ in acetate-buffered methanol[333]. The structure of this complex is thought to be

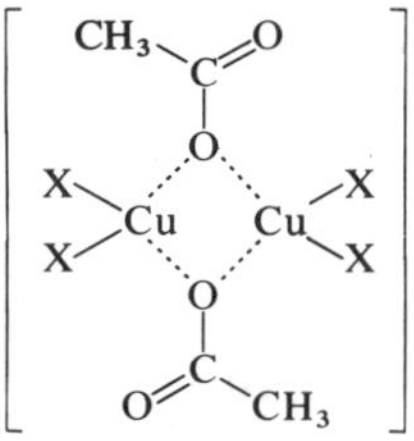

where X represents methanol molecules.

In fact this species is a heteropolynuclear complex, for the formation of which there are many examples. When an aqueous solution of chromium(III) citrate, malate or tartarate containing a large excess of hydroxycarboxylic acid at a pH of about 3 is treated with indium(III) or tin(II), heteropolynuclear species are formed but no information is available on the stabilities of these complexes[334]. The addition of nickel(II) and lanthanum(III) to alizarin fluorine blue (1,2-dihydroxyanthraquinon-3-ylmethylamine-*N*,*N*-diacetic acid) yields a heterobinuclear complex[335]; its apparent stability constant has been determined but more work is necessary for the complete characterisation of the equilibrium behaviour of this interesting complex. Heteropolynuclear complexes are formed when $Hg(CN)_2$ reacts with coordinatively inert complexes[336]. The driving force of these reactions is the residual affinity of mercury(II) in the parent complex and the donor ability of the coordinated cyanides in the inert complexes in question. Stability constants of these complexes have been obtained spectrophotometrically and potentiometrically but it is not known whether one or two cyanide groups of the inert complex serve as bridging ligands.

An interesting class of polynuclear complexes is formed as a consequence of the expansion of the coordination sphere of the central ion.

Bis-(β-diketone)metal(II) complexes polymerise to form trimers or tetramers, some oxygen atoms being shared by two metal atoms[337, 338]. In certain cases a very extensive rearrangement of the coordination sphere occurs. Although it is evident that an equilibrium mixture is always present in the solution, the stability data are not at present available.

The formation of polynuclear complexes involving peptides[339, 340], proteins[340, 341] and polymethacrylic acid[342] has also been studied and is evidently an extremely important and promising subject. A new potentiometric method (potentiostatic titration) developed by Österberg[340] seems a valuable technique in the quantitative study of these systems.

1.5.4 Metal complexes of biological importance

In the living organisms there are a number of metal ions which form a great variety of complexes with the different species containing donor atoms. These potential ligands are aminoacids, peptides, proteins, nucleic acids, different metabolites, vitamins, etc. The complexes are vitally important in the strict meaning of the word as regards enzyme action, metabolism and transport phenomena, and moreover in the effect of many drugs, pharmaceuticals, bacteriostatics, fungicides, etc. A prerequisite of the complete understanding of these and many other biological processes and phenomena is the clarification of the complicated equilibria involved. This conclusion is evident considering the great number of relevant studies, many of which have already been mentioned particularly in connection with mixed ligand complexes. In this section some further examples are given to illustrate the broadness of this intriguing field.

Most of the results were achieved in connection with the complexes of aminoacids and their derivatives[343–351]. A critical survey of the stability constants of glycine complexes has been published[351]. Particularly interesting problems are the mode of coordination of hydroxyamino acids and the stabilities of complexes of aromatic aminoacids. The stabilities of a number of aminoacid ester–metal complexes were studied to throw light on the mechanism of their hydrolyses[352, 353]. A systematic investigation of complexes of peptides has been made by Österberg[340]. Infrared and potentiometric studies have elucidated the equilibria involving copper(II) and tri- and tetraglycine[402].

The complex-forming properties of adrenaline[355] and DOPA[356] (L- β-) 3,4-dihydroxyphenyl(alanine) have been studied. There are a number of natural macrocyclic complex systems where the reversible uptake of ligands occurs only at the axial positions (octahedral configuration). Typical and the most thoroughly studied compound is Vitamin B_{12} [357]. The heterocyclic bases adenine, guanine, cytosine, uracil and thymine have many donor atoms and their complex-forming properties may be biologically important.

Quantitative equilibrium studies have been made[358–361] for some of their metal complexes including mixed ligand complexes. Data were obtained for the behaviour of their important phosphate derivatives[362, 363]. However, at present we know nothing about the complex equilibria involving these substances *in vivo* or their biological consequences. A little more is known about the role of the stabilities of certain metal complexes in their antifungal effects and some correlation has been found between the stability constants and their efficiencies. It could be concluded from the stability studies and from the *in vivo* experiments that a 1:1 chelate is the active toxicant[297].

1.5.5 New central ions and ligands

The number of central ions is evidently limited by the periodic system. However, a great number of organometallic ions have been prepared, the original properties of the central ion of which are influenced substantially by the formation of the carbon–metal bond(s) and because the bond between the organic group and the metal ion is not ruptured in the reversible uptake of ligands, they may be considered as new central ions. Assigning a formal oxidation number to the central ion is arbitrary and the comparison of mercury(II) and methylmercury(II) ions is rather difficult. Of course these species may also be regarded as mixed ligand complexes, the organic group(s) being considered as one kind of ligand. The stabilities of different complexes of a number of organometallic cations have been studied intensively and a review[364] has been published providing tabulated stability data.

It is more obvious that the number of investigated systems has been increased by the purposeful preparation of new multidentate ligands. Only a few of them are mentioned here; the reader should consult with the coming Supplement of Stability Constants[365].

Complexes of polyaminopolycarboxylic acids have been especially thoroughly studied. Triethylenetetraminehexaacetic acid has as many as ten donor atoms; it therefore yields very stable complexes and exhibits a tendency to form polynuclear species[366–368]. Ligands containing phosphonate groups have been synthesised and the stability constants of their complexes determined[369]. Aromatic aminopolycarboxylic acid complexes have been systematically studied[370]. Such investigations furnish important data towards the derivation of relationships between the electronic structures of the substituents and the stabilities of the complexes. The list of sulphur-containing ligands too is continuously increasing. For example, the stabilities of complexes of bithional structure(I), fenticlor structure(II)[371], commercially available bacteriostatic compounds, and of the derivatives of dithio-oxamide[372] have been determined.

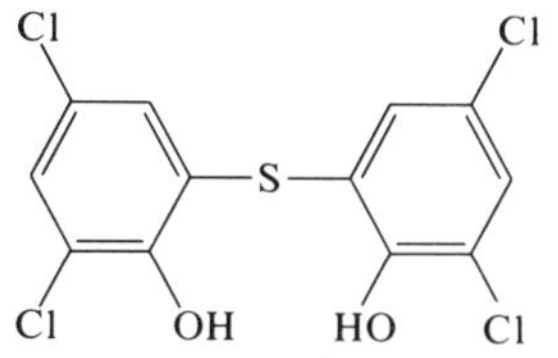

I

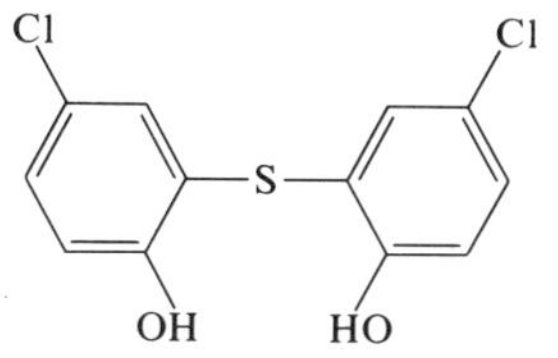

II

1.6 COMPLEX EQUILIBRIA IN NON-AQUEOUS MEDIA

An overwhelming part of the work quoted in the previous sections referred to aqueous solutions. There is an infinite number of possibilities of working with other media, from molten salts to a great variety of organic and inorganic solvents. The study of complex equilibria in non-aqueous solution is of twofold importance: first it provides a possibility for deeper understanding of complex equilibria *in general* by the comparison of different solvent systems; secondly there are many systems of practical importance requiring elucidation of the complex equilibria involved. The very concentrated aqueous solutions mean a transition stage between dilute aqueous solutions and molten salt systems, and their description may be easier by approaching the problem from the molten salt side[373, 374]. An excellent survey on the complex equilibria in different non-aqueous solvents has been given by Gutmann[375]. Most of the work has led to qualitative results; there are only a few systems for which reliable stability constants are available. Interestingly, relatively many enthalpy values have been determined, making possible the introduction of the concept of donicity[375, 376].

Silver(I) halide systems in different organic solvents of high dielectric constants have been studied most thoroughly[377–382]. A striking result of these investigations is that generally the formation of the anionic complexes is much more preferred in comparison with the aqueous medium. Experiments with dimethyl sulphoxide–water[381] and propylene carbonate–water[379] mixtures have revealed that a small amount of water results in quite a large decrease of the constants. This can be explained by specific solvation effects and by change in the solvent structure. On the other hand these results stress the importance of the controlling of the water content of these 'non-aqueous' media. One must consider that it is practically impossible to decrease the water content of such solvents below $10^{-4}-10^{-3}$ mol l^{-1}. Chloro-complexes of gold(I) complexes in acetonitrile have also been studied and the stability data of copper(I), silver(I) and gold(I) complexes have been compared[383]. Qualitative differences between the stability patterns of these complexes (e.g. $K_1/K_2<1$ in the case of Au^I, while $K_1/K_2>1$ in the case of Cu^I) await clarification. Surprising results have also been obtained for the stability constants of mercury(II) halide complexes in dimethyl formamide[384]. All the stability constants are much greater than the corresponding constants referred to water, and the increase of the stabilities of the third and fourth complexes is especially pronounced. Furthermore, for the fourth stepwise constant the order $Cl>Br>I$ was found, which is just the opposite of that in aqueous solution. Similarly, an unusual stability order was found for tetrahalonickelate(II) complexes in dimethyl sulphone[385]. These results show convincingly that solvation effects have to be considered carefully in the explanation of stability behaviour.

1.7 CORRELATIONS BETWEEN THE STABILITY CONSTANTS OF METAL COMPLEXES AND EXTERNAL AND INTERNAL FACTORS

There is an increase of demand with the increasing number of data for relationships among the stability constants of related complexes and between the stability constants and characteristic properties of their constituents. Furthermore, one should find how the variation of different external factors such as pressure, dielectric constant, etc. influence the stability constants. As the whole problem has recently been discussed this survey is restricted to treat some basic ideas only.

1.7.1 The effect of pressure and solvents on the stability constant

The effect of a change in the dielectric constant of the medium on the stability constants was thoroughly studied and a few data are now available on the effect of pressure. Marshall and Quist offered a basically new approach to the problem by defining the complete equilibrium constant,

$$K^0 = \frac{[M(H_2O)_m][L(H_2O)_l]}{[ML(H_2O)_k]} = \frac{K}{C^n_{H_2O}}$$

K is the concentration equilibrium constant, C_{H_2O} is the water concentration in mol l^{-1} and n is the number of water molecules becoming free in the complexation reaction ($n = m+l-k$). For non-aqueous solvents consisting of one donor and one inert solvent analogous 'complete constants' can be defined. A change in C_{H_2O} may occur as a consequence of pressurisation or of replacing water partly by an inert solvent. Marshall[387] believes that the introduction of 'complete constants' can dismiss the concept of activity. The great number of experimental data referring to ion pairs and obtained mainly at high temperatures and pressures could be quantitatively described by the 'complete constants', considering the change in C_{H_2O}. This approach has been severely criticised by Gilkerson[388] and Matheson[389]. It is evident from this dispute that Marshall's treatment, as any other description of a complicated system, is a simplification of the real situation and n should better be regarded an adjustable parameter than the effective change in hydration, nevertheless one may expect that it will be useful from both practical and theoretical points of view.

1.7.2 Linear free energy relationships

In most cases a parallelism is observed between the proton affinity and metal ion affinity of a certain ligand. There are examples, however, for the opposite behaviour[393] and back coordination was assumed as being responsible for this. It was convincingly pointed out[394, 395] that this explanation is not correct because the inverse relationship is observed with ligands where there is no opportunity for π-bonding and several different factors lead to this behaviour.

A critical analysis of linear free energy relationships among stability constants has been made by Nieboer and McBryde[396], who derived the following equations:

$$\log K_{ML} = B \log K_{M'L} + (\log K_{ML'} - B \log K_{M'L'})$$
$$\log K_{ML} = C \log K_{ML'} + (\log K_{M'L} - C \log K_{M'L'})$$

M′ and L′ are reference metal ion and ligand, respectively. B is the rate of change of stability constants in a series of complexes ML of the metal ion M with the stability constant of the related series of complexes M′L of the reference metal ion. Similarly, C is the rate of change of stability constants in a series of complexes ML of the ligand L with the stability constants of the related series of complexes ML′ of the reference ligand L′. In testing these equations with data for a great number of complex systems, two kinds of behaviour were observed. In one there was a mutual dependence of B on the particular metal ions compared and of C on the particular ligands compared. In the other one all values of B and C were found to equal unity irrespective of metal ion and ligand compared. At present there is no explanation of this empirical correlation. As a matter of course such linear relationships may be expected to occur among complexes of structurally analogous ligands and metal ions of similar electron structure.

1.7.3 Correlation between the properties of metal ions (and ligands) and the formation enthalpies and entropies

1.7.3.1 Linear relationships between ΔH and ΔS

It was found in many cases[390–392] that there is a linear correlation between ΔH and ΔS for a series of complexes, that is

$$\Delta H = a + b\,\Delta S$$

Such correlations are sometimes regarded[391] as evidence for a common structure throughout the series of complexes. Considering that $\Delta G = \Delta H - T\Delta S$, Fay and Purdie[392] pointed out that such a linear correlation is based on certain restrictions on ΔG (i.e. $\Delta G = 0$ or constant throughout the series) and ΔG itself does not give information on the structure of the species, this linear relationship between ΔH and ΔS is also independent of the structure of the species in question.

1.7.3.2 Correlation between the bonding type and the formation enthalpies and entropies

As is well known the distinction between class a (hard) and class b (soft) metal ions is based on the stability order of complexes formed with different ligands. The different types of bonding, however, are more clearly reflected in the enthalpy and entropy values of complex formation reactions. On the basis of the available data Ahrland[397, 398] has given some useful qualitative rules. It seems that the driving force of complex formation in the case of essentially electrostatic bonding is the large gain of entropy, while in the case of essentially covalent bonding the large decrease of enthalpy overcompensates the unfavourable entropy change. The more negative ΔH, the greater the covalency of the bonding.

A more sophisticated and quantitative treatment has been given by Nancollas[399, 400] and Anderegg[401], who splits both ΔG and ΔH into an electrostatic and a covalent part.

References

1. Sillén L. G. and Martell, A. E. (1964). *Stability Constants of Metal-ion Complexes*, Special Publication No. 17. (London: Chemical Society)
2. Eigen, M. and Tamm, K. (1962). *Z. Elektrochem.*, **66**, 107
3. Guggenheim, E. A. and Stokes, R. A. (1969). *Equilibrium Properties of Aqueous Solutions of Single Strong Electrolytes.* (Oxford: Pergamon Press)
4. Prue, J. E. (1966). *Ionic Equilibria.* (Oxford: Pergamon Press)
5. Nancollas, G. A. (1966). *Interactions in Electrolyte Solutions.* (Amsterdam: Elsevier)
6. Beck, M. T. (1970). *Chemistry of Complex Equilibria.* (London: Van Nostrand Reinhold Co.)
7. Prue, J. E. (1970). *Proc. 3rd. Symposium Coord. Chem. Vol. 1*, 25. (Budapest: Akadémiai Kiadó)
8. Prue, J. E. and Read, A. J. (1966). *J. Chem. Soc. A*, 1812
9. Ginstrup, O. (1970). *Acta. Chem. Scand.*, **24**, 875
10. Huston, R. and Butler, J. N. (1969). *Anal. Chem.*, **41**, 1695
11. Butler, J. N. and Huston, R. (1970). *Anal. Chem.*, **42**, 676
12. Guggenheim, E. A. and Turgeon, J. C. (1955). *Trans. Faraday Soc.*, **51**, 747
13. Burnett, M. G. (1970). *J. Chem. Soc. A*, 2480
14. Bond, A. M. (1970). *J. Phys. Chem.*, **74**, 331
15. Ilcheva, L. and Beck, M. T. (1970). *Proc. 3rd Symp. Coord. Chem.*, 89. (Budapest: Akadémiai Kiadó)
16. Choppin, G. R., Kelly, D. A. and Ward, E. H. (1967). *Solvent Extraction Chemistry*, 46. (Amsterdam: North-Holland)
17. Hartings, S. H., Franklin, J. L., Schiller, J. C. and Matser, F. A. (1952). *J. Amer. Chem. Soc.*, **75**, 2900
18. Prue, J. E. (1965). *J. Chem. Soc.*, 7534
19. Person, W. B. (1965). *J. Amer. Chem. Soc.*, **87**, 167
20. Norheim, G. (1969). *Acta Chem. Scand.*, **23**, 2808
21. Heric, E. L. (1969). *J. Phys. Chem.*, **73**, 3496
22. Butler, J. N. and Huston, R. (1970). *J. Phys. Chem.*, **74**, 2976
23. Butler, J. N. and Huston, R. (1970). *Anal. Chem.*, **42**, 1308
24. Durst, R. A. (1969). *Ion-selective Electrodes.* (Washington: National Bureau of Standards)
25. Butler, J. N. and Huston, R. (1967). *J. Phys. Chem.*, **71**, 4479
26. Coyley, D. R. and Butler, J. N. (1968). *J. Phys. Chem.*, **72**, 1017
27. Hall, F. M. and Slater, I. J. (1968). *Aust. J. Chem.*, **21**, 2663
28. Budevsky, O. and Platikanova, E. (1967). *Talanta*, **14**, 901
29. DeAlneida Neves, E. F. and Sant'Agostino, L. (1970). *Anal. Chim. Acta*, **49**, 591
30. Eisenman, G. (1967). *Glass Electrodes for Hydrogen and Other Cations.* (New York: Manuel Dekker)
31. McClure, J. E. and Rechnitz, G. A. (1966). *Anal. Chem.*, **38**, 136
32. Rechnitz, G. A. and Liu, Z. F. (1967). *Anal. Chem.*, **39**, 1406
33. Gardner, G. L. and Nancollas, G. A. (1969). *Anal. Chim. Acta*, **41**, 202
34. Scibone, G., Mantella, R. and Danesi, P. R. (1970). *Anal. Chem.*, **42**, 844
35. Pungor, E. and Havas, I. (1966). *Acta. Chim. Hung.*, **50**, 77
36. Hirata, H. and Dati, K. (1970). *Talanta*, **17**, 883
37. Pungor, E. (1967). *Anal. Chem.*, **39**, 28A
38. Buchanan, E. B. and Seago, J. L. (1968). *Anal. Chem.*, **40**, 5A
39. Ruzicka, J. and Tjell, C. (1970). *Anal. Chim. Acta*, **49**, 346
40. Rechnitz, G. A. and Lin, Z. F. (1968). *Anal. Chem.*, **40**, 696
41. Huston, R. and Butler, J. N. (1970). *Anal. Chem.*, **41**, 200
42. Watters, J. I., Kalliney, S. and Machen, R. C. (1969). *J. Inorg. Nucl. Chem.*, **31**, 3823
43. Shatkay, A. (1967). *Anal. Chem.*, **39**, 1056
44. Shatkay, A. (1967). *J. Phys. Chem.*, **71**, 3858
45. Rechnitz, G. A. and Hseu, T. M. (1969). *Anal. Chem.*, **41**, 111
46. Nakayama, F. S. and Rasnick, B. A. (1967). *Anal. Chem.*, **39**, 1022
47. Stefanac, Z. and Simon, W. (1966). *Chimia*, **20**, 436
48. Shemyakin, M. M., Ovchinnikov, Yu. A., Ivanov, V. T., Antonov, V. K., Shkrob, A. M., Mikhaeleva, I. I., Evstratov, A. V. and Malenkov, G. G. (1967). *Biochem. Biophys. Res. Comm.* **29**, 834

49. Thompson, M. E. and Ross, J. W. (1966). *Science,* **154,** 1643
50. Pioda, L. A. R., Stankova, V. and Simon, W. (1969). *Anal. Lett.,* **2,** 665
51. Lal, S. and Christian, G. D. (1970). *Anal. Lett.,* **3,** 11
52. Krull, I. H., Mask, C. A. and Cosgrave, R. E. (1970). *Anal. Lett.,* **3,** 43
53. Mohan, M. S. and Rechnitz, G. A. (1970). *J. Amer. Chem. Soc.,* **92,** 5839
54. Frant, M. and Ross, J. W. (1966). *Science,* **154,** 1553
55. Srinisavan, K. and Rechnitz, G. A. (1968). *Anal. Chem.,* **40,** 1818
56. Baumann, E. (1969). *J. Inorg. Nucl. Chem.,* **31,** 3155
57. Elgquist, B. (1970). *J. Inorg. Nucl. Chem.,* **32,** 937
58. Bond, A. M. and Heffer, G. (1970). *Inorg. Chem.,* **9,** 1021
59. Mesmer, R. E. and Baes, C. F. (1969). *Inorg. Chem.,* **8,** 618
60. Aziz, A. and Lyle, S. J. (1969). *Anal. Chim. Acta,* **47,** 49
61. Rhyl, T. (1969). *Acta Chem. Scand.,* **23,** 2667
62. Baumann, E. W. (1970). *J. Inorg. Nucl. Chem.,* **32,** 3823
63. Hume, D. N. and Grossino, S. L. (1970). *Proc. 13th I.C.C.C., Cracow-Zakopane,* Vol. 1., 75
64. Burger, K. and Pintér, B. (1966). *Hung. Sci. Instr.,* No. **8,** 11
65. Hseu, T. M. and Rechnitz, G. A. (1968). *Anal. Lett.,* **1,** 23
66. Dobbelstein, T. N. and Diehl, H. (1969). *Talanta,* **16,** 1341
67. Coetzee, C. J. and Freiser, H. (1968). *Anal. Chem.,* **40,** 2071
68. Coetzee, C. J. and Freiser, H. (1969). *Anal. Chem.,* **41,** 1128
69. Hirsch, R. F. and Portock, J. D. (1969). *Anal. Lett.,* **2,** 295
70. Baldwin, W. G. (1969). *Arkiv Kemi,* **31,** 407
71. Brosset, C. and Orring, J. (1943). *Svensk Kemisk Tidskr.,* **55,** 101
72. Nordén, B. (1967). *Acta Chem. Scand.,* **21,** 2435
73. Nordén, B. (1967). *Acta Chem. Scand.,* **21,** 2457
74. Shulman, V. M., Kremazyova, T. V., Larionov, S. V. and Dubinskij, V. I. (1967). *Proc. 10th I.C.C.C.,* 63. (Tokyo: Chemical Society of Japan)
75. Irving, H. M., Miles, M. G. and Pettit, L. D. (1967). *Anal. Chim. Acta,* **38,** 475
76, Light, T. S. and Fletcher, K. S. (1967). *Anal. Chem.,* **39,** 70
77. Beck, W. H., Bottom, A. E. and Covington, A. E. (1968). *Anal. Chem.,* **40,** 501
78. Irving, H. M. N. H. and Mahnot, U. S. (1968). *J. Inorg. Nucl. Chem.,* **30,** 1215
79. Norberg, K. (1966). *Talanta,* **13,** 745
80. McBryde, W. A. E. (1967). *Can. J. Chem.,* **45,** 2093
81. Nagypál, I. and Gergely, A. (1971). *Magy. Kém. Foly.,* in print
82. Carpéni, G., Double, G., Ferrani, G., Rocque, C., Asensi, P. and Ngoc, G. N. (1969). *Electrochim. Acta,* **14,** 587
83. Synnott, J. C. and Butler, J. N. (1969). *Anal. Chem.,* **41,** 189
84. Baucke, F. G. K. and Tobias, C. W. (1969). *J. Electrochem. Soc.,* **116,** 34
85. Marple, L. W. (1967). *Anal. Chem.,* **39,** 844
86. Manning, C. W. and Purdy, W. C. (1970). *Anal. Chim. Acta,* **51,** 124
87. Kratochvil, B., Lorah, E. and Garber, C. (1969). *Anal. Chem.,* **41,** 1793
88. Bond, A. M. (1969). *J. Electroanal. Chem.,* **23,** 277
89. Bond, A. M. (1969). *J. Electroanal. Chem.,* **23,** 269
90. Momoki, K., Sato, H. and Ogawa (1967). *Anal. Chem.,* **39,** 1072
91. Momoki, K., Ogawa, H. and Sato, H. (1969). *Anal. Chem.,* **41,** 1826
92. Tanaka, N., Ogino, K. and Sato, G. (1966). *Bull. Chem. Soc. Japan,* **39,** 366
93. Tanaka, N. and Yamada, A. (1967). *Z. Anal. Chem.,* **224,** 117
94. Katakis, D. (1965). *Anal. Chem.,* **37,** 876
95. Kankare, J. J. (1970). *Anal. Chem.,* **42,** 1322
96. Lingane, P. J. and Hugus, Z. Z. (1970). *Inorg. Chem.,* **9,** 1015
97. Coleman, P. J., Varga, L. P. and Mastin, S. H. (1970). *Inorg. Chem.,* **9,** 757
98. Ramette, R. W. (1967). *J. Chem. Educ.,* **44,** 647
99. Wallin, T. (1966). *Arkiv Kemi,* **26,** 13
100. Marcus, Y. (1967). *Israel J. Chem.,* **5,** 143
101. Larsson, R. and Miezis, A. (1968). *Acta Chem. Scand.,* **22,** 3261
102. Taylor, K. A., Long, T. V. and Plane, R. A. (1967). *J. Phys. Chem.,* **47,** 138
103. Spiro, T. G. (1965). *Inorg. Chem.,* **4,** 731
104. Keller, O. L. and Chetham-Strode, A. (1966). *Inorg. Chem.,* **5,** 367
105. Davis, A. R. and Plane, R. A. (1968). *Inorg. Chem.,* **7,** 2565

106. Miller, J. T. and Irish, D. E. (1967). *Can. J. Chem.*, **45,** 147
107. Clarke, J. H. R. and Woodward, L. A. (1968). *Trans. Faraday Soc.*, **64,** 1041
108. Tobias, R. S., Sprague, M. J. and Glass, G. E. (1968). *Inorg. Chem.*, **7,** 1714
109. Krishnan, K. and Plane, R. A. (1966). *Inorg. Chem.*, **5,** 852
110. Gruen, E. C. and Plane, R. A. (1967). *Inorg. Chem.*, **6,** 1123
111. Krishnan, K. and Plane, R. A. (1968). *J. Amer. Chem. Soc.*, **90,** 3195
112. Loehr, T. M. and Plane, R. A. (1968). *Inorg. Chem.*, **7,** 1708
113. Maroni, V. A. and Spiro, T. G. (1966). *J. Amer. Chem. Soc.*, **88,** 1410
114. Maroni, V. A. and Spiro, T. G. (1968). *Inorg. Chem.*, **7,** 188
115. Griffith, W. P. and Wickins, T. D. (1967). *J. Chem. Soc. A*, 675
116. Carreira, L. A., Maroni, V. A., Swaine, J. W. and Plumb, R. C. (1968). *J. Chem. Phys.*, **45,** 2216
117. Matwiyoff, N. A., Darley, P. E. and Movius, W. G. (1968). *Inorg. Chem.*, **7,** 2173
118. Movius, W. G. and Matwiyoff, N. A. (1968). *J. Phys. Chem.*, **72,** 2063
119. Supran, L. D. and Sheppard, N. (1967). *Chem. Commun.*, 832
120. O'Brien, J. F. and Alei, A. (1970). *J. Phys. Chem.*, **74,** 743
121. Haraguchi, A. and Fujiwara, S. (1969). *J. Phys. Chem.*, **73,** 3467
122. Matwiyoff, N. A. and Arley, P. E. (1968). *J. Phys. Chem.*, **72,** 2659
123. Foreman, M. J., Gorton, J. and Foster, R. (1970). *Trans. Faraday Soc.*, **66,** 2120
124. Lin, W. and Tsay, S. (1970). *J. Phys. Chem.*, **74,** 1037
125. Al-Baldawi, S. A. and Gough, R. E. (1969). *Can. J. Chem.*, **47,** 1417
126. Burlemacchi, L. and Tiezzi, E. (1969). *J. Inorg. Nucl. Chem.*, **31,** 2159
127. Frankel, L. S., Stengle, T. R. and Langford, C. H. (1967). *J. Inorg. Nucl. Chem.*, **29,** 243
128. Katz, J. J., Strain, H. H., Leussing, D. L. and Sougherty, R. C. (1968). *J. Amer. Chem. Soc.*, **90,** 784
129. Aochi, Y. O. and Sawyer, D. T. (1966). *Inorg. Chem.*, **5,** 2085
130. Smith, B. B. and Sawyer, D. T. (1969). *Inorg. Chem.*, **8,** 1154
131. Enman, D. L. and Sawyer, D. T. (1970). *Inorg. Chem.*, **9,** 204
132. Smith, B. B. and Sawyer, D. T. (1969). *Inorg. Chem.*, **8,** 379
133. Kula, R. J. (1966). *Anal. Chem.*, **38,** 1382
134. Smith, B. B. and Sawyer, D. T. (1968). *Inorg. Chem.*, **7,** 2020
135. Kula, R. J. (1967). *Anal. Chem.*, **39,** 1171
136. Whidby, J. F. and Leyden, D. E. (1970). *Anal. Chim. Acta.*, **51,** 25
137. Pratt, L. and Smith, B. B. (1969). *Trans. Faraday Soc.*, **65,** 1703
138. Sudmeier, J. L. and Senzel, A. J. (1968). *Anal. Chem.*, **40,** 1693
139. Ho, F. F. L. and Reilley, C. N. (1969). *Anal. Chem.*, **41,** 1855
140. Thwaites, J. D. and Sacconi, L. (1966). *Inorg. Chem.*, **5,** 1029
141. Thwaites, J. D. and Sacconi, L. (1966). *Inorg. Chem.*, **5,** 1036
142. Holm, R. H., Chakravorty, A. and Theriot, L. J. (1966). *Inorg. Chem.*, **5,** 625
143. Dillon, K. B. and Rossotti, F. J. C. (1966). *Chem. Commun.*, 768
144. Baker, B. C. and Sawyer, D. T. (1968). *Anal. Chem.*, **40,** 1945
145. McCormick, D. B., Sigel, H. and Wright, L. D. (1969). *Biochim. Biophys. Acta*, **184,** 318
146. Sigel, H. Griesser, R. and McCormick, D. B. (1969). *Arch. Biochem. Biophys.*, **134,** 217
147. McCain, D. C. and Myers, R. J. (1968). *J. Phys. Chem.*, **72,** 4115
148. Burlemacchi, L. and Tiezzi, E. (1969). *J. Phys. Chem.*, **73,** 1588
149. Burlemacchi, L. and Tiezzi, E. (1968). *J. Mol. Struct.*, **2,** 261
150. Burlemacchi, L., Martini, G. and Tiezzi, E. (1970). *J. Phys. Chem.*, **74,** 3980
151. Sancier, K. M. (1968). *J. Phys. Chem.*, **72,** 1317
152. Burlemacchi, L., Martini, G. and Tiezzi, E. (1970). *J. Phys. Chem.*, **74,** 1809
153. Boas, J. F., Pilbrow, J. R. and Smith, T. D. (1969). *J. Chem. Soc. A*, 723
154. Boas, J. F., Dunhill, R. H., Pilbrow, J. R., Srivastava, R. C. and Smith, T. D. (1969). *J. Chem. Soc. A*, 94
155. Johansson, S. (1965). *Arkiv Kemi*, **24,** 189
156. Arnek, R. and Kakolowicz, W. (1967). *Acta Chem. Scand.*, **21,** 1449
157. Brunetti, A. P., Lim, M. C. and Nancollas, G. H. (1968). *J. Amer. Chem. Soc.*, **90,** 5120
158. Harris, P. C. and Moore, T. E. (1968). *Inorg. Chem.*, **7,** 656
159. Christiansen, J. J., Izatt, R. M. and Hansen, L. D. (1965). *Rev. Sci. Instr.*, **36,** 779
160. Simeon, V. and Tkalcec, M. (1968). *Lab. Pract.*, **17,** 1353
161. Barnes, D., Laye, P. G. and Pettit, L. D. (1969). *J. Chem. Soc. A*, 2073
162. Arnek, R. (1970). *Arkiv Kemi*, **32,** 55

163. Gerding, P. (1966). *Acta Chem. Scand.*, **20,** 79
164. Gerding, P. (1968). *Acta Chem. Scand.*, **22,** 2255
165. Gerding, P. and Jönsson, I. (1968). *Acta Chem. Scand.*, **22,** 2247
166. Gerding, P. (1969). *Acta Chem. Scand.*, **23,** 1695
167. Kennedy, M. B. and Lister, M. W. (1966). *Can. J. Chem.*, **44,** 1709
168. Fay, D. P. and Purdie, N. (1969). *J. Phys. Chem.*, **74,** 3462
169. Grenthe, I. and Williams, D. R. (1967). *Acta Chem. Scand.*, **21,** 347
170. Pearson, K. H., Baker, J. R. and Goodrich, J. D. (1969). *Anal. Lett.*, **2,** 577
171. Walter, J. L. and Rosalie, S. M. (1966). *J. Inorg. Nucl. Chem.*, **28,** 2969
172. Stack, W. F. and Skinner, H. A. (1967). *Trans. Faraday Soc.*, **63,** 1136
173. Thornton, H. C. R. and Skinner, H. A. (1969). *Trans. Faraday Soc.*, **65,** 2044
174. Williams, D. R. (1970). *J. Chem. Soc. A,* 1550
175. Nancollas, G. H. and Poulton, D. J. (1969). *Inorg. Chem.*, **8,** 680
176. Milburn, R. M. (1967). *J. Amer. Chem. Soc.*, **89,** 54
177. McAuley, A., Nancollas, G. H. and Torrance, K. (1967). *Inorg. Chem.*, **6,** 136
178. Simeon, V., Svigir, B. and Paulic, N. (1969). *J. Inorg. Nucl. Chem.*, **31,** 2085
179. Uhlig, E. and Martin, A. (1969). *J. Inorg. Nucl. Chem.*, **31,** 2781
180. Brunetti, A. P., Nancollas, G. H. and Smith, P. N. (1969). *J. Amer. Chem. Soc.*, **91,** 4680
181. Carson, A. S., Laye, P. G. and Smith, P. N. (1968). *J. Chem. Soc. A,* 1384
182. Degischer, G. and Nancollas, G. H. (1970). *Inorg. Chem.*, **9,** 1259
183. Brenner, A. (1965). *J. Electrochem. Soc.*, **112,** 611
184. Christensen, J. J., Izatt, R. M., Hansen, L. D. and Partridge, J. A. (1966). *J. Phys. Chem.*, **70,** 2003
185. Christensen, J. J., Wrathall, D. P. and Izatt, R. M. (1968). *Anal. Chem.*, **40,** 175
186. Christensen, J. J., Wrathall, D. P., Oscarson, J. O. and Izatt, R. M. (1968). *Anal. Chem.*, **40,** 1713
187. Wanders, A. C. M. and Zwietering, T. N. (1969). *J. Phys. Chem.*, **73,** 2076
188. Izatt, R. M., Eatough, D., Snow, R. L. and Christensen, J. J. (1968). *J. Phys. Chem.*, **72,** 1208
189. Arnek, R. (1970). *Arkiv Kemi,* **32,** 81
190. Christensen, J. J., Rytting, J. H. and Izatt, R. M. (1969). *J. Chem. Soc. A,* 861
191. Gerding, P. (1968). *Acta Chem. Scand.*, **22,** 1283
192. Christensen, J. J., Izatt, R. M. and Hansen, L. D. (1967). *J. Amer. Chem. Soc.*, **89,** 213
193. Christensen, J. J., Izatt, R. M., Wrathall, D. P. and Hansen, L. D. (1969). *J. Chem. Soc. A,* 1212
194. Metzger, W. H., Brenner, A. and Salmon, H. I. (1967). *J. Electrochem. Soc.*, **114,** 131
195. Brenner, A. and Metzger, W. A. (1968). *J. Electrochem. Soc.*, **115,** 258
196. Frei, V. and Solcova, A. (1967). *Collect. Czech. Chem. Commun.*, **32,** 1815
197. Frei, V. and Solcova, A. (1965). *Collect. Czech. Chem. Commun.*, **30,** 961
198. Folkesson, B. and Larsson, R. (1968). *Acta Chem. Scand.*, **22,** 1953
199. Antikainen, P. J. and Tevaner, K. (1970). *J. Inorg. Nucl. Chem.*, **32,** 1915
200. Larsson, R., Mason, B. F. and Norman, B. J. (1966). *J. Chem. Soc. A,* 301
201. Larsson, R., Marcusson, B. and Nunziata, G. (1970). *Proc. 3rd. Symp. Coord. Chem. Vol. I.,* 53/Budapest: Akadémiai Kiadó
202. Crescenzi, V., Quadrifoglio, F. and Pispisa, B. (1968). *J. Chem. Soc. A,* 2175
203. Johansson, L. (1968). *Coordin. Chem. Rev.*, **3,** 293
204. Haight, G. P. and Johansson, L. (1968). *Acta Chem. Scand.*, **22,** 961
205. Preer, J. R. and Haight, G. P. (1966). *Inorg. Chem.*, **5,** 656
206. Ahrland, S. and Rawsthorne, J. (1970). *Acta Chem. Scand.*, **24,** 1578
207. Johansson, L. (1970). *Acta Chem. Scand.*, **23,** 1572
208. Dyrssen, D., Liljenzin, J. O. and Rydberg, J. (1967). *Solvent Extraction Chemistry.* (Amsterdam: North Holland)
209. Kertes, A. S. and Marcus, Y. (1969). *Solvent Extraction Research.* (New York: Wiley-Interscience)
210. Marcus, Y. and Kertes, A. S. (1970). *Ion Exchange and Solvent Extraction of Metal Complexes.* (New York: Wiley-Interscience)
211. Rydberg, J. (1969). *Acta Chem. Scand.*, **23,** 647
212. Reinhardt, H. and Rydberg, J. (1969). *Acta Chem. Scand.*, **23,** 2773
213. Andersson, C., Andersson, S. O., Liljenzin, J. O., Reinhardt, H. and Rydberg, J. (1969). *Acta. Chem. Scand.*, **23,** 2181

214. Elding, L. J. (1966). *Acta Chem. Scand.*, **20,** 2559
215. Espenson, J. H. and Helzer, S. R. (1969). *Inorg. Chem.*, **8,** 1051
216. Kolski, G. B. and Margerum, D. W. (1968). *Inorg. Chem.*, **7,** 2239
217. Swaddle, T. W. and Guastalla, G. (1969). *Inorg. Chem.*, **8,** 1604
218. Wells, C. F. (1970). *Proc. 3rd Symp. Coord. Chem. Vol. 1,* 387. (Budapest: Akadémiai Kiadó)
219. Durham, D. A. and Beck, M. T. (1970). *Proc. 3rd Symp. Coord. Chem., Vol. 1,* 415, (Budapest: Akadémiai Kiadó)
220. Poon, C. K. and Tobe, M. L. (1967). *J. Chem. Soc. A.* 2069
221. Diebler, H. (1970). *Z. Elektrochem.*, **74,** 268
222. Strehlow, H. and Knoche, W. (1969). *Z. Elektrochem.*, **73,** 427
223. Matusch, M. and Strehlow, H. (1969). *Z. Elektrochem.*, **73,** 982
224. Jones, J. P. and Margerum, D. W. (1969). *Inorg. Chem.*, **8,** 1486
225. Yasuda, M. (1968). *Bull. Chem. Soc. Japan,* **41,** 139
226. Hildebrand, J. H. and Scott, R. L. (1962). *Regular Solutions.* (Englewood Cliffs: Prentice Hall)
227. Irving, H. M. N. H. and Smith, J. S. (1969). *J. Inorg. Nucl. Chem.*, **31,** 3163
228. Irving, H. M. N. H. and Smith, J. S. (1969). *J. Inorg. Nucl. Chem.*, **31,** 159
229. Dézsi, I., Vértes, A. and Komor, M. (1968). *Inorg. Nucl. Chem. Lett.*, **4,** 649
230. Vértes, A., Komor, M., Dézsi, I., Burger, K., Suba, M. and Gelencsér, P. (1970). *Proc. 3rd Symp. Coord. Chem. Vol. 1,* 477. (Budapest: Akadémiai Kiadó)
231. Johannson, G. (1968). *Acta Chem. Scand.*, **22,** 399
232. Johansson, G. and Olin, Å. (1968). *Acta Chem. Scand.*, **22,** 3197
233. Wertz, D. L. and Kruh, R. F. (1970). *Inorg. Chem.*, **9,** 595
234. Bacon, W. E. and Brown, G. H. (1969). *J. Phys. Chem.*, **73,** 4163
235. Stryker, L. J. and Matijevic, E. (1969). *J. Phys. Chem.*, **73,** 1484
236. Sillén, L. G. and Warnquist, B. (1969). *Arkiv. Kemi,* **31,** 315
237. Sillén, L. G. and Warnquist, B. (1969). *Arkiv. Kemi,* **31,** 341
238. Arnek, R., Sillén, L. G. and Wahlberg, O. (1969). *Arkiv. Kemi,* **31,** 353
239. Brauner, P., Sillén, L. G. and Whiteker, R. (1969). *Arkiv. Kemi*
240. Sillén, L. G. and Warnquist, B. (1969). *Arkiv. Kemi,* **32,** 377
241. Hume, D. N. (1971). *Proc. 3rd Symp. Coord. Chem. Vol. 2.* (Budapest: Akadémiai Kiadó)
242. Sayce, I. G. (1968). *Talanta,* **15,** 1397
243. Tobias, S. R. and Yasuda, M. (1963). *Inorg. Chem.*, **2,** 1307
244. Perrin, D. D. and Sayce, I. G. (1967). *J. Chem. Soc. A,* 82
245. Sayce, I. G. (1971). *Talanta,* in print
246. Lund, W. (1969). *Anal. Chim. Acta,* **45,** 109
247. Varga, L. P. (1969). *Anal. Chem.*, **41,** 323
248. Thun, H., Verbeek, F. and Vanderlen, W. (1967). *J. Inorg. Nucl. Chem.*, **29,** 2109
249. Unwin, E. A., Beimer, R. G. and Fernando, Q. (1967). *Anal. Chim. Acta,* **39,** 95
250. Romary, J. K., Donnelly, D. L. and Andrews, A. L. (1967). *J. Inorg. Nucl. Chem.*, **29,** 1805
251. Nagypál, I., Gergely, A. and Jékel, P. (1969). *J. Inorg. Nucl. Chem.*, **31,** 3447
252. Perrin, D. D. and Sayce, I. G. (1967). *Talanta,* **14,** 833
253. Ingri, N., Kakolowicz, W., Sillén, L. G. and Warnquist, B. (1967). *Talanta,* **14,** 1261
254. Elgquist, B. (1969). *Talanta,* **16,** 1502
255. Dyrssén, D., Jagner, D. and Wengelin, F. (1968). *Computer Calculation of Ionic Equilibria and Titration Procedures.* (Stockholm: Almquist and Wiksell)
256. Beck, M. T. (1968). *Coordin. Chem. Rev.*, **31,** 91
257. Beck, M. T. (1972). *Chemistry of Outer Sphere Complexes.* (Budapest: Akadémiai Kiadó)
258. Ogino, K. and Saito, U. (1967). *Bull. Chem. Soc. Japan,* **40,** 826
259. Ogino, K. (1969). *Bull. Chem. Soc. Japan,* **42,** 442
260. Atkinson, G. and Kor, S. K. (1967). *J. Phys. Chem.*, **71,** 673
261. Atkinson, G. and Petrucci, S. (1966). *J. Phys. Chem.*, **70,** 3122
262. Elder, A. and Petrucci, S. (1970). *Inorg. Chem.*, **9,** 19
263. Larsson, R. and Johansson, L. (1965). *Proc. Symp. Coord. Chem. Tihany,* 31. (Budapest: Akadémiai Kiadó)
264. Olsen, I. and Bjerrum, J. (1967). *Acta. Chem. Scand.*, **21,** 1112
265. Larsson, R. and Nunziata, G. (1970). *Acta Chem. Scand.*, **24,** 1878
266. Langford, C. H. and Warner, R. M. (1967). *J. Amer. Chem. Soc.*, **89,** 3141
267. Chan, S. C. (1966). *J. Chem. Soc. A,* 1124

268. Miller, W. A. and Watts, D. W. (1967). *J. Amer. Chem. Soc.*, **89,** 6858
269. Duffy, N. V. and Earley, J. E. (1967). *J. Amer. Chem. Soc.*, **89,** 272
270. Kelm, H. and Harris, G. M. (1967). *Inorg. Chem.*, **6,** 706
271. Bott, H. L., Poe, A. J. and Shaw, K. (1970). *J. Chem. Soc. A*, 1744
272. Spiro, T. G., Revesz, A. and Lee, J. (1968). *J. Amer. Chem. Soc.*, **90,** 4000
273. Larsson, R. (1962). *Acta Chem. Scand.*, **16,** 2305
274. Mironov, V. E., Lubamirova, K. N. and Ragulin, G. K. (1970). *Zh. Fiz. Khim.*, **44,** 416
275. Ebsworth, E. A. W. and Hughes, R. G. (1967). *J. Inorg. Nucl. Chem.*, **29,** 1799
276. Garbett, K. and Gillard, R. D. (1968). *J. Chem. Soc. A*, 1725
277. Cummins, K. and Jones, T. P. (1970). *Chem. Commun.*, 638
278. Elsbrand, H. and Beattie, J. K. (1968). *Inorg. Chem.*, **7,** 2468
279. Marcus, Y. and Eliezer, I. (1969). *Coordin. Chem. Rev.*, **41,** 273
280. Fridman, Ya. D. (1971). *Proc. 3rd Symp. Coord. Chem. Vol. 2.* (Budapest: Akadémiai Kiadó)
281. Sigel, H. (1967). *Chimia*, **21,** 489
282. Martell, A. E. (1971). *Proc. 3rd Symp. Coord. Chem. Vol.* **2**. (Budapest: Akadémiai Kiadó)
283. Zangen, M. (1969). *J. Inorg. Nucl. Chem.*, **31,** 867
284. Maslowska, J. (1969). *Roczniki Chemii*, **43,** 15
285. Gaizer, F. and Beck, M. T. (1966). *J. Inorg. Nucl. Chem.*, **28,** 503
286. Srivastava, S. C. and Newman, L. (1967). *Inorg. Chem.*, **6,** 762
287. Newman, L. and Srivastava, S. C. (1970). *Proc. 3rd Symp. Coord. Chem. Vol. 1*, 131. (Budapest: Akadémiai Kiadó)
288. Bounsall, E. J. and Poe, A. J. (1966). *J. Chem. Soc. A*, 286
289. Krishnan, K. and Plane, R. A. (1967). *Inorg. Chem.*, **6,** 85
290. Fridman, Ya.D. and Veresova, R. A. (1968). *Zh. Neorg. Khim.*, **13,** 762
291. Kanemura, Y. and Watters, J. J. (1967). *J. Inorg. Nucl. Chem.*, **29,** 1701
292. L'Heureux, G. A. and Martell, A. E. (1966). *J. Inorg. Nucl. Chem.*, **28,** 481
293. Sigel, H. and Griesser, R. (1967). *Helv. Chim. Acta*, **50,** 1842
294. Sigel, H., Becker, K. and McCormick, D. B. (1967). *Biochim. Biophys. Acta*, **148,** 655
295. Condike, S. F. and Martell, A. E. (1969). *J. Inorg. Nucl. Chem.*, **31,** 2455
296. Schulman, S. G. and Gershon, H. (1969). *J. Inorg. Nucl. Chem.*, **31,** 2467
297. Gershon, H., Schulman, S. G. and Olney, D. (1969). *Contrib. Boyce Thompson Inst.*, **24,** 167
298. Näsänen, R. and Koskinen, M. (1967). *Suomen Kemistilehti B*, **40,** 23
299. Näsänen, R. and Koskinen, M. (1967). *Suomen Kemistilehti B*, **40,** 108
300. Perrin, D. D., Sayce, I. G. and Sharma, V. S. (1967). *J. Chem. Soc. A*, 1755
301. Perrin, D. D. and Sharma, V. S. (1969). *J. Chem. Soc. A*, 2060
302. Freeman, H. C. and Martin, R. P. (1969). *J. Biol. Chem.*, **244,** 4823
303. Sharma, V. S. and Schubert, J. (1969). *J. Chem. Educ.*, **46,** 506
304. Farona, F., Perry, D. C. and Kuska, H. A. (1968). *Inorg. Chem.*, **7,** 2415
305. Griesser, R. and Sigel, H. (1970). *Inorg. Chem.*, **9**, 1238
306. Burger, K. and Pintér, B. (1967). *Magy. Kém. Foly.*, **73,** 209
307. Huber, P. R., Griesser, R., Prijs, B. and Sigel, H. (1969). *European J. Biochem.*, **10,** 283
308. Sigel, H. (1968). *Angewandte Chemie. Int. Ed.*, **7,** 137
309. Leussing, D. L. and Bai, K. S. (1968). *Anal. Chem.*, **40,** 575
310. Sigel, H. and McCormick, D. B. (1970). *Accounts Chem. Res.*, **3,** 201
311. Advani, A. T., Barnes, D. S. and Pettit, L. D. (1970). *J. Chem. Soc. A*, 2691
312. Advani, A. T., Irving, H. M. N. H. and Pettit, L. D. (1970). *J. Chem. Soc. A*, 2699
313. Simeon, V. and Weber, O. A. (1966). *Croat. Chem. Acta*, **38,** 161
314. Morris, D. J. and Martin, R. B. (1970). *J. Inorg. Nucl. Chem.*, **32,** 2891
315. Barnes, D. S. and Pettit, L. D. (1970). *Chem. Commun.*, 1000
316. Gillard, R. D. (1967). *Inorg. Chim. Acta Rev.*, **1,** 69
317. Ohtaki, H. (1967). *Inorg. Chem.*, **6,** 808
318. Hietanen, S. and Sillén, L. G. (1968). *Acta Chem. Scand.*, **22,** 265
319. Aveston, J. (1966). *J. Chem. Soc. A*, 1599
320. Hentz, C. F. and Johnson, J. S. (1966). *Inorg. Chem.*, **5,** 1337
321. Moolenar, R. J., Evans, J. C. and McKeever, L. D. (1970). *J. Phys. Chem.*, **74,** 3629
322. Aveston, J. (1969). *J. Chem. Soc. A*, 273
323. Pajdowski, L. (1966). *J. Inorg. Nucl. Chem.*, **28,** 433

324. Baron, V. (1961). *Coordin. Chem. Rev.*, **6,** 65
325. Harris, J. I. and Tobias, R. S. (1969). *Inorg. Chem.*, **8,** 2259
326. Pellerito, L. and Tobias, R. S. (1970). *Inorg. Chem.*, **9,** 953
327. Carpéni, G. (1969). *J. Chim. Phys.*, **66,** 327
328. Zompa, L. J. and Bogucki, R. F. (1966). *J. Amer. Chem. Soc.*, **88,** 5186
329. Perrin, D. D. and Sharma, V. S. (1966). *J. Inorg. Nucl. Chem.*, **28,** 1271
330. Schugar, H., Walling, C., Jones, R. B. and Gray, H. B. (1967). *J. Amer. Chem. Soc.*, **89,** 3712
331. Spiro, T. G., Pape, L. and Saltman, P. (1967). *J. Amer. Chem. Soc.*, **89,** 5555
332. Spiro, T. G., Bates, L. and Saltman, P. (1967). *J. Amer. Chem. Soc.*, **89,** 5559
333. Sigwart, C., Hemmerich, P. and Spence, J. T. (1968). *Inorg. Chem.*, **7,** 2545
334. Srivastava, R. C. and Smith, T. D. (1968). *J. Chem. Soc. A*, 2192
335. Leonard, M. A. and Nagi, F. I. (1969). *Talanta*, **16,** 1104
336. Beck, M. T. and Porzsolt, Cs.E. (1971). *J. Coord. Chem.*, **1,** in press
337. Buckingham, D. A., Gorges, R. C. and Henry, J. T. (1967). *Aust. J. Chem.*, **20,** 28
338. Bishop, J. J., Cotton, F. A., Eiss, R. and Hugel, R. P. (1967). *Nature (London)*, **214,** 111
339. Österberg, R. and Sjöberg, B. (1968). *Acta Chem. Scand.*, **22,** 689
340. Österberg, R. (1970). *Proc. 3rd Symp. Coord. Chem., Vol. I*, 221. (Budapest: Akadèmiai Kiadó)
341. Tsangaris, J. M., Chang, J. W. and Martin, R. B. (1969). *Arch. Biochem. Biophys.*, **103,** 53
342. Gustafson, R. L. and Lirio, J. A. (1965). *J. Phys. Chem.*, **72,** 1127
343. Birch, C. G. and Manahan, S. E. (1967). *Anal. Chem.*, **39,** 1182
344. Clark, E. R. and Martell, A. E. (1970). *J. Inorg. Nucl. Chem.*, **32,** 911
345. Katzin, L. J. and Gulyás, E. (1969). *J. Amer. Chem. Soc.*, **91,** 6940
346. Wellman, K. M., Mecca, T. G., Mungall, W. and Hare, C. R. (1968). *J. Amer. Chem. Soc.*, **90,** 805
347. Anderson, K. P., Greenhalgh, W. O. and Butler, E. A. (1967). *Inorg. Chem.*, **6,** 1056
348. Petit-Ranel, M. M. and Paris, M. R. (1968). *Bull. Soc. Chim. France*, 2791
349. Janjic, T. J., Pfendt, L. B. and Celap, M. (1970). *Z. Anorg. Allg. Chem.*, **378,** 83
350. Martin, R. P. and Mosoni, L. (1970). *Bull. Soc. Chim. France*, 2917
351. Gergely, A. (1969). *Acta Chim. Hung.*, **59,** 309
352. Porter, L. J., Perrin, D. D. and Hay, R. W. (1969). *J. Chem. Soc. A*, 118
353. Hopgood, D. and Angelici, R. J. (1968). *J. Amer. Chem. Soc.*, **90,** 2508
354. Österberg, R. (1966). *Thesis*, Göteborg
355. Jameson, R. F. and Neillie, W. F. S. (1966). *J. Inorg. Nucl. Chem.*, **28,** 2667
356. Gorton, J. E. and Jameson, R. F. (1968). *J. Chem. Soc. A*, 2615
357. Firth, R. A., Hill, H. A. O., Pratt, J. M., Thorp, R. G. and Williams, R. J. P. (1969). *J. Chem. Soc. A*, 381
358. Tewari, K. C., Lee, J. and Li, N. C. (1970). *Trans. Faraday Soc.*, **66,** 2069
359. Pradel, J. and Martin, R. P. (1970). *Compt. Rend.*, **270,** 1863
360. Feldman, I., Jones, J. and Cross, R. (1967). *J. Amer. Chem. Soc.*, **89,** 49
361. Khan, M. M. T. and Martell, A. E. (1966). *J. Amer. Chem. Soc.*, **88,** 668
362. Phillips, R. C., George, P. and Rutman, R. J. (1966). *J. Am. Chem. Soc.*, **88,** 2631
363. Sigel, H. (1968). *Europ. J. Biochem.*, **3,** 530
364. Tobias, R. S. (1966). *Organometal. Chem. Rev.*, **1,** 93
365. Sillén, L. G. and Martell, A. E. (1971). *Stability of Metal-ion Complexes, Supplement No.* **1,** (London: The Chemical Society)
366. Harju, L. (1970). *Anal. Chim. Acta*, **50,** 475
367. Yingst, A. and Martell, A. E. (1969). *J. Amer. Chem. Soc.*, **91,** 6927
368. Hoyer, E., Wagler, D. and Anton, J. (1969). *Z. Phys. Chem.*, **241,** 65
369. Rajan, K. S., Murase, I. and Martell, A. E. (1969). *J. Amer. Chem. Soc.*, **91,** 4408
370. Uhlig, E. and Walther, D. (1969). *Vortragsberichte zum Symp. Koordinationschemie der Übergangselemente, Vol. 3*, 150 (Jena: Chemische Geselschaft)
371. Fogg, A. G., Gray, A. and Burnus, D. T. (1970). *Anal. Chim. Acta*, **51,** 265
372. van Poucke, L. C. and Herman, M. A. (1968). *Anal. Chim. Acta*, **42,** 467
373. Braunstein, J. (1968). *Inorg. Chim. Acta Rev.*, **2,** 19
374. Braunstein, J., Alvarez-Funes, A. R. and Braunstein, H. (1966). *J. Phys. Chem.*, **70,** 2732
375. Gutmann, V. (1968). *Coordination Chemistry of Non-aqueous Solutions.* (Berlin: Springer)
376. Gutmann, V. (1969). *Record. Chem. Progr.*, **30,** 169

377. Alexander, R., Ko, E. C. F., Mac, Y. C. and Parker, A. J. (1967). *J. Amer. Chem. Soc.*, **89**, 3703
378. Butler, J. N. (1968). *J. Phys. Chem.*, **72**, 3288
379. Butler, J. N. (1967). *Anal. Chem.*, **39**, 1799
380. Butler, J. N. (1968). *J. Electrochem. Soc.*, 115, 446
381. Synnott, J. C. and Butler, J. N. (1969). *J. Phys. Chem.*, **73**, 1470
382. Luehrs, D. C., Iwamoto, R. T. and Kleinberg, J. (1966). *Inorg. Chem.*, **5**, 201
383. Bravo, O. and Iwamoto, R. T. (1969). *Inorg. Chim. Acta*, **3**, 663
384. Matsui, Y. and Date, Y. (1970). *Bull. Chem. Soc. Japan*, **43**, 2052
385. Liu, C. H., Newman, L. and Hasson, J. (1968). *Inorg. Chem.*, **7**, 1868
386. Marshall, W. L. and Quist, A. S. (1967). *Proc. Nat. Acad. Sci.*, **58**, 901
387. Marshall, W. L. (1970). *J. Phys. Chem.*, **74**, 346
388. Gilkerson, W. R. (1970). *J. Phys. Chem.*, **74**, 746
389. Matheson, R. A. (1969). *J. Phys. Chem.*, **73**, 3625
390. Kennedy, M. B. and Lister, M. W. (1966). *Can. J. Chem.*, **44**, 1709
391. Mesmer, R. E. and Baes, C. F. (1968). *J. Phys. Chem.*, **72**, 4720
392. Fay, D. P. and Purdie, N. (1969). *J. Phys. Chem.*, **73**, 195
393. Sigel, H. and Kaden, T. (1966). *Helv. Chim. Acta*, **49**, 1619
394. Yingst, A. and McDaniel, D. H. (1966). *J. Inorg. Nucl. Chem.*, **28**, 2919
395. Nieboer, E. and McBryde, W. A. E. (1967). *Inorg. Nucl. Chem. Lett.*, **3**, 569
396. Nieboer, E. and McBryde, W. A. E. (1970). *Can. J. Chem.*, **48**, 2549
397. Ahrland, S. (1968). *Structure and Bonding*, **5**, 118
398. Ahrland, S. (1967). *Helv. Chim. Acta*, **50**, 306
399. Nancollas, G. H. (1970). *Coordin. Chem. Rev.*, **5**, 379
400. Degischer, G. and Nancollas, G. H. (1970). *J. Chem. Soc. A*, 1125
401. Anderegg, G. (1968). *Helv. Chim. Acta*, **51**, 1856
402. Kim, M. K. and Martell, A. E. (1966). *J. Amer. Chem. Soc.*, **88**, 914
403. Bond, A. M. and O'Donnell, J. (1970). *J. Electroanal. Chem.*, **26**, 137
404. Haynes, L. V. and Sawyer, D. T. (1967). *Inorg. Chem.*, **6**, 2146

2
Stereochemical Non-rigidity

E. L. MUETTERTIES
E. I. du Pont Nemours and Company, Delaware

2.1 INTRODUCTION

Molecules and ions are not rigid point group collections of nuclei and electrons. Vibrational modes impart an intrinsic flexibility to the nuclear array. These motions, if sufficiently large, may permute nuclear positions and hence lead to stereoisomerisation. Molecular species that undergo such rearrangements are called stereochemically non-rigid[1,2]. Alternative terms or modifiers commonly encountered in organic[3] and organometallic[4] chemistry are valence tautomerism, degenerate tautomerism, and fluxional. Stereochemical non-rigidity applies to any discrete molecular array in which vibrational motion allows a rapid intramolecular rearrangement. The net process may comprise (1) stereochemical change, e.g., a racemisation, or (2) permutation of like nuclei, i.e., the form of the Hamiltonian is invariant to the transformation.

It is more precise to characterise these processes first by the permutation class or set and then to relate the permutation type to a physical process(es) or operation(s)[5–9]. Classically well-defined operations are the C—C bond rotation in ethane and its derivatives, the inversion process in ammonia, and pseudorotation in cyclopentane. In all these operations, the ground state and the intermediate or transition state can be characterised as an idealised polygon or polyhedron[2, 5, 7, 10, 11] which are designated *polytopal*[7, 8] isomers. This applies to (a) coordination compounds wherein the nuclei bonded to the 'central' nucleus describe the vertices of an idealised polytope, (b) polyhedral boranes and metal clusters with extensive multicentre bonding, where the core nuclei describe the vertices of an idealised polytope, (c) conformational problems in ring structures, (d) bond rotamers, and (e) complex stereochemistry involving tertiary or higher-order structure as especially found in biochemistry. This formalism can be applied to the mechanistic analysis of any problem in stereochemical non-rigidity.

The consequences of high stereochemical non-rigidity can be profound: inability to separate enantiomers by classical techniques, loss of stereospecificity in reactions proceeding through relatively long-lived intermediates, and inadequacy of resonance techniques for stereochemical determination. On the more positive side, an understanding of such rearrangements can lead to a thorough delineation of reaction mechanism as has been done by Westheimer and co-workers in organophosphate chemistry[12].

Nuclear magnetic resonance studies have had an enormous impact on the knowledge and understanding of dynamic stereochemistry. This structural probe has greatly extended the established classes of stereochemically non-rigid molecules or ions and added a substantive element of mechanistic comprehension. The following critique of advances during the period 1965–1970 is limited in specifics to inorganic and organometallic chemistry, but there is a carry-over in generalities to all of chemistry. Organisation follows systematics previously established[2, 5]. Coordination compounds (ML_N), boron hydrides (B_NH_y), and metal clusters (M_NL_y) are discussed as N-atom aggregates and specific aspects of organometallics are analysed separately.

2.2 ANALYSIS OF STEREOCHEMICAL PROBLEMS

About 1968, there began to appear a large number of papers which dealt with so-called topological and non-topological analyses of stereochemical problems in intramolecular or polytopal rearrangements[5–9, 13–21]. As yet no complete, abstract mathematical model has been presented, and therefore, a critical review of this facet of stereochemical non-rigidity is inappropriate. However, new terminology has developed, and this is appearing in many papers on stereochemical non-rigidity. Accordingly, a brief summary of the analysis follows.

A stereochemical family, F, for a specific x-atom problem comprises all stereoisomers of all polytopal isomers G_A, G_B, G_C[5–9, 16]. Since specific rearrangements in stereochemically non-rigid molecules or ions may be formally considered in terms of a ground-state polytopal form G_A and a transition-state polytopal form G_B, the analysis may be reduced to a consideration of the G_A and G_B relationship, i.e., the operation required for conversion of a G_A to a G_B stereoisomer[4–9].

In an ML_N coordination compound wherein all ligands are different, the number of stereoisomers in G_A or G_B is $N!/h_r$, where N is the number of ligands and h_r is the order of the rotational sub-group of the point group for idealised polytopal (G_A and G_B) forms. The connectivity, δ, for a G_A (or G_B) polytopal form is defined as the number of *new* stereoisomers of G_B (or G_A) generated from *a* stereoisomer of G_A (or G_B)[5–9, 16] in one step, where a step is the product of the operations o and o′

$$ML_x(G_A) \overset{o}{\rightleftharpoons} ML_x(G_B) \overset{o'}{\rightleftharpoons} ML_x'(G_A).$$

The stereochemical system (G_A–G_B) may be open or closed in a graphical or topological sense. Closure simply means that traverse of any stereoisomer in either polytopal form is allowed in a series of steps based on the specified operations. Necessary conditions for closure are that the product of the isomer count, I, and the connectivity, δ, of one polytopal system (G_A) be equal to that of the other system (G_B), i.e., $I_A\delta_A = I_B\delta_B$. Furthermore, one of the operators (including pure rotations) required to define the system must produce permutations (of nuclear positions) that are odd if the group of all permutations representing the isomer system is even–odd. If no further invariant is imposed by the operators, the system will be closed[5–9, 16].

Specific elements in the analyses of mathematical or stereochemical significance are cycles, $\mathscr{C}$, which are circular paths traversing an isomer a and its enantiomer $\bar{a}$ with no isomer traversed more than once in the circuit. A sub-cycle, C, is also a circular path, but no enantiomeric isomers are traversed, i.e., it cycles isomers a, b, $\bar{c}$, but not $\bar{a}$, $\bar{b}$, c. The racemisation chain, c_E is non-circular and is the shortest path from isomer a to isomer $\bar{a}$[5–9, 16].

Graphical, topological, and matrix representations of stereochemical problems are of substantial value in consideration of mechanistic arguments, and examples are cited in the following sections[5–9, 13–21]. The more general treatment not yet fully published[16] treats the stereoisomers as an N-dimensional vector allowing a representation of the rearrangement operators by $N \times N$ matrices. The transformation of vector coordinates permits analysis of the geometrical invariants associated with the rearrangement operators. A theorem has been proved regarding the specific invariants the rearrangement operator may impose in order that it generate a closed stereochemical system[16].

In the general four-atom case, a ground-state tetrahedral form and transitional planar form comprise one of the simplest stereochemical systems. There are two isomers (enantiomers) in tetrahedral form and three in square-planar form. These are interconverted by the torsional twist rearrangement R (Figure 2.1), operating on edges 12, 13, and 14 of the tetrahedron:

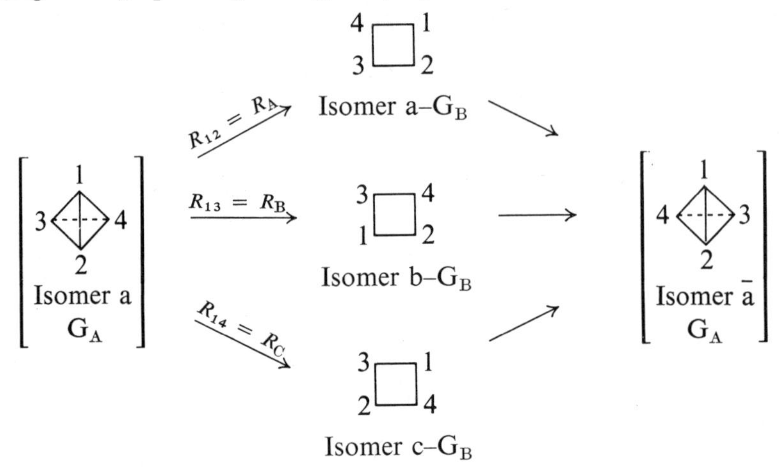

The $I\delta$ products are equivalent here: $(2 \times 3)_{G_A} = (3 \times 2)_{G_B}$. Rotations required to define the system are the twofold *rotations* on edges 12(r_2), 13 (r'_2), and 14 (r''_2); these produce not new stereoisomers but equivalent configurations. The two threefold rotations (about the face 234) are r_3 and r_3^2. The rotational operator group is then E, r_2, r'_2, r''_2, r_3, r_3^2, r_2r_3, $r_2r_3^2$, r'_2r_3, $r'_2r_3^2$, r''_2r_3, and $r''_2r_3^2$. Combining these with the three rearrangement operations R_A, R_B, and R_C, an operator group of order 24 is formed comprising the rotational group operators and R_A, R_B, R_C, R_Ar_2, $R_Ar_2r_3$, $R_Ar_2r_3^2$, R_Ar_3, $R_Ar_3^2$, R_Br_3, $R_Br_3^2$, R_Cr_3, and $R_Cr_3^2$. This group is isomorphic to the stereochemical system for G_A, with a torsional twist mechanism, which has two stereoisomers, and each of these has 12 equivalent configurations (generated by rotational operators). If a rectangular intermediate or transition state were alternatively considered, the group order remains the same but elements change because now there would be six distinct rearrangement operators: R_A, R_B, R_C, plus R_D, R_E, R_F, where the latter three operate on edges 23, 34, and 42.

For any general case in which the stereochemical system is closed, the order of the operator or permutational group is for an ML_N or M_N system simply N!. The groups have been identified for the general five-atom (of order 120) and six-atom (of order 720) cases[16].

2.3 THREE-ATOM SYSTEM—QUASI-FOUR-ATOM CASE

The favoured idealised polytopal form for the three-atom system is the plane and for quasi-tetrahedral :ML_3 coordination compounds the trigonal-pyramid. Stereochemical non-rigidity is primarily relevant to the pyramidal coordination compounds since planar MXYZ species have no stereoisomers. Conceptually, the pyramidal :ML_3 system is best visualised in potential energy form. In Figure 2.2, an idealised potential energy diagram for an $:ML_3$ molecule or ion is presented in relation to the LML angle. Only molecules like $:CH_3^-$, $:NH_3$, and $:OH_3^+$ stabilise in pyramidal form where ostensibly an electron or a non-bonding pair of electrons resides in a directed orbital. In these cases, the planar excited state may be a maximum, as depicted in Figure 2.2, on the potential energy surface. Interconversion of isomers (enantiomers exist for pyramidal MLL′L″) is a pyramidal conversion, and these 'isomers' have been designated invertomers[22]. For the conversion,

$$\mathrm{L\,L'\,L''\,M{:}} \rightleftharpoons \mathrm{{:}M\,L\,L'\,L''}$$

two processes operate competitively. One is the inversion involving deformation of the LML angles to 120 degrees (if all ligands are identical) at which point there is an equal probability, on angular compression, to form either invertomer in the pyramidal form. Factors affecting this process are

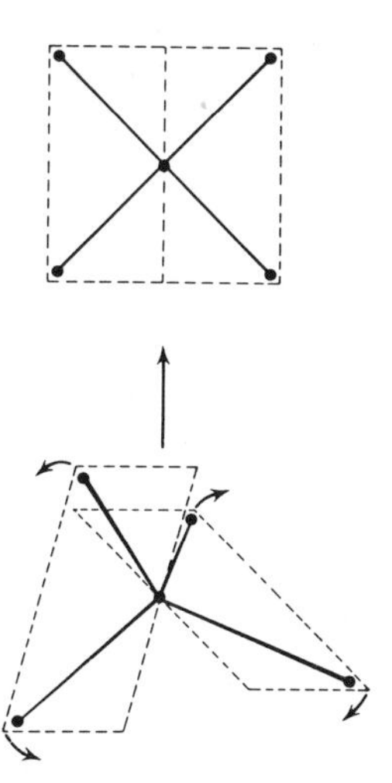

Figure 2.1 An operation, a digonal twist, for interconversion of planar and tetrahedral forms

considered below. The second process is quantum-mechanical tunnelling with penetration of the walls of the energy wells. Tunnelling contributes substantially to the overall inversion rate only when the mass of the ligands is very low, specifically when at least one of the ligands is hydrogen or deuterium and the invertomers are not diastereomers.

It is extremely difficult to accurately assess barriers to intramolecular rearrangements by quantum-mechanical calculations, although one of the simplest problems in this area, the ammonia inversion, has been calculated with some accuracy by a molecular orbital approximation[23]. Typically the problem has been approached, for the last 30 years, empirically or mechanistically with approximated potential functions which weight the mass contributions of the central M nucleus and peripheral ligands and the value of the equilibrium LML angle(s). Such functions[24–26] anticipate or rationalise some well-established trends in inversion barriers. For example, the barrier increases in a given periodic group with increase in atomic number of the central nucleus as in the NR_3, PR_3, and AsR_3 series and the OR_3^+ and SR_3^+ series. However, the effect of the R group cannot be reduced to simple considerations of mass as especially noted by Rauk, Allen, and Mislow[22] who attempted to evaluate steric, conjugative–hyperconjugative, electronegativity, delocalisation of lone pairs by $(p\text{–}d)_\pi$ conjugation, angular constraint, and other effects on the inversion barrier. Such a detailed mechanistic approach will not be comprehensively discussed here (see Reference 22) because a sound theoretical treatment is still not in hand. The most recent advances

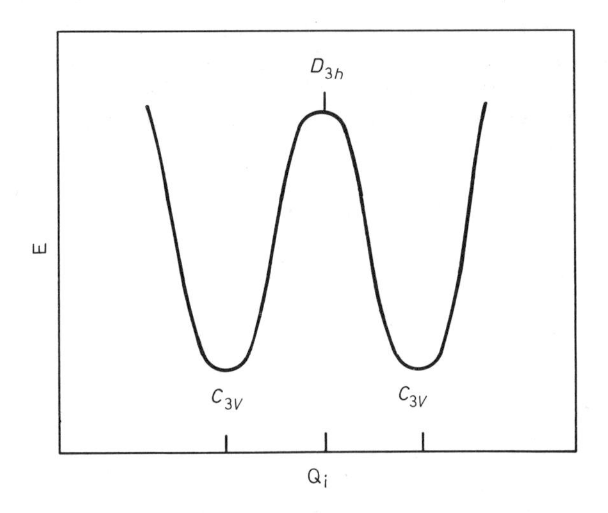

Figure 2.2 Potential energy diagram of an $:MX_3$ pyramidal molecule as a function of the equilibrium interbond angles

in experimental delineation of barriers is concisely assayed below. Because of the extensive literature on pyramidal inversions, a comprehensive review is not possible here. The reviews by Koeppl *et al.*[26] and by Rauk *et al.*[22] provide a more detailed analysis.

For first row pyramidal molecules, the relevant species are of the type CR_3^-, $\cdot CR_3$, NR_3, and OR_3^+. Inversion barriers for ammonia and simple amines fall in a relatively narrow range of 4–8 kcal mol^{-1}. Fairly accurate values from microwave or vibrational spectroscopic data are

NH_3	5.8 kcal mol^{-1}	(27)
CH_3NH_2	4.8 kcal mol^{-1}	(28)
$(CH_3)_2NH$	4.4 kcal mol^{-1}	(29)
$(CH_3)_3N$	7.46 kcal mol^{-1}	(25, 26)

Angular constraints *if sufficiently severe* do stabilise the pyramidal form, and significantly higher barriers have been found in such strained molecules. Thus with *N*-methylpyrrolidine[30] the barrier of 8 kcal mol^{-1} obtained from n.m.r. studies is essentially the same as for trimethylamine, but in the three-membered ring aziridine system (2.1) the parent member aziridine (R=H) has a barrier,

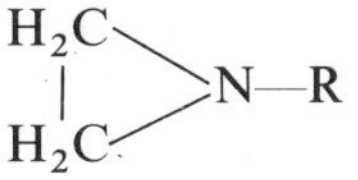

$$\begin{array}{l} H_2C \\ \;\;| \qquad \rangle N\!-\!R \\ H_2C \end{array} \qquad (2.1)$$

estimated *at the lower limit*, of 11.6 kcal mol^{-1} [31]. The steric effect on barriers is nicely illustrated in the aziridine series; the inversion barrier falls with increasing bulk of R, i.e., $CH_3 > C_2H_5 \gg t\text{–}C_4H_9$, with a value of 19 kcal mol^{-1} reported for R = CH_3 [32–35]. Diastereomeric *N*-chloroaziridine derivatives have been actually *isolated* by Felix and Eschenmosser[35].

Electronegative groups greatly increase the inversion barrier. For example, the most reliable estimate for the inversion barrier in NF_3 is 56–59 kcal mol^{-1} [36]. In the carbon free radical system, $\cdot CH_3$ is planar but $\cdot CF_3$ is pyramidal with a calculated inversion barrier of 27.4 kcal mol^{-1} [37]. Complexation of polyamines, e.g., 1,4,7,10-tetrazadecane, with transition-metal ions render the nitrogen nuclei, after a hydrogen abstraction, rather resistant to inversion[38, 39].

Nuclear magnetic resonance studies of oxonium salts R_3O^+ have demonstrated a barrier increase with angular constraints. Thus, the activation energy for inversion in the isopropyloxonium salt of oxirane (ethylene oxide) is 10 ± 2 kcal mol^{-1}, whereas inversion is too fast to be determined by n.m.r. techniques in the six-membered tetrahydropyran system[40]. No other good experimental data are available for first-row-based pyramidal molecules. Nevertheless, calculations and other inferential information suggest that carbanions and oxonium salts will have barriers little different from their nitrogen analogues.

It has long been recognised that an increase in the atomic number of the central nucleus in a pyramidal molecule or ion raises the inversion barrier. For example, the inversion barriers, $\Delta G^\ddagger$, for trialkylamines are 7–8 kcal mol^{-1} and for the analogous phosphines 29–36 kcal mol^{-1} [41, 42]. Configurational stability in organophosphines and arsines has been demonstrated in the separation of the enantiomers. The inversion barrier in acyclic phosphines is quite insensitive to variation in structural parameters[41, 42]. However, in diphosphines and diarsines the barriers, $\Delta G^\ddagger$, are slightly lower, ~22.5–24 kcal mol^{-1} [43, 46], ostensibly due to delocalisation of lone pair electrons into empty d orbitals by $(p\text{–}d)_\pi$ conjugation. A rather striking illustration of the high barrier to inversion and to bond rotation in arsines was the isolation of isomers of $(C_6F_5)_2AsAs(C_6H_5)_2$ by sublimation at 140 and 230 °C [47]. Dramatic increases in inversion rates for phosphines have been effected by Mislow and coworkers through substituent effects that increase the stability of the planar form. The inversion barrier ($\Delta G^\ddagger_{25°}$) for

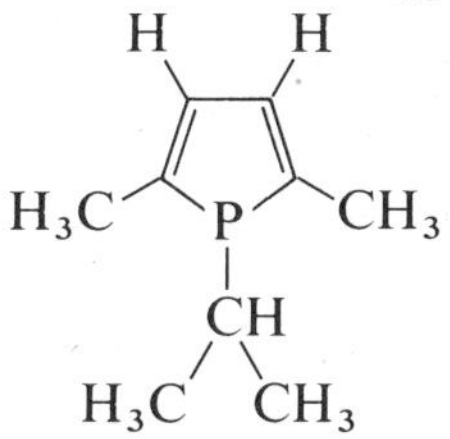

H H

H_3C P CH_3 (2.2)

CH

H_3C CH_3

is 16 kcal mol^{-1}. Barrier reduction was ascribed to $(p\text{–}p)_\pi$ delocalisation and increased aromaticity of the phosphine in the planar transition state[42]. Similarly, $(p\text{–}d)_\pi$ conjugation in the silyl substituted phosphine

$$C_6H_5\text{—}P\begin{matrix}\diagup CH(CH_3)_2 \\ \diagdown Si(CH_3)_3\end{matrix} \tag{2.3}$$

is apparently the cause of a barrier reduction to 18.9 kcal mol^{-1} ($\Delta G^{\ddagger}_{62°}$)[48].

Tricoordinate sulphur is configurationally stable at 25 °C as shown for sulphoxides[49], sulphinate esters[50], sulphinamides[51], sulphinimines[52], and sulphites[53]. Inversion barriers ($\Delta H^{\ddagger}$) in sulphoxides are in the region 35–42 kcal mol^{-1} [49, 54]. Various claims to low inversion barriers for tricoordinate sulphur compounds should be re-examined as pointed out by Rauk *et al.*[22]. This same comment applies to some phosphorus and arsenic inversion values[22]. The relatively high barrier in tricoordinate sulphur inversion may be of substantial consideration in stereochemical analysis of solution structure for biological materials, particularly iron–sulphur-based enzymes. Some inversion values have been reported for sulphur in model metal complexes. The activation energy barriers for inversion about sulphur in

$Cl_2Pt[S(CH_2\text{—}C_6H_5)_2]_2$ (2.4) and $[(\pi\text{-}C_5H_5)(OC)Fe(\mu\text{-}SC_6H_5)_2Fe(CO)(\pi\text{-}C_5H_5)]$ (2.5)

were found to be 18.0 [55] and ~31 [56] kcal mol^{-1}, respectively.

2.4 FOUR-ATOM SYSTEM

The tetrahedron is the prevailing polytopal form in four-atom systems. In coordination chemistry, the tetrahedral form is invariably of lowest energy if the central nucleus has a d^0- or d^{10}- electronic configuration. With partially filled d-shells, an alternative form, the plane (square or rectangle), may be of comparable energy to, or of lower energy than, the tetrahedron. This is especially true for complexes of metals with d^7-, d^8-, or d^9-electronic configurations. Factors affecting relative population of tetrahedral and planar forms include steric, electronic and solvation or packing effects as well as the character of the ligand and the charge on the metal complex. Generalisations on the importance of these factors have been concisely reviewed by Gerlach and Holm[57].

The energy separation between tetrahedral and planar forms in a d^0- or d^{10}-coordination compound is so large that it is improbable that the excited planar state will be significantly populated. The gap should generally be larger than the energy required for bond scission, homolytic or heterolytic, e.g., simply consider the depressed attractive interactions and elevated repulsive interactions of planar methane relative to those for tetrahedral methane. Hund[58] estimated back in 1927 the frequency of permutation of hydrogen atom positions in methane (by tunnelling) to be once every 10^9 years. It would be of interest to obtain some non-empirical calculations for the binding energy of methane in tetrahedral and in planar forms.

In transition-metal complexes where there may be population of planar and tetrahedral or quasi-tetrahedral forms, all evidence to date points to very rapid rearrangements, $>10^5\ s^{-1}$. Most of the meaningful rate data relate to d^8-nickel(II) complexes. The first series of nickel complexes to be investigated were those based upon the aminotroponeimine ion, (2.6)[59]. With R = H, the only form populated is the spin-paired (s = 0) plane. On replacement of H by alkyl or aryl groups there is thermal population of the triplet

(2.6)

state (s = 1) (1.6% with R = CH_3 and 84% with R = $C_6H_5CH_2$)[59]. The geometry of the triplet state is not known. It cannot be planar, but the extent that digonal twisting (Figure 2.1) takes place is unknown. The geometry could be tetrahedral (D_{2d} symmetry) or only slightly non-planar. X-ray studies of representative singlet and triplet state species is critically needed in a series of analogous compounds such as the aminotroponeiminates or other series[57, 60, 61] such as those based on ligands (2.7) and (2.8).

In all nickel chelates investigated to date[57–67], the rate of interconversion of singlet and triplet states is too fast for n.m.r. study. Rates exceed $10^3\ s^{-1}$. Ostensibly, the interconversion proceeds by a digonal twist mechanism

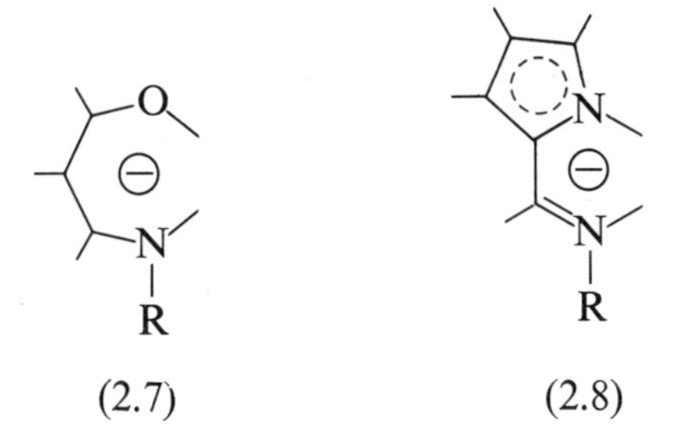

(2.7) (2.8)

(Figure 2.1). Interconversion rates for singlet and triplet states have been obtained for a series of bis(diarylalkylphosphine)nickel(II) halides. In studies by two different groups[68–70], the steric and electronic effects on the rearrangement rates were sought from the following substituent variance in (2.9) and (2.10):

$$\underset{\substack{R = CH_3,\ C_2H_5,\ n\text{-}C_3H_7,\ n\text{-}C_4H_9 \\ X = Cl,\ Br,\ I}}{[(C_6H_5)_2\overset{\overset{R}{|}}{P}]_2NiX_2} \quad (2.9) \qquad \underset{\substack{Z = H,\ R,\ OR,\ Cl,\ CF_3,\ NR_2 \\ X = Cl,\ Br,\ I}}{[(p\text{-}ZC_6H_4)\overset{\overset{CH_3}{|}}{P}]_2NiX_2} \quad (2.10)$$

Consistently, the rates for halide variation followed the order I > Cl > Br, e.g., 29, 2.6, and $0.45 \times 10^5\ s^{-1}$ (298 K) respectively, for the nickel iodide, chloride, and bromide complexes based on the $(C_6H_5)_2(CH_3)P$ ligand. There was, however, no obvious rationale to the very small rate and activation parameter variation with changes in R and in Z for the two series. This probably reflects the complex interplay of steric and electronic factors on the rearrangements. Generally, rate data fell in the range of 10^2 to $10^3\ s^{-1}$ at -50 °C and 10^5 to $10^6\ s^{-1}$ at 25 °C. Sample $\Delta H^‡$ and $\Delta S^‡$ data in series 2.9 are:

	$\Delta H^‡$ ± 4 kcal mol^{-1}	$\Delta S^‡$ ± 10 cal deg^{-1} mol^{-1}
$[(C_6H_5)_2PCH_3]_2NiBr_2$	11	6
$[(p\text{-}ClC_6H_4)_2PCH_3]_2NiBr_2$	11	4
$[(p\text{-}CH_3OC_6H_4)_2PCH_3]_2NiBr_2$	13	9
$[(p\text{-}CH_3OC_6H_4)_2PCH_3]_2NiCl_2$	10	5

No experimental evidence has been presented for polytopal rearrangements in polyhedral boranes and metal clusters. It has been suggested[71] that rearrangements may occur in metal clusters such as $Co_4(CO)_{12}$ (Figure 2.3) by a bending of carbonyl groups from bridging to terminal positions. For the cobalt cluster, the rearrangement rate at 25 °C must be less than $\sim 10^2\ s^{-1}$ since the necessary non-equivalence of cobalt nuclei for the ground state structure was evident in the ^{57}Co n.m.r. spectrum[72].

2.5 FIVE-ATOM SYSTEM

In coordination chemistry, stereochemically non-rigid ML_5 molecules or ions are a pervasive theme[2,5]. This facet of 5-coordination is explicable in consideration of ground-state geometries and the relative ease of interconversion of the two predominant geometrical forms. Additionally, there is a special class of transition-metal hydrides in which the hydrogen nucleus exerts little stereochemical effect and the molecules are quite non-rigid. This special class is discussed first.

In $RhH[P(C_6H_5)_3]_4$, the phosphorus atoms describe, within experimental error, a regular tetrahedron with the rhodium atom at the centre. The hydrogen atom, which was not located through x-ray diffraction data, must lie on the three-fold axis or be randomly disordered in the crystal[73]. Similarly, the related compound $RhH[P(C_6H_5)_3]_3As(C_6H_5)_3$ in a hemibenzene-solvated crystal has arsenic and phosphorus at tetrahedral positions, and the hydrogen ligand is supposedly *trans* to both arsenic and phosphorus[74]. In $CoH(PF_3)_4$ [75], there is a regular-tetrahedron of PF_3 groups. If in these hydrides one of the phosphorus ligands is replaced by a less bulky group such as carbon monoxide, the geometry departs from tetrahedral toward trigonal-bipyramidal. In those hydride complexes with a regular-tetrahedral arrangement of ligands, there should be little barrier to hydrogen atom tunnelling of triangular faces[77]. In fact, the proton resonance in $HRh[P(C_6H_5)_2C_2H_5]_2$ [76] is a simple quintet of doublets which does not change to a complex spectrum at low temperatures. The above authors[73–75] reported these results as being consistent with a square pyramidal geometry, which is highly unlikely based on the x-ray studies. Similar behaviour was observed for $HCo(PR_3)_4$ complexes, although the proton quintet broadens at low temperatures due to quadrupole relaxation effects[77].

For near-spherically-symmetrical central nuclei, the favoured disposition of five equivalent nuclei or ligands should be, and is, the trigonal-bipyramid[2, 5, 78]. However, square-pyramidal form should be only *slightly* higher

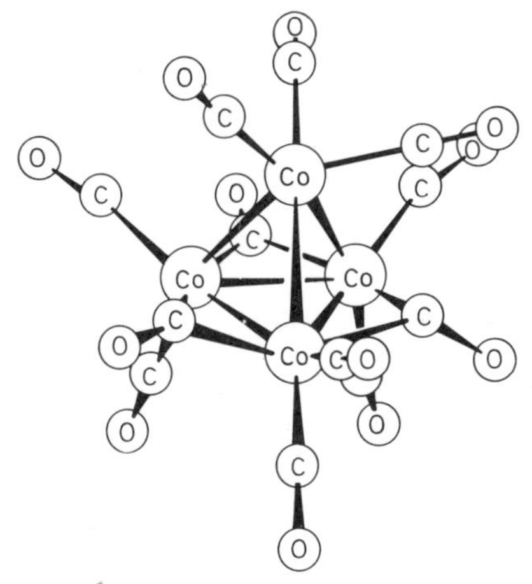

Figure 2.3 Structure of $Co_4(CO)_{12}$

in energy[2, 5, 78]. With non-equivalent ligands, the potential energy surface may well be more amorphous with forms intermediate between the limits of trigonal-bipyramidal and square- or tetragonal-pyramidal geometry very slightly favoured in special cases. Nevertheless, the majority of complexes with mixed ligands have near trigonal-bipyramidal forms. Multidentate ligands compound this delicate problem of ground-state geometry; however, most non-transition element complexes with multidentate ligands have the bonding nuclei at positions approximating a trigonal-bipyramid[2, 5, 78]. Even for the transition group, the chelates are almost evenly divided between trigonal-bipyramidal and square-pyramidal for established solid-state structures.

All data, primarily accumulated in the 1960s[2, 5, 78–80], indicate that the energy difference between trigonal-bipyramidal and square-pyramidal forms is small. In complexes with identical ligands the energy difference is very small (6 kcal mol^{-1} or less). The subtlety of energy differentiation is well illustrated in the crystal structure of the pentacyanonickelate(II) anion in the form of the chromium(III) (tris-ethylenediamine)$_3^{3+}$ salt[81]. Two polytopal forms of $Ni(CN)_5^{3-}$ are present: a near-perfect square-pyramid and a distorted trigonal-bipyramid.

There are some stereochemical rules critical to an understanding of the mechanistic and kinetic aspects of rearrangements in 5-coordinate complexes. In trigonal-bipyramidal complexes with mixed ligands, the more electronegative nuclei or ligands are found at the axial positions[2, 79, 80, 82] where ostensibly better overlap obtains. Steric factors may override these considerations, and very bulky ligands are best accommodated at equatorial positions. Ring strain in chelate complexes can overcome the electronegativity consideration. All established exceptions to the electronegativity rule are found in chelate systems in which ring strain in stereochemically determinant as in molecules like $[X_3PNR]_2$ (2.11), where the nitrogen nuclei are necessarily at equatorial and apical positions to minimise $(PN)_2$ ring strain[83]. Another apparent example is (2.12)

(2.11) (2.12)

in which ring constraints require the oxygen atoms from each ring to be equatorial and axial, leaving the more electronegative fluorine atom equatorial, although a near-square-pyramidal geometry cannot be excluded as a possibility[84].

The limiting idealised polytopal forms for 5-coordinate complexes are obviously quite different with respect to symmetry elements, see (2.13) and (2.14). However, the *similarities* are quite evident in viewing the two forms from the respective C_2 and C_4 axes as in (2.15) and (2.16). In fact, only a small bending mode is required to traverse either geometry as illustrated in Figure 2.4. This rearrangement process is referred to as the Berry mecha-

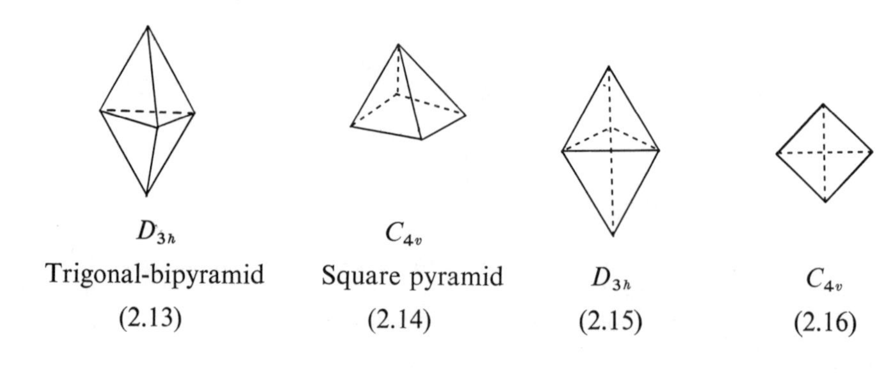

D_{3h} Trigonal-bipyramid (2.13) | C_{4v} Square pyramid (2.14) | D_{3h} (2.15) | C_{4v} (2.16)

nism*[85]. This mechanism accounts for all aspects of dynamical stereochemistry in 5-coordinate molecules or ions as well as reaction intermediates with the established exception of the quasi-tetrahedral MHL_4 aggregates mentioned earlier. The possibility of alternatives to the Berry mechanism must, however, be seriously considered in any mechanistic analysis.

Most data relevant to stereochemical non-rigidity of 5-coordinate compounds first came from n.m.r. studies of phosphorus fluorides[79,80,82]. A concise exposition[2] of the dynamic stereochemical principles is available from the nuclear magnetic resonance studies of phosphorus pentafluoride and its derivatives. The ^{19}F n.m.r. spectrum of the pentafluoride is a doublet[2,86]

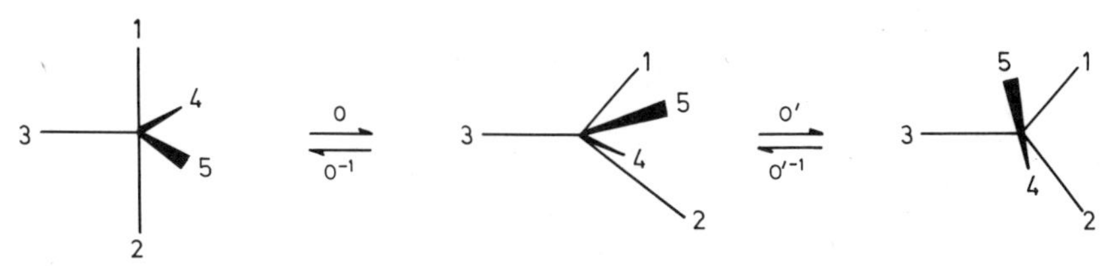

Figure 2.4 The Berry rearrangement mechanism for 5-coordinate complexes. In a single step, two axial and two equatorial positions are permuted.

(PF spin–spin coupling) from ~ -197 °C to at least 60 °C. The apparent magnetic equivalence of fluorine nuclei in this trigonal-bipyramidal molecule presumably reflects a very low barrier to the Berry rearrangement. *All* ML_5 molecules investigated by n.m.r. have shown apparent magnetic equivalence of ligand nuclei, e.g., AsF_5 [79,80], $P(C_6H_5)_5$ [80], $Sb(CH_3)_5$ [80], and $Fe(CO)_5$ [87].

Monosubstitution in PF_5 should not substantially alter the barrier to the Berry rearrangement: one equatorial position is effectively held fixed in a single rearrangement step, and the fluorine atom positions are permuted without traverse of the high-energy, alternative trigonal-bipyramidal stereoisomers. Hence an APF_4 molecule should also display an invariant ^{19}F doublet spectrum. The CH_3PF_4 molecule which has the CH_3-group at an equatorial position[88] yields a ^{19}F doublet spectrum invariant from < -120–100 °C [79]. This is the case for all such molecules investigated to date with but one important exception. Aminophosphoranes, R_2NPF_4, show fluorine atom equivalence at 25 °C and the expected two sets (PF splitting) of doublets and triplets (FF splitting) at low temperatures[80,89] for an equatorially-substituted fluorophosphorane. Rearrangement barriers are low, *c.* 6 kcal mol^{-1}. The definitive work in this class comes from the ^{31}P study of Whitesides and Mitchell[90], who showed the character of the transitional ^{31}P line shape for $(CH_3)_2NPF_4$ *to be consonant only with the Berry mechanism*[84] *or a permutational equivalent.* The value of n.m.r. in mechanistic studies is exemplified by this study (see Figure 2.5). Note that irrespective of the nitrogen coordination geometry—planar or pyramidal—rotation about the PN bond must be rapid to achieve ultimate fluorine atom

* The Berry mechanism is often referred to as pseudorotation, which the process is. However, any polytopal rearrangement for any x-atom system may be viewed as a pseudorotation[2,8]. The terminology is imprecise and should be dropped.

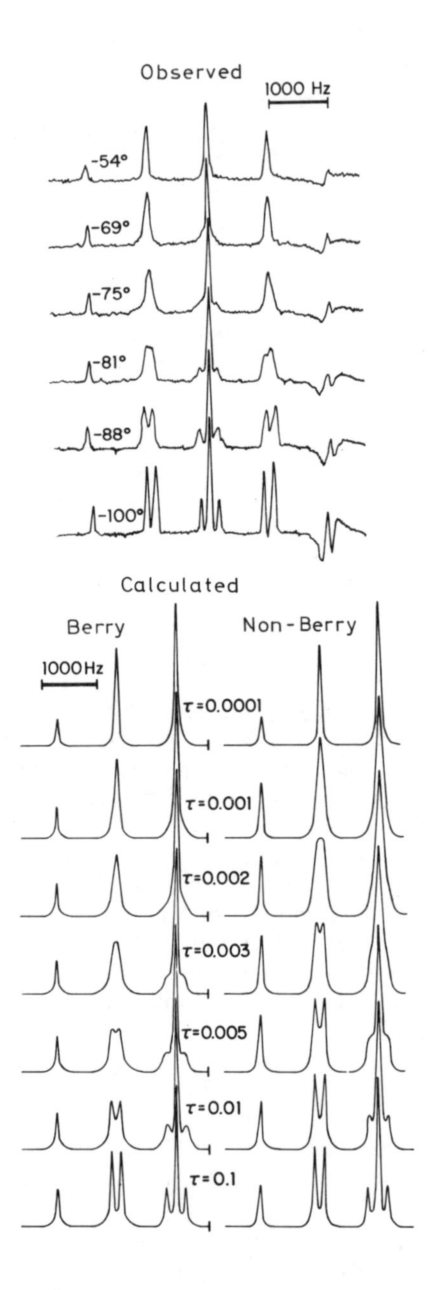

Figure 2.5 The observed ^{31}P n.m.r. spectra of $(CH_3)_2NPF_4$ as a function of temperature. The observed spectrum is to be compared with the calculated spectra shown in the lower half of the figure based on a Berry mechanism and on the alternative non-Berry type of mechanism. The complete spectra are not reproduced for the calculated sets; only about 60% of the calculated spectra are depicted. The central line is associated with a transition which is unaffected by a Berry rearrangement. As can be clearly seen in the observed spectra, the central line undergoes no significant line-width change. In contrast, a non-Berry rearrangement would lead to broadening of the central line as is illustrated in the lower right-hand portion of the figure.

equivalence via the Berry mechanism. It was suggested[80] that the PN rotation process is relatively activated due to a degree of multiple PN bonding and that this contributes to the higher barrier to equilibration of fluorine environments in aminofluorophosphoranes relative to the other tetrafluorophosphoranes. A similar phenomenon is operative in alkyl- and aryl-thiophosphoranes, $RSPF_4$, where fluorine atom non-equivalence is evident below $\sim -60\,°C$ [91].

The three possible stereoisomers in an A_2PF_3 molecule are (2.17–2.19), with increasing energy in going from (2.17) to (2.18) to (2.19). The energy gap should be directly related to the magnitude of the F—A electronegativity

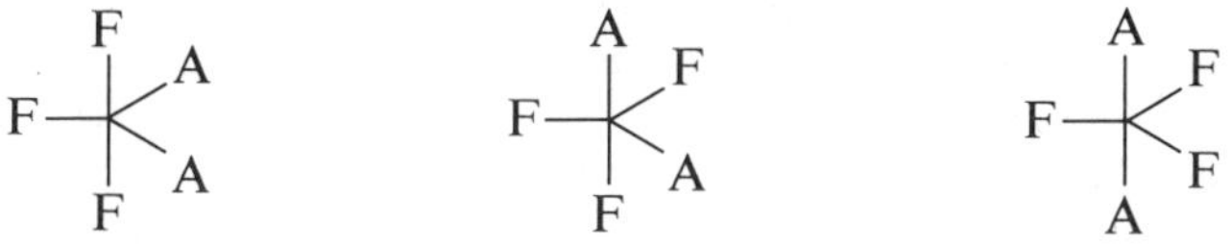

(2.17) (2.18) (2.19)

difference with the (2.18) to (2.19) gap being the larger of the two. Isomer (2.17) is the only detectable form in A_2PF_3 species[79, 80, 82]. Gaseous $(CH_3)_2PF_3$ has near-trigonal-bipyramidal form with equatorial methyl groups[88]. There is no evidence for intramolecular rearrangement in the ^{19}F spectra of dialkyl- or diarylfluorophosphoranes at high temperatures. A Berry rearrangement in an R_2PF_3 molecule would require traverse of energetically unfavourable isomeric states like (2.18). In contrast, Cl_2PF_3 and Br_2PF_3, in which the F—A electronegativity differences are much smaller than in the organophosphoranes, are less stereochemically rigid. Fluorine atom non-equivalence required for the ground-state form (2.17) is evident in the n.m.r. spectra of the chloride and bromide only at $-60\,°C$ and below[79, 82]. Rearrangement barriers are low, $\sim 6\ kcal\ mol^{-1}$ [82]. Gorenstein[92] has shown that in alkoxyphosphoranes the barriers due to intermediate (O *v.* C) electronegativity effects are $\sim 10–17\ kcal\ mol^{-1}$. The observations were based on n.m.r. studies[93–101] of non-rigid cyclic oxyphosphoranes primarily of the type (2.20)–(2.22). The enthalpies and free energies of activation for a Berry

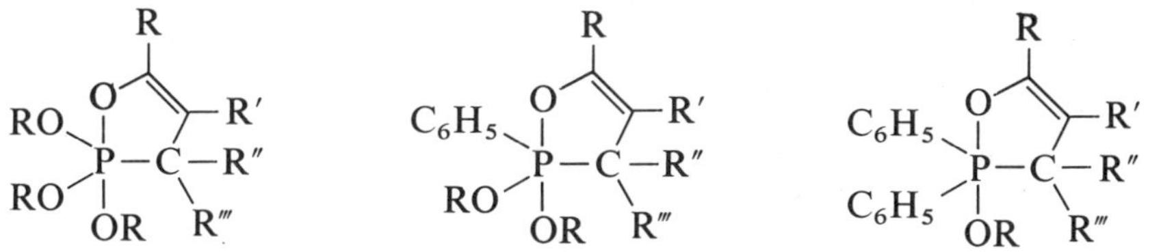

(2.20) (2.21) (2.22)

type of rearrangement both fell in a narrow range of $\sim 10–17\ kcal\ mol^{-1}$. Ring-strain effects on pathways for the Berry mechanism are far more severe. Consider the *cis* (2.23) and *trans* (2.24) forms of the phosphorane derived from dimethyl phenylphosphonite and benzylidene acetylacetone. The rearrangement mixes the apical (starred) methoxy group of the *cis* isomer with the equatorial (starred) methoxy group of the *trans* isomer, and similarly the equatorial methoxy group of the *cis* with the apical methoxy of

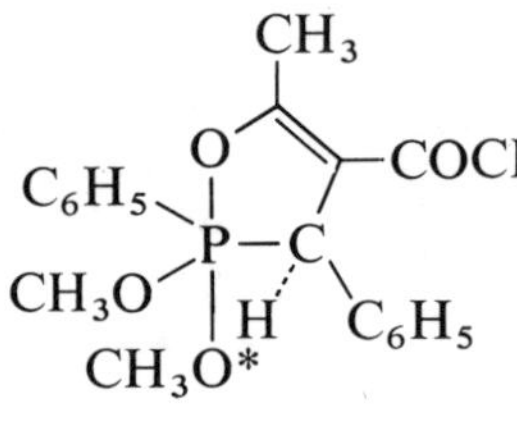

cis

(2.23)

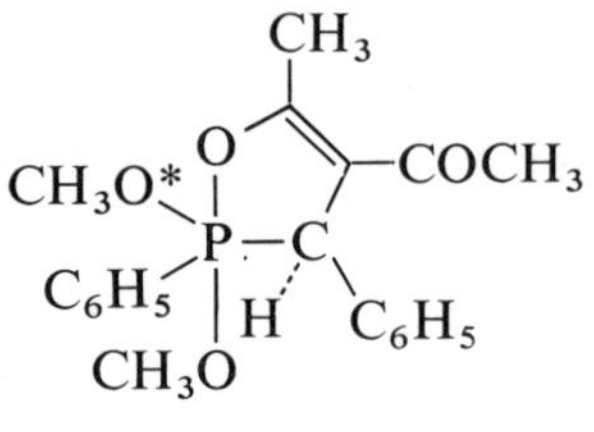

trans

(2.24)

the *trans*. This is evident in the 25 °C proton n.m.r. spectrum. Equilibration of methoxy and starred methoxy groups would require a Berry rearrangement which traverses the stereoisomer with the —C—C=C—C— ring system at diequatorial positions—a highly strained configuration. Strain-imposed barriers of this type are 20 kcal mol^{-1} or greater.

Effect of ring strain on rearrangement barriers is also evident from the n.m.r. studies of (2.25) and (2.26). Like all R_2PF_3 molecules, the ^{19}F n.m.r. of (2.25) consists of two sets of doublets and triplets and is temperature-invariant[79]. The expected fluorine non-equivalence for (2.26) is found only at low temperatures. Presumably ring strain in the PC_4 ring in (2.26) is

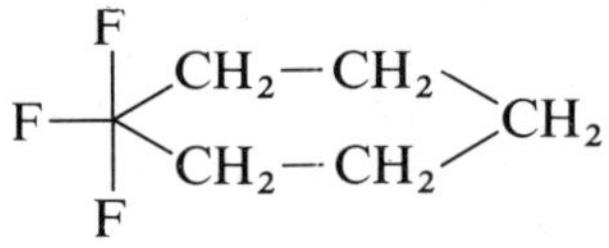

(2.25)

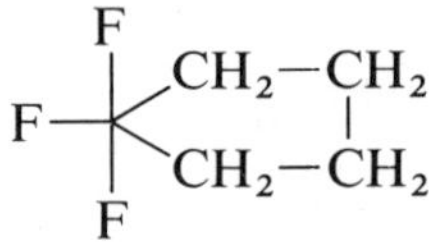

(2.26)

sufficient to raise the energy level of isomer form (2.17) to the point that isomer form (2.18) can be rapidly traversed and fluorine atom equivalence achieved with greater facility than in (2.25) and its acyclic R_2PF_3 analogues.

As noted above, pentaphenylphosphorane appears to be stereochemically non-rigid based on the low temperature n.m.r. equivalence of the aryl hydrogen atoms[80]. Similarly, pentamethylantimony shows n.m.r. equivalence of methyl hydrogens to quite low temperatures[80]. The barrier to a Berry rearrangement in this type of molecule has been increased by introducing steric or ring constraints. Consider first the case of the bis-bitolylphosphorane (2.27) with an equatorial phenyl group:

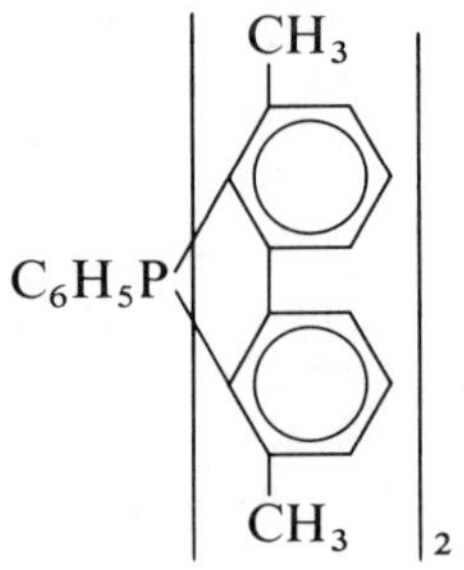

(2.27)

This compound, which is dissymmetric, can racemise in one step via the Berry process without going through a high-energy transition state derived from any kind of ring strain associated with the bidentate ligand. Racemisation in this system is very fast (*vide infra*) except where a bulky group, e.g., 1-naphthyl or 9-anthryl, is substituted for the phenyl ligand[102]. On the other hand, removal of the two methyl groups as in (2.28) changes the problem

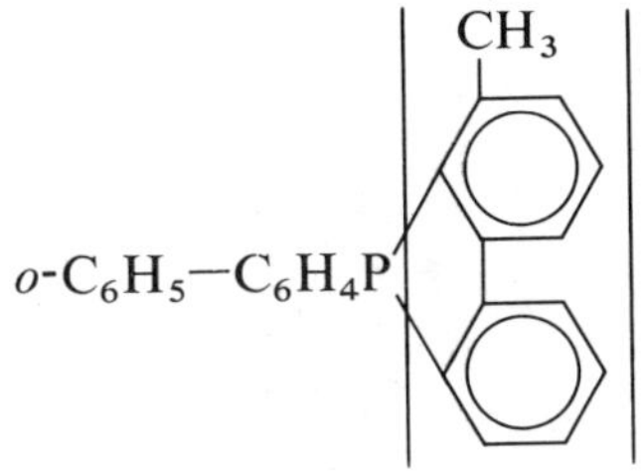

(2.28)

completely. Here it is not possible to racemise via the Berry process without traversing a stereoisomer in which one of the bidentate moieties spans equatorial positions. The latter would be a high-energy state due to ring strain, and it is of note that Hellwinkel[103] has succeeded in isolating enantiomers for this particular structure. A related study of considerable mechanistic importance came from an n.m.r. investigation of $o\text{-}C_6H_4[CH(CH_3)_2]\text{-}P(p,p'\text{-bitolyl})_2$. At room temperature, the isopropyl methyl groups as well as the bitolyl methyl groups are non-equivalent, consistent with the structure shown in Figure 2.6. This non-equivalence is lost at 130 °C. The rates of isopropyl-CH_3 and of tolyl-CH_3 proton equilibration are independent sensors of enantiomer interconversion and of axial–equatorial vertex positional exchange, respectively, and within experimental error these rates are identical. Only a Berry rearrangement *appears* to be consistent with these data[104]. A related molecule in which configurational stability has been achieved through steric factors for phosphorus as well as for nitrogen inversion is (2.29) with the substituted naphthyl group at an equatorial position[105].

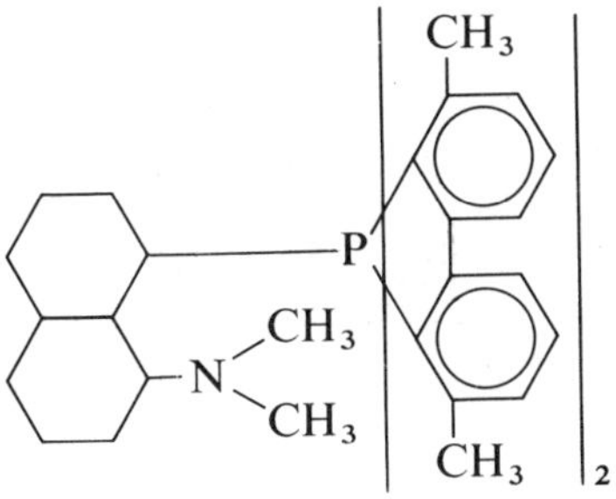

(2.29)

The above data on stereochemically non-rigid 5-coordinate rearrangements relate primarily to phosphorus(V) chemistry; however, the principles established from the extensive and quantitative studies of this element should extend to other elements, and all available information bears this out. The arsenic fluoride system of AsF_5–$RAsF_4$–R_2AsF_3 follows precisely the analogous phosphorus system with respect to structure and relative stereochemical non-rigidity[79, 80]. A claim[106] of very low barriers to rearrangements in a series of arsenic(V)–glycol complexes is anomalous and of questionable validity; further mechanistic studies are required. The data for antimony and bismuth are quite limited, but the basic analogy holds; compounds such as $Sb(CH_3)_5$ and $RSbF_4$ are stereochemically non-rigid[80], whereas R_3SbF_2 and R_3BiF_2 do not rearrange on the n.m.r. time-scale[80]. Penta-coordinate

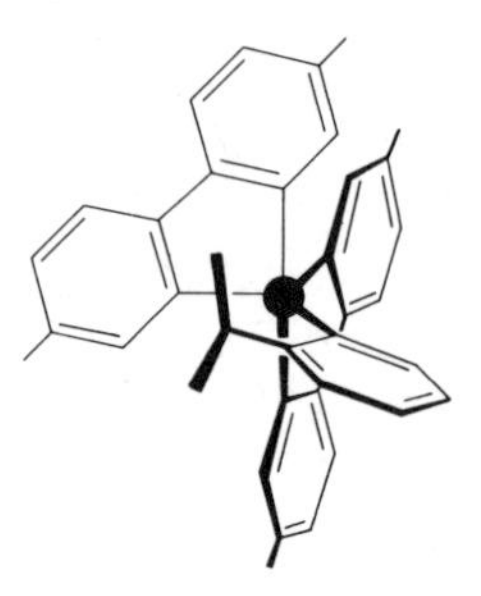

Figure 2.6 Probable structure of o-$C_6H_4[CH(CH_3)_2]P(p,p'$-bitolyl$)_2$. The o-isopropylphenyl substituent is at an equatorial position

silicon compounds are now well-established and trigonal-bipyramidal geometry prevails. In the series SiF_5^-, $RSiF_4^-$, and $R_2SiF_3^-$, which is isoelectronic to the fluorophosphoranes, the relative stereochemical non-rigidity is qualitatively the same as in the phosphorus series; only the $R_2SiF_3^-$ ions possess sufficient configurational stability to exhibit on the n.m.r. time-scale the non-equivalence of the axial and equatorial fluorine atoms[107].

Quasi-5-coordinate molecules such as sulphur tetrafluoride, $:SF_4$, should be subject to relatively fast Berry-type rearrangement, but there are no definitive data for sulphur tetrafluoride or its derivatives. Fast fluorine exchange does occur in $:SF_4$ [108, 109], but this is primarily intermolecular fluorine exchange. If a polytopal rearrangement does take place in $:SF_4$, it must have a barrier greater than 4 kcal mol^{-1} [109]. Fast intramolecular rearrangements have been postulated in $:S(i\text{-}C_3F_7)_2F_2$ and $:S(i\text{-}C_3F_7)(CF_3)F_2$ to account for S—F fluorine atom n.m.r. equivalence[80]. In contrast, fast exchange is not observed[80] in $:S(i\text{-}C_3F_7)F_3$, $:S(C_6H_5)F_3$; these are roughly analogous to X_2PF_3 fluorophosphoranes which generally show configurational stability on the n.m.r. time-scale. From stereochemical and mechanistic studies[110] of base hydrolysis of alkoxy-3-methyl thietanium cations (2.30), it has been suggested that the barrier to Berry rearrangements

R_2

R_1

S^+—:   (2.30)

OR

may be higher in quasi-5-coordinate sulphur(IV) compounds than in formally analogous phosphorus compounds. This thesis may be correct, but there are no supportive hard data. Quasi-5-coordinate molecules like

:C̈lF$_3$ and :B̈rF$_3$ should have a relatively high barrier to a Berry type of rearrangement. A sulphur(VI) derivative, $CF_3N{=}SF_4$, has been shown by ^{19}F n.m.r. studies to undergo rapid intramolecular fluorine positional exchange with an activation barrier of ~6 kcal mol^{-1} [80].

Substantive information on polytopal rearrangements in 5-coordinate transition elements is quite limited. More detailed knowledge of the factors affecting rearrangement barriers is critically needed to better understand reaction mechanisms of 4- and 6-coordinate complexes that involve a 5-coordinate [2] reaction intermediate. The studies of Clark and co-workers[111,112] on PF_3 derivatives of iron(0) and cobalt(I) provide the only data of value. For example, the compound $CF_3Co(CO)_3PF_3$ shows a complex ^{19}F n.m.r. spectrum at low temperatures, < -30 °C, that is characteristic for the presence of the two stereoisomers, (2.31) and (2.32). Above

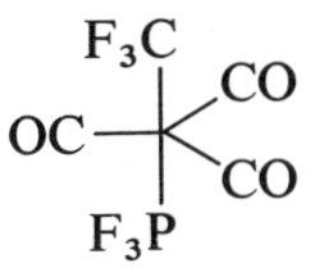

F$_3$C, CO, OC, CO, F$_3$P (2.31)

F$_3$C, CO, F$_3$P, CO, OC (2.32)

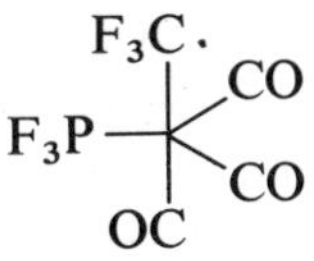

-30 °C, the ^{19}F spectrum broadens and then sharpens to give an averaged CF resonance and an averaged PF resonance with retention of spin-coupling. Retention of the coupling definitively establishes a polytopal rearrangement. The mechanism is presumably the Berry process. The activation energy for the process is about 11 kcal mol^{-1} [111].

Hydrolysis of 4-coordinate phosphate esters (phosphates, phosphinates, and phosphonates) proceeds via a 5-coordinate intermediate. The intermediates possess sufficient lifetimes that Berry-type rearrangements occur. With knowledge of the restraints on the rearrangement as outlined above with respect to electronegativity and ring strain, the hydrolysis rates, stereochemical facets and the actual course of hydrolytic cleavage can be rationalised for a vast body of data. The general rules outlined by Westheimer[100] are of substantial predictive value. For example, in the hydrolysis of ethylene phosphate, the rate is ~10^8 greater than for dimethyl phosphate, and oxygen (^{18}O) exchange is 20% faster than the hydrolysis[97,113,114]. The presumed intermediate, (2.33), achieves a large decrease in energy due to

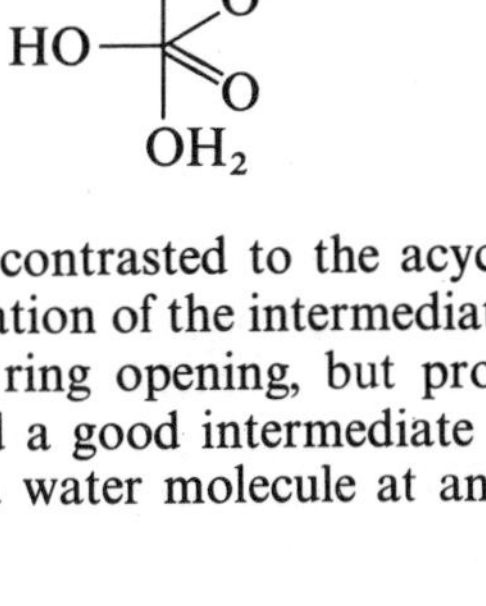

CH$_2$, CH$_2$, O, O, HO, O, OH$_2$ (2.33)

decreased ring strain (as contrasted to the acyclic analogues or large ring cyclic analogues). Protonation of the intermediate at a ring oxygen provides a good intermediate for ring opening, but protonation at the equatorial HO group does not yield a good intermediate for oxygen exchange (^{18}O) studies) since release of a water molecule at an equatorial position would violate the principle of microscopic reversibility. However, a rapid Berry rearrangement to (2.34) allows oxygen exchange because the OH group is now axial.

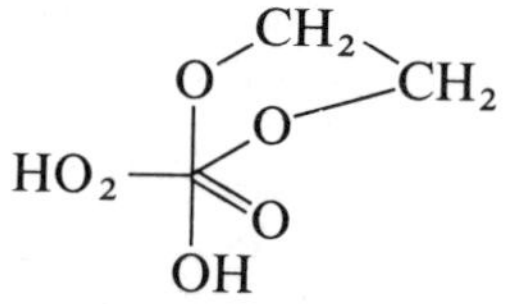

CH_2 CH_2 O O HO_2 O OH (2.34)

The phostonate ester, (2.35), hydrolyses rapidly with almost exclusive ring opening; the OCH_3 group is not substantially attacked[95]. Again this is most

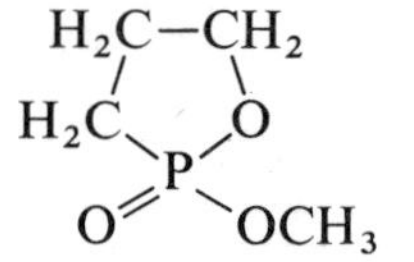

H_2C-CH_2 H_2C O P O OCH_3 (2.35)

readily understood in terms of stereochemical rules for a 5-coordinate species and a Berry rearrangement. The initial intermediate in the hydrolysis of the ester would be (2.36):

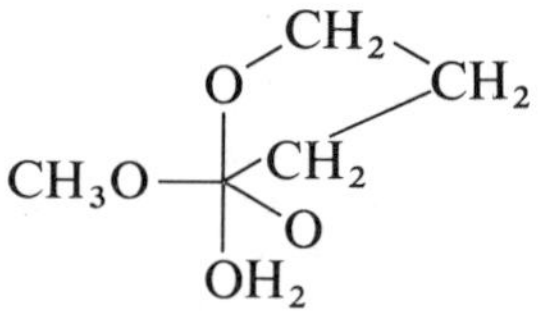

CH_2 CH_2 O CH_3O CH_2 O OH_2 (2.36)

Protonation at the ring oxygen allows for ring cleavage. Facile cleavage of the OCH_3-moiety would require a Berry rearrangement to stereoisomers of the type (2.37) and (2.38), both of which are high-energy states because of

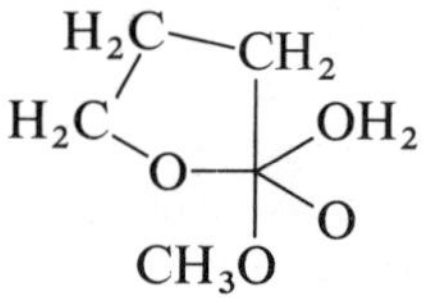

H_2C-CH_2 H_2C OH_2 O O CH_3O

(2.37)

CH_3O CH_2-CH_2 H_2O O CH_2 O

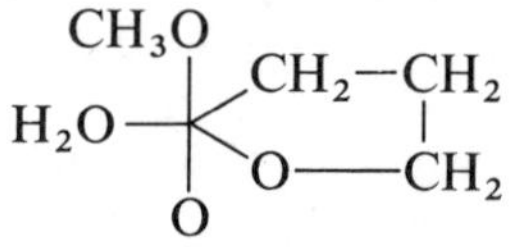

(2.38)

unfavourable disposition of the electropositive CH_2 group and because of ring-strain increase ($\sim$120 degrees rather than $\sim$90 degrees) respectively. Accordingly, the rate of rearrangement in the intermediate should be low, and hence the almost exclusive ring cleavage[95,100].

These basic studies and generalisations have been greatly reinforced by elegant studies of other types of phosphorus(V) esters in work by Westheimer and co-workers[93–97,110,115–117], Mislow and co-workers[18,118,119], Ramirez and co-workers[101], Frank and Usher[120], and by others[121–124].

Application of 5-coordinate stereochemical rules and the possibility of polytopal rearrangements in reaction intermediates from reaction of 4- and 6-coordinate complexes derived from the transition-metal group has been relatively neglected. Speculations have been presented for rearrangements in reaction intermediates in some reactions of 4-coordinate complexes[2, 125, 126] and 6-coordinate cobalt complexes[127], but there are no definitive data on rearrangements in such reaction intermediates.

Analysis of the stereochemical problems in 5-coordinate rearrangements has been a subject of considerable recent interest. For the general case in which a central metal atom is attached to five different nuclei or groups, there are 20 and 30 stereoisomers in the trigonal-bipyramid and square-pyramid forms, respectively. The connectivity of the trigonal-bipyramid is three. This can be easily visualised by referring back to Figure 2.4. One Berry process can be performed at any one of the equatorial positions. Each of these would generate a unique square-pyramidal isomer. The square-pyramid is only two-connective. The connected graph requirement is satisfied: $I\delta$ products are 20×3 and 30×2 for the trigonal-bipyramid and square-pyramid, respectively. The one-step rearrangement for a given isomer is shown in Figure 2.7. Isomer B can be generated in six different configurations, all of which are equivalent and are related by symmetry elements of the D_{3h} point group (for the idealised trigonal-bipyramid). The operations referenced to Figure 2.7 and the associated permutations are:

		Parity
$R^+ \rightarrow$	$\begin{pmatrix} 12345 \\ 45321 \end{pmatrix} = (1425)$	odd
$C_3R^+ \rightarrow$	$\begin{pmatrix} 12345 \\ 31245 \end{pmatrix}(1425) = (132)(1425) = (14)(253)$	odd
$C_3^2R^+ \rightarrow$	$\begin{pmatrix} 12345 \\ 23145 \end{pmatrix}(1425) = (123)(1425) = (143)(25)$	odd
$R^- \rightarrow$	$\begin{pmatrix} 12345 \\ 54312 \end{pmatrix} = (1524)$	odd
$C_3R^- \rightarrow$	$\begin{pmatrix} 12345 \\ 23145 \end{pmatrix}(1524) = (123)(1524) = (153)(24)$	odd
$C_3^2R^- \rightarrow$	$\begin{pmatrix} 12345 \\ 31245 \end{pmatrix}(1524) = (132)(1524) = (15)(243)$	odd

With these different operations (only those that include R^+ or R^- lead to actual stereoisomerism), there are 120 elements in this system. Considering the isomer system properties, the following can be delineated. Complete graph analysis shows that the stereochemical system is closed; all isomers can be reached from any given isomer by the operations postulated. The total number of operations[16] which are distinct between all isomer configurations is 120 or 5!. The racemisation chain, c_E, ($A \rightarrow \bar{A}$) is of order five. For each isomer there are 12 distinct chains of order five, and there are, of course, ten different chains, that is different A—$\bar{A}$ pairs. The smallest sub-cycle, **C** (non-trivial A—A circuit) is of order six. There are others of order 10, 14, etc. The racemisation cycle $\mathscr{C}_E$(A—$\bar{A}$—A) is of order 10. Each cycle connects two enantiomers or more, and there are 12 possible cycles per pair. Hence there are 120 cycles[16].

The stereochemical or system properties may be related to group properties. The set of all operations corresponding to all paths between any isomer

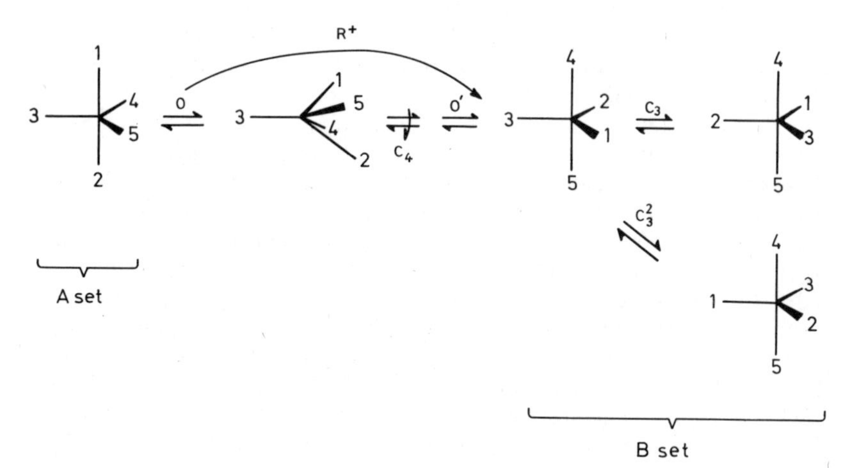

Figure 2.7 The operations involved in the Berry rearrangement of a trigonal-bipyramid traversing the square-pyramidal intermediate or transition state. Also shown are the threefold rotational operations which generate equivalent configurations in the isomer B set. Three other equivalent configurations would be similarly generated with the R^- operation which is distinguished from the R^+ operation by a counter C_4 rotation of the square-pyramid

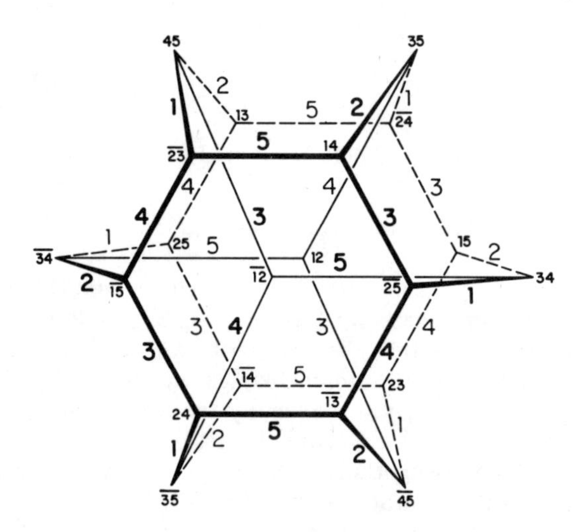

Figure 2.8 A 20-vertex graph representation of the general case in the five-atom family based on the Berry rearrangement[18]. This particular graph representation has D_{3d} symmetry. (There are many isomorphic graphs that have been employed to depict this particular system.) Isomers are denoted by simply labelling axial substituents; enantiomers are related by the notations: 12 and $\overline{12}$, etc. Proper sub-groups for this symmetric S_5 group are readily identified from the figure. These are isomorphic to the isomer sub-cycles of order six. These proper sub-groups comprise all of the six vertex hexagons or puckered hexagons

in the system and any equivalent configuration of any other isomer in the system is isomorphic to the symmetric group S_5 which possesses 120 elements. There is a proper sub-group of order six generated by the permutation (12)(345), and this is isomorphic to a sub-cycle of isomerisation. Cycles present in the isomer system are not proper sub-groups of the symmetric group S_5 [16].

Topological representation of this group cannot be presented in three dimensions without recourse to interpenetrating figures[5–9]. A number of isomorphic graphs have been ingenuously employed[13–15, 18] and are helpful in revealing mechanistic problems associated with stereoisomerism in 5-coordinate compounds or 5-coordinate reaction intermediates of reasonable lifetime as has been demonstrated by De Bruin *et al.*[18] (Figure 2.8). An alternative representation is a tabular or matrix display[8, 20]. Any of these representations are extremely useful since stereochemical problems can be literally solved at a glance, whereas paper and pencil solutions are laborious and errors are too readily made. Most importantly, mechanistic distinctions are incisively evident by inspection[5–9]. An example[18] of the graphical representation is given in Figure 2.8.

Intramolecular rearrangements in five-atom polyhedral boranes and in metal clusters are possible and mechanistically should be similar to those in 5-coordinate complex chemistry[10, 11]. To date, there are no data on rearrangements in these classes of molecular aggregates[11].

2.6 SIX-ATOM SYSTEM

The possibility of an intramolecular rearrangement in a 6-coordinate complex was considered long ago[128–132], but only recently have any definitive data on such rearrangements been presented.

Stereochemical non-rigidity has been unequivocally established for a class of transition-metal dihydrides of the general formula MH_2L_4, where M is iron or ruthenium(II) and L is a phosphine, phosphite, phosphonite, or phosphinite[133–135]. These complexes in solution have almost exclusively a *cis* configuration with the exception of those based on phosphinite ligands, e.g., $C_6H_5P(OC_2H_5)_2$ and $C_6H_5P(OCH_3)_2$, which have *cis* and *trans* forms with a *cis*:*trans* ratio of 3:1 or larger. The analysis of the solution-state structures was rigorous and was based on ^1H and ^{31}P n.m.r. analysis. The iron complexes are configurationally stable only below 25 °C. Intramolecular rearrangement is rapid ($\sim$500 s^{-1}) at $\sim$25 °C and limiting high-temperature ^1H and ^{31}P spectrum consisting of quintets and triplets were observed at $\sim$50–80 °C. Retention of spin correlation in these transitions, the averaging of coupling constants in the high-temperature forms, and invariance of spectra to dilution and to added ligand rigorously establish a polytopal rearrangement. Illustrations of the temperature-dependent n.m.r. spectra for *cis-trans* $FeH_2[C_6H_5P(OC_2H_5)_2]_4$ and for *cis* $FeH_2\{C_6H_5P[OCH(CH_3)_2]_4\}$ are given in Figures 2.9 and 2.10.

The activation energy for rearrangement in $H_2Fe[P(OC_2H_5)_3]_4$ is 12 kcal mol^{-1} [135]. Rearrangement barriers for the iron complexes vary surprisingly little with large changes in character of the ligand. It does appear,

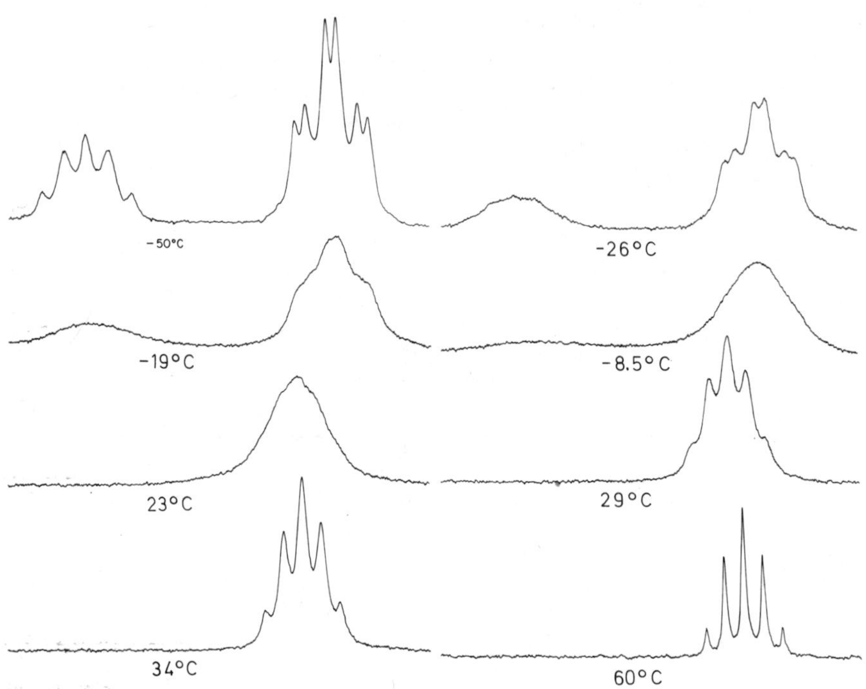

Figure 2.9 Temperature-dependence of the 220 MHz ^1H hydride spectrum of $H_2Fe[C_6H_5P(OC_2H_5)_2]_4$

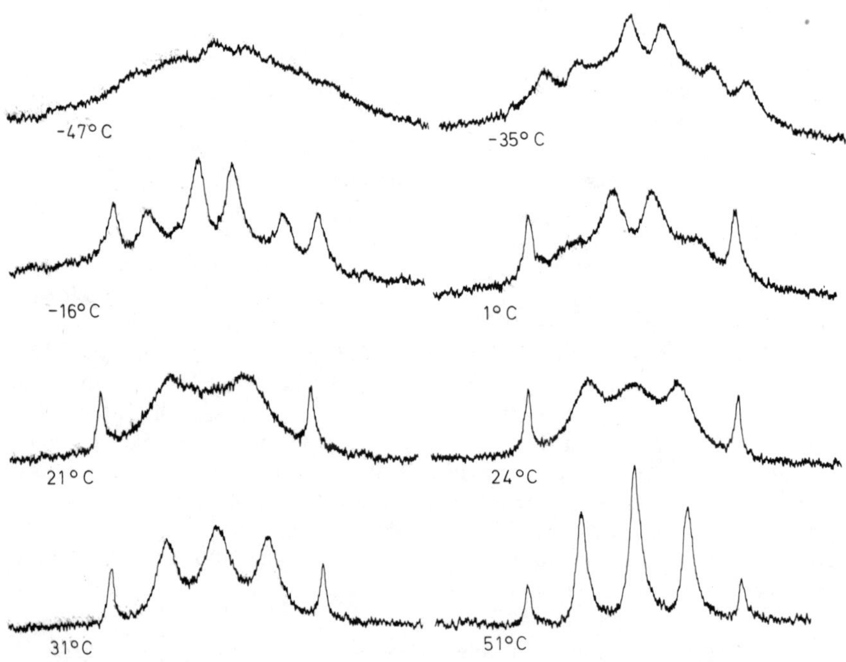

Figure 2.10 Temperature-dependence of the 220 MHz ^1H hydride spectrum of $H_2Fe[C_6H_5P(OCH(CH_3)_2)_2]_4$

however, that the complex with the ligand that has the smallest PX_3 cone angle, $H_2Fe[P(OCH_2)_3CC_2H_5]_4$, has the highest barrier and the one with the largest ligand cone angle, $H_2Fe[(C_6H_5)_2PCH_3]_4$, the lowest barrier.

All but one of the ruthenium complexes examined to date have exclusive *cis* stereochemistry [133–138]. Again, spectra are temperature-dependent (Figure 2.11). The high-temperature limiting proton quintet requires temperatures well in excess of 150 °C, but this has not been spectrally demonstrated because decomposition and dissociation rates are high above 150–200 °C. Nevertheless, near-retention of line shape for the central two lines in the proton

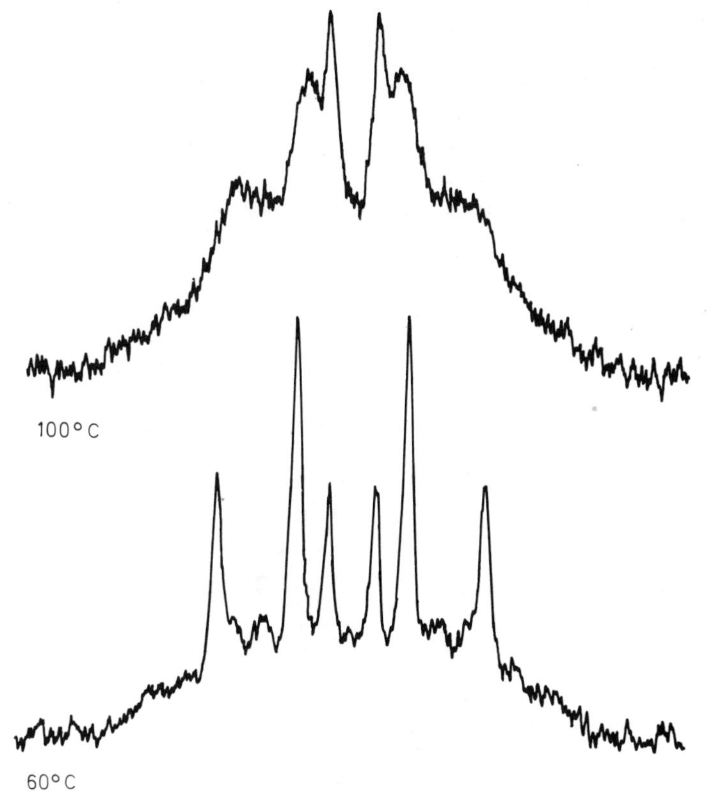

Figure 2.11 Temperature-dependence of the 220 MHz ^1H hydride spectrum of $H_2Ru[C_6H_5P(C_2H_5)_2]_4$. The barrier for the related $H_2Ru[C_6H_5P(CH_3)_2]_4$ complex is higher, ostensibly because of less significant ligand–ligand repulsions

spectra (Figure 2.11) throughout the transition clearly establishes a polytopal rearrangement. The inner lines stay sharp for the *cis*-ruthenium complexes, in contrast to the outer lines for the *cis*-iron set (Figure 2.10); this simply reflects differences in relative signs and magnitudes of the coupling constants for the two sets. The $H_2Ru[C_6H_5P(OC_2H_5)_2]_4$ complex has *cis* and *trans* forms present in the solution state, and the general temperature behaviour (n.m.r.) is similar to that of the iron analogue (Figure 2.9). Two significant properties of the ruthenium complexes are the relative invariance of rearrangement barriers to ligand change and the much larger barrier values for the set as compared to the iron set. As found for the iron complexes, increasing steric bulk of the ligand does *appear* to lower the rearrangement barrier. The barrier for the complex based on the $C_6H_5P(CH_3)_2$ ligand is higher than that for the complex from the bulkier $C_6H_5P(C_2H_5)_2$ ligand.

Ligand-ligand non-bonding repulsions in the H_2ML_4 complexes should distort the coordination polyhedron from an octahedron toward a tetrahedral

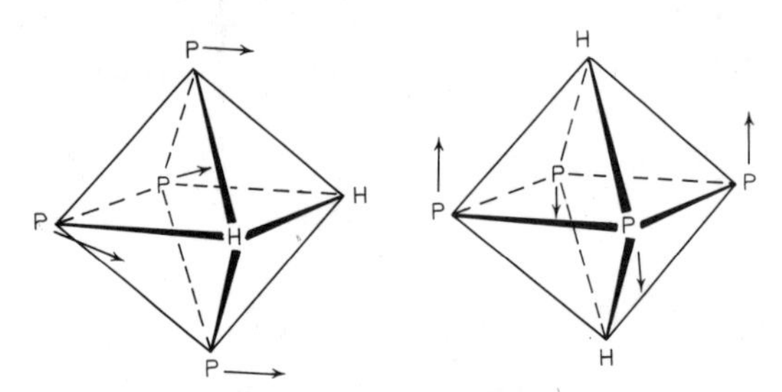

Figure 2.12 Structures of H_2FeL_4 isomers. Arrows indicate probable distortions from idealised geometries

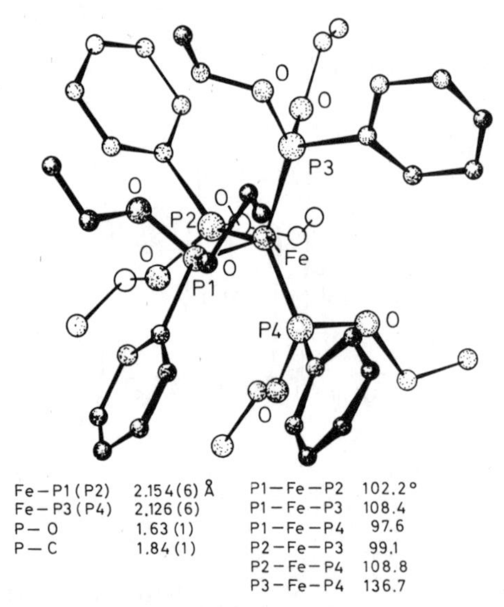

Fe—P1 (P2)	2.154 (6) Å	P1—Fe—P2	102.2°
Fe—P3 (P4)	2.126 (6)	P1—Fe—P3	108.4
P—O	1.63 (1)	P1—Fe—P4	97.6
P—C	1.84 (1)	P2—Fe—P3	99.1
		P2—Fe—P4	108.8
		P3—Fe—P4	136.7

Figure 2.13 Structure of $H_2Fe[C_6H_5P(OC_2H_5)_2]_4$. The hydride hydrogen atoms are presumed to be vicinal as indicated by their *trans* influence on the Fe–P(1) and Fe–P(2) bond distances. The shortest non-hydrogen contacts between atoms on different phosphorus atoms involve oxygen contacts with methylene or benzene carbon atoms

disposition of phosphorus nuclei, as illustrated in Figure 2.12. The extent of this distortion in the ground state was established by a crystal structure determination of $H_2Fe[C_6H_5P(OC_2H_5)_2]_4$ [134]. The structure is illustrated in Figure 2.13. The iron–phosphorus, i.e., FeP_4, geometry is midway between the idealised angles for octahedral and tetrahedral geometry with r.m.s. angular deviation favouring tetrahedral geometry. Thus, these dihydrides are related, in a geometrical-deformation sense, to the monohydrides of the type $HM(PR_3)_4$, described in the five-atom section, which have near-tetrahedral geometry. The distortion modes (Figure 2.12) illustrated for *cis* and *trans* isomers could, through higher vibrationally-excited states, provide a mechanism for achieving spin-equivalence of hydrogen and of phosphorus nuclei: population of a P_4M pseudo-tetrahedral transition state in which hydrogen atoms tunnel face-edge positions. In such a mechanism, rearrangement barriers would reflect the extent of PMP angle deformation required to approach tetrahedral values and the hydrogen atom motion. Line-shape analysis of transitional n.m.r. spectra[134, 135] show that the permutation exemplified by this mechanism is the only *single* one that can account for the observed data.

This class of stereochemically non-rigid 6-coordinate dihydrides is somewhat unique, and their susceptibility to rearrangement stems from the gross distortion imposed by the bulky PR_3 ligands and ease of hydrogen motion. Generally, 6-coordinate molecules are quite rigid. The relative rigidity is aptly illustrated in the n.m.r. study of $C_6H_5SF_5$ [133]. The ^{19}F n.m.r. spectrum of this fluoride is an AB_4 pattern which does not change up to 215 °C. At least in this molecule, the barrier to any kind of rearrangement must be >30–40 kcal mol^{-1}. This reflects the rather pervasive stability of the octahedral form in 6-coordinate molecules or ions.

The possibility of a specific kind of rearrangement, a trigonal twist, has been considered[128–132, 138] for chelate structures of the type $M(chel)_3$. Recently a very careful study[139] was made of the kinetics of racemisation and of *cis-trans* isomerisation in the cobalt complex (2.39):

CH₃–C(–O)=CH–C(–O)–CH(CH₃)₂ chelated to Co, [ ]₃ (2.39)

The analysis quite clearly eliminates the twist type of mechanism as the dominant isomerisation or racemisation process and indicates that a bond rupture mechanism proceeding through a 5-coordinate transition state is the most plausible process. This conclusion is consistent with earlier inferences on isomerisation mechanism in 6-coordinate chelate complexes[132].

A class of compounds that should be more susceptible to trigonal twist rearrangements are sulphur-based chelates. In particular, the demonstrated

M[S–C(R)=C(R)–S]₃ (2.40)

M[P(R)–C(R″)=C(R″)–P(R′)]₃ (2.41)

range of ground-state structures for dithiolate complexes (2.40) from octahedral to intermediate to trigonal-prismatic[140–144],* suggests the possibility of facile polytopal rearrangement via a trigonal twist. Decisive demonstration of this phenomenon may require synthesis of analogues[2],* like (2.41) which have ideal nuclei for n.m.r. studies – although preliminary data for some sulphur-based chelates, iron dithiocarbamates, do indicate a relatively fast intramolecular twist rearrangement[146].

Rearrangements in polyhedral boranes and metal clusters is a little-explored area in the six-atom family. *Cis* to *trans* isomerisation at 250 °C has been established for the octahedral carborane $C_2B_4H_6$, but the mechanism is unknown[147]. Barriers to polytopal rearrangements in this class should generally be higher than for coordination compounds[5, 11].

2.7 SEVEN-ATOM SYSTEM

A large number of idealised geometries may be realistically considered for the ground-state structure of a 7-coordinate complex. Some of these are the pentagonal-bipyramid (D_{5h}), capped octahedron (C_{3v}), monocapped trigonal prism (C_{2v}), and variations of a square base-trigonal cap (C_s) (Figure 2.14)[148].

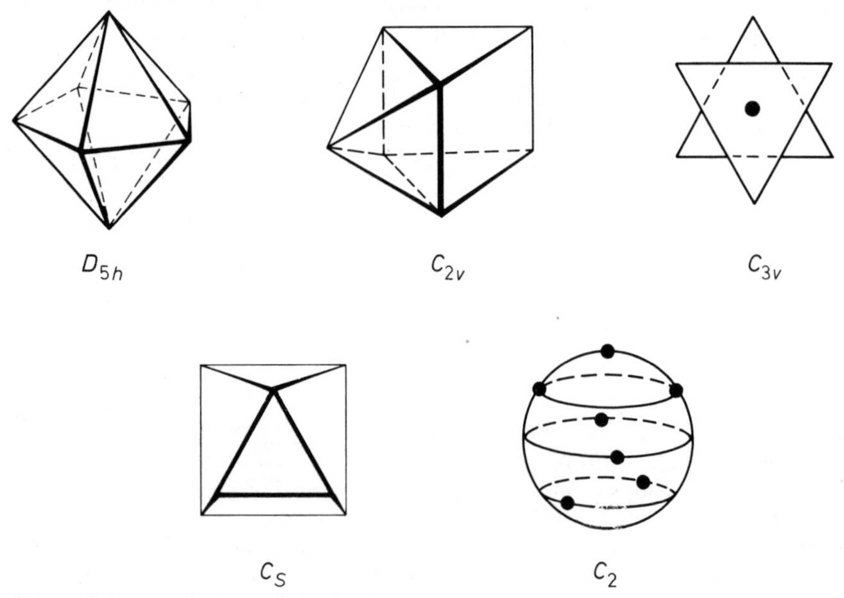

Figure 2.14 Idealised geometries for the seven-atom system

All these geometries have been established in the solid state for different 7-coordinate complexes[148–151]. Unless unusual constraints are operative as in a chelate complex, the energy differentiation among the various polytopal forms should be very small[148–151]. More significantly, in relation to stereochemical non-rigidity, the bending required to deform any one of the

* Interligand R—R repulsion may greatly raise the trigonal prismatic energy level.

idealised forms into another is quite small, e.g., see Figure 2.15 [2,5]. Accordingly, 7-coordinate complexes should generally be stereochemically non-rigid, comparable to or even less rigid than 5-coordinate complexes. Available experimental data are limited and non-definitive, but do point to relatively non-rigid structures. For example, the IF_7 ^{19}F n.m.r. resonance is a broad multiplet (I—F spin–spin coupling) over a broad temperature range. This spectroscopic equivalence of fluorine atoms is consistent with a very rapid intramolecular rearrangement[152,153]. The same data and possible

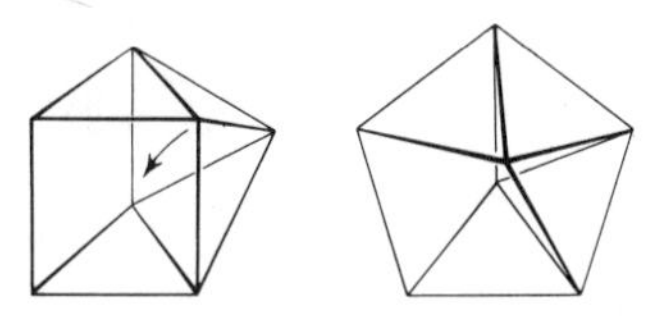

Figure 2.15 A possible mechanism for the conversion of the C_{2v} monocapped-trigonal prism to the D_{5h} pentagonal-bipyramid by a relatively small bending mode

conclusions apply to ReF_7 [152]. The 7-coordinate osmium hydride[2], $H_4Os[C_6H_5P(OC_2H_5)_2]_3$, has a simple quartet proton n.m.r. spectrum at 25 °C. This quartet begins to broaden at lower temperatures until a single broad peak prevails at ~ -100 °C. This *may be indicative* of a transition region due to an intermediate rate of polytopal rearrangement[2]. Similar inferential data apply to 7-coordinate chelate structures[154].

2.8 EIGHT-ATOM FAMILY

As in 7-coordination, the number of reasonable polytopal forms in the eight-atom case is quite large. However, most coordination compounds generally approximate the D_{4d} square antiprismatic, D_{2d} trigonal–dodecahedral, and C_{2v} bicapped trigonal prismatic forms (Figure 2.16)[158–160]. Energy differentiation among these forms should not be large (barring unusual steric or electronic factors) and interconversions of forms require quite minor bending modes[5,11,148], e.g., see Figures 2.17 and 2.18. There are no definitive data on intramolecular rearrangements in coordination compounds; however, all discrete 8-coordinate molecules or ions examined by n.m.r. have shown spectroscopic equivalence of ligand nuclei which suggest rapid intramolecular rearrangements. These include $Mo(CN)_8^{4-}$ (^{13}C) [1], $ReH_5[P(C_6H_5)_3]_3$ (^1H) [161], $H_5Ir[C_6H_5P(C_2H_5)_2]_3$ (^1H and ^{31}P) [2], and numerous chelates[154,162–164].

2.9 NINE-ATOM SYSTEM

Most 9-coordinate complexes have a geometry very similar to the symmetrically tricapped trigonal prism (D_{3h}) but an alternative form, the monocapped square antiprism (C_{4v}) has also been observed[5,11,148] (Figures 2.19 and 2.20). Rearrangements in 9-coordinate structures should require very little activation. Two possible rearrangement mechanisms are illustrated in Figures

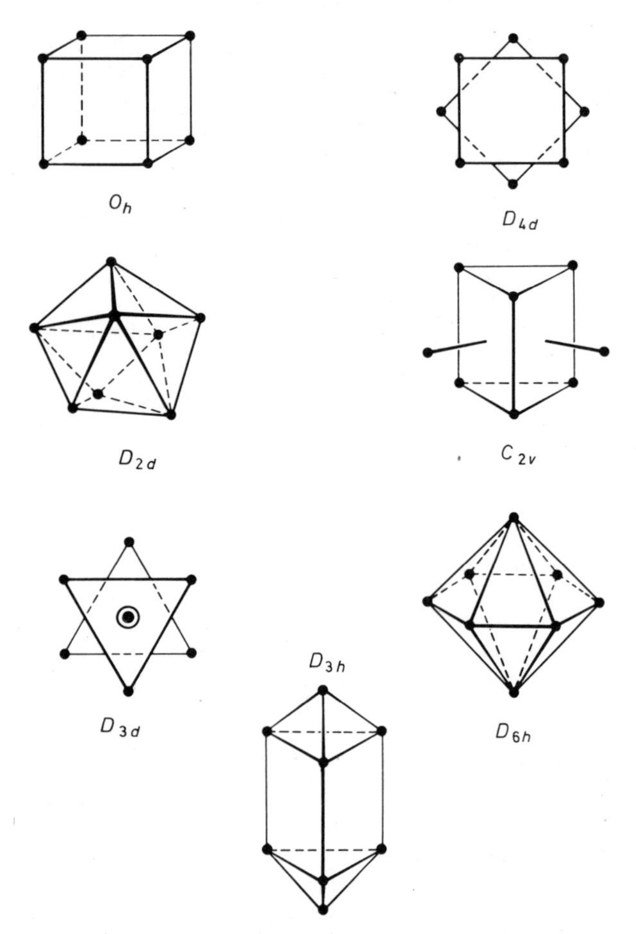

Figure 2.16 Idealised geometries for the eight-atom system

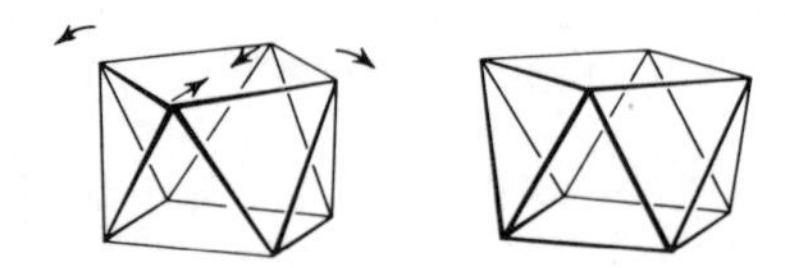

Figure 2.17 Bending mode for the interconversion of the square antiprismatic form to the C_{2v} bicapped trigonal prismatic form

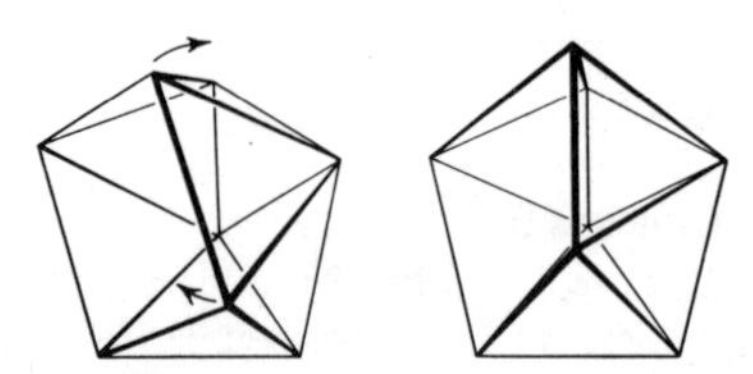

Figure 2.18 Bending mode for the conversion of the C_{2v} bicapped trigonal-prismatic form to the D_{2d} trigonal-dodecahedral form

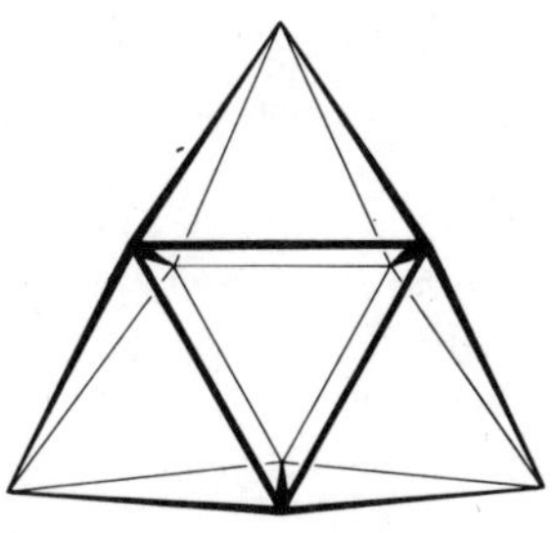

Figure 2.19 Favoured, idealised geometry for the nine-atom system, the symmetrically tricapped trigonal prism of D_{3h} symmetry

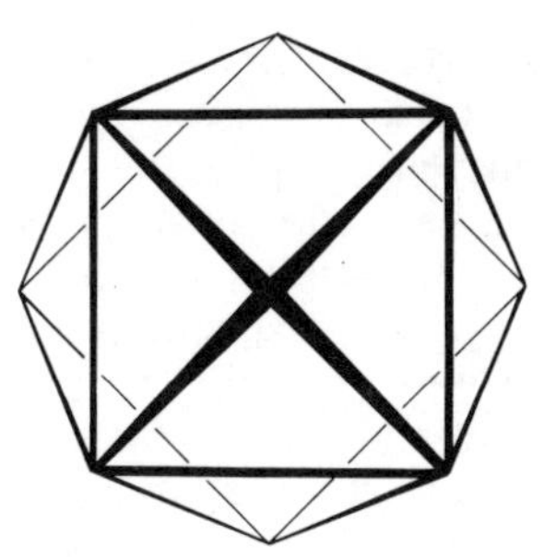

Figure 2.20 An alternative polyhedral form for the nine-atom system, the monocapped square antiprism of C_{4v} symmetry

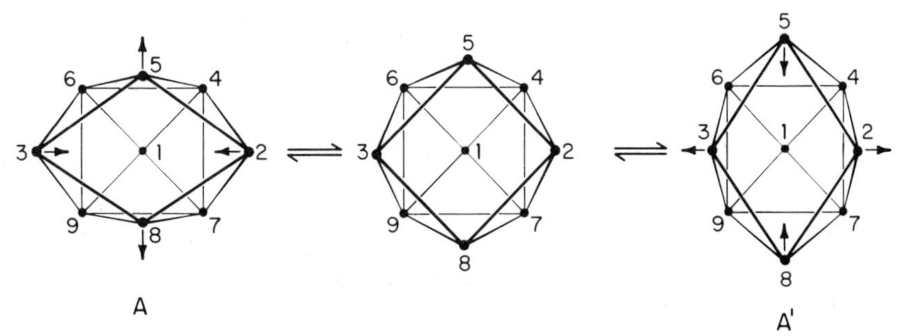

Figure 2.21 Possible rearrangement mechanism for a nine-atom complex with the symmetrically tricapped trigonal prismatic ground-state geometry which traverses the monocapped square antiprism as an intermediate or transition state

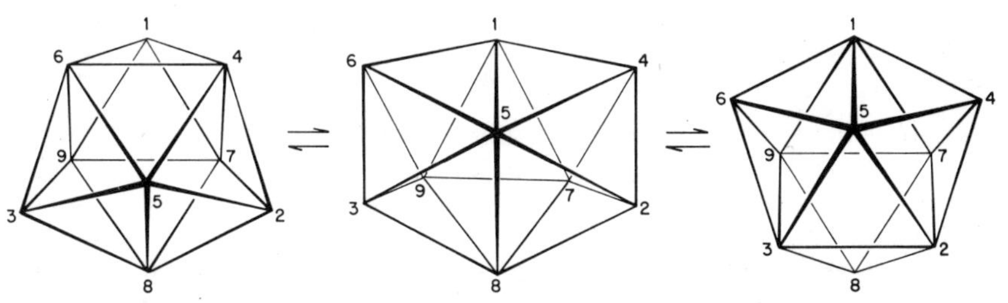

Figure 2.22 Alternative rearrangement mechanism in the nine-atom system for a complex based on the tricapped trigonal-prismatic form which traverses a reaction intermediate or transition state of C_{2v} symmetry

2.21 and 2.22 [5,9]. The fact that a single Re—H proton n.m.r. resonance is observed[2,165] for the D_{3h} ReH_9^{2-} ion argues strongly for stereochemical non-rigidity in this ion.

The boron nuclei in the polyhedral borane ion $B_9H_9^{2-}$ also have the D_{3h} symmetrically tricapped trigonal-prismatic arrangement. The ^{11}B n.m.r. spectrum (two doublets of intensities 2 and 1) of the ion in solution is unaltered up to temperatures above 200 °C [166]. The barrier to rearrangement is greater than 30–40 kcal.

2.10 TEN- TO TWELVE-ATOM SYSTEM

N.M.R. study of polytopal rearrangements in 10–12-coordinate complexes has proved to be difficult because rearrangement barriers are intrinsically very low, the number of discrete complexes of adequate stability is small, and typically the central nuclei have large quadrupole moments[2,5] exacerbating problems associated with n.m.r. studies. On the other hand, rearrangements in ten-, eleven-, and twelve-atom polyhedral boranes have been well studied, and considerable information concerning mechanism is now available[10,11]. The ground-state structures and the rearrangement mechanisms for polyhedral boranes should serve as realistic models for analogous coordination compounds (i.e., M_NH_x and ML_N) [2,5,78,148] as delineated in definitive reviews[5,11,148]. Qualitative differences will reside only in the rearrangement barriers with those for coordination compounds being an order of magnitude, or more, lower than those for polyhedral boranes[5,11].

Clearly, the results of empirical calculations[148] and experimental structural

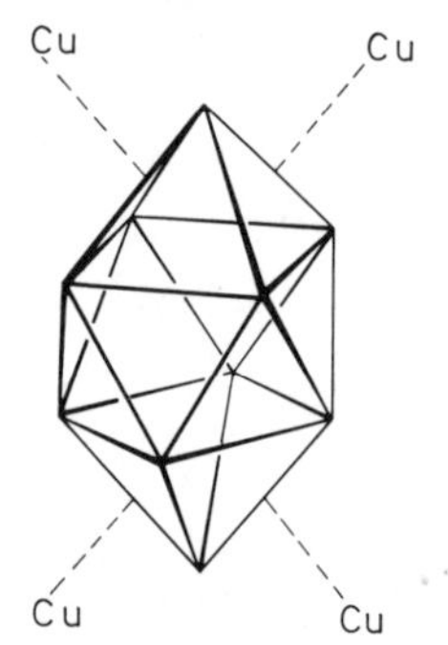

Figure 2.23 The favoured polyhedral form for the ten-atom system, the symmetrically bicapped antiprism of D_{4d} symmetry as found for the copper salt of the $B_{10}H_{10}^{2-}$ ion. The terminal B—H hydrogen atoms are not depicted

data indicate that the favoured ground-state geometries are the D_{4d} symmetrically bicapped square antiprism for the ten-atom system (alternatives are described in a review[148]), the octadecahedron for the eleven-atom system, and the icosahedron for the twelve-atom system. These favoured polytopal forms are shown in Figures 2.23–2.25. Alternative twelve-atom forms are illustrated in Figure 2.26. Structural references for the favoured forms are the $B_{10}H_{10}^{2-}$ ion [167,168] and $B_8H_8(CCH_3)_2$ for the ten-atom system[169], $B_9H_9(CCH_3)_2$ and $B_{11}H_{11}^{2-}$ for the eleven-atom system[170,171], and the

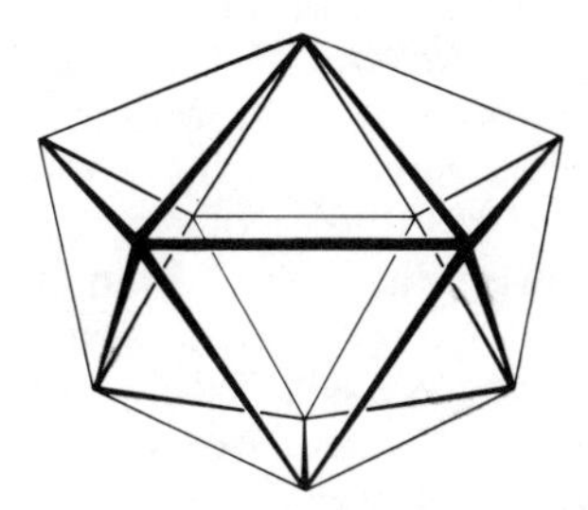

Figure 2.24 The favoured polyhedral form for the eleven-atom system, the octadecahedron of C_{2v} symmetry

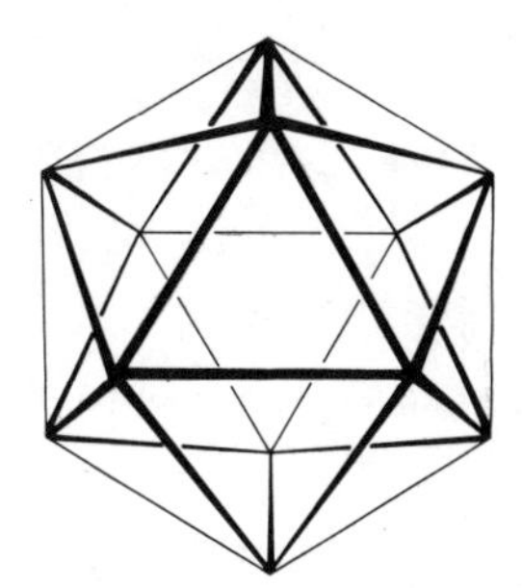

Figure 2.25 The favoured polyhedral form for the twelve-atom system, the icosahedron of I_h symmetry

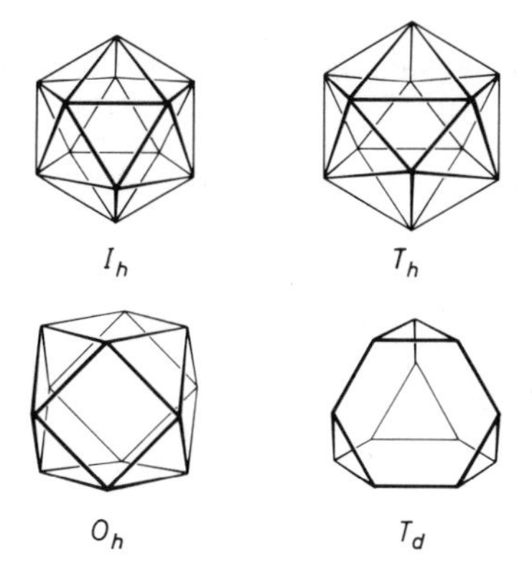

Figure 2.26 Possible idealised geometries for the twelve-atom system

$B_{12}H_{12}^{2-}$ ion[168, 172], various carboranes[11] and metalloboranes[11] for the twelve-atom system. Specific discussion of coordination compounds for these systems is presented in reviews[5, 148].

To illuminate the following discussion of polytopal rearrangements, the recommended numbering conventions for nuclear positions in the ten-, eleven-, and twelve-atom boranes are shown in Figures 2.27–2.29.

The mechanism of rearrangements in the ten-atom polyhedral boranes is reasonably well explicated by studies of rearrangements in derivatives of $B_{10}H_{10}^{2-}$. The most extensive data[173] comes from the bis-trimethylammonium derivative, $B_{10}H_8[N(CH_3)_3]_2$. Five of the seven possible isomers of this disubstituted B_{10}-derivative were isolated in pure form. The only two isomers never isolated were two of the three vicinally-substituted derivatives, namely, the 1,2 and 2,6(9) isomers (see Figure 2.27 for numbering convention).

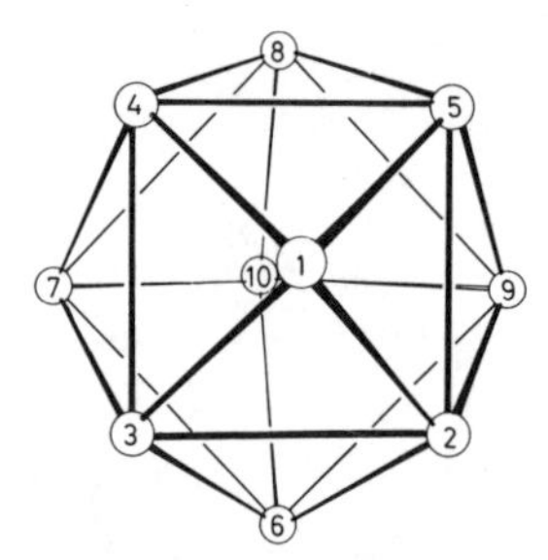

Figure 2.27 Numbering convention for the nuclear positions in the ten-atom system based on the symmetrically bicapped square antiprism

All five of these isomers undergo isomerisation at 300 °C in a melt to give a mixture of five isomers. The only two isomers not detected in the reaction mixture were the vicinal 1,2 and 2,6(9) isomers. The composition of a thermally-equilibrated isomeric mixture of bis-substituted trimethylammonium derivatives is given in Table 2.1. Relative concentrations are given for

Table 2.1 Composition of thermally-equilibrated isomeric mixtures of $B_{10}H_8[N(CH_3)_3]_2$ [173]

Isomer	*Relative concentration* 300 °C	325 °C	*Composition of a statistical mixture*
1, 10	1.0	1.0	1
2, 7(8)	4.19	4.31	8
2, 4	1.75	1.51	4
1, 6	3.90	4.85	8
2, 3*	0.043	0.017	8†

*This may be a mixture of the 2,3, 1,2, and 2,6(9) isomers.
†This number should be 24 if the 1,2 and 2,6(9) isomers are included here.

isomerisation at 300 °C and 325 °C and are compared with the composition of a statistical mixture. Thermodynamic data obtained for the isomerisation are: 2,7(8) $\rightleftharpoons$ 1,10: $\Delta G = -1.63$ kcal mol^{-1}, $\Delta H = +0.77$ kcal mol^{-1}, and $\Delta S = 4.19$ e.u. at 300 °C [173]. The accuracy of these values is relatively low because of the narrow temperature range explored and because of the thermal reactivity of the derivatives. Essentially, all of the entropy term is accounted for by a statistical factor (see Table 2.1). Solution studies of the intramolecular rearrangements showed that at 230 °C none of the *non-vicinal* isomers underwent detectable isomerisation within a period of 30 min. However, under these conditions the 2,3 isomer, the only vicinal isomer available, rearranged exclusively to the 1,6 isomer. Attempts to find conditions under which the 1,10, 1,6, 2,4, and 2,7(8) isomers would undergo

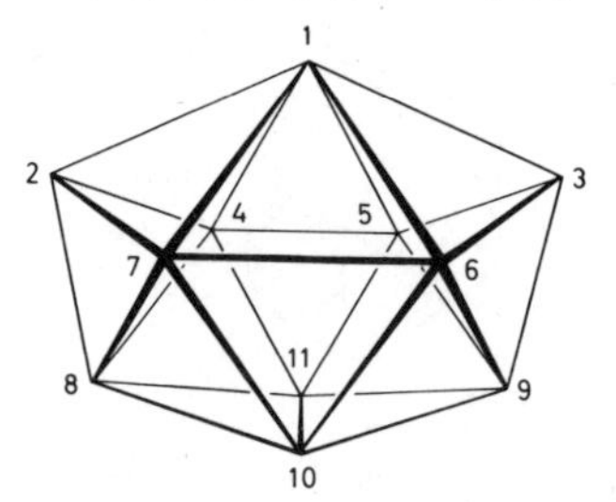

Figure 2.28 The recommended numbering convention for nuclear positions in the eleven-atom system based on the C_{2v} octadecahedron

selective isomerisation failed. The activation energy for rearrangement of 2,3 to 1,6 was found to be 37 kcal mol^{-1} and the entropy of activation, +2.4 e.u. [173].

The substitution in the above trimethylammonium system was exo-polyhedral in nature; hence, the rearrangement cannot be rigorously characterised as an intramolecular rearrangement. However, since the 2,3 isomerisation is very clean and selective and since all five available isomers lead to an identical mixture of isomers at 300 °C and 325 °C, any mechanism other than an intramolecular rearrangement is unlikely. Mecha-

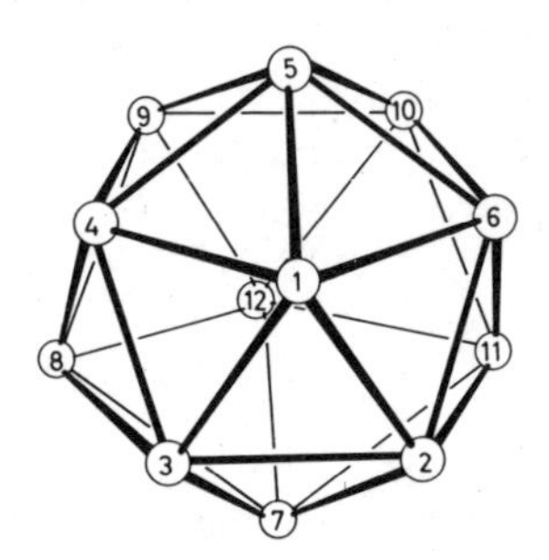

Figure 2.29 The recommended numbering convention for the nuclear positions in the twelve-atom system based on the icosahedron

nistically, the *facile* rearrangement of the 2,3 vicinal isomer can be readily rationalised. Firstly, there should be a lowering of the barrier to B_{10}-polyhedral rearrangement simply due to the non-bonding repulsions of the bulky vicinal trimethylamine groups. As shown in Figure 2.30, a simple stretching of two boron–boron equatorial interactions leads to an intermediate state (or in Lipscomb's terminology[10], the S or square-type configuration) that can then either return to the original configuration or specifically yield the

1,6 isomer, as is experimentally observed. These experiments support the mechanism, but they are not definitive in themselves. Studies of 1,2- and 1,6-$(CH_3)_2SB_{10}H_8N(CH_3)_3$ establish the mechanism outlined in Figure 2.30 as a realistic one[173, 174].

None of the relatively high-energy vicinal isomers, with the exception of the 2,3 isomer, can proceed through the above mechanism directly to a

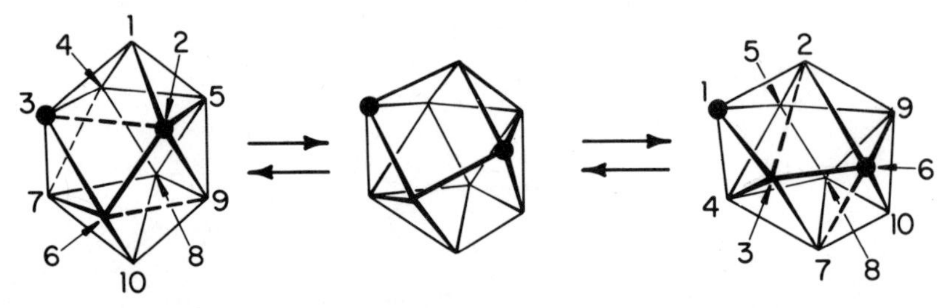

Figure 2.30 Proposed mechanism for the rearrangement of ten-atom polyhedral boranes with a ground-state geometry approximating the bicapped square antiprism. A simple stretching of the two boron–boron interactions leads to an intermediate state that can either return to the original configuration or yield a new isomer. The above figure specifically illustrates the isomerisation of a 2,3 isomer to the 1,6 isomer

non-vicinal isomer. Either of the two other vicinal isomers, the 1,2 and the 2,6(9), must first go through another high-energy state, the 2,3 isomer, before the substituents can become non-vicinal. Experimentally, this was confirmed in qualitative studies of the 1,2- and 1,6-$(CH_3)_2SB_{10}H_8N(CH_3)_3$ isomers. The vicinal 1,2 isomer undergoes rearrangement no more readily than does the 1,6 isomer[173, 174].

Although there are no quantitative data on the rearrangement of the various isomers in the B_8C_2-carborane series, Tebbe and co-workers have shown that the 1,6 isomer of $B_8H_8(CCH_3)_2$ is converted at 350 °C within a 22-h period to the 1,10 isomer[169].

Thermal rearrangements of ten-atom metalloboranes have also been described. The cobalt carbonyl, $Co[B_7C_2H_9]_2^{2-}$, in the form of the caesium salt, rearranges from the isomer shown in Figure 2.31 to the isomer in which the carbon atoms are at apical positions as illustrated in Figure 2.32. This rearrangement occurs quantitatively at 315 °C within 24 h [175].

No rearrangement of a closed polyhedral eleven-atom structure has as yet been reported. The ^{11}B n.m.r. spectrum of $B_{11}H_{11}^{2-}$ is unchanged to temperatures well above 250 °C; thus, the barrier to rearrangement in this system must be over 25–30 kcal mol^{-1} [170]. Rearrangements for eleven-atom polyhedral borane *fragments* have been established; for example, the anion 1,2-$B_9C_2H_{12}^-$ and its derivatives thermally rearrange at temperatures of about 350 °C to the corresponding 1,7 systems[176]. The structures of these two stereoisomers is given in Figures 2.33 and 2.34.

The most probable mechanism for rearrangements in closed polyhedral eleven-atom systems is presented in Figure 2.35 [5].

Although a number of attempts have been made to experimentally delineate polyhedral rearrangements in derivatives of the $B_{12}H_{12}^{2-}$ ion, there is no evidence of such stereoisomerisation. Studies have been extended

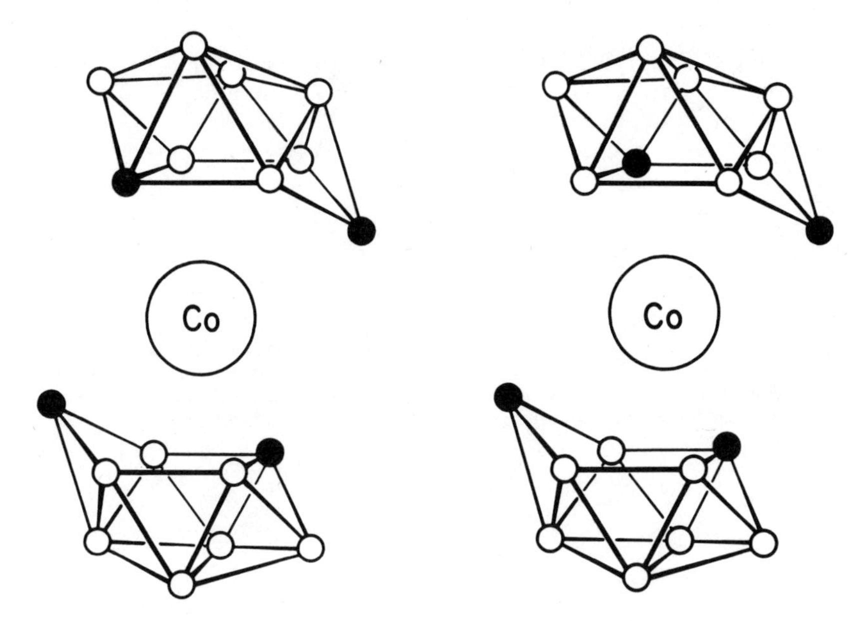

Figure 2.31 The anti and syn forms for the bis-π(2)-1,6-dicarbazapylcobalt(III) ion isomer. The carbon atoms are the black circles. Terminal hydrogen atoms are not depicted.

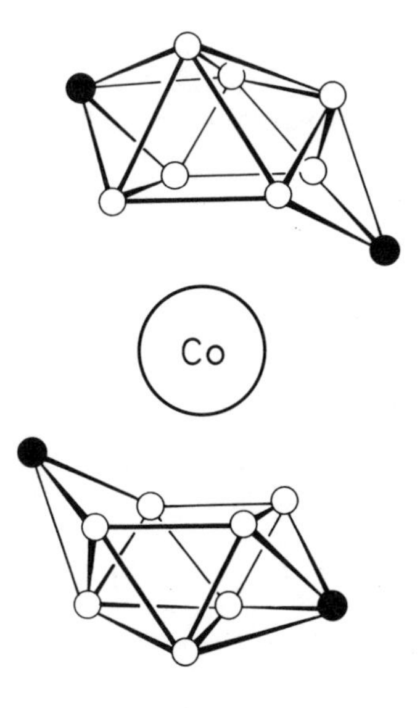

Figure 2.32 The bis-apical (carbon nuclei) isomer of the bis-dicarbazapylcobalt ion obtained by the thermal rearrangement of the isomer shown in the preceding figure

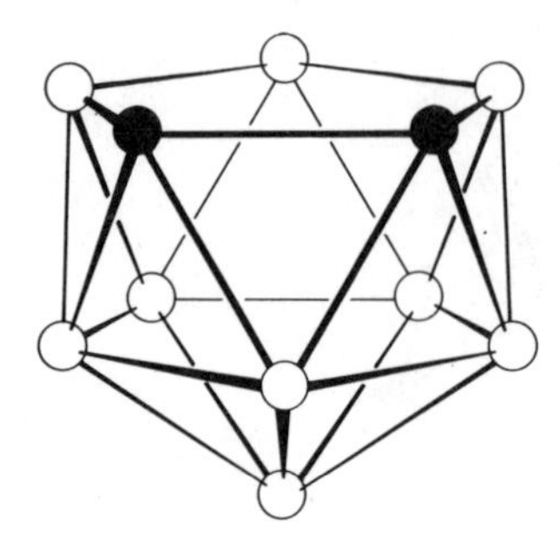

Figure 2.33 One of the stereoisomers in the eleven-atom fragment ion $B_9C_2H_{12}$, namely, the 1,2 isomer in which the carbon atoms are vicinal. The hydrogen nuclei are not depicted in this drawing. Each polyhedral nucleus has one terminal hydrogen bond. The twelfth hydrogen atom is associated with the open pentagonal face of the polyhedral fragment

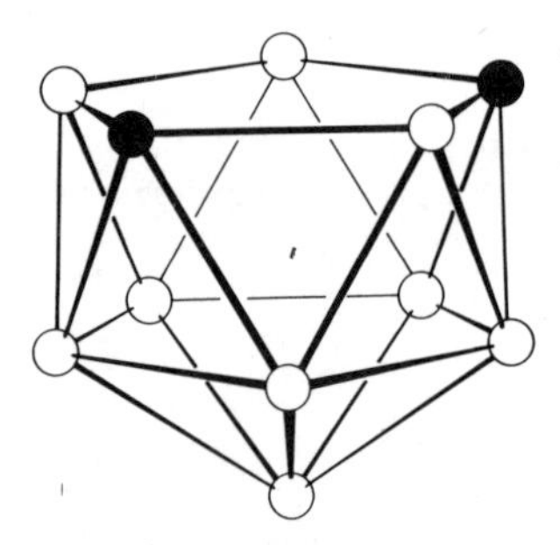

Figure 2.34 Another stereoisomer established in the $B_9C_2H_{12}^-$ series, namely, the 1,6 isomer in which both carbon atoms are at the open pentagonal face but non-vicinal to each other

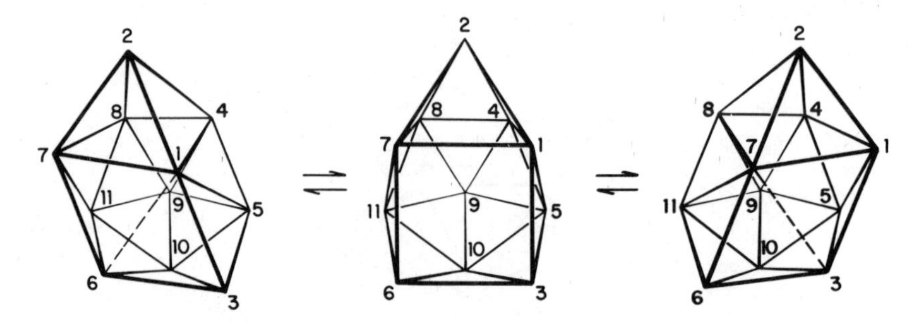

Figure 2.35 A possible mechanism for the rearrangement in the polyhedral eleven-atom system based on a C_{2v} octadecahedral ground state

to temperatures in the 400–500 °C region, indicating that rearrangements in this system have relatively high activation energies[11,173]. On the other hand, studies of the carborane system, that is, $B_{10}C_2H_{12}$ and its C-substituted derivatives, have clearly shown that the 1,2 isomer (refer to Figure 2.29) rearranges quite cleanly to the 1,7 isomer. For the parent carborane this rearrangement begins at about 425 °C [177,178]. The activation energy for the rearrangement of the 1,2 to the 1,7 isomer is 62 kcal mol^{-1} and the entropy of activation is 7 e.u. at 455 °C [179]. This barrier to rearrangement in the parent carborane can be substantially reduced by the introduction of vicinal, bulky groups on the carbon atoms. For example[179], reaction conditions for conversion of 1,2- to 1,7-carboranes as a function of the substituents on the carbon nuclei are as follows: 425 °C for 2 days for the parent carborane, 300 °C for 10 days for the bis-trimethylsilyl derivative, 300 °C for 4 days for the bis-dimethylphenylsilyl derivative, and 260 °C for 2 days for the bis-chlorodiphenylsilyl and bis-methyldiphenylsilyl derivatives. Rate data obtained for the thermal rearrangement of the 1,2 isomer of the bis-methyl-diphenylsilyl derivatives yielded an enthalpy of activation of 45 kcal mol^{-1} and an entropy of activation of −1 e.u. at 280 °C as compared with the previously cited 62 kcal mol^{-1} and 7 e.u. at 455 °C for the parent carborane. The ortho-isomer, 1,2, of $B_{10}CPH_{11}$, which is an icosahedral borane with ten borons, one carbon, and a phosphorus comprising the framework,

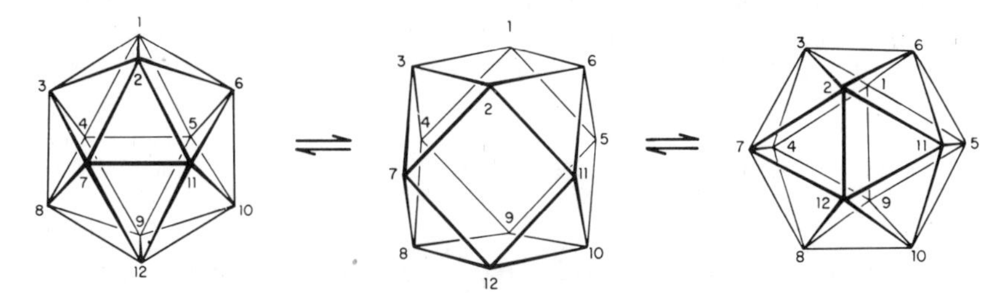

Figure 2.36 A possible mechanism for rearrangement in the icosahedral twelve-atom system based on traverse of an intermediate cube-octahedral form. Antipodal relationship of vertices must be maintained in this process. Hence a 1,2 or 1,7 isomer cannot be rearranged to a 1,12 isomer

undergoes thermal rearrangement at 480 °C to give primarily the 1,7(C,P) isomer[180]. Hawthorne and co-workers have also described thermal rearrangements in icosahedral metalloborane species[181], e.g., $Ni[1,2\text{-}B_9C_2H_{11}]_2$.

One of the first mechanistic proposals for rearrangements in the twelve-atom system was that of Lipscomb[10,183,184] and co-workers. The proposal was an ingenuous and aesthetically-pleasing one based on the traverse of icosahedral and cubic octahedral* forms. The icosahedron may be distorted as shown in Figure 2.36 to give a cube-octahedron. The intermediate or transition state may then return to the 1,2 isomer or to a 1,7 isomer. This process may well be a low-energy path for rearrangement in the icosahedral boranes, but it is not sufficient to account for all of the experimental results. It is not possible to convert either the 1,2 or 1,7 isomer through the cube-

* These are the two highest symmetry polyhedra having twelve vertices.

octahedral intermediate or transition state to the symmetrical 1,12 isomer because antipodal relationship of vertices must be maintained throughout. It has been experimentally shown that these isomers can be rearranged at 600 °C to the 1,12 isomer[185]. Hence, either the cube octahedral intermediate is not operative in any of these rearrangements, or else an additional higher-energy process is possible. The subject of rearrangements in icosahedral

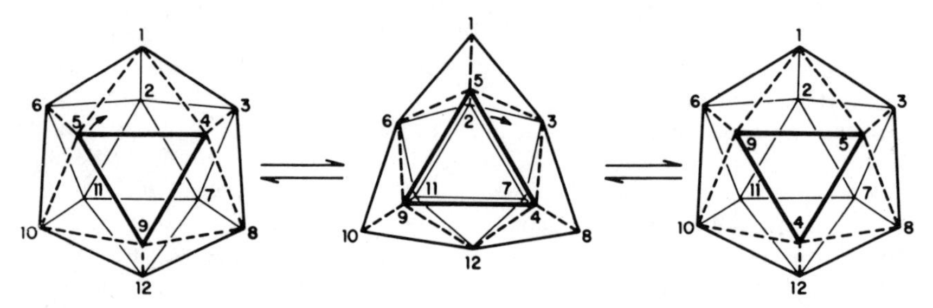

Figure 2.37 An alternative mechanism for rearrangement in the icosahedral twelve-atom system based on a face-rotational process

boranes is now under intensive study[186]. In later studies, an alternative to the cube-octahedral process was proposed by Lipscomb and co-workers. They suggested that the cube-octahedral process may be accompanied by trigonal face-rotation in the intermediate[186]. Another alternative mechanism is the face-rotation process[5,11] (Figure 2.37).

2.11 REARRANGEMENTS IN BORANES AND CLUSTERS VIA LIGAND TRAVERSE OF TERMINAL AND BRIDGING POSITIONS

A large number of boron hydrides and metal clusters have both terminal- and bridge-bonded ligands. A facile mechanism available to these molecular aggregates, by which permutation of ligand positions may occur, comprises a bending of ligands from terminal to bridging positions. An example of a

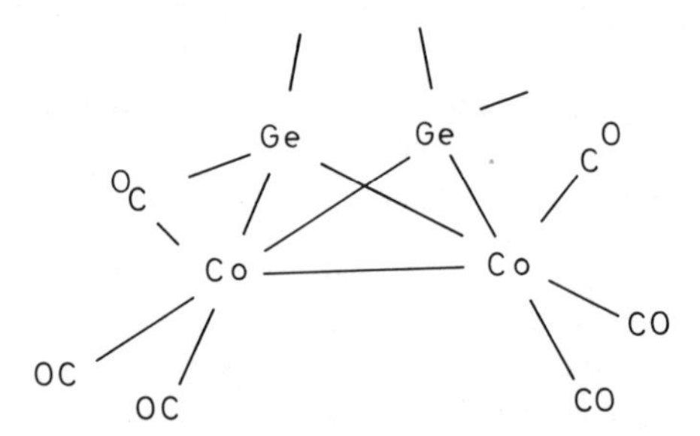

Figure 2.38 Probable structure of $[(CH_3)_2Ge]_2Co_2(CO)_6$

possible mechanism for rearrangements in $Co_4(CO)_{12}$ leading to cobalt atom and to CO group environmental equivalence was earlier outlined in the four-atom section and illustrated in Figure 2.3. This mechanism applies to all metal cluster carbonyls, e.g., $Fe_3(CO)_{12}$ and $Rh_6(CO)_{16}$, and to metal cluster halides, e.g., $Ta_6Cl_{12}^{2+}$ and $Mo_6Cl_{14}^{2-}$. Carbonyl positional exchange should be relatively rapid in binuclear carbonyls, such as $Co_2(CO)_8$, by interchange of bridge and terminal carbonyl ligands. Recently it was reported[187] that a germanium derivative of the cobalt carbonyl underwent some kind of intramolecular rearrangement. The geometry of the germanium derivative was not established, but the most probable form is given in Figure 2.38. At low temperatures, the expected two environments for the methyl proton are evident in the n.m.r. spectrum. However, as temperature is raised, these peaks merge to give a single sharp proton resonance. A somewhat speculative mechanistic proposal was presented[187]. A more probable mechanism for equilibration of the germanium methyl environments would be through an intermediate or transition state in which the methyl germanium groups went to terminal-type bonding positions as postulated above for the rearrangement mechanisms in metal clusters. The only substantive experimental information on this class of non-rigid compounds comes from n.m.r. studies of boron hydrides. The paradigm is the n.m.r. behaviour of the $B_3H_8^-$ ion. This ion has the structure outlined in Figure 2.39. The B_3 skeleton

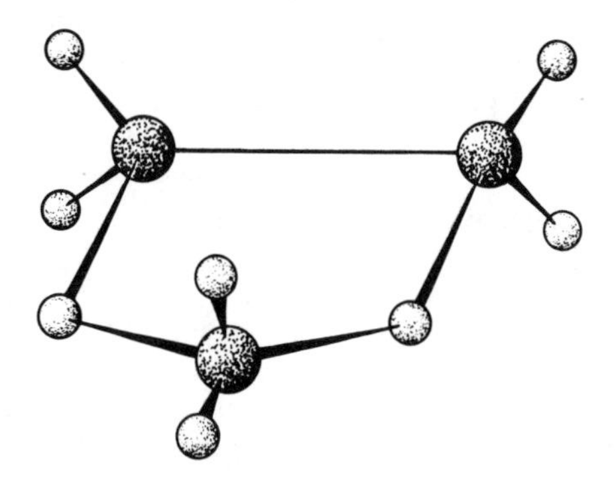

Figure 2.39 The structure of the triborohydride ion, $B_3H_8^-$

describes an isosceles triangle and two edges are bridged by hydrogen nuclei. Inconsistent with this ground-state structure ('static' with respect to the n.m.r. time-scale) are the ^{11}B and ^1H n.m.r. spectra. At ~25 °C (above and well below), the ^{11}B spectrum is a nonet and the ^1H spectrum a decet. This denotes n.m.r. equivalence of all B and of all H nuclei[188]. Each B nucleus is equally spin-coupled with *all* eight H(s = 1/2) nuclei and each H nucleus is equally spin-coupled with all three B(s = 3/2) nuclei. Ostensibly, the explanation[189, 190] is that there is a rapid exchange of bridge-hydrogen nuclei with terminal-hydrogen nuclei. Low temperature ^1H n.m.r. data (220 Mc) indicate that the dynamical process for the equivalence of H nuclei is very fast, even at −137 °C[191].

There is also evidence of bridge–terminal hydrogen exchange in metal derivatives of $B_3H_8^-$ ion. A complex of this type for which relatively extensive information is available is the bis-phosphine copper(I) derivative of $B_3H_8^-$. The structure of this metalloborane has been established and is depicted in

Figure 2.40. The ^{31}P n.m.r. spectrum of this complex is the expected AB pattern, consistent with the structure shown in Figure 2.39, but this AB pattern is discernible only at low temperatures[192, 193]. At about −70 °C, the ^{31}P n.m.r. AB pattern broadens and then merges at about −60 °C to a broad, single peak that sharpens on further temperature increase. Mecha-

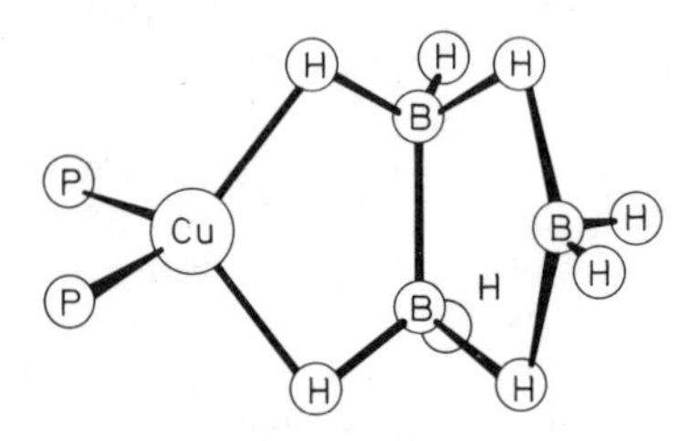

Figure 2.40 The structure of the bis-phosphinocopper(I) derivative of $B_3H_8^-$. The three organic groups attached to each of the phosphorus nuclei are not depicted

nistic stúdies *suggest* that this rearrangement mechanism involves a transitional state or intermediate in which there is a unidentate Cu—H—B coordination. Similar evidence for rearrangement of this type has been indicated from a low-temperature proton n.m.r. study of this sytem[191].

2.12 ORGANOMETALLIC COMPOUNDS

Stereochemical non-rigidity in organometallic compounds was first noted in 1956[194, 195] from an n.m.r. study of σ-cyclopentadienyl metal complexes, but this was ignored for about a decade. Within the last five years, a very large number of publications has appeared on this subject, and four[4] general classes of non-rigid organometallic complexes have been denoted:

(a) complexes of acyclic olefins, e.g., allene, the allyl radical and ion and some α,ω-polymers,
(b) complexes with a cyclopolyene σ-bonded to a metal atom,
(c) complexes with a cyclopolyene bonded to a metal atom through its π-system,
(d) compounds with one or more polyenes π-bonded or σ- and π-bonded to several metal atoms.

Examples of some of these are discussed below, especially from the mechanistic view. Unfortunately, definitive data bearing on reaction mechanism are few in number, although many ingenuous studies have been devised. Cotton[4] has suggested an hypothesis of 'least action' which comprises 1,2 shifts of the metal atoms with respect to the organic ligand, and this is probably the dominant mechanism in the rearrangement of organometallic complexes.

Most π-allyl metal complexes show in the n.m.r. spectrum at room temperature the expected non-equivalence of the syn and anti protons in the π-allyl ligand. This non-equivalence is lost at higher temperatures or by the introduction of certain kinds of ligands such as phosphines. The mechanism for equilibration of the syn and anti proton environments was a subject of dispute for quite some time. However, relatively definitive experimental data bearing on mechanism are now available, and the consensus is that the mechanism comprises traverse of an intermediate or transition state in which the allyl moiety is σ-bonded to the metal:

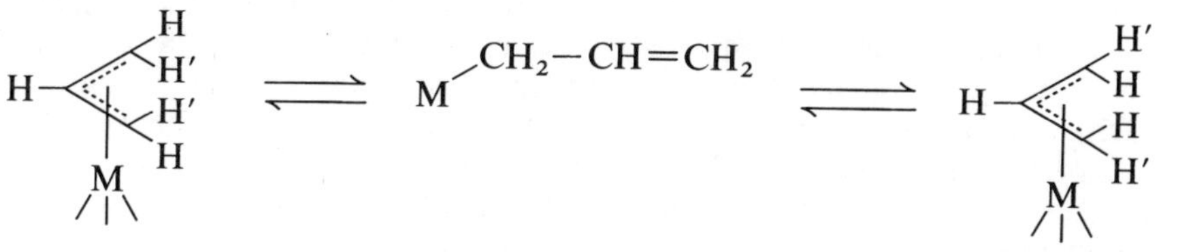

$$(\pi\text{-allyl})M \rightleftharpoons M-CH_2-CH=CH_2 \rightleftharpoons (\pi\text{-allyl})M$$

The literature in this area is extensive and will not be reviewed in detail here. The best discussion of the details for the rearrangement of π-allyl metal complexes may be found in an article by Powell and Shaw[196].

Stereochemical non-rigidity appears to be a general property for σ-cyclopentadienyl derivatives of transition and non-transition elements. The barrier to rearrangement is generally raised by methyl substitution in the ring and by fusion of the ring with an aromatic system as exemplified by the indenyl ligand. Some of the fluxional molecules[194, 195, 197–206] in this class are depicted below, (2.42) to (2.47):

$(\pi\text{-}C_5H_5)Fe(CO)_2(\sigma\text{-}C_5H_5)$
(2.42)

$(\pi\text{-}C_5H_5)Cr(CO)_2(\sigma\text{-}C_5H_5)$
(2.43)

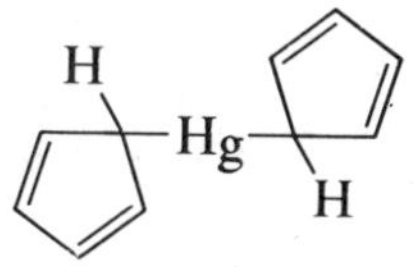

$(\sigma\text{-}C_5H_5)Hg(\sigma\text{-}C_5H_5)$
(2.44)

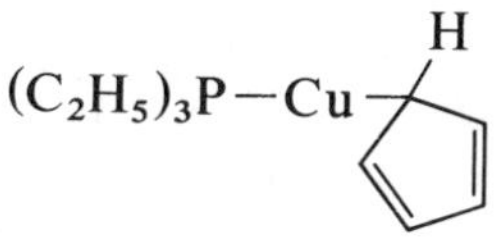

$(C_2H_5)_3P-Cu(\sigma\text{-}C_5H_5)$
(2.45)

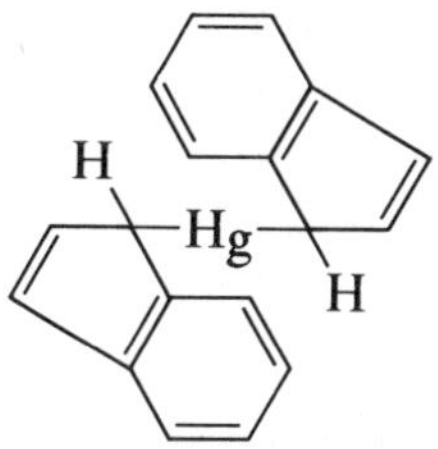

$(\sigma\text{-}C_9H_7)Hg(\sigma\text{-}C_9H_7)$
(2.46)

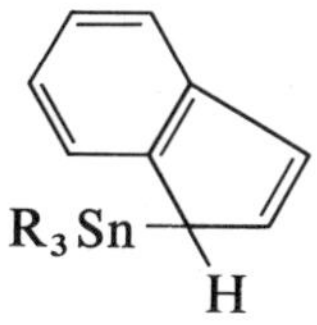

$R_3Sn(\sigma\text{-}C_9H_7)$
(2.47)

The x-ray crystal structure of $(\pi\text{-}C_5H_5)Fe(CO)_2(\sigma\text{-}C_5H_5)$ established the σ-bonded character of one of the cyclopentadienyl ligands[205]. The temperature-dependence of the proton n.m.r. spectrum of this molecule was studied over the temperature range of $+30$–-100 °C. The limiting low-temperature spectrum showed the expected non-equivalence of the σ-cyclopentadienyl hydrogen atoms. Above -80 °C, the three main resonances broaden and finally merge to a single broad peak slightly below room temperature. The broad peak sharpens to give a single peak at about $+30$ °C.

Line-shape analysis of the transition region of the n.m.r. spectrum was consistent with either a succession of 1,2 shifts or a succession of 1,3 shifts. A unique decision between these two possibilities cannot be made, but again invoking the principle of least motion, it would seem most probable that the 1,2 shift is the dominant mechanism. It is notable that another study of the non-rigid complex $(C_2H_5)_3PCuC_5H_5$ based on an analysis of the n.m.r. transitional region was reported to be consistent only with a 1,3 shift. Unfortunately, the assignment of individual resonances was not uniquely achieved, and therefore there is no unambiguous decision here between the 1,2 or 1,3 type of shift[201]. A later study[206] of mercury derivatives (*vide infra*) suggested the possibility that the 1,3 shift does prevail in the isoelectronic copper complex.

A study of the 1-indenyl derivative, $(\pi\text{-}C_5H_5)Fe(CO)_2$(1-indenyl), analogous to the iron compound discussed above, showed that it was stereochemically rigid on the n.m.r. time-scale. In this molecule a 1,2 shift would require traverse of an isoindenyl structure which should represent a significantly higher energy state. This was then assumed to be further evidence for the dominance of 1,2 shifts in these cyclopentadienyl-type complexes[197].

There apparently is no significant perturbation of barriers in these σ-cyclopentadienyl complexes on substitution with other metal ions because the molecule $(\pi\text{-}C_5H_5)Cr(Mo)_2C_5H_5$, which is isoelectronic to the iron compound discussed above, displays an n.m.r. behaviour analogous to the iron derivative[197].

Stereochemical non-rigidity has been established for a number of σ-cyclopentadienyl derivatives of non-transitional metals. Excellent examples come from studies of silicon, germanium, and tin complexes[202, 203]. The studies of Davison and Rakati argue strongly for 1,2 shifts in these Group IV derivatives of the σ-cyclopentadienyl ligand. 1-Indenyl derivatives were examined, and in these cases a fluxional behaviour was discerned from the temperature-dependence of the proton n.m.r. spectra. The barriers for rearrangement were relatively high, *c.* 13 kcal mol^{-1}, and ostensibly this was due to the fact that rearrangements in these systems proceed through 1,2 shifts with population of the relatively high-energy state involving isoindenyl forms. The spin-tickling n.m.r. experiments of Cotton and Marks with $CH_3SiCl_2(\sigma\text{-}C_5H_5)$ further support the 1,2 shift hypothesis[204].

Bis-(σ-cyclopentadienyl)mercury is also a non-rigid molecule[195, 199, 200]. Replacement of the hydrogen atoms by methyl groups to give $Hg[\sigma\text{-}C_5(CH_3)_5]_2$ substantially raises the rearrangement barrier, and the expected non-equivalence of CH_3 protons in a static σ-bonded structure is evident in the proton n.m.r. spectrum at 25 °C. Curiously, the proton n.m.r. spectrum of this molecule was not explored at elevated temperatures so it is not known whether the molecule is non-rigid within the time-scale afforded by resonance techniques[198]. The indenyl mercury analogue is non-rigid on the n.m.r. time-scale, and the data indicate that the instantaneous structure is a bis-1-monohaptoindenyl) form[206]. Arguments were advanced in this molecule for a prevailing 1,3 shift[206].

In 1966, a number of papers appeared describing the proton n.m.r. spectra of cyclo-octatetraene iron tricarbonyl as a function of temperature[207–209]. Different conclusions were reached by all three groups concerning the stereochemistry of this molecule in solution. The geometry for the complex in the solid state had been established and is shown in Figure 2.41[210]. The main reason different conclusions were reached was the assumption that the proton n.m.r. spectrum observed at about −145 °C was a limiting spectrum. This was later shown not to be the case, and the better resolved and nearly-limiting spectrum, obtained by Grubbs *et al.*, showed that the solution stereochemistry for this complex is that found for the solid state (Figure 2.41)[211]. Further work by Anet[212] on the substituted complex, $CH_3C_8H_7Fe(CO)_3$, showed that the complex was of the 1,3-diene metal type with the methyl group at an inner position of the bound diene, and that the general stereochemistry was analogous to the parent complex as established in the

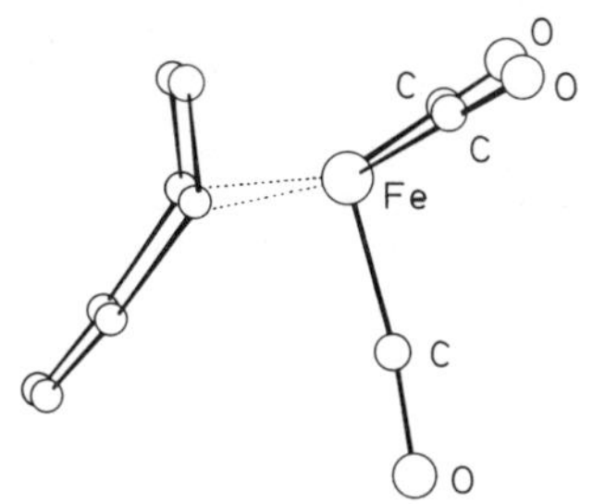

Figure 2.41 The structure of the cyclo-octatetraene iron tricarbonyl molecule as established by x-ray for the solid state

solid-state structure determination (Figure 2.41). The definitive work on the ground-state geometries of molecules of this type in the iron group came from a study of the ruthenium derivative of cyclo-octatetraene by Cotton and co-workers[145, 182]. For this molecule the limiting low-temperature n.m.r. spectrum was observed at substantially higher temperatures than for the analogous iron system, and this spectrum was fully consistent with the geometry (Figure 2.41) established for the iron complex. The transitional n.m.r. line-shapes for the ruthenium system, which is also non-rigid, was studied in some detail. In the high-temperature limiting form, all protons of the cyclo-octatetraene ligand are equivalent due to some kind of shift mechanism. The observed n.m.r. transition line-shapes were compared with computer-simulated spectra for 1,2, 1,3, 1,4 and random shift mechanisms. Of this group only the computed spectra for a 1,2 shift were in close agreement with the observed spectrum. Thus, in this one case it would appear that it is necessary to invoke only one process, that is, a sequence of 1,2 shifts, in order to account for the observations. This does not, however, rule out the possibility that *several* mechanisms are operative in this particular rearrangement.

A number of rather specialised cases of stereochemical non-rigidity in organometallic systems have been discussed by Cotton[4].

References

1. Muetterties, E. L. (1965). *Inorg. Chem.*, **4**, 769
2. Muetterties, E. L. (1970). *Accounts Chem. Res.*, **3**, 266
3. Doering, W. von E., Roth, W. R. (1963). *Angew. Chem., Int. Ed. Engl.*, **2**, 115

4. Cotton, F. A. (1968). *Accounts Chem. Res.*, **1,** 257
5. Muetterties, E. L. (1970). *Rec. Chem. Progr.*, **31,** 51
6. Muetterties, E. L. (1968). *J. Amer. Chem. Soc.*, **90,** 5097
7. Muetterties, E. L. (1969). *ibid.*, **91,** 1636
8. Muetterties, E. L. (1969). *ibid.*, **91,** 4115
9. Muetterties, E. L., Storr, A. T. (1969). *ibid.*, **91,** 3098
10. Lipscomb, W. N. (1966). *Science*, **153,** 373
11. Muetterties, E. L., Knoth, W. H. (1968). *Polyhedral Boranes*, (New York: Marcel Dekker, Inc.)
12. Westheimer, F. H. (1968). *Accounts Chem. Res.*, **1,** 70
13. Balaban, A. T., Farcasiu, D., Banica, R. (1966). *Rev. Roum. Chim.*, **11,** 1205
14. Lauterbur, P. C., Ramirez, F. (1968). *J. Amer. Chem. Soc.*, **90,** 6722
15. Dunitz, J. O., Prelog, V. (1968). *Angew. Chem., Int. Ed. Engl.*, **7,** 725
16. Storr, A. T. Paper presented at Chirality in Chemistry Conference in Elmau, Germany, October 19, 1970. Storr, A. T., Muetterties, E. L., to be published
17. Gielen, M., Nasielski, J. (1969). *Bull. Soc., Chim. Belges.*, **78,** 339
18. De Bruin, K. E., Naumann, K., Zon, G., Mislow, K. (1969). *J. Amer. Chem. Soc.*, **91,** 7031
19. Gielen, M., Mayence, G., Topart, J. (1969). *J. Organomet. Chem.*, **18,** 1
20. Gielen, M., de Clercq, M., Nasielski, J. (1969). *ibid.*, **18,** 217
21. Ugi, I., Marquarding, D., Klusacek, H., Gokel, G., Gillespie, P. (1970). *Angew Chem., Int. Ed. Engl.*, **9,** 703
22. Rauk, A., Allen, L. C., Mislow, K. (1970). *Angew. Chem., Int. Ed. Engl.*, **9,** 400
23. Rauk, A., Allen, L. C., Clementi, E. (1970). *J. Chem. Phys.*, **52,** 4133
24. Kincaid, J. F., Henriques, F. C., jr. (1940). *J. Amer. Chem. Soc.*, **62,** 1474
25. Weston, R. F. (1954). *ibid.*, **76,** 2645
26. Koeppl, G. W., Sagatys, D. S., Krishnamurthy, G. S., Miller, S. I. (1967). *ibid.*, **89,** 3396
27. Swalen, J. D., Ibers, J. A. (1962). *J. Chem. Phys.*, **36,** 1914
28. Tsuboi, M., Hirakawa, A. Y., Tamagake, K. (1967). *J. Mol. Spect.*, **22,** 272
29. Wollrab, J. E., Laurie, V. W. (1968). *J. Chem. Phys.*, **48,** 5058
30. Lambert, J. B., Oliver, W. L., jr. (1969). *J. Amer. Chem. Soc.*, **91,** 7774
31. Kemp, M. K., Flygare, W. H. (1968). *ibid.*, **90,** 6267
32. Brois, S. J. (1967). *ibid.*, **89,** 4242
33. Brois, S. J. (1969). *Trans. N. Y. Acad. Sci.*, **31,** 931
34. Gutowsky, H. S. (1958). *Ann. N. Y. Acad. Sci.*, **70,** 786
35. Felix, D., Eschenmosser, A. (1968), *Angew Chem., Int. Ed. Engl.*, **7,** 224
36. Gordon, M. S., Fischer, H. (1968). *J. Amer. Chem. Soc.*, **90,** 2471
37. Morokuma, K., Pedersen, L., Karplus, M. (1968). *J. Chem. Phys.*, **48,** 4801
38. Buckingham, D. A., Marzilli, P. A., Sargeson, A. M. (1969). *Inorg. Chem.*, **8,** 1595
39. Buckingham, D. A., Marzilli, L. G., Sargeson, A. M. (1969). *J. Amer. Chem. Soc.*, **91,** 5227
40. Lambert, J. B., Johnson, D. H. (1968). *ibid.*, **90,** 1349
41. Baechler, R. D., Mislow, K. (1970). *ibid.*, **92,** 3090
42. Egan, W., Tang, R., Zon, G., Mislow, K. (1970). *ibid.*, **92,** 1442
43. Lambert, J. B., Jackson, G. F., III, Mueller, D. C. (1970). *ibid.*, **92,** 3093
44. Lambert, J. B., Mueller, D. C. (1966). *ibid.*, **88,** 3669
45. Lambert, J. B., Jackson, G. F., III, Mueller, D. C. (1968). *ibid.*, **90,** 6401
46. Lambert, J. B., Jackson, G. F., III. (1968). *ibid.*, **90,** 1350
47. Green, M., Kirkpatrick, D. (1967). *Chem. Commun.*, 57; (1968). *J. Chem. Soc. (A)*, 483
48. Baechler, R. D., Mislow, K. (1970). *J. Amer. Chem. Soc.*, **92,** 4758
49. Rayner, D. R., Miller, E. G., Bickart, P., Gordon, A. J., Mislow, K. (1966). *ibid.*,
50. Mislow, K., Green, M. M., Laur, P., Melillo, J. T., Simmons, T., Ternay, A. L., jr. (1965). *ibid.*, **87,** 1958
51. Nudelman, A., Cram, D. J. (1968). *ibid.*, **90,** 3869
52. Day, J., Cram, D. J. (1965). *ibid.*, **87,** 4398
53. Lauterbur, P. C., Pritchard, J. G., Vollmer, R. L. (1963). *J. Chem. Soc.*, 5307
54. Rayner, D. R., Gordon, A. J., Mislow, K. (1968). *J. Amer. Chem. Soc.*, **90,** 4854

55a. Turley, P. C., Haake, P. (1967). *ibid.*, **89,** 4617

55b. Haake, P., Turley, P. C. (1967). *ibid.*, **89,** 4611

56. Dekker, M., Know, G. R., Robertson, C. G. (1969). *J. Organomet. Chem.*, **18,** 161

57. Gerlach, D. H., Holm, R. H. (1969). *Inorg. Chem.*, **8,** 2292; (1970). *J. Amer. Chem. Soc.*, **9,** 588; (1969). **91,** 3457
58. Hund, F. (1927). *Z. Physik.*, **43,** 805
59. Eaton, D. R., Phillips, W. D. (1965). *Advan. Magn. Resonance*, **1,** 103
60. Holm, R. H., Everett, G. W., jr., Chakravorty, A. (1966). *Progr. Inorg. Chem.*, **7,** 83
61. Holm, R. H., Chakravorty, A., Theriot, L. J. (1966). *Inorg. Chem.*, **5,** 625
62. O'Connor, M. J., Ernst, R. E., Holm, R. H. (1968). *J. Amer. Chem. Soc.*, **90,** 4561
63. O'Connor, M. J., Ernst, R. E., Schoenborn, J. E., Holm, R. H. (1968). *ibid.*, **90,** 1744
64. Everett, G. W., jr., Holm, R. H. (1968). *Inorg. Chem.*, **7,** 776
65. Everett, G. W., jr., Holm, R. H. (1965). *J. Amer. Chem. Soc.*, **87,** 2117; (1966) **88,** 2442
66. Chakravorty, A., Holm, R. H. (1964). *Inorg. Chem.*, **3,** 1010
67. Holm, R. H., Chakravorty, A., Dudek, G. O. (1964). *J. Amer. Chem. Soc.*, **86,** 379
68. LaMar, G. N., Sherman, E. O. (1969). *Chem. Commun.*, 809
69. LaMar, G. N., Sherman, E. O. (1970). *J. Amer. Chem. Soc.*, **92,** 2691
70. Pignolet, L. H., Horrocks, W. DeW., jr., Holm, R. H. (1970). *ibid.*, **92,** 1855
71. Cotton, F. A. (1966). *Inorg. Chem.*, **5,** 1083
72. Lucken, E. A. C., Noack, K., Williams, D. F. (1967). *J. Chem. Soc., A.* 148
73. Baker, R. W., Pauling, P. (1965). *Chem. Commun.*, 1495
74. Baker, R. W., Ilmaier, B., Pauling, P. J., Nyholm, R. S. (1970) *Chem. Commun.* 1077
75. Frenz, B. A., Ibers, J. A. (1970). *Inorg. Chem.*, **9,** 2403
76. Dewhirst, K. C., Keim, W., Reilly, C. A. (1968). *Inorg. Chem.*, **7,** 546
77. Meakin, P., Jesson, J. P., Tebbe, F. N., Muetterties, E. L. (1971). *J. Amer. Chem. Soc.*, **93,** 1791
78. Muetterties, E. L., Schunn, R. A. (1966). *Quart. Rev. Chem. Soc.*, **20,** 245
79. Muetterties, E. L., Mahler, W., Schmutzler, R. (1963). *Inorg. Chem.*, **2,** 613
80. Muetterties, E. L., Mahler, W., Packer, K. J., Schmutzler, R. (1964). *ibid.*, **3,** 1298
81. Raymond, K. N., Corfield, P. W. R., Ibers, J. A. (1968). *ibid.*, **7,** 1362
82. Mahler, W., Muetterties, E. L. (1965). *ibid.*, **4,** 1520
83. Hoard, L. G., Jacobson, R. A. (1966). *J. Chem. Soc.* 1203
84. Doak, G. O., Schmutzler, R. (1970). *Chem. Commun.*, 476
85. Berry, R. S. (1960). *J. Chem. Phys.*, **32,** 933
86. Gutowsky, H. S., McCall, D. W., Slichter, C. P. (1953). *J. Chem. Phys.*, **21,** 279
87. (a) Cotton, F. A., Danti, A., Waugh, J. S., Fessenden, R. W. (1958). *J. Chem. Phys.*, **29,** 1427; (b) Bramley, R., Figgis, B. N., Nyholm, R. S. (1962). *Trans Faraday Soc.*, **58,** 1893
88. Bartell, L. S., Hansen, K. W. (1965). *Inorg. Chem.*, **4,** 1777
89. Tebbe, F. N., Muetterties, E. L. (1968). *ibid.*, **7,** 172
90. Whitesides, G. M., Mitchell, H. L. (1969). *J. Amer. Chem. Soc.*, **91,** 5374
91. Peake, S. C., Schmutzler, R. (1969). *Chem. Commun.*, 1662; (1970) *J. Chem. Soc., A*1049
92. Gorenstein, D. (1970). *J. Amer. Chem. Soc.*, **92,** 644
93. Gorenstein, D., Westheimer, F. H. (1970). *ibid.*, **92,** 634; (1967). *J. Amer. Chem. Soc.*, **89,** 2762
94. Gorenstein, D., Westheimer, F. H. (1967). *Proc. Nat. Acad. Sci. (U.S.)*, **58,** 1747
95. Dennis, E. A., Westheimer, F. H. (1966). *J. Amer. Chem. Soc.*, **88,** 3431; (1966) **88,** 3432
96. Kluger, R., Westheimer, F. H. (1969). *ibid.*, **91,** 4143
97. Kluger, R., Covitz, F., Dennis, E. A., Williams, L. D., Westheimer, F. H. (1969). *ibid.*, **91,** 6066
98. Ramirez, F., Pilot, J. F., Madan, D. P., Smith, C. P. (1968). *ibid.*, **90,** 1275
99. Ramirez, F. (1968). *Accounts Chem. Res.*, **1,** 168
100. Ramirez, F. (1966). *Bull. Soc. Chim. France*, 2443
101. Ramirez, F., Smith, C. P., Pilot, J. F. (1968). *J. Amer. Chem.*, **90,** 6726
102. Hellwinkel, D. (1966). *Angew. Chem. Intern. Ed.*, **5,** 725
103. Hellwinkel, D. (1966). *Chem. Ber.*, **99,** 3628, 3642, 3660, 3668
104. Whitesides, G. M., Bunting, W. M. (1967). *J. Amer. Chem. Soc.*, **89,** 6801
105. Hellwinkel, D., Wilfinger, H. J. (1969). *Tetrahedron Letters*, **39,** 3423
106. Goldwhite, H. (1970). *Chem. Commun.*, 651
107. Klanberg, F., Muetterties, E. L. (1968). *Inorg. Chem.*, **7,** 155
108. Muetterties, E. L., Phillips, W. D. (1959). *J. Amer. Chem. Soc.*, **81,** 1084
109. Muetterties, E. L., Phillips, W. D. (1967). *J. Chem. Phys.*, **46,** 2861
110. Tang, R., Mislow, K. (1969). *J. Amer. Chem. Soc.*, **91,** 5644
111. Udovich, C. A., Clark, R. J. (1969). *ibid.*, **91,** 526
112. Udovich, C. A., Clark, R. J., Haas, H. (1969). *Inorg. Chem.*, **8,** 1066.

113. Bruice, T. C., Benkovic, S. (1966). *Bioorganic Mechanisms,* Vol. 2, 1–109, (New York: W. A. Benjamin, Inc.)
114. Haake, P. C., Westheimer, F. H. (1961). *J. Amer. Chem. Soc.,* **83,** 1102
115. Usher, D. A., Dennis, E. A., Westheimer, F. H. (1965). *ibid.,* **87,** 2320
116. Kluger, R., Kerst, F., Lee, D., Dennis, E. A., Westheimer, F. H. (1967). *J. Amer. Chem. Soc.,* **89,** 3918
117. Kluger, R., Kerst, F., Lee, D., Westheimer, F. H. (1967). *ibid.,* **89,** 3919
118. DeBruin, K. E., Mislow, K. (1969). *ibid.,* **91,** 7393
119. DeBruin, D. E., Zon, G., Naumann, K., Mislow, K. (1969). *ibid.,* **91,** 7027
120. Frank, D. S., Usher, D. A. (1967). *ibid.,* **89,** 6360
121. Hamer, N. K. (1968). *Chem. Commun.,* 1399
122. Haake, P., Cook, R. D., Koizumi, T., Ossip, P. S., Schwarz, W., Tyssee D. A. (1970). *J. Amer. Chem. Soc.,* **92,** 3828
123. Cremer, S. E., Trivedi, B. C. (1969). *J. Amer. Chem. Soc.,* **91,** 7200
124. Benkovic, S. J., Schray, K. J. (1969). *ibid.,* **91,** 5653
125. Fackler, J. P., jr., Fetchin, J. A., Seidel, W. C. (1969). *ibid.,* **91,** 1217
126. Odell, A. L., Raethel, H. A. (1969). *Chem. Commun.,* 87
127. Buckingham, D. A., Olsen, I. I., Sargeson, A. M. (1968). *J. Amer. Chem. Soc.,* **90,** 6654
128. Gehman, W. G. (1954). *Ph.D. Thesis,* The Pennsylvania State University
129. Seiden, L. (1957). *Ph.D. Thesis,* Northwestern University
130. Bailar, J. C., jr. (1958). *J. Inorg. Nucl. Chem.,* **8,** 165
131. Rây, P., Dutt, N. K. (1943). *ibid.,* **20,** 81
132. Fay, R. C., Piper, T. S. (1964). *Inorg. Chem.,* **3,** 348
133. Tebbe, F. N., Meakin, P., Jesson, J. P., Muetterties, E. L. (1970). *J. Amer. Chem. Soc.,* **92,** 1068
134. Meakin, P., Guggenberger, L. J., Jesson, J. P., Gerlach, D. H., Tebbe, F. N., Peet, W. G., Muetterties, E. L. (1970). *ibid.,* **92,** 3482
135. Meakin, P., Jesson, J. P., Muetterties, E. L. (1971). *ibid.,* **93,**
136. Ligand L = $P(OCH_3)_3$, $P(OC_2H_5)_3$, $P[OCH(CH_3)_2]_3$, $C_6H_5P(OC_2H_5)_2$, $(C_6H_5)_2POCH_3$, $(C_6H_5)_2PCH_3$, $C_6H_5P(CH_3)_2$, and $C_6H_5P(C_2H_5)_2$
137. Dewhirst, K. C., Keim, W., Reilly, C. A. (1968). *Inorg. Chem.,* **7,** 546
138. Springer, C. S., jr., Sievers, R. E. (1967). *ibid.,* **6,** 852
139. Gordon, J. G., II, Holm, R. H. (1970). *J. Amer. Chem. Soc.,* **92,** 5319
140. Eisenberg, R., Ibers, J. A. (1965). *ibid.,* **87,** 3776; (1966). *Inorg. Chem.,* **5,** 411
141. Eisenberg, R., Stiefel, E. I., Rosenberg, R. C., Gray, H. B. (1966). *J. Amer. Chem. Soc.,* **88,** 2874
142. Smith, A. E., Schrauzer, G. N., Mayweg, V. P., Heinrich, W. (1965). *ibid.,* **87,** 5798
143. Stiefel, E. I., Dori, Z., Gray, H. B. (1967). *ibid.,* **89,** 3353
144. Bernal, I., Sequeira, A. Paper presented before the *American Crystallographers Association,* Minneapolis, Minn., August 1967
145. Cotton, F. A., Davison, A., Musco, A. (1967). *ibid.,* **89,** 6796
146. Pignolet, L. H., Holm, R. H. (1970). *J. Amer. Chem. Soc.,* **92,** 1791
147. Onak, T., Drake, R. P., Dunks, G. B. (1964). *Inorg. Chem.,* **3,** 1686
148. Muetterties, E. L., Wright, C. M. (1967). *Quart. Rev. Chem. Soc., London,* **21,** 109, provide a comprehensive review of this subject but also see references 149–151 for theoretical analyses
149. Britton, D. (1963). *Canad. J. Chem.,* **41,** 1632
150. Claxton, T. A., Benson, G. C. (1966). *ibid.,* **44,** 157
151. Thompson, H. B., Bartell, L. S. (1968). *Inorg. Chem.,* **7,** 488
152. Muetterties, E. L., Packer, K. J. (1964). *J. Amer. Chem. Soc.,* **86,** 293
153. Gillespie, R. J., Quail, J. W. (1964). *Canad. J. Chem.,* **42,** 2671
154. Muetterties, E. L., Alegranti, C. W. (1969). *J. Amer. Chem. Soc.,* **91,** 4420
155. Gavin, R. M., jr., Bartell, L. S. (1968). *J. Chem. Phys.,* **48,** 2460
156. Bartell, L. S., Gavin, R. M., jr. (1968). *ibid.,* **48,** 2466
157. Klanberg, F., Eaton, D. R., Guggenberger, L. J., Muetterties, E. L. (1967). *Inorg. Chem.,* **6,** 1271
158. See analyses and reviews by Hoard and Silverton[159], Kepert[160], and Muetterties and Wright[148]
159. Hoard, J. L., Silverton, J. V. (1963). *Inorg. Chem.,* **2,** 235
160. Kepert, D. L. (1965). *J. Chem. Soc.,* 4736
161. Malatesta, L., Freni, M., Valenti, V. (1964). *Gazzetta,* **94,** 1278

162. Muetterties, E. L., Wright, C. M. (1965). *J. Amer. Chem. Soc.,* **87,** 4706
163. Pinnavaia, T. J., Fay, R. C. (1966). *Inorg. Chem.,* **5,** 223
164. Adams, A. C., Larsen, E. M. (1966). *ibid.,* **5,** 228
165. Ginsberg, A. P. (1964). *ibid.,* **3,** 567
166. Klanberg, F., Muetterties, E. L. (1966). *ibid.,* **5,** 1955
167. Dobrott, R. D., Lipscomb, W. N. (1962). *J. Chem. Phys.,* **37,** 1779
168. Muetterties, E. L., Merrifield, R. E., Miller, H. C., Knoth, W. H., Downing, J. R. (1962). *J. Amer. Chem. Soc.,* **84,** 2506
169. Tebbe, F. N., Garrett, P. M., Young, D. C., Hawthorne, M. F. (1966). *ibid.,* **88,** 609
170. Klanberg, F., Muetterties, E. L. (1966). *Inorg. Chem.,* **5,** 1955
171. Tsai, C., Streib, W. E. (1966). *J. Amer. Chem. Soc.,* **88,** 4513
172. Wunderlich, J. A., Lipscomb, W. N. (1960). *ibid.,* **82,** 4427
173. Hertler, W. R., Knoth, W. H., Muetterties, E. L. (1964). *ibid.,* **86,** 5434
174. Knoth, W. H., Hertler, W. R., Muetterties, E. L. (1965). *Inorg. Chem.,* **4,** 280
175. George, T. A., Hawthorne, M. F. (1969). *J. Amer. Chem. Soc.,* **91,** 5475
176a. Hawthorne, M. F. (1968). *Pure and Applied Chemistry,* **17,** 195
176b. Garrett, P. M., Tebbe, F. N., Hawthorne, M. F. (1964). *J. Amer. Chem. Soc.,* **86,** 5016
177. Grafstein, D., Dvorak, J. (1963). *Inorg. Chem.,* **2,** 1128
178. Zakharkin, L. I., Kalinin, V. N. (1966). *J. Gen. Chem. USSR, Eng. Transl.,* **36,** 376
179. Salinger, R. M., Frye, C. L. (1965). *Inorg. Chem.,* **4,** 1815
180. Little, J. L., Moran, J. T., Todd, L. J. (1967). *J. Amer. Chem. Soc.,* **89,** 5485
181. Warren, L. F., jr., Hawthorne, M. F. (1970). *ibid.,* **92,** 1157
182. Bratton, W. K., Cotton, F. A., Davison, A., Musco, A., Faller, J. W. (1967). *Proc. Nat. Acad. Sci. (U.S.),* **58,** 1324
183. Kaczmarczyk, A., Dobrott, R. D., Lipscomb, W. N. (1962). *Proc. Nat. Acad. Sci., (U.S.),* **48,** 729
184. Hoffmann, R., Lipscomb, W. N. (1963). *Inorg. Chem.,* **2,** 231
185. Papetti, S., Heying, T. L. (1964). *J. Amer. Chem. Soc.,* **86,** 2295
186. Kaesz, H. D., Bau, R., Beall, H. A., Lipscomb, W. N. (1967). *ibid.,* **89,** 4128
187. Adams, R. D., Cotton, F. A. (1970). *ibid.,* **92,** 5003
188. Muetterties, E. L., Phillips, W. D. (1962). *Advances in Inorganic and Radiochemistry,* Vol. 4, 206, (New York: Academic Press)
189. Lipscomb, W. N. (1959). *ibid.,* Vol. 1, 132
190. Phillips, W. D., Miller, H. C., Muetterties, E. L. (1959). *J. Amer. Chem. Soc.,* **81,** 4496
191. Beall, H., Bushweller, C. H., Dewkett, W. J., Grace, M. (1970). *ibid.,* **92,** 3484
192. Muetterties, E. L., Peet, W. G., Wegner, P. A., Alegranti, C. W. (1970). *Inorg. Chem.,* **9,** 2447
193. Muetterties, E. L., Alegranti, C. W. (1970). *J. Amer. Chem. Soc.,* **92,** 4114
194. Wilkinson, G., Piper, T. S. (1956). *J. Inorg. Nucl. Chem.,* **2,** 32
195. Piper, T. S., Wilkinson, G. (1956). *ibid.,* **3,** 304
196. Powell, J., Shaw, B. L. (1967). *J. Chem. Soc., A*1839
197. Cotton, F. A., Musco, A., Yugupsky, G. (1967). *J. Amer. Chem. Soc.,* **89,** 6136
198. Floris, B., Illuminati, G., Ortaggi, G. (1969). *Chem. Commun.,* 492
199. Maslowsky, E., Nakamoto, K. (1968). *ibid.,* 257
200. Nesmeyanov, A. N., Federov, L. A., Materikova, R. B., Fedin, E. I., Kocketkova, N. S. (1968). *Tetrahedron Letters,* 3753
201. Whitesides, G. M., Fleming, J. S. (1967). *J. Amer. Chem. Soc.,* **89,** 2855
202. Rakati, J. C., Davison, A. (1969). *Inorg. Chem.,* **8,** 1164
203. Davison, A., Rakati, J. C. (1970). *J. Organometal. Chem.,* **23,** 407; (1970). *Inorg. Chem.,* **9,** 289
204. Cotton, F. A., Marks, T. J. (1970). *Inorg. Chem.,* **9,** 2802
205. Bennet, M. J., Cotton, F. A., Davison, A., Faller, J. W., Lippard, S. J., Morehouse, S. M. (1966). *J. Amer. Chem. Soc.,* **88,** 4371
206. Cotton, F. A., Marks, T. J. (1969). *ibid.,* **91,** 3178
207. Kreiter, C. G., Maasbol, A., Anet, F. A. L., Kaesz, H. D., Winstein, S. (1966). *ibid.,* **88,** 3444
208. Cotton, F. A., Davison, A., Faller, J. W. (1966). *ibid.,* **88,** 4507
209. Keller, C. E., Shoulders, B. A., Pettit, R. (1966). *ibid.,* **88,** 4760
210. Dickens, B., Lipscomb, W. N. (1962). *J. Chem. Phys.,* **37,** 2084
211. Grubbs, R., Breslow, R., Herber, R., Lippard, S. J. (1967). *J. Amer. Chem. Soc.,* **89** 6864
212. Anet, F. A. L. (1967). *ibid.,* **89,** 2491

3
Substitution Reactions of the Light Elements

J. C. LOCKHART
University of Newcastle

3.1 INTRODUCTION

The work on the light elements is presented under headings for the individual element. The major portion deals with the elements lithium and boron. Kinetic work in this area presents many practical problems since most of the systems involving the light elements are air- and moisture-sensitive, but kinetics are being tackled by increasing numbers of workers trained in the appropriate glove-box and vacuum techniques. The increasing interest in this area also reflects the formidable array of physico-chemical techniques which have been made available to the chemist in the last decade.

Welcome advances in our knowledge of homolytic substitution reactions have been made, especially for organo derivatives of lithium and of boron. This work owes much to the phenomenon of electron spin–nuclear spin-coupling (as detected by e.s.r. spectra) and to the more recently recognised, but related, phenomenon of chemically induced nuclear dynamic polarisation[1] (CINDP observed in n.m.r. spectra). The predictions of rate for several bimolecular reactions of diatomic molecules made by Noyes[2] in 1966 have been tested for a number of reactions including as a propagating step a homolytic substitution. Some interesting calculations of potential energy surfaces for simple hydrogen-exchange systems have been made – these suggest energy *minima* rather than energy barriers in the course of reaction, and may help to explain some curious isotope effects which have been observed[3].

3.2 LITHIUM

Three main types of substitution reaction on lithium have been studied mechanistically. In the first, some rather unusual substitutions have been found in the chemistry of lithium alkyls and aryls. Reaction between a lithium alkyl (aryl) and another metal alkyl (aryl) (e.g. of aluminium, boron, lithium, zinc, cadmium) gives rise to substitution of the alkyl (aryl) from the secondary source for the original organo group on lithium (equation (3.1))[4].

$$x(\mathrm{LiR})_n + n\mathrm{MR}'_x \rightarrow x(\mathrm{LiR}')_n + n\mathrm{MR}_x \tag{3.1}$$

This type of reaction is often described as redistribution, scrambling, or exchange[4, 5]. Studies of this kind of reaction have been almost exclusively undertaken by n.m.r. spectroscopic techniques. The second main type of substitution is again by alkyl or aryl groups on a lithium alkyl, but this time the reaction effectively exchanges the organic moiety on an organolithium with that from an organohalide (equation (3.2)). This reaction is of considerable

$$(\mathrm{LiR})_n + n\mathrm{R}'\mathrm{X} \rightarrow (\mathrm{LiR}')_n + n\mathrm{RX} \tag{3.2}$$

importance in the synthesis of new organolithium reagents. The systems depend very strongly on the nature of R,R′ *and* X, not to mention solvent; side-reactions (e.g. Wurtz coupling, equation (3.3)) are sometimes extensive; and intermediates ranging from free radicals R· to arynes are known. The third is the exchange of hydrogen for lithium (equation (3.2), X = H) which is essentially quantitative[6] for acidic hydrocarbons (R′X).

$$LiR + R'X \rightarrow LiX + RR' \quad (3.3)$$

3.2.1 Substitution of alkyl (aryl) for alkyl (or aryl) on lithium by organometallics

The alkyl and aryl derivatives of lithium are usually polymeric $(LiR)_n$ where n may be 2 (as in phenyl lithium, R = Ph), 4 (as in t-butyl lithium, R = Bu^t, in certain solvents) or 6 (as in methyl lithium, R = Me). The structures suggested for tetramer and hexamer are shown in (1) and (2) and imply

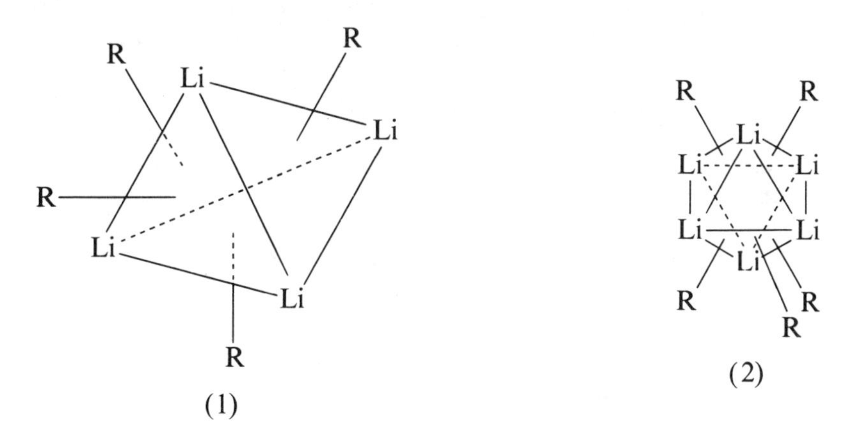

triangular arrays of lithium atoms with an alkyl group above the plane of the three lithium atoms and equi-distant from each[4]. The tetramer can be thought of as four such units edge-linked to form a tetrahedron of lithium atoms (1), while the hexamer has six such units edge-linked to form an octahedron of lithium atoms (2). In mixtures of any two organolithium reagents $(LiR)_n$ and and $(LiR')_n$ the R and R′ groups are found to transfer between lithium atoms, the reaction apparently going to a statistical product composition (as is common in scrambling reactions)[5] and by mechanisms involving molecular dissociation.

T. L. Brown and his co-workers have largely been responsible for studies of these systems, having used ^1H and ^7Li n.m.r. spectroscopy to determine the nature and concentrations of species in reaction mixtures. A review of this work appeared in 1968 [4]. This is an area where complex species have to be identified in the analysis of the reaction, and accurate kinetic studies are mostly restricted to reactions rapid on the n.m.r. time-scale, for which accurate line-shape analyses can be made. The outline of the reaction scheme can often be established, but the fine details of mechanism await fresh investigation. Typical of the studies by Brown and co-workers of the substitution of one alkyl group for another is the reaction shown in equation (3.4). The original paper[7] assumed (4) was a tetramer, but it is now known[4, 8] to be a hexamer, while the structures of the (several) mixed species (6) are not known for certain[8]. The ^7Li n.m.r. spectra

$$\underset{(3)}{Li_4Bu^t_4} + \underset{(4)}{[Li(CH_2SiMe_3)]_6} \rightleftharpoons \underset{(5)}{Li_4(Bu^t)_3(CH_2SiMe_3)} + \underset{(6)}{Li_nBu_x(CH_2SiMe_3)_{n-x}} \quad (3.4)$$

of this intermolecular reaction at various stages gave considerable information about the mechanism. The half-life for reaction of (3) and (4) in cyclopentane at room temperature (about 28 °C) was about 6–8 h [4]. The process is first-order in t-butyl lithium concentration, independent of the concentration of compound (4), and has an activation energy[4] of 24 ± 6 kcal mol^{-1}. The ^7Li spectrum at first consisted of two signals, one for the tetramer (3) and one for the hexamer (4). The species (6) appeared first, apparently giving two new ^7Li signals, and the t-butyl-rich mixed compound (one new signal) only latterly. Integration of signal areas[8] allowed the concentration of (3) to be determined as a function of time, and the (first-order) rate constant for the first half of the reaction was *c.* $6–16 \times 10^{-6}\ s^{-1}$ at 20 °C. The results are consistent with a slow dissociative step involving (3) and a similar rapid step for (4). The mechanism proposed is set out in equations (3.5–3.7).

$$Li_4(Bu^t)_4 \xrightleftharpoons{slow} 2Li_2(Bu^t)_2 \quad (3.5)$$

$$xLi_6(CH_2SiMe_3)_6 \xrightleftharpoons{fast} 6Li_x(CH_2SiMe_3)_x \text{ etc.} \quad (3.6)$$

$$Li_2(Bu^t)_2 + Li_x(CH_2SiMe_3)_x \xrightleftharpoons{fast} \underset{(6)}{Li_{x+2}(CH_2SiMe_3)_x(Bu^t)_2} \quad (3.7)$$

Species like (6), rich in trimethyl-silylmethyl groups are built up first on this mechanism, and only by subsequent dissociation and re-association could give rise to t-butyl-rich mixtures. This is the observed order of formation of mixed products (6). The reaction is slower in cyclopentane than in toluene and is accelerated by added triethylamine. At 80 °C in toluene the ^7Li signal of (3) is distinct from the four collapsed signals corresponding to the rest of the components of reaction (3.4).

Despite the variety of lithium environments possible in the mixed compounds (for example a lithium atom in (5) can have the environment (7)

	R			R′	
	Li			Li	
R		R	R		R
	(7)			(8)	

or (8), only three new ^7Li signals are observed in the reaction mixture. If the reaction mixture is cooled to −25 °C, these signals collapse and four new signals appear[4, 7]. This is best explained by an intramolecular switching of alkyl groups between environments like (7) or (8), a process rapid on the n.m.r. time-scale at room temperature and slow at −25 °C. Lithium alkyls, in general, are subject to configurational instability of this type, and n.m.r. spectra must always be interpreted with this in mind[4].

A critical experiment to examine the nature of the slow dissociation of t-butyl lithium, postulated as equation (3.5), was devised[8]. This was a study of the equilibration of equimolar quantities of isotopically enriched[6] $LiBu^t$ and $Li_n(Bu^t)_x(CH_2SiMe_3)_{n-x}$ in cyclopentane (cf. the n.m.r. kinetics of equation (3.4)). If the dissociation of t-butyl lithium goes via a free t-butyl carbanion or free radical, we would not expect to find lithium isotope-scrambling at a comparable rate, since the tetrahedral lithium skeleton should remain intact. Only a dissociation of a lithium-containing fragment (e.g. dimer or monomer) can cause lithium isotope exchange. The sequence A–E in Figure 3.1. shows

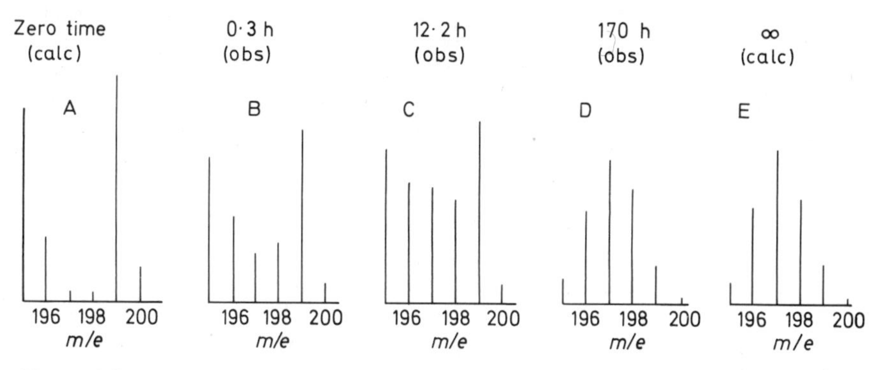

Figure 3.1 Mass spectra calculated and observed for equimolar mixture of $Bu^{t6}Li$ (96 % ^6Li) and $Bu^{t7}Li$ (99.9 % ^7Li). (From Darensbourg, Kimura, Hartwell and Brown[8]).

the mass spectral pattern calculated and observed at various times for such mixtures in the *m/e* region 195–200. Parent ions are not observed for the t-butyl lithium tetramer and this region comprises the highest observed fragment, $[Li_4Bu^t_3]^+$. There is clearly a process of rate comparable to the exchange in equation (3.4), which results in complete scrambling of the lithium isotopes. No actual kinetic parameters were evaluated[8].

The substitution of methyl for ethyl on lithium occurs in ether solution between the tetrameric species Li_4Me_4 and Li_4Et_4. The rate-determining step was again thought to be dissociation of methyl lithium to reactive dimers[4, 9]

$$Li_4Me_4 \xrightleftharpoons{\text{slow}} 2(Li_2Me_2) \tag{3.8}$$

as in equation (3.8). In this reaction only three mixed species are expected. Four ^7Li signals are observed at $-80\,°C$, corresponding to the four possible lithium environments in the limit of slow intramolecular switching, and these collapse to a one-line spectrum above $-50\,°C$. The signals for ethyl-rich species collapse first with increasing temperature. An activation energy for equation (3.8) (the supposed rate-determining step in this exchange) has been obtained in another way, and is given[4] as 11 kcal mol^{-1}, with an activation entropy of -6 cal deg^{-1} mol^{-1}.

Another angle on dynamic stereochemistry in these systems is provided by the neohexyl lithium system (9). Witanowski and Roberts[10] have observed in the ^1H n.m.r. spectrum of this compound in ether a change with increasing

$$\begin{array}{c} \quad\;\; Me \;\; H_B \;\; H_A \\ Me-\underset{|}{\overset{|}{C}}-\underset{|}{\overset{|}{C}}-\underset{|}{\overset{|}{C}}-Li \\ \quad\;\; Me \;\; H_{B'} \; H_{A'} \end{array}$$

(9)

temperature consistent with an AA′BB′ system changing to an A_2B_2 system by a process involving inversion at the metal-bearing carbon. The methylene groups of the neohexyl system in a stable compound (e.g. a halide) should provide an AA′BB′ spectrum. However, a labile metal system like (9) provides a mechanism for rapid inversion at the metal-bearing carbon, thus removing the magnetic non-equivalence of A and A′ and B and B′. The activation energy of the exchange has been estimated at 15 ± 2 kcal mol^{-1} from an analysis of line-shape to provide exchange lifetimes at different temperatures, and a dissociative mechanism, possibly S_E1 was suggested. This activation energy is of the same order of magnitude as given for mixed alkyl exchanges in the Me–Et and Bu^t–CH_2SiMe_3 lithium systems, where the interpretation was that exchange occurred via a rate-determining dissociation of tetramer methyl- or t-butyl-lithium to reactive dimer. The state of aggregation of neohexyl lithium in ether is not known. Further work is clearly necessary to determine if there are indeed these quite different processes available (S_E1 and molecular dissociation). Clearly the inversion of the metal-bearing carbon in the neohexyl lithium cannot be explained without separation of the organic moiety from lithium. A carbanion (S_E1) or free radical mechanism could also be envisaged for the supposed inter- and intra-molecular types of alkyl exchange found in the t-butyl–silyl system between the several different lithium compounds present (slow) and the several lithium local environments present (fast). However, the mass spectral data of Figure 3.1. tend to rule this out. On similar grounds, the line-shape in ^1H n.m.r. spectra is also governed by inversion rate for compound (10).

$$\begin{array}{c} \quad H_X \;\; H_A \\ Et-\underset{|}{\overset{|}{C}}-\underset{|}{\overset{|}{C}}-Li \\ \quad Me \;\; H_{A'} \end{array}$$

(10)

The metal-bearing methylene group gives an AB(X) signal at low temperatures changing to an A_2(X) signal on heating[11]. The rate of this inversion is first-order in (10) in pentane, toluene or ether, and is unaffected (in pentane) by the presence of t-butyl lithium, but is increased by addition of base (with accompanying increase in $\Delta H^\ddagger$ and $\Delta S^\ddagger$) [11].

Methyl exchange[12,13] occurs between ether solutions of methyl lithium and the complex Li_2MMe_4 or Li_3MMe_5, where M is Cd, Zn or Mg, and is faster in the sequence Cd > Zn > Mg. The two zinc complexes are indistinguishable in terms of their ^{1}H n.m.r. spectra, so the system was considered as a two-site exchange, and an approximate activation energy of 8.5 ± 3 kcal mol^{-1} estimated for methyl exchange. The rate-determining step could thus be the previously-proposed dissociation of methyl lithium tetramer, and a suggested mechanism is given by equation (3.8) followed by equation (3.9). Since lithium

$$*Me_2Li_2 + Li_2ZnMe_4 \rightleftharpoons Li_2ZnMe_2^*Me_2 + Li_2Me_2 \quad (3.9)$$

exchange (measured from ^{7}Li line-shape data) between the sites has the same rate as alkyl exchange, the mechanism must allow for the LiMe to transfer as a unit. The extension of this work to cadmium analogues[13] in ether brought no advances in interpretation. Methyl exchange here is faster than lithium, so the mechanism is slightly different. The comparison of activation parameters for exchange of methyl lithium and the same metal complexes in THF is made in Table 3.1. Lithium exchange and methyl exchange occur at the same rate and substantially the same activation energy for all three systems (Cd, Zn and Mg), as shown in the Table 3.1 [13]. Much reliance is placed

Table 3.1 Activation energies determined from n.m.r. line-shape data for exchange of MeLi with $Li_2MMe_{x(+2)}$ in THF. Data in kcal mol^{-1} ($\pm$ 2 kcal mol^{-1}) from Seitz and Little[13]

	M = Mg	Zn	Cd
Exchange of ^{7}Li	12.4	10.7	11.0
Exchange of Me	9.2	9.0	10.8

on the dissociation of tetramer to dimer as a basic mechanism without much strong evidence for it. Significant advances in interpretation here await a more positive demonstration of such a step (equation (3.8)). It would also be helpful to investigate these systems thoroughly for the presence of free radicals.

In the related exchange of phenyl lithium with Li_2MPh_4 or $LiMPh_3$ where M is Mg or Zn, the overall stoichiometry of the complexes is different, and the exchange processes seem to be different in character[14]. N.M.R. spectroscopic studies show that ^{7}Li exchange occurs between phenyl lithium and the complex Li_2MPh_4, and is considerably faster than phenyl exchange between the two species. It has been suggested that lithium exchanges via an ion-pair formed as in equation (3.10), while phenyl

$$Li_2MPh_4 \xrightleftharpoons{\text{slow}} Li^+ \cdot LiMPh_4^- \quad (3.10)$$

substitution requires a different (slower) mode of break-down, equation (3.11).

$$Li_2MPh_4 \xrightleftharpoons{\text{slow}} MPh_2 + Li_2Ph_2 \quad (3.11)$$

The rate of phenyl exchange is independent of the absolute concentrations of the species. Phenyl exchange is slightly slower in the zinc than in the magnesium system, consistent with a more stable zinc complex and hence slower rate-determining stage in equation (3.11).

3.2.2 Substitution of halogen for alkyl (or aryl) on lithium

The exchange reaction between alkyl or aryl lithium and alkyl or aryl halides results in substitution of a new group on lithium. The reaction is in competition with Wurtz coupling and disproportionation reactions (see equations (3.10), (3.11) and 3.12))

$$\mathrm{Bu^sLi + Pr^iBr \xrightarrow{\text{pentane, 25°C, 20 h}} C_6H_{14} + C_7H_{16} + C_8H_{18}} \qquad (3.12)$$

to such an extent that sometimes no metal–halogen interchange is observed as in equation (3.12). Early work[15] purported to measure carbanion stabilities as a function of the equilibrium constant K for reaction in equation (3.13), in systems in which it was demonstrated that side reactions (equation (3.3)) did not occur.

$$\mathrm{RLi + R'I \overset{K}{\rightleftharpoons} R'Li + RI} \qquad (3.13)$$

The kinetics of exchange in equation (3.14) have been measured and exhibit a complicated form[16]. In general, they refer to salt-free and salt-containing organolithium reactions in which no side-reaction (e.g. via arynes to biaryl derivatives) is able to compete kinetically[17]. In mixtures in which the organolithium contains an equimolar quantity of residual lithium halide (from the synthesis $\mathrm{RX + 2Li \rightarrow LiX + RLi}$), the rate of exchange of aryl groups is first-order in aryl lithium and first-order in aryl iodide in both forward and reverse directions (equation (3.14)), with an activation energy of 16.8 kcal mol^{-1} and $\Delta S^{\ddagger}_{277} = -23.6$ cal deg^{-1} mol^{-1}*. The variation of rate with substituent Y *para* to the bromide in the aryl bromide followed the Hammett linear free

$$\mathrm{PhLi + Y{-}C_6H_4{-}Br \overset{K}{\rightleftharpoons} Y{-}C_6H_4{-}Li + PhBr} \qquad (3.14)$$

energy relationship $\log k/k_0 = \sigma\rho$ with a high ρ value of 4.0. This high ρ value shows that electron-withdrawing substituents on the aryl bromide and electron-donating substituents on the aryl lithium (reverse reaction, equation (3.14)) increase the reaction rate. A more complex behaviour was noted in mixtures made from pure aryl lithium (i.e. no residual lithium halide). The rate is again first-order in aryl lithium and aryl bromide, but is faster than in salt-containing systems. The overall second-order rate constant is linearly dependent on initial concentrations (Table 3.2). For salt-free systems it is probable that the degree of association of the aryl lithium is relevant to the rate. The

* The effect of lithium halide on the organolithium is thought to be structure-breaking, although there is not enough information to be sure of this. Mixtures of lithium bromide with lithium methyl in THF or ether do, in fact, show signs of lithium exchange. The ^{7}Li signal of mixtures at 0 °C is a singlet, at −100 °C is two lines, and the methyl lithium line sharpens before that of the salt on cooling. Retention of configuration occurs for the organic portion in these exchanges[18].

mechanism shown in equations (3.15)–(3.16) allows for this. The kinetic equation (3.17) can be derived if a random distribution of aryl residues is assumed and if the forward rates k_{15} and k_{16} and the reverse rate k_{-15} and k_{-16} are assumed to be equal. The rate constants derived from this expression (equation (3.17)) are reasonably constant for changing reagent concentration (see Table 3.2).

$$(PhLi)_2 + Me\text{-}C_6H_4\text{-}Br \rightleftharpoons PhBr + PhLi \cdot Me\text{-}C_6H_4\text{-}Li \qquad (3.15)$$

$$(PhLi)_2Me\text{-}C_6H_4\text{-}Li + Me\text{-}C_6H_4\text{-}Br \rightleftharpoons PhBr + (Me\text{-}C_6H_4\text{-}Li)_2 \qquad (3.16)$$

$$dx/dt = k_{17}/2a[(a^2-x^2)(b-x)-(a^2-x_e^2)(b-x_e)(2ax^2-x^3)/(2ax_e^2-x_e^3)] \qquad (3.17)$$

Subsequent to this work, clear evidence has been obtained for the presence of free radicals in a multitude of R—Li—R′—X systems. These are certainly implicated in the Wurtz coupling processes (e.g. equation (3.12)) and possibly also in the metal–halogen exchange. Screttas and Eastham[19] noted a broad continuous ultraviolet–visible absorption band in the spectrum of butyl-lithium dissolved in tetrahydrofuran, and suggested it was due to charge transfer. They proposed that the primary reaction process for alkyl lithiums in presence of base was the generation of radicals according to equation (3.18).

$$Bu_2Li_2THF + \text{solvent} \rightarrow (Bu_2Li_2THF)\text{solvent}^+ + e(\text{solvent})^- \qquad (3.18)$$

Free radicals have now been observed by a variety of means – inhibition of reaction by reagents known to devour free radicals, e.s.r. spectra and chemically induced nuclear dynamic polarisation (CINDP), which is

Table 3.2 Variation of K_{obsd} and K_{17} for salt-free systems of PhLi and *p*-bromotoluene (0.3 M) in ether at 25 °C [16]

Molarity of PhLi	$k_{obsd.}$*	k_{17}†
0.58	0.73	1.19
0.29	0.55	1.28
0.23	0.47	1.11
0.10	0.38	1.28

* Calculated for a simple second-order reaction (l. mol^{-1} h^{-1})
† Calculated for equation (3.17)

observed when nuclear spins in reacting molecules become dynamically coupled to an unpaired electron while traversing the path from reagent to product[1]. This method provides spectacular n.m.r. spectra (see e.g.

Figure 3.2) of anomalous intensities for nuclei which are transiently part of a free radical. The majority of such CINDP studies have been on alkyl systems and provide evidence for free radical substitution on lithium.

A typical study using the phenomenon of CINDP is that of Lepley[20] who examined the ^{1}H n.m.r. spectra of reaction mixtures containing alkyl lithium and alkyl iodide in hexane solution (see Figures 3.2 and 3.3). The normal

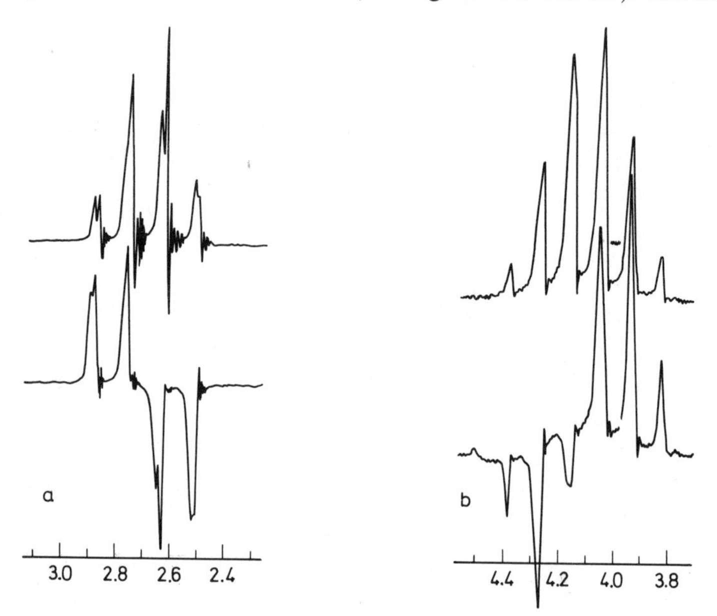

Figure 3.2 A comparison of the normal absorption (upper) and nuclear polarised (lower) proton magnetic resonance bands for the iodomethylene quarter (a) of ethyl iodide and the iodomethine sextet; (b) of 2-iodobutane. Nuclear polarisation of these bands occurs during halogen–metal exchange. (Reprinted from *Chemical Communications*, 1969, p. 64, by permission of the Chemical Society and Professor A. R. Lepley)

absorption spectra for the methine septet of sec-butyl iodide and the methylene quartet for ethyl iodide are shown in the upper trace. The reaction mixture of ethyl lithium and sec-butyl iodide produced the spectra shown in the lower trace (Figure 3.2) after about 5 min. Earlier scans showed duplicated strong polarisation of the newly formed ethyl iodide, but the polarisation of the *sec*-butyl iodide (the reagent) only slowly increased to that shown. The stimulated emission and enhanced absorption for the methylene hydrogen nuclei have a reversed phase relationship to those for the methine hydrogen. The lithomethylene hydrogen nuclei showed no polarisation in these reactions, and in the related reaction (Figure 3.3) of n-butyl lithium and i-propyl iodide had disappeared within 4 min. The lithium compound probably does not take direct part in the polarisation process. It is clear that polarisation occurs in the radical obtained from the alkyl lithium and is transferred to alkyl halide, at least in part, by a radical path. Several studies of this type have been made and nuclear polarisation observed in the systems ethyl lithium–ethyl iodide[21], n-butyl lithium–*sec*-butyl iodide[21], ethyl lithium–iodobenzene[21], n-butyl lithium–n-butyl iodide[22], ethyl lithium–α,α-dichlorotoluene. Reaction schemes such as those depicted in equations (3.19)–(3.22) have been put forward.

$$\mathrm{RLi + R'I \rightarrow R\cdot + R'\cdot + LiI} \qquad (3.19)$$

$$\mathrm{R^{*}\cdot + R'I \rightarrow R'I^{*} + R\cdot} \qquad (3.20)$$

or

$$\mathrm{R^{*}\cdot + R'I \rightarrow RR'I \rightarrow R^{*}I + R'\cdot} \qquad (3.21)$$

The rapid disappearance of the alkyl lithium signals is consistent with some

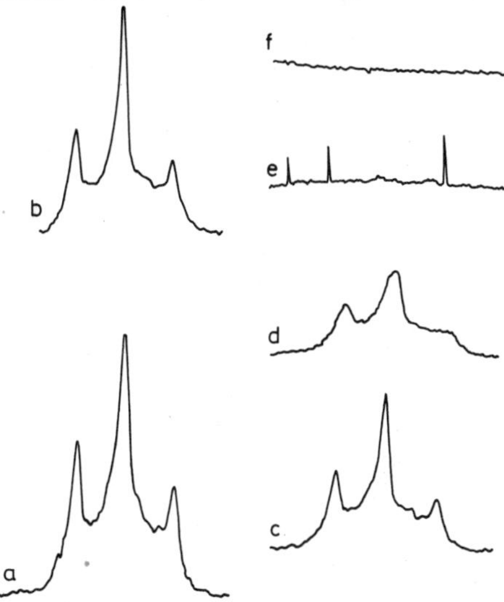

Figure 3.3A sequence displaying the disappearance of the lithiomethylene proton magnetic resonance multiplet at $\delta - 0.82$ p.p.m.; (a) before halide addition and with scans started at 0.50. (b) 1.25 (c) 2.00 (d) 3.00 (e) and 4.00 (f) min after the addition of 150 μl of *sec*-butyl iodide to 625 μl of 1.5 M n-butyl lithium in hexane. (Reprinted from *Chemical Communications*, 1969, p. 64, by permission of the Chemical Society and Professor A. R. Lepley)

of the free radicals R· being trapped by RLi to form a complexed radical (equation (3.22)), which might take part in alkyl transfer.

$$\mathrm{R'\cdot + RLi \rightarrow R'[RLi]\cdot} \qquad (3.22)$$

Lepley has also tried to measure relative concentrations of polarised and unpolarised alkyl halides as a function of time during the reaction of i-propyl iodide with n-butyl lithium in hexane[23] (Figure 3.4). The rates of appearance and disappearance can thus be estimated, and it would clearly be of enormous value to compare these with rates of alkyl–halogen interchange obtained by another route, so that the exact relationship between the polarisation and exchange mechanism can be more fully elucidated. It is important to note that these reactions finally result in radical-coupling processes and disproportionation. Screttas and Eastham[19] had previously suggested the efficacy of basic solvents in promoting free radical species. Basic solvents probably make these reactions uncontrollably fast, and hexane or benzene is used to give measurably slow reactions.

A free radical mechanism for lithium–halogen interchange would still be consistent with earlier stereochemical studies in which tripticyl bromide exchanged with butyl lithium[26]. Front-side attack of the reagent is required

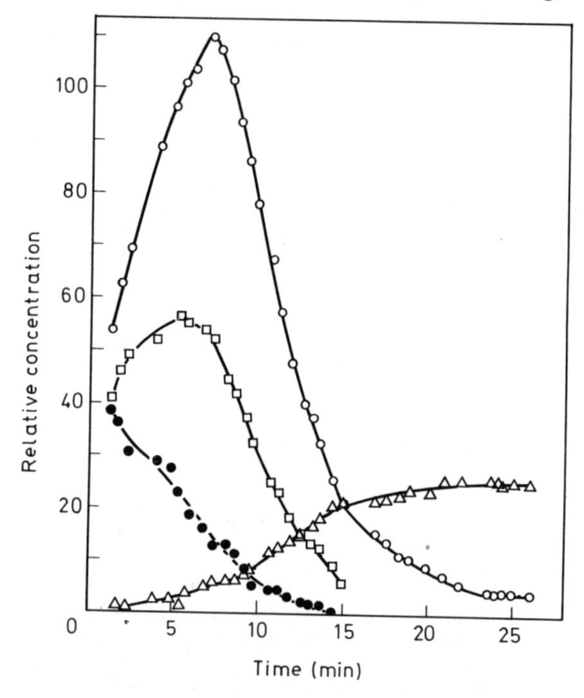

Figure 3.4 Relative variation with time for the polarised (*) and normal forms of 2-iodopropane (PrI, ●; PrI*, □) and 1-iodobutane (BuI, Δ; Bu*I, ○) from the reaction of 150 μl of 2-iodopropane with 0.6 ml of 1.5 M n-butyl lithium in hexane. These values were calculated from the statistical integral intensities for normal spectra and the optimum emission–absorption spectra of the individual species. (Reprinted from Lepley[23]. Copyright (1970) by the American Chemical Society. Reprinted by permission of the copyright owner and Professor A. R. Lepley)

and it could occur via a four-centre radical mechanism. 2-Iodo-octane is partly racemised by interchange of halogen with *sec*-butyl lithium, and the route may be partly free radical[27].

Russell and Lamson have examined some of these systems using e.s.r. spectra to study the free radicals formed *in situ*[28]. They studied the reaction depicted in equation (3.23). The systems were dissolved in benzene containing base (ether or tetramethylethylenediamine) equimolar with lithium compound. An e.s.r. signal could be obtained within 0.03 s of mixing in a flow system. This study detects a different part of the reaction from the CINDP experiment. The radical detected was found to be R· or R′· depending on the

Table 3.3 Radical for which e.s.r. signals were detected 0.03 s after mixing in the reaction mixture shown in equation (3.23)[28]

R *in* RLi	R X	*Radical observed*
Bu^n	MeI	$Me\cdot + Bu^n\cdot$
Bu^n	EtI	$Et\cdot$
Bu^n	Pr^nI	$Pr^n\cdot$
Bu^n	Pr^iI	$Pr^i\cdot$
Bu^n	Bu^nI	$Bu^n\cdot$
Bu^n	Bu^tI	$Bu^t\cdot$
Bu^n	Bu^iBr	$Bu^n\cdot$
Bu^n	Bu^tBr	$Bu^n\cdot$
Bu^n	$PhCH_2Br$	$Bu^n\cdot$
Bu^s	EtBr	$Bu^s\cdot + Et\cdot$
Bu^n	AllylBr	allyl·

nature of the halide X, as shown in Table 3.3. A series of steps (equations (3.24)–(3.26)) was postulated to explain the overall pattern of products. Equation (3.25) is effectively homolytic substitution on lithium. In the limiting case where $k_{24} \gg k_{25}$ the concentration

$$RLi + R'X \rightarrow R\cdot + R'\cdot + LiX \tag{3.23}$$

$$R\cdot + R'X \xrightarrow{k_{24}} RX + R'\cdot \tag{3.24}$$

$$R'\cdot + RLi \xrightarrow{k_{25}} R'Li + R\cdot \tag{3.25}$$

$$R\cdot + R'\cdot \rightarrow \text{alkanes, alkenes} \tag{3.26}$$

of R stays small. Metal–halogen interchange occurs and R′ builds up. This step is likely to be fastest where the halide is iodide (a prediction based on the relative bond strengths of different halides attached to alkyl carbon) as in fact is observed (Table 3.3). In the other limit, where $k_{25} \gg k_{24}$, only R should be obtained initially. Of course, extensive metal–halogen interchange would be expected to lead to mixtures of R and R′ but these were not observed. It is not clear that this would follow from the mechanism in equations (3.24) and (3.25) unless k_{23} is very much less than k_{24} or k_{25}, and the RI formed is only a small fraction of the R′I.

3.2.3 Lithiation of hydrocarbons

Investigations of this topic do not provide a substantial test for any particular mechanism. A 4-centre mechanism has previously been proposed for reactions of the general lithium reagent RLi with hydrocarbons R′H, equation (3.27) providing one example[29]. In this particular reaction (lithiation of anisole), anomalous product distributions have been observed. The kinetic isotope

$$C_6H_5(OMe)H + Bu^nLi \longrightarrow C_6H_4(OMe)Li + Bu^nH \quad (3.27)$$

effect for reaction of anisole-2-d was found to be $k_H/k_D = 6.0–8.2$ for both Bu^n and Bu^tLi, so the rate-determining step *removes* ring hydrogen. Radical intermediates are now proposed although there is no direct evidence. There is, however, a step of low *steric* requirement involving the alkyl group since Bu^tLi actually reacts faster than Bu^nLi in the reaction depicted in equation (3.27). The Bu^t radical is the more stable and might account for it. Considerable information about the reaction of triphenylmethane with organolithium reagents (equation (3.28)) has been amassed[6, 30]. Table 3.4

Table 3.4 The metalation of triphenylmethane and triphenyldeuteriomethane with organolithium reagents in tetrahydrofuran[30]

R	*Order in* RLi(H)*	*Order in* RLi(D)*	k_H/k_D	*State of aggregation of* RLi
allyl	1.07±0.05	0.93±0.10	7.1	monomer and dimer
benzyl	0.90±0.27	0.90±0.16	4.3	monomer
vinyl	0.24±0.04	0.30±0.16	6.7	tetramer
phenyl	0.64±0.07	0.69±0.04	3.6	dimer
methyl	0.28±0.08	0.25±0.05	6.2	tetramer
n-butyl	0.33±0.05	0.27±0.08	8.9	tetramer

* H refers to Ph_3CH, D to Ph_3CD

shows the effect on reactivity of the following changes; (i) Ph_3CD studies to determine

$$Ph_3CH + RLi \rightarrow Ph_3CLi + RH \quad (3.28)$$

kinetic isotope effects, (ii) substituents on the aryl rings, and (iii) variation of R. Reaction in THF at 22 °C gave the general order of reactivity Bu > vinyl > Me > benzyl > phenyl. The order in organolithium reagent depends on R, and no doubt reflects the degree to which the RLi aggregate (monomer, dimer, or tetramer) is broken down in the transition state. Substantial changes in the isotope effect show that the methine C—H bond is broken in the rate-determining step, and that the precise nature of the transition state changes with different R groupings. For each individual R group, Hammett $\sigma\rho$-plots were obtained by varying substituents on the triaryl methane ring, with ρ values between 2.2 and 3.0, suggesting an appreciable build-up of negative charge on the Ph_3C residue. These results considerably extend those of Pocker and Exner[31].

3.3 BERYLLIUM

The coordination sphere of beryllium is anomalous among M^{2+} species in that it is much less labile although fast reaction techniques are still required for kinetic studies[32].

When the solvated ion $Be(DMF)_4^{2+}$ is dissolved in DMF, the solvent in the primary coordination sphere can be distinguished from the bulk solvent by the use of n.m.r. spectroscopy[33]. The rate constant for exchange from bulk to complexed environment was measured by line-shape analysis and is given as 310 s^{-1} at 25 °C. The mixed complex $Be(DMF)_2(acac)^+$ ($k = 22\ s^{-1}$ at 25 °C) is less labile. Enthalpies and entropies of activation for solvent exchange are 14.6 kcal mol^{-1} and 2.6 cal deg^{-1} mol^{-1} for the $Be(DMF)_4^{2+}$ ion, and 13.9 kcal mol^{-1} and −6 cal deg^{-1} mol^{-1} for $Be(DMF)_2(acac)^+$, and a dissociative mechanism is assumed. The main difference in reactivity between the two is thus the entropy term.

Another dissociative process has been observed by n.m.r. spectroscopy[34]. The ^{19}F signals in aqueous B—F systems represent the species BeF_4^{2-}, BeF_3^- and F^-, and the rate constants for F transfer at 33 °C between these three sites have been determined as

$$k_{29} = 23.8\ s^{-1} \qquad k_{-29} = 3\times10^2\ l\ mol^{-1}\ s^{-1}$$
$$k_{30} = 4.3\times10^2\ l\ mol^{-1}\ s^{-1} \qquad k_{31} = 4.2\times10^2\ l\ mol^{-1}\ s^{-1}$$

$$BeF_4^{2-} \underset{k_{-29}}{\overset{k_{29}}{\rightleftharpoons}} BeF_3^- + F^- \tag{3.29}$$

$$BeF_4^{2-} + {}^*F^- \overset{k_{30}}{\rightleftharpoons} Be^*F_4^{2-} + F \tag{3.30}$$

$$BeF_3^- + {}^*F^- \overset{k_{31}}{\rightleftharpoons} Be^*F_3^- + F^- \tag{3.31}$$

Baldwin and Stranks[35] measured the rate of formation of $Be(H_2O)_3F^+$ by a conductimetric stopped-flow technique in acid concentrations greater than 0.01 M. There appear to be two paths for the reaction of the beryllium tetra-aquo ion, each with the same activation energy of 8.9±0.8 kcal mol^{-1}. The path given in equation (3.32) has a second-order rate constant of 724±95 l mol^{-1} s^{-1}, and that in equation (3.33) a rate constant of 73±31 mol^{-1} s^{-1}.

$$Be(H_2O)_4^{2+} + F^- \rightleftharpoons Be(H_2O)_3F^+ \tag{3.32}$$

$$Be(H_2O)_4^{2+} + HF \rightleftharpoons Be(H_2O)_3F^+ + H^+ \tag{3.33}$$

A dissociative ion-pair mechanism S_N1 IP is suggested in which dissociation of water molecule is rate-determining, and precedes association of the fluoride or HF to form an ion-pair.

Pearson and Moore[36] had earlier measured the hydrolysis rates for $Be(acac)_2$ in acid solutions at 25 °C and 0.25 M ionic strength. Both rings were hydrolysed at the same rate. The reaction rate was first-order in $Be(acac)_2$, with a curious non-linear dependence on $[H^+]$, interpreted in terms of a rate-determining one-ended dissociation of acetylacetonate from beryllium to give the half-bonded protonated ligand. The activation energy for the ring-opening step was calculated as 10 kcal mol^{-1}.

3.4 BORON

3.4.1 Free radical reactions at boron centres

In the last few years very considerable progress has been made in examining homolytic substitution reactions at boron, usually as part of an overall chain mechanism involving the boron substrate. Homolytic substitution on boron by radicals such as RO· [37], $RO_2^{\cdot}$ [38], Br· [39], Me_2N· [40], and $RCH_2CH{=}CHO$· [41], has now been observed in solution, and occasionally, for some of these radicals, in the gas phase. Some of the innovations which appear to have led to this remarkable spread of knowledge about radical processes on boron are (a) the speculative use of stereochemistry, which helped to destroy the 1,3-dipolar addition mechanism for autoxidation of boron[42–44], (b) the use of powerful radical scavengers such as galvinoxyl to inhibit rates of chain reactions which had previously been unrecognised for lack of an induction period[42–44], and (c) the use of e.s.r. spectroscopy to detect and identify radicals formed in the reactions[45, 46].

3.4.1.1 *Substitution of alkoxy radicals on boron*

Krusic and Kochi discovered an S_H2 process (equation (3.34)) on boron when the e.s.r. spectrum of a reaction mixture of trimethylboron and irradiated t-butyl peroxide was observed. No hyperfine structure due to

$$Me_3B + Bu^tO\cdot \rightarrow Me\cdot + Me_2BOBu^t \qquad (3.34)$$

coupling with boron could be observed, so it seems unlikely that any stable intermediate based on 4-coordinate boron is formed[45]. Davies and Roberts have used the same method to detect radicals derived from organometallics in general[46].

Davies and co-workers have now measured the absolute rates for such homolytic substitutions on boron using some novel competition methods[37]. In one of these, Bu^tO· radicals were allowed to compete for equimolar cyclopentane (equation (3.35)) and tributylboron (equation (3.36)). The radicals cyclopentyl and butyl produced in these reactions were detected by e.s.r. spectroscopy, and their relative concentrations determined by integration of e.s.r. signal intensities. Since the rate constant for equation (3.35) is already known, the absolute rate constant for equation (3.36) is readily obtained.

$$C_5H_{10} + RO\cdot \rightarrow C_5H_9\cdot + ROH \quad (3.35)$$

$$Bu_3B + RO\cdot \rightarrow Bu\cdot + ROBBu_2 \quad (3.36)$$

The other method utilised the chain reaction of t-alkyl hypochlorites with tributylboron, which has as propagating steps the homolytic substitution depicted in equations (3.37) and (3.38). For R = Pr^i the self-decomposition process in equation (3.39) has a comparable (known) rate. G.l.c.-analysis of

$$RCMe_2O\cdot + BBu_3 \rightarrow Bu\cdot + RCMe_2OBBu_2 \quad (3.37)$$

$$Bu\cdot + RCMe_2OCl \rightarrow BuCl + RCMe_2O\cdot \quad (3.38)$$

$$Pr^iCMe_2O\cdot \rightarrow Pr^i + Me_2CO \quad (3.39)$$

the product ratio $[BuCl]/[Me_2C{=}O]$ enabled the relative rate constants (k_{37} and k_{39}) to be obtained. The rate constants determined for the homolytic substitution processes determined by the two methods are comparable in magnitude:

$$k_{37} = 4.5 \times 10^7 \text{ l mol}^{-1} \text{ s}^{-1} \text{ at } 40\,°\text{C (g.l.c.)}$$
$$k_{36} = 7.3 \times 10^6 \text{ l mol}^{-1} \text{ s}^{-1} \text{ at } 40\,°\text{C (e.s.r.)}$$

3.4.1.2. *Autoxidation of alkyl boron compounds*

The autoxidation of boron compounds in solution, formerly assumed to proceed by a polar 1,3 shift, is now thought to proceed via a rapid free radical reaction of long chain length. Davies and roberts[42, 44] and Allies and Brindley[43] employed stereochemical criteria in autoxidation processes. Optically active phenylethyl-boronic acid (equation (3.40)) reacted in a

$$\underset{\text{(opt. active)}}{MeCH(Ph)B(OH)_2} + O_2 \rightarrow \underset{\text{(racemic)}}{MeCH(Ph)OOB(OH)_2} \quad (3.40)$$

first-order reaction with $k = 0.0020 \pm 0.0002 \text{ s}^{-1}$ at 20 °C, to give the *racemic* peroxy-compound[42]. A polar, 1,3 shift mechanism could not account for the stereochemical result. Similar arguments have been used by Davies and Roberts in the autoxidation of epimeric mixtures of trinorbornylboranes, which, no matter what the starting ratio of *endo* to *exo*, always leads to the same proportion (76% *exo* and 24% *endo*) of norbornylperoxy links, suggesting a common norbornyl radical in the process[44]. Allies and Brindley described the extensive racemisation of di(isopinocampheyl)butylboranes during autoxidation[43]. In each of these studies clear-cut inhibition of reaction by potent radical scavengers such as galvinoxyl was observed, in accord with a radical process of long chain length. The chain-propagating steps suggested are given in equations (3.41) and (3.42).

$$R\cdot + O_2 \rightarrow RO_2\cdot \tag{3.41}$$

$$RO_2\cdot + R'_3B \xrightarrow{S_H2} RO_2BR'_2 + R'\cdot \tag{3.42}$$

K.U. Ingold has also observed such an S_H2 displacement in a liquid phase autoxidation of *sec*-butylboronic anhydride $(Bu^sBO)_3$, which can either be self-initiated or started with t-butyl peroxide[38]. The mechanism suggested is

initiation $$\rightarrow R\cdot \tag{3.43}$$

propagation $$R\cdot + O_2 \rightarrow RO_2\cdot \tag{3.44}$$

$$RO_2\cdot + [RBO]_n \xrightarrow{k_p} [ROOBO]_n + R\cdot \tag{3.45}$$

termination $$R\cdot + R\cdot \xrightarrow{2k_t} \text{non-radical species} \tag{3.46}$$

The rate of oxidation is given by equation (3.47). Since the termination rate is known, the rate of the propagation step (equation (3.45)) which in S_H2 can be obtained from equation (3.47).

$$-d[O_2]/dt = k_p[Bu^sBO]R_i^{\frac{1}{2}}/(2k_t)^{\frac{1}{2}} \tag{3.47}$$

The gas-phase autoxidation of triethylboron is an auto-catalytic chain process giving a variety of hydrocarbon products, CO and acetaldehyde[47].

3.4.1.3 Reaction of trialkylboranes with bromine

Reaction of trialkylboranes with dry bromine in the dark results in substitution of bromide for alkyl on boron (equation (3.48))[39]. In methylene chloride $d[RBr]/dt \approx -d[R_3B]/dt \approx -d[Br_2]/dt$. However, in CCl_4 this is not so (see Figure 3.5). The butyl bromide is formed in a step slower than loss of bromine or borane, possibly as depicted in the overall scheme in equations (3.49)–(3.53).

$$Br_2 + Bu_3B \rightarrow Bu_2BBr + BuBr \tag{3.48}$$

$$Br_2 + BBu_3 \xrightarrow{\text{slow}} Br\cdot + BBrBu_2 + Bu\cdot \tag{3.49}$$

$$BuB{-}\overset{|}{\underset{\underset{H}{|}}{C}}{-} + Br\cdot \rightarrow Bu_2B{-}\overset{|}{\underset{|}{C}}\cdot + HBr \tag{3.50}$$

$$Bu_2B{-}\overset{|}{\underset{|}{C}}\cdot + Br_2 \rightarrow Bu_2B{-}\overset{|}{\underset{\underset{Br}{|}}{C}}{-} + Br\cdot \tag{3.51}$$

$$\mathrm{Bu_2B{-}\overset{|}{\underset{\underset{Br}{|}}{C}}{-}} + \mathrm{HBr} \rightarrow \mathrm{Bu_2BBr} + \mathrm{H\overset{|}{\underset{\underset{Br}{|}}{C}}{-}} \tag{3.52}$$

$$\mathrm{Bu{-}B}\lt + \mathrm{HBr} \rightarrow \mathrm{BuH} + \mathrm{BrB}\lt \tag{3.53}$$

The reaction of bromine with *sec*-butylboron cannot be the direct process shown in equation (3.48), but might create bromodibutyl boron in two ways—

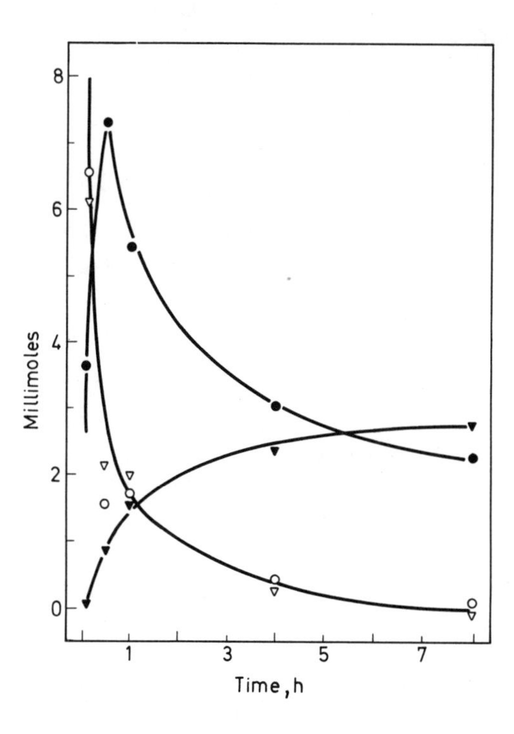

Figure 3.5 The dark reaction of tri-sec-butylborane (10 m mol) with bromine (10 m mol) in 20 ml of carbon tetrachloride solvent: ○, Br_2; ∇, R_3B; ●, total HBr; ▼, 2-bromobutane. (Reprinted from Lane and Brown[39], copyright (1970) by the American Chemical Society. Reprinted by permission of the copyright owner)

equation (3.49), suggested as a slow step, and by reaction of HBr with an α-brominated trialkyl boron (equation (3.52)). If this latter step were favoured in methylene chloride, it could account for the differing rates of consumption and production of materials in carbon tetrachloride (Figure 3.5). Cyclohexane and chloroform gave the same result as carbon tetrachloride.

The related gas-phase reaction of triethyl boron with bromine was found to proceed by direct free radical substitution (by photolytically generated Br·) and H· abstraction to give HBr and a radical (equation (3.50))[48]. A competitive kinetic technique showed that the free radical reaction was very slow in comparison to that depicted in equation (3.50).

3.4.1.4 *Substitution of dimethylamine radicals on boron*

The dimethylamine radical, generated photolytically from tetramethyl tetrazene was found to substitute for Bu· on tributylboron. The butyl radical was detected through its e.s.r. spectrum. The hydrogens of the dialkylamine group in the product gave appropriate ^{1}H n.m.r.

$$Me_2N\cdot + BBu_3 \xrightarrow{S_H2} Me_2NBBu_2 + Bu\cdot \quad (3.54)$$

spectra. It was found that disubstitution could be effected[40]. The monosubstitution (equation (3.54)) was also observed in chlorobenzene at 35 °C when the radical $Me_2N\cdot$ was generated from Me_2NCl, but here two processes were found to be in competition – equation (3.54) and the 1,2 polar reaction in equation (3.55). The homolytic process (equation (3.54)) could be totally inhibited by galvinoxyl, enabling the polar

$$Bu_3B + Me_2NCl \rightarrow BuNMe_2 + Bu_2BCl \quad (3.55)$$

process (equation (3.55)) to occur alone. The homolytic process is probably part of a long chain-propagating sequence in which the butyl radical formed abstracts chlorine from Me_2NCl to regenerate the $Me_2N\cdot$ radical[49].

3.4.1.5 *Free radical reactions of α,β-unsaturated carbonyls with trialkylboron. Precursors for saturated aldehydes*

Another interesting avenue opened up through exploitation of the hydroboration reaction by H. C. Brown and his collaborators, is the reaction of α,β-unsaturated carbonyls (e.g. acrolein) with trialkyl-boranes (equation (3.56)). The primary product apparently is the enolborinate (11), which on hydrolysis gave the aldehyde RCH_2CH_2CHO and R_2BOH. This reaction is now found to be inhibited by galvinoxyl, and is probably a radical chain process (equation 3.56)).

$$R\cdot + CH_2{=}CH\ CHO \rightarrow RCH_2\dot{C}HCHO$$

$$\updownarrow$$

$$\underset{(11)}{RCH_2CH{=}CHOBR_2} + R\cdot \xleftarrow[S_H2]{R_3B} R\dot{C}H_2CH{=}CHO\cdot \quad (3.56)$$

Some reactions of this type proceed spontaneously[41], but where this is not so the reaction may be induced photochemically or by introduction of acyl peroxides[50], or by admission of small quantities of oxygen[51] (possibly autoxidation, see Section 3.4.1.2).

3.4.1.6 *Miscellaneous*

Reactions of organoboranes with neutral and alkaline H_2O_2 are now thought to involve radicals, since hydrocarbons are produced[52, 53]. Stereochemical results for epimeric tri-2-norbornyl-boranes, however, suggest a polar process is also operative, since a different epimer composition in the final product is obtained for each different epimeric mixture of reagents[53].

3.4.2 Hydrolysis and alcoholysis of boron substrates

Interest continues in the hydrolysis reactions of amine-boranes and amino-haloboranes with evidence for a number of different mechanisms with (a) attack of H^+ and separation of H^-, (b) B—N fission and (c) B–halogen fission as rate-determining steps. The activation parameters which have appeared in the literature over the past 5 years are given in Table 3.5. The borane systems tend to have unusual kinetic isotope effects, some of which are shown in Table 3.6.

Table 3.5 Activation parameters for some hydrolysis reactions of 4-co-ordinate boranes and haloboranes

Compound	$\Delta H^‡$ (kcal mol^{-1})	$\Delta S^‡$ cal deg^{-1} mol^{-1}	*Solvent*	*Reference*
$NaBH_4$	7.8±0.9	−6±3	H_2O	61
$Et_3N \cdot BH_3$	24.3	+2	50% dioxan	59
quinuclidine-BH_3	25.3	+5.6		
pyridine-BH_3	22.7	−9.9	Pr^nOH	71
2-picoline-BH_3	22.9	−7.5		
3-picoline-BH_3	23.7	−7.9		
4-picoline-BH_3	24.9	−5.1		
2,6-lutidine-BH_3	20.9	−8.3		
$Me_3N \cdot BH_2I$	17.2	−11.9	67% dioxan	55
$Me_3N \cdot BHI_2$	18.8	−14.0		
$Me_3N \cdot BH_2Br$	17.6	−14.5		
$Me_3N \cdot BH_2Cl$	19.8	−19.2		
pyridine-$PhBH_2$	8.9		H_2O	54
pyridine-BCl_3	12		25% EtOH	58
2-picoline-BCl_3	13			
4-picoline-BCl_3	10			

R. E. Davis and co-workers have continued their examination of kinetic isotope effects with the hydrolysis of pyridine adducts of $ArBH_2$ where Ar is a phenyl group with a *para*-substituent[54]. The rate expression is $-d[\text{borane}]/dt = k[H^+][\text{borane}]$, and the substituent effects on rate are correlated with Hammett σ values, the reaction constant ρ being −1.4.

Increased electron density at the boron centre is required for rapid reaction. The isotope effects obtained using $ArBD_2$ analogues are expressed as k_H/k_D in Table 3.6. They *increase* from less than unity as the stretching frequency $\nu_{B-H(D)}$ drops. There is no correlation with B—H(D) bending modes. This suggests (a) a delicately balanced series of electronic effects, (b) a hydridic leaving-group on boron, (c) an attacking proton and (d) a

Table 3.6 Isotope effects in hydrolysis kinetics of some boranes

Compound	*Solvent*	*Hydrolysis rate* $10^3\,k\,(s^{-1})$	k_{H_2O}/k_{D_2O}	k_H/k_D
$Me_3N{\cdot}BH_2I$ [55]	67% dioxan	4.0±0.2 (25 °C)	1.8	—
$Me_3N{\cdot}BH_2I$ [57]	H_2O	4.6 (25 °C)	—	1.24
$Me_3N{\cdot}BHI_2$ [55]	67% dioxan	0.085 (25 °C)	1.8	—
$Me_3N{\cdot}BHI_2$ [55]	67% dioxan	0.071 (25 °C)	—	1.09
$Me_3B{\cdot}BDI_2$ [55]	67% dioxan	0.066 (25 °C)	—	—
$NaBH_4$ [61]	H_2O	—	3.0–4.9	0.7
$NaBH_3CN$ [60]	H_2O	—	0.7	—
pyridine $ArBH_2$ [54] :X=Cl*	H_2O	0.11† (30 °C)	2353‡	0.92
X=H	H_2O	0.244† (30 °C)	2345‡	0.966
X=Me	H_2O	1.05† (30 °C)	2331‡	1.147
X=MeO	H_2O	5.44† (30 °C)	2326‡	1.406

* Ar=× *p*-X—C_6H_4
† at pH 6.86
‡ stretching frequency ν_{B-H} in cm^{-1}

linear $B\ldots\overset{\delta-}{H}\ldots\overset{\delta+}{H}$ configuration as factors in the transition state. The curious isotope effects may be explained if the *secondary* (bond-stiffening) effect is more sensitive to substituent than the *primary* (bond-stretching) effect, so that the primary effect predominates for *p*-MeO, with increasing secondary isotope effect for more electron-withdrawing substituents.

For the haloborane derivatives $Me_3N{\cdot}BH_2Cl(Br,I)$ and $Me_3N{\cdot}BHI_2$, hydrolysis rates in 67% dioxane are independent of $[H^+]$ up to 0.3 M, $[OH^-]$ up to 0.1 M [55]. The order of rates Cl>Br>I reflects ionic bond strengths[56], while the slight isotope effect (Table 3.6) on replacing H by D in the BHI_2 or BH_2I derivative suggests B—H bond-fission is not important[55, 57]. The small solvent isotope effect k_{H_2O}/k_{D_2O}, and the independence of $[OH^-]$, also suggest that the attacking nucleophile is not very important. Accordingly a rate-determining B—X fission, perhaps to give an intermediate boron cation, is suggested.

The boron trichloride adducts of substituted pyridines show a correlation of the lowest activation energy for complex hydrolysis with the greatest B—N bond dissociation energy, so that B—N fission is not operative. This is suggestive of electron donation to the site of reaction, which may then be a rate-determining B—Cl fission[58].

Bridgehead nitrogen has almost no effect on hydrolysis parameters[59] (compare quinuclidine and triethylamine boranes, Table 3.5).

Borane cations[57] are normally inert to hydrolysis, but Miller has observed hydrolysis of the cation containing ethyl *N,N*-dimethylglycinate and trimethylamine as tertiary amines. Hydrogen evolution follows ester hydrolysis and intramolecular cyclisation, with loss of trimethylamine to give an isolatable intermediate

$$\overline{Me_2NCH_2CO_2BH_2}$$
(12)

The rate of hydrolysis of the zwitterion (12) is first-order in concentration of [12] and in $[OH^-]$, second-order overall.

Comparison of the acid-catalysed hydrolysis of cyanoborohydride with early work on borohydride itself results in a more complete statement of the nature of the intermediate[60, 61]. The rate is 10^8-times slower for BH_3CN^- because of the electron-withdrawing CN^- group. Reaction in D_2O (followed by ^1H n.m.r. spectroscopy) indicated an exchange of D with BH_3CN^- (not observed in BH_4^-) at a rate 15-times that of hydrolysis. The intermediate for acid hydrolysis of BH_3CN must then incorporate D from the general acid present in a position equivalent to the leaving hydride. This might be written as in equation (3.57). If the reverse reaction k_- is faster than the hydrolysis

$$CNBH_3^- \underset{k_-}{\overset{D^+ \quad k_+}{\rightleftharpoons}} \left[\begin{array}{c} NC \quad H \\ H{-}B \vdots \\ H \quad D \end{array} \right] \xrightarrow[-H_2]{slow} \text{products} \tag{3.57}$$

step, exchange of D occurs (as an Al hydrolysis). Presumably for BH_4^-, $k_{(hydrolysis)}$ is *faster* than k_-.

$$3HOCHMeCHMeOH + [PhBO]_3 \rightleftharpoons 3PhB\begin{matrix} O{-}CHMe \\ | \\ O{-}CHMe \end{matrix} + 3H_2O \tag{3.58}$$

The reversible formation of cyclic esters of phenylboronic anhydride trimer is a rapid reaction proceeding with B—O rather than C—O bond fission, since no DL → *meso* conversion occurred when the reaction was performed with a mixture of *meso*- and DL-butane-2,3-diol[62] (equation (3.58)).

3.4.3 Nucleophilic substitution on boron by amines

Dissociation mechanisms are frequently proposed (or disposed of) in this kind of reaction for 4-coordinate boron adducts by comparison of activation energy with the thermodynamic bond dissociation energy for the adduct. This practice is to be encouraged provided people use a solution-phase value for ΔH (usually derivable by a thermochemical cycle) to compare with solution-phase activation energies[63].

The trimethylamine–trimethylborane system permits two substitution processes, electrophilic substitution on nitrogen (see Section 3.5.1) and nucleophilic substitution on boron (equations (3.59), (3.60))[64].

In excess trimethylamine the activation energy for the nucleophilic

$$\underset{(a)}{Me_3N \cdot BMe_3} \xrightleftharpoons{S_N1} \underset{(b)}{Me_3N} + \underset{(c)}{BMe_3} \quad (3.59)$$

$$Me_3N^* + BMe_3 \xrightleftharpoons{fast} Me_3N^*BMe_3 \quad (3.60)$$

substitution was *c*. 17 kcal mol^{-1}, (about equal to the gas-phase dissociation energy 17.6 kcal mol^{-1}) and $\Delta S^‡$ was 15 cal deg^{-1} mol^{-1} at -24 °C. The exchange lifetime for (a), estimated from ^{1}H n.m.r. line-shape parameter analysis was independent of concentration, and that for (c) was proportional to [a]/[b]. These facts support a dissociative process (equation (3.59)) as the rate-determining stage.

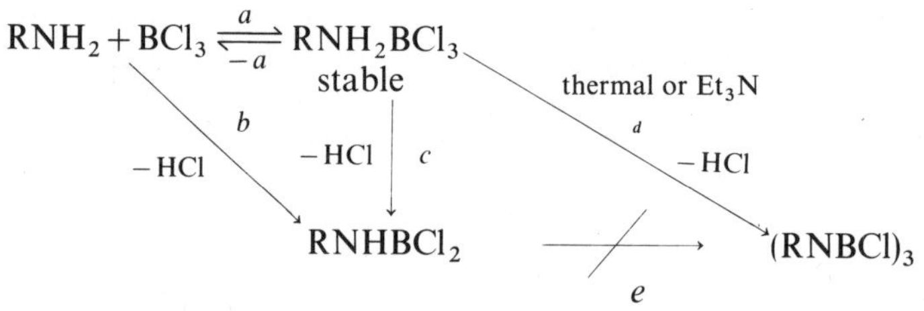

$$RNH_2 + BCl_3 \underset{-a}{\overset{a}{\rightleftharpoons}} \underset{\text{stable}}{RNH_2BCl_3}$$

$RNH_2 + BCl_3 \xrightarrow[-HCl]{b} RNHBCl_2$; $RNH_2BCl_3 \xrightarrow[-HCl]{c} RNHBCl_2$; $RNH_2BCl_3 \xrightarrow[-HCl]{\text{thermal or } Et_3N,\ d} (RNBCl)_3$; $RNHBCl_2 \not\xrightarrow{e} (RNBCl)_3$

$$RNH_2 + MeCN \cdot BX_3 \rightarrow RNHBX_2 + MeCN + HX \quad (3.61)$$

The general study of the reactivity of aromatic amines towards boron halides has made clear three distinct categories of reaction; (i) for *o*-nitro-amines the path *b* in the above scheme leads directly to the substitution product[56, 63], (ii) for other *o*-substituted amines, the path $a+c$ leads to substitution via an intermediate of some stability[65], and (iii) for other aromatic amines the path is $a+d$ and the substitution product $RNHBCl_2$ is never observed[66]. Mechanisms have been suggested for several reactions in category (i). The adducts of acetonitrile with boron trichloride and tribromide undergo substitution processes in acetonitrile with 2,4-dinitroaniline and 2,4-dinitronaphthylamine (equation (3.61))[63]. The activation energy in the bromide reactions was within experimental error the same as the solution value for the dissociation enthalpy of the bromide adduct, while for the chloride reaction it was much less. These facts, and the sign of $\Delta S^‡$ (positive for bromide, but negative for chloride), are consistent with adduct dissociation as the rate-determining step for bromide systems, but association of adduct and nucleophile in the chloride system. The reaction of dinitro-naphthylamine with phenyldichloroborane was thought to be S_N2[56].

3.4.4 Exchange at boron centres with silanes and boranes

The steric course of a number of reactions substituting alkoxy or dialkyl-amino groups on silicon for halogen on boron halides has been studied for optically active silane centres, $R'R''R'''Si^*NR_2$ and $R'R''R'''Si^*OR$[67]. This sometimes provides information on the stereochemistry for the boron reagent also, but not invariably. Rather than tag the boron with a stereochemical label, it should be possible to put a stereochemical label on the linking atom of the departing substituent, but this has not often been done. The polar 4-centre mechanism often postulated in such systems would be expected to lead to *retention* of configuration of silane, but in fact this is not the most common result so far obtained. Inversion of configuration is more common and may be explained by a trigonal-pyramid configuration at

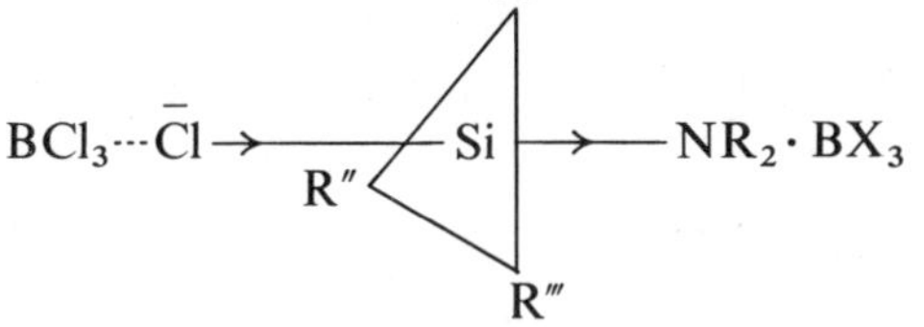

$$BCl_3\cdots\bar{Cl}\rightarrow Si(R')(R'')(R''')\rightarrow NR_2\cdot BX_3$$

inversion
(13)

$$\text{Me, Ph, Np}\ Si\cdots H\cdots BCl_2\cdots Cl\cdots Si$$

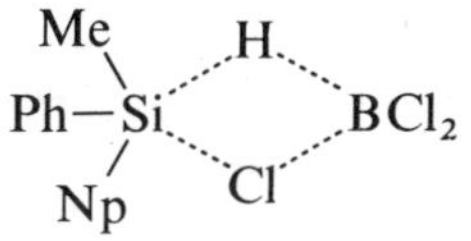

retention
(14)

silicon with entering and leaving groups axial. The course of reaction at boron is not defined for reactions leading to inversion (13). Exchange of methylphenylnaphthylsilyl methoxide[67] or hydride[68] with boron trichloride in non-polar solvents results in retention of configuration at silicon, and points to the 4-centre mechanism (14) which specifies more clearly the role of boron.

Substitution of alkoxy groups from boric or phenylboronic esters on boric esters (redistribution) may be observed by ^1H n.m.r. spectroscopic investigation of, for example, the reversible redistribution reaction, equation (3.62). Reactions are faster for the stronger Lewis acids of the set, and are retarded by donor solvents, all of which is not only consistent with a 4-centre mechanism but also with others[69].

$$B(OPr^i)_3 + B(OMe)_3 \rightleftharpoons B(OPr^i)_2(OMe) + B(OPr^i)(OMe)_2 \qquad (3.62)$$

3.4.5 A photochemical reaction

Mixtures of methyl bromide and borazine on irradiation at wavelength 1849 Å produce monobromoborazine. Neither free bromine atoms nor free methyl radicals are required for the reaction, but CH_3D is produced when deuterated borazine is used with CH_3Br. The reaction must proceed through some excited state of borazine[70].

3.5 NITROGEN, OXYGEN AND FLUORINE

Substitution reactions of nitrogen, oxygen and fluorine are not normally indexed under substitution or under the individual element, but rather under the name of the compound containing them (e.g. ether, amine).

Consequently it is difficult to be comprehensive in this section. The following topics seem relevant to this area.

3.5.1 Substitution on nitrogen

Substitution of excess BM_3 on Me_3NBMe_3 is thought to proceed by a dissociative process (equation (3.63), (3.64)) for much the same reasons as the related nucleophilic substitution on boron (Section 3.4.3). The activation energy was 18.0 kcal mol^{-1}, the activation entropy was positive (17.0 cal deg^{-1} mol^{-1} at -22 °C) and the concentration dependence of the exchange lifetimes is consistent with the scheme:

$$Me_3N{\cdot}BMe_3 \underset{\text{fast}}{\overset{\text{slow}}{\rightleftharpoons}} Me_3N + BMe_3 \tag{3.63}$$

$$^*BMe_3 + Me_3N \rightleftharpoons Me_3N^*BMe_3 \tag{3.64}$$

Although there is a considerable amount of work on exchange of H^+ with amines, alcohols etc., which is technically electrophilic substitution on nitrogen and oxygen, most of this is probably best considered as organic chemistry and is not reviewed here. Typical of this work are the n.m.r. kinetic studies of Grunwald and co-workers[72, 73]. There are some important papers on fundamental reactions such as that depicted in equation (3.65)

$$N'H_3 + [NH_3H]^+ \rightleftharpoons [N'H_3H]^+ + NH_3 \tag{3.65}$$

for which an exchange rate constant of 6.9×10^9 s^{-1} at 25 °C was measured by ^1H n.m.r. spectroscopic methods. ($\Delta H^\ddagger = 1.6 \pm 0.5$ kcal mol^{-1} and $\Delta S^\ddagger = -8 \pm 2$ cal deg^{-1} mol^{-1})[74]. The reaction mixture consisted of ammonium halide in very pure liquid ammonia. Also in liquid ammonia[75], the exchange between water and ammonia is 'slow' on the n.m.r. timescale; there are several possible paths and the overall rate is given by equation (3.66) where k_a is 7.4×10^4 s^{-1}, k_b is 0.84×10^4 s^{-1} M^{-2}, k_c is 3.5×10^6 s^{-1} M^{-1} and $k_d = 0.058$ s^{-1} $M^{-(n-1)}$, at 29.6 °C.

$$\text{rate} = k_a[OH^-] + k_b[OH^-][H_2O]^2 + k_c[NH_4^+][H_2O] + k_d[H_2O]^n \tag{3.66}$$

3.5.2 Some reactions of diatomic molecules

Some predictions were made by Noyes[76] in 1966 of possible pathways and rates of reactions of diatomic molecules. Data are now available to test some of these predictions. For instance, the kinetics of the halogen interchange between fluorine and chlorine for which the rate expression is $d[ClF]/dt = k[F_2][Cl_2]^{\frac{1}{2}}$ has been determined in the range 80–120 °C in the gas phase; k was found to be 4.6×10^8 $e^{-19,800/RT}$ $(l\ mol^{-1})^{\frac{1}{2}}$ s^{-1}. This does not follow Noyes' prediction[77]. It appears to be a chain reaction, as shown in the scheme:

$$\begin{array}{llll} F_2 & \xrightarrow{k_a} 2F\cdot + M & (38\ \text{kcal mol}^{-1}) & \text{initiation} \\ F\cdot + Cl_2 & \xrightarrow{k_b} ClF + Cl\cdot & (-3\ \text{kcal mol}^{-1}) & \\ Cl\cdot + F_2 & \xrightarrow{k_c} ClF + F\cdot & (-23\ \text{kcal mol}^{-1}) & \\ Cl\cdot + F\cdot + M & \xrightarrow{k_d} ClF & (-61\ \text{kcal mol}^{-1}) & \\ F\cdot + F\cdot + M & \xrightarrow{k_e} F_2 & (-38\ \text{kcal mol}^{-1}) & \text{termination} \\ Cl\cdot + Cl\cdot + M & \xrightarrow{k_f} Cl_2 & (-58\ \text{kcal mol}^{-1}) & \end{array}$$

The isotopic exchange of oxygen (equation (3.67)) for which Noyes

$$^{16}O\text{–}^{16}O + ^{18}O\text{–}^{18}O \rightleftarrows 2\,^{16}O\text{–}^{18}O \tag{3.67}$$

predicted an activation energy of 51 kcal mol^{-1} has been examined by Carroll and Bauer[78] in a shock-tube investigation. Two empirical rate expressions were found to fit the exchange data:

$$d[^{34}O_2]/dt = 4 \times 10^{12} \exp[-(41 \pm 4)/RT]\,[^{16}O\text{–}^{16}O][^{18}O\text{–}^{18}O]\,M^{-1}\,s^{-1}$$

or

$$d[^{34}O_2]/dt = 7 \times 10^{10} \exp[-(39 \pm 4)/RT][^{16}O\text{–}^{16}O + ^{18}O\text{–}^{18}O]^2 M^{-1}\,s^{-1}$$

The reaction is second-order with respect to total oxygen, but the activation energy (39–44 kcal mol^{-1}) is too low to agree with Noyes' postulate. The mechanism possibly involves a two-level vibrational excitation and may be a 4-centre process.

3.5.3 Potential energy surfaces

Attention is being focused on the computation of potential energy surfaces for simple reactions, due, no doubt, to the increasing availability of digital computers[79]. Some of these are concerned with reaction of hydride with the molecules HF [80], H_2O [81] and NH_3 [82], which are technically to be considered as displacement reactions on *hydrogen* (e.g. equation (3.68)). The calculations, by the LCAO–MO–SCF method using Gaussian basis functions, indicate that the potential energy surface for such a

$$H^- + HF \rightarrow F^- + H_2 \tag{3.68}$$

reaction has no classical activation *barrier*, but in fact a minimum, indicating a weak complex, e.g. H_2F^-, H_3O^-, NH_4^- in the reaction path. The gas phase reaction of H^- with H_2O occurs at the encounter-controlled rate [81]. The unusual isotope effects which have been observed in practice for some of these reactions show that hydrogen does not transfer in the rate-determining step, which is consistent with these calculations.

References

1. Fischer, H. and Bargon, J. (1969). *Accounts Chem. Res.*, **2,** 110
2. Noyes, R. M. (1966). *J. Amer. Chem. Soc.*, **88,** 4318
3. Ritchie, C. D. and King, H. F. (1968). *J. Amer. Chem. Soc.*, **90,** 825, 833, 838
4. Brown, T. L. (1968). *Accounts Chem. Res.*, **1,** 23
5. Lockhart, J. C. (1970). *Redistribution Reactions* (New York: Academic Press)
6. West, P., Waack, R. and Purmort, J. I. (1970). *J. Amer. Chem. Soc.*, **92,** 840
7. Hartwell, G. E. and Brown, T. L. (1966). *J. Amer. Chem. Soc.*, **88,** 4625
8. Darensbourg, M. Y., Kimura, B., Hartwell, G. E. and Brown, T. L. (1970). *J. Amer. Chem. Soc.*, **92,** 1236
9. Seitz, L. M. and Brown, T. L. (1966). *J. Amer. Chem. Soc.*, **88,** 2174
10. Witanowski, M. and Roberts, J. D. (1966). *J. Amer. Chem. Soc.*, **88,** 737
11. Fraenkel, G., Dix, D. T. and Carlson, M. (1968). *Tetrahedron Lett.*, **5,** 579
12. Seitz, L. M. and Brown, T. L. (1967). *J. Amer. Chem. Soc.*, **89,** 1607
13. Seitz, L. M. and Little, B. F. (1969). *J. Organometal. Chem.*, **18,** 227
14. Seitz, L. M. and Brown, T. L. (1967). *J. Amer. Chem. Soc.*, **89,** 1602
15. Applequist, D. E. and O'Brien, D. F. (1963). *J. Amer. Chem. Soc.*, **85,** 743
16. Winkler, H. J. S. and Winkler, H. (1966). *J. Amer. Chem. Soc.*, **88,** 969
17. *Idem. Ibid.*, 964
18. Waack, R., Doran, M. A. and Baker, E. B. (1967). *Chem. Commun.*, 1291
19. Screttas, C. G. and Eastham, J. F. (1966). *J. Amer. Chem. Soc.*, **88,** 5668
20. Lepley, A. R. (1969). *Chem. Commun.*, 64
21. Ward, H. R., Lawler, R. G. and Cooper, R. A. (1969). *J. Amer. Chem. Soc.*, **91,** 748
22. Lepley, A. R. and Landau, R. L. (1969). *J. Amer. Chem. Soc.*, **91,** 748
23. Lepley, A. R. (1969). *J. Amer. Chem. Soc.*, **91,** 749
24. Ward, H. R., Lawler, R. G. and Loken, H. Y. (1968). *J. Amer. Chem. Soc.*, **90,** 7360
25. Ward, H. R., Lawler, R. G., Loken, H. Y. and Cooper, R. A. (1969). *J. Amer. Chem. Soc.*, **91,** 4928
26. Letsinger, R. L. (1950). *J. Amer. Chem. Soc.*, **72,** 4842
27. Wittig, G. and Schollkopf, U. (1958). *Tetrahedron*, **3,** 91
28. Russell, G. A. and Lamson, D. W. (1969). *J. Amer. Chem. Soc.*, **91,** 3967
29. Shirley, D. A. and Hendrix, J. P. (1968). *J. Organometal. Chem.*, **11,** 217
30. West, P., Waack, R. and Purmort, J. I. (1969). *J. Organometal. Chem.*, **19,** 267
31. Pocker, Y. and Exner, J. H. (1968). *J. Amer. Chem. Soc.*, **90,** 6764
32. Eigen, M. (1963). *Seventh International Conference on Coordination Chemistry*, Plenary Lecture No. 6 (London: Butterworths)
33. Matwiyoff, N. A. and Movius, W. G. (1967). *J. Amer. Chem. Soc.*, **89,** 6077
34. Feeney, J., Haque, R., Reeves, L. W. and Yue, C. P. (1968). *Canad. J. Chem.*, **46,** 1389
35. Baldwin, W. G. and Stranks, D. R. (1968). *Austral. J. Chem.*, **21,** 2161
36. Pearson, R. G. and Moore, J. W. (1966). *Inorg. Chem.*, **5,** 1528
37. Davies, A. G., Griller, D., Roberts, B. P. and Tudor, R. (1970). *Chem. Commun.*, 640
38. Ingold, K. U. (1969). *Chem. Commun.*, 911
39. Lane, C. F. and Brown, H. G. (1970). *J. Amer. Chem. Soc.*, **92,** 7212
40. Davies, A. G., Hook, S. C. W. and Roberts, B. P. (1970). *J. Organometal. Chem.*, **22,** C37
41. Kabalka, G. W., Brown, H. C., Suzuki, A., Honma, S., Arase, A. and Itoh, M. (1970). *J. Amer. Chem. Soc.*, **92,** 710
42. Davies, A. G. and Roberts, B. P. (1967). *J. Chem. Soc. B.*, 17
43. Allies, P. G. and Brindley, P. B. (1967). *Chem. Ind.*, 319
44. Davies, A. G. and Roberts, B. P. (1969). *J. Chem. Soc. B.*, 311
45. Krusic, P. J. and Kochi, J. K. (1969). *J. Amer. Chem. Soc.*, **91,** 3942
46. Davies, A. G. and Roberts, B. P. (1969). *Chem. Commun.*, 699
47. Grotewold, J., Lissi, E. A. and Scaiano, J. C. (1969). *J. Chem. Soc. B.*, 475
48. Grotewold, J., Lissi, E. A. and Scaiano, J. C. (1969). *J. Organometal. Chem.*, **19,** 431
49. Davies, A. G., Hook, S. C. W. and Roberts, B. P. (1970). *J. Organometal. Chem.*, **23,** C11
50. Brown, H. C. and Kabalka, G. W. (1970). *J. Amer. Chem. Soc.*, **92,** 712
51. Brown, H. G. and Kabalka, G. W. (1970). *J. Amer. Chem. Soc.*, **92,** 714
52. Bigley, D. B. and Payling, D. W. (1970). *J. Chem. Soc. B.*, 1811
53. Davies, A. G. and Tudor, R. (1970). *J. Chem. Soc.*, 1815
54. Davis, R. E. and Kenson, R. E. (1967). *J. Amer. Chem. Soc.*, **89,** 1384

55. Lowe, J. R., Uppal, S. S., Weidig, C. and Kelly, H. C. (1970). *Inorg. Chem.,* **9,** 1423
56. Lockhart, J. C. (1966). *J. Chem. Soc. A.,* 809
57. Miller, N. E. (1970). *J. Amer. Chem. Soc.,* **92,** 4564
58. Heaton, G. S. and Riley, P. N. K. (1966). *J. Chem. Soc. A.,* 952
59. Kelly, H. C. and Underwood, J. A. (1969). *Inorg. Chem.,* **8,** 1202
60. Kreevoy, M. M. and Hutchins, J. E. C. (1969). *J. Amer. Chem. Soc.,* **91,** 4329
61. Davis, R. E., Bromels, E. and Kibby, C. L. (1962). *J. Amer. Chem. Soc.,* **84,** 885
62. Lockhart, J. C. (1968). *J. Chem. Soc. A.,* 869
63. Blackborrow, J. R. and Lockhart, J. C. (1968). *J. Chem. Soc. A.,* 3015
64. Cowley, A. H. and Mills, J. L. (1969). *J. Amer. Chem. Soc.,* **91,** 2911
65. Bartlett, R. K., Turner, H. S., Warne, R. J., Young, M. A. and Lawrenson, I. J. (1966). *J. Chem. Soc. A.,* 479
66. Blackborrow, J. R. and Lockhart, J. C. (1969). *J. Chem. Soc. A.,* 816
67. Sommer, L. H., Citron, J. D. and Parker, G. A. (1969). *J. Amer. Chem. Soc.,* **91,** 4729
68. Attridge, C. J., Haszeldine, R. N. and Newlands, M. J. (1966). *Chem. Commun.,* 911
69. Heyes, R. and Lockhart, J. C. (1968). *J. Chem. Soc. A.,* 326
70. Oertel, M. and Porter, R. F. (1970). *Inorg. Chem.,* **9,** 904
71. Ryschkewitsch, G. E. and Birnbaum, E. R. (1965). *Inorg. Chem.,* **4,** 575
72. Grunwald, E. and Ku, A. Y. (1968). *J. Amer. Chem. Soc.,* **90,** 29
73. Grunwald, E., Lipnick, R. L. and Ralph, E. K. (1969). *J. Amer. Chem. Soc.,* **91,** 4333
74. Clutter, D. R. and Swift, T. J. (1968). *J. Amer. Chem. Soc.,* **90,** 601
75. Alei, M. and Florin, A. E. (1969). *J. Phys. Chem.,* **73,** 857, 863
76. Noyes, R. M. (1966). *J. Amer. Chem. Soc.,* **88,** 4311, 4318
77. Fletcher, E. A. and Dahneke, B. E. (1969). *J. Amer. Chem. Soc.,* **91,** 1603
78. Carroll, H. F. and Bauer, S. H. (1969). *J. Amer. Chem. Soc.,* **91,** 7727
79. Kraus, M. (1970). *Amer. Rev. Phys. Chem.,* **21,** 39
80. Ritchie, C. D. and King, H. F. (1968). *J. Amer. Chem. Soc.,* **90,** 825, 833 and 838
81. Paulson, J. F. (1966). *Ion-Molecule Reactions in the Gas Phase,* Advances in Chemistry Series No. 58, Ch. 3, (Washington D.C.: Amer. Chem. Soc.)

4
Oxidative Addition

A. J. DEEMING
University College, University of London

4.1 INTRODUCTION

Most work to date on the organo-chemistry of the transition metals has been concerned with the preparation of complexes of widely varying types and with their structural elucidation, which in itself has been an interesting and challenging aspect of recent chemistry. With this approach to the field, oxidative addition has provided a convenient preparative route to complexes with many different and often novel types of organic or inorganic ligands. Mechanistic studies of the reactions of organometallic complexes, however, have largely been confined to ligand substitutions and to sorting out the reaction sequences in homogeneous catalytic processes. Many such processes have been shown to involve oxidative addition and reductive elimination steps which are, of course, the subjects of this chapter, but generally, however, kinetic studies of such catalytic processes have provided only information about the basic sequence of steps in a process and little about the oxidative additions themselves. For instance, in the well-known hydrogenation of olefins catalysed by [$RhCl(PPh_3)_3$] the rate-determining step is not the addition of H_2 to the rhodium(I) complex but is believed to be a subsequent substitution whereby the olefin enters the coordination shell[1]. Similarly, using [$IrCl(CO)(PPh_3)_2$] as the hydrogenation catalyst the kinetics of the oxidative addition of H_2 cannot be determined from the overall kinetics[2,3].

Most of the early ideas on the mechanism of oxidative addition were based on structural studies on the adducts, for example, the *cis*-addition of H_2 supported a one-step addition and such a mechanism was sometimes tacitly assumed for other addendum molecules. However, since Chock and Halpern[4] reported their work in 1966 on the kinetics of the addition of H_2, O_2 and MeI to complexes of the type [$IrX(CO)(PPh_3)_2$] (X = halide) several more mechanistic studies have been reported. Nevertheless, mechanistic studies in this field are embryonic and there are still many unsettled issues, such as whether or under what conditions alkyl halide addition to square-planar d^8-metal complexes is a one- or a two-step process, and for the majority of oxidative addition reactions the mechanisms are still unknown.

Several reviews on oxidative addition[5–8] or aspects of it[9–12] have appeared, the best general one being that of Collman and Roper[5].

There have been attempts to classify the reactivity of metal complexes[14,15] in the way that the reactivity of carbon has been classified in organic chemistry. Indeed, metals appear to show many of the distinct behaviour patterns that carbon also exhibits in its various states. For instance, the oxidative addition reactions of d^8- and d^{10}-complexes have been likened to the addition reactions of carbenes and pursuing the analogy d^7-complexes could be likened to alkyl radicals in their ability to convert to diamagnetic species by abstraction of radicals. Such analogies may be useful for the generation of ideas but are probably only superficial and, as Ugo *et al.* have pointed out[13], carbenes act as electrophiles adding preferentially to olefins with electron-donating substituents, whereas nucleophilic behaviour is the main characteristic of the reactivity of d^8- and d^{10}-complexes.

A brief outline of possible types of oxidative addition will now be given before examining the mechanisms observed for different addendum molecules in detail.

4.1.1 Oxidation of d^8- to d^6-complexes

It is Vaska's compound $[IrCl(CO)(PPh_3)_2]$ that most readily comes to mind in this context, and oxidative addition of molecules XY to complexes of the type $[ML_4]$ may occur by one of the two limiting mechanisms shown in Figure 4.1, resulting in *cis*- or *trans*-adducts of the type $[ML_4XY]$. If XY is H_2, the process is of the one-step type but for other molecules the problem is largely open. A two-electron oxidation of a coordinatively saturated molecule on the other hand only allows the metal to expand its coordination shell by one, and so for 5-coordinate complexes preliminary dissociation precedes the oxidative addition or an ionic mechanism takes place without dissociation, the ionic species being the final product in certain cases (see Figure 4.1).

4.1.2 Oxidation of d^{10}- to d^8-complexes

The relationship between species existing in solutions of d^{10}-complexes, notably complexes of Ni^0, Pd^0 and Pt^0, is not completely understood, but it has been shown that equilibria can occur between the coordinatively saturated complex ML_4 and species formed by dissociation of L:

$$ML_4 \rightleftharpoons ML_3 \rightleftharpoons ML_2$$

It appears that $Ni[P(OEt)_3]_4$ is barely dissociated in solution[16] whereas the main species formed on dissolving $Pt(PPh_3)_4$ is the complex $Pt(PPh_3)_3$[17], although solutions containing mainly $Pt(PPh_3)_2$ may also be obtained by dissolving $[Pt(PPh_3)_2(C_2H_4)]$, for example, and flushing out the ethylene from the system[17]. Similar solutions can be prepared by irradiating $[Pt^{II}$ $(oxalato)(PPh_3)_2]$ to give $Pt(PPh_3)_2$ and carbon dioxide[19]. Dicoordinate nickel compounds have been prepared *in situ* by decomposing the nitrogen

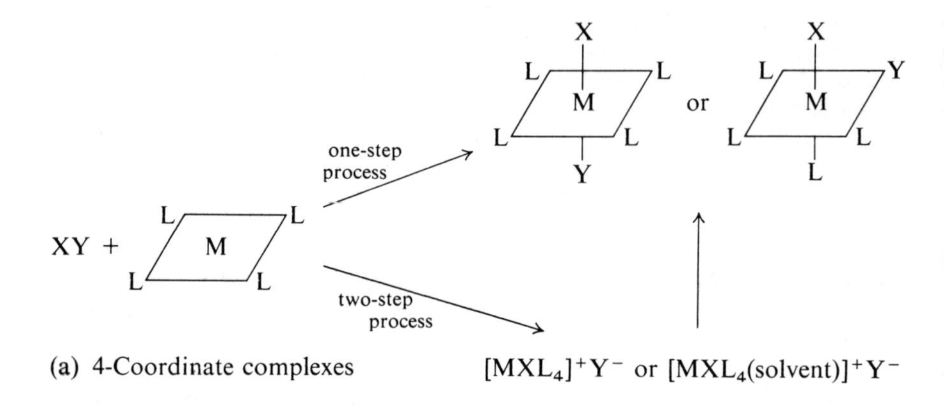

(a) 4-Coordinate complexes

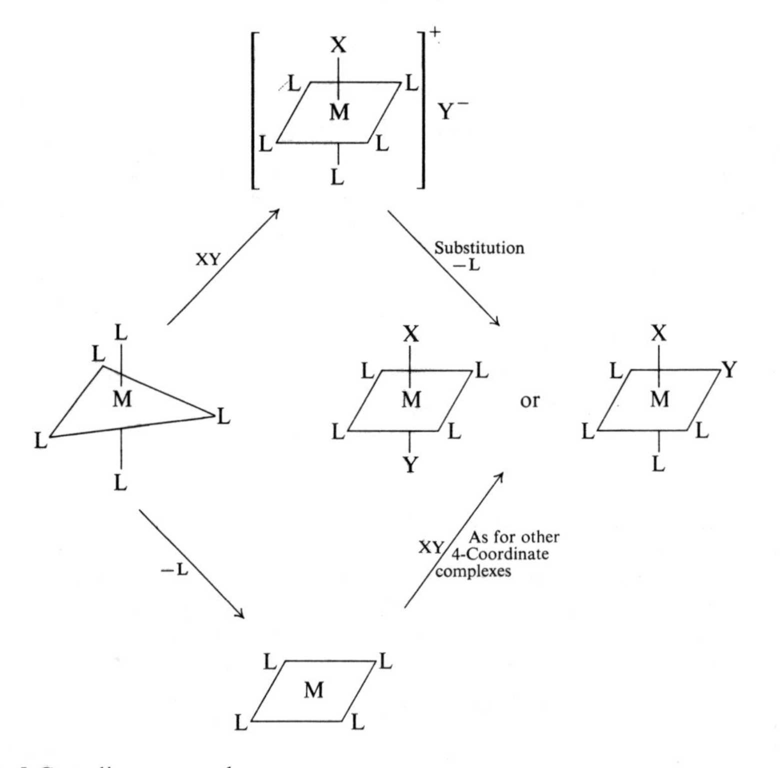

(b) 5-Coordinate complexes

Figure 4.1 Oxidative addition to d^8-metal complexes

complex $\{[P(C_6H_{11})_3]_2Ni\}_2N_2$[18]. Nevertheless, even though a certain species may predominate under certain conditions in solution, it is sometimes difficult to establish the actual reactive species in a particular reaction. This complication has made an examination of oxidative addition difficult, but it is probable that both one- and two-step mechanisms occur as for d^8-complexes.

4.1.3 Oxidation of d^7- to d^6-complexes

These reactions go largely by free-radical mechanisms as shown below, with the initial step being a free-radical abstraction from the addendum molecule by the d^7-complex; although a different mechanism has been proposed for the addition of hydrogen (see Section 4.7.3).

$$\underset{d^7}{ML_5} + XY \rightarrow \underset{d^6}{ML_5X} + Y^{\bullet}$$

$$\underset{d^7}{ML_5} + Y^{\bullet} \rightarrow \underset{d^6}{ML_5Y}$$

4.2 ADDITION OF HALOGENS

5-Coordinate (saturated) d^8-complexes often undergo halogenation with the expulsion of one of the ligands originally coordinated. An example of this is the loss of carbon monoxide from $[Os(CO)_3(PPh_3)_2]$ on reaction with halogens[20], and confirmation that a preliminary dissociation of CO is not the first step in the reaction was the isolation of an ionic intermediate which then gave the final product by ligand substitution:

$$X_2 + [Os(CO)_3(PPh_3)_2] \rightarrow [OsX(CO)_3(PPh_3)_2]^+X^- \rightarrow [OsX_2(CO)_2(PPh_3)_2] + CO$$

Similarly, in the reaction of bromine with iron pentacarbonyl to give $[FeBr_2(CO)_4]$ two intermediates have been observed in the i.r. spectra of solutions at low temperatures[21]. The following is an interpretation of spectral changes observed on raising the temperature from -80 °C.

$$[Fe(CO)_5] + Br_2 \rightarrow [FeBr(CO)_5]^+Br^- \rightarrow [FeBr(COBr)(CO)_4] \rightarrow [FeBr_2(CO)_4] + CO$$

In this case the substitution reaction is preceded by nucleophilic attack of bromide ion at a carbonyl rather than at the metal centre as is more usual, although attack at the carbon atom of a carbonyl by other nucleophiles is well established, for example see Reference 22. So the actual oxidation of the metal in these two examples occurs during the reaction: $M + X_2 \rightarrow MX^+ + X^-$. Since the kinetics of this type of reaction remain to be studied, one can only speculate at this stage concerning the mechanism of this step. The problem is analogous to that of the formation of bromonium ions during the bromination of olefins and the reaction may be imagined to go in one step with a transition state of type δ^+M----X----$X\delta^-$ or by the formation of the Lewis acid-base adduct M→X—X as an intermediate prior to heterolysis of the X_2 molecule. Such problems are still to be resolved.

With square-planar d^8-complexes there is the further possibility of a one-step three-centred transition state for the addition of halogen since there are now two available coordination sites. However, there is only evidence available at the moment for a mechanism analogous to that described above, that is, oxidation of the metal associated with heterolysis of the halogen molecule as in the example below:

$$[Pt^{II}(CN)_4]^{2-} + Br_2 \rightleftharpoons trans\text{-}[Pt^{IV}(CN)_4Br(H_2O)]^- + Br^-$$

$$trans\text{-}[Pt^{IV}(CN)_4Br(H_2O)]^- + Br^- \xrightarrow{\text{R.D.S.}} trans\text{-}[Pt^{IV}(CN)_4Br_2]^{2-}$$

Although the mechanism of the first step is unknown, these workers[23] favour the rapid formation of an intermediate adduct with a bromine molecule before any significant cleavage of the Br—Br bond.

The stereochemistry of addition of X_2 to square-planar complexes might have bearing on the mechanism of addition but, although both *cis*-[25] and *trans*-additions[24] have been reported, in no case has the stereochemistry of addition been proved to be kinetically controlled.

Halogens may also be added oxidatively to complexes containing olefins, which may or may not be coordinated, with one of the halogen atoms adding to the olefin to give a metal–carbon σ-bond. For example[26, 27]

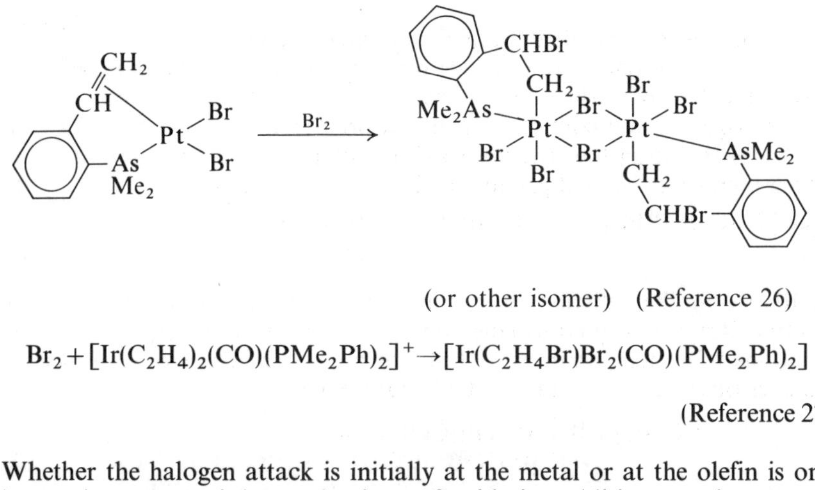

(or other isomer) (Reference 26)

$$Br_2 + [Ir(C_2H_4)_2(CO)(PMe_2Ph)_2]^+ \rightarrow [Ir(C_2H_4Br)Br_2(CO)(PMe_2Ph)_2]$$

(Reference 27)

Whether the halogen attack is initially at the metal or at the olefin is one interesting aspect of the mechanisms of oxidative addition reactions as yet to be resolved.

4.3 ADDITION OF METAL HALIDES

Many halides of transition or post-transition metals have been added oxidatively to d^8- or d^{10}-complexes but virtually nothing is known of the mechanisms of these additions. One can only argue by analogy. For example, mercuric halide has been shown to give Lewis acid-base adducts with transition-metal bases, and the structure of $[(\pi\text{-}C_5H_5)(CO)_2Co \rightarrow HgCl_2]$ has been confirmed by an x-ray study[28], and it is quite possible that the formation of such an adduct is the first step in the mercuric halide addition to $[Os(CO)_3(PPh_3)_2]$[29], a reaction which gives as its final product $[Os(HgX)(CO)_3(PPh_3)_2]^+HgX_3^-$. This species is analogous to the intermediate formed in the addition of X_2 to the same osmium compound[20], but here there is no subsequent substitution of carbon monoxide probably because of the formation of HgX_3^- which has no non-bonding electrons on the mercury available for coordination to the osmium. Analogous species are probably intermediates in the reaction of HgX_2 ($X = Cl$ or I) with $[Ir(CO)_2(PMe_2Ph)_3]^+$[30]. The doubly-charged cationic intermediate $[Ir(HgX)(CO)_2(PMe_2Ph)_3]^{2+}$ undergoes attack by X^- ($X = I$) or by MeO^- ($X = Cl$, solvent = methanol) to give the species $[Ir(HgI)I(CO)(PMe_2Ph)_3]^+$ or $[Ir(HgCl)(CO_2Me)(CO)(PMe_2Ph)_3]^+$ respectively which were isolated.

4.4 ADDITION OF ALKYL HALIDES

4.4.1 Some possible mechanisms

Ionic mechanisms are not only feasible but also quite likely in the oxidative addition of alkyl halides and a variety of mechanisms may be envisaged. Reactions can involve transition states of either type (A) or (B), that is either (A) three-centred transition states of low polarity in which both addendum atoms interact with the metal simultaneously, or (B) dipolar transition states very similar indeed to those accepted for classical S_N2 reactions of alkyl halides with non-metallic nucleophiles.

$$M\overset{R}{\underset{X}{\vdots}} \qquad\qquad \delta^+\, M\cdots R\cdots X\, \delta^-$$

(A) (B)

There is no reported evidence for free-radical mechanisms in the case of two-electron oxidations of the metal; however, one-electron oxidations of Co^{II} complexes with alkyl halides to give Co^{III} alkyl compounds have been shown to involve alkyl radicals and these reactions will be discussed later. Although there are only two basic mechanistic types to be considered there are several possible modifications. For example, a transition state of type (B) may result in complete ionisation with the liberation of free halide ion into solution or an ion-pair may be formed which rapidly recombines. Alternatively, the complex may react with an alkyl halide which is coordinated to another metal through the halogen atom, a type of coordination shown to be possible[61]. Nevertheless, all mechanisms so far proposed are either of type (A) or (B) and the main criteria that have been used in attempts to distinguish them are as follows:

(a) Type (A) should involve retention of configuration at the carbon atom, while for (B) inversion should be observed.

(b) The values of activation parameters of type (B) reactions and their solvent-dependence should be characteristic of reactions that involve molecules of low polarity (for neutral complexes) reacting to give highly polar transition states, for example, as in the Menschutkin reaction ($R_3N + RX \rightarrow R_4N^+X^-$).

(c) It may be possible to detect ionic intermediates, for example free halide generated from the alkyl halide, which would support mechanism (B).

4.4.2—Coordinatively saturated d⁸-complexes

The oxidative addition of alkyl halides to 5-coordinate (coordinatively saturated) d^8-complexes will be considered first because the situation is fairly clear with only evidence for mechanisms of type (B). In the reactions of alkyl halides with coordinatively saturated complexes, such as those being discussed here, processes of type (A) would involve 7-coordinate intermediates with 20 valence electrons (two in excess of those allowed by the inert gas electron configuration rule) and it is possible to appreciate why only mechanisms of type (B) have been reported. As mentioned above, free-radical mechanisms involving an initial one-electron oxidation to give a d^7-metal intermediate might be envisaged, but no such intermediates have been observed nor their presence inferred, and such mechanisms probably do not occur.

4.4.2.1 π-cyclopentadienyl complexes

The reaction of RX with the coordinatively saturated complex $[(\pi\text{-}C_5H_5)Ir(CO)(PPh_3)]$ has been shown to give the ionic species $[(\pi\text{-}C_5H_5)IrR(CO)(PPh_3)]X$ [31], whereas the corresponding reactions for cobalt or rhodium[32] gave the neutral acyl complexes $[(\pi\text{-}C_5H_5)MX(COR)(PPh_3)]$.

$$[(\pi\text{-}C_5H_5)M^{I}(CO)(PPh_3)] \xrightarrow[\text{Oxidative addition}]{\text{MeI}} [(\pi\text{-}C_5H_5)M^{III}(CO)(Me)(PPh_3)]^+ I^-$$

$$\xrightarrow{\text{Carbonyl insertion, For M = Co,Rh}} [(\pi\text{-}C_5H_5)M^{III}(COMe)(I)(PPh_3)]$$

M = Co, Rh, Ir

It has been shown that the preliminary reaction is the same for each metal[33]. There is a second-order reaction to give halide ion and a cationic alkyl complex which is stable for iridium, but for cobalt or rhodium undergoes rapid attack by halide ion with migration of the alkyl group on to the carbonyl (i.e. carbonyl insertion) to give the neutral acyl complexes which were isolated.

The proposed cationic alkyl–rhodium intermediate could be isolated by addition of $NaBPh_4$ to a methanol solution of $[(\pi\text{-}C_5H_5)RhBr(COMe)(PMe_2Ph)]$, and as expected the complex $[(\pi\text{-}C_5H_5)RhMe(CO)(PMe_2Ph)]^+$ which was precipitated reacts rapidly with halide ion to regenerate the acyl complex. This provided evidence that such methyl complexes could indeed be the reaction intermediates.

The kinetics of the initial oxidative addition of MeI were measured in dichloromethane and gave activation parameters as given in Table 4.1.

Table 4.1 Activation parameters for reactions with MeI in dichloromethane[33]

Reacting complex	$\Delta H^\ddagger$ (kcal mol^{-1})	$\Delta S^\ddagger$ (cal deg^{-1} mol^{-1})
$(C_5H_5)Co(CO)(PPh_3)$	12 ± 1	-31 ± 3
$(C_5H_5)Rh(CO)(PPh_3)$	10 ± 1	-36 ± 3
$(C_5H_5)Ir(CO)(PPh_3)$	$\sim(8–9)$	$\sim -(35–40)$

The quite large negative entropies of activation and the considerable solvent-dependence of the reaction rate are consistent with the expected highly polar transition state. The reaction is accelerated in methyl cyanide and decelerated in toluene relative to the rate in dichloromethane, although activation parameters were only reported for the reaction in the latter solvent. Ethyl iodide reacts 400–1200 times more slowly than methyl iodide (probably for steric reasons) and the reaction rate is accelerated on replacement of PPh_3 by other tertiary phosphines in the order $PPh_3 < PMePh_2 < PMe_2Ph$, which is the same order as increasing basicity of the ligand. However, the deceleration of the reaction on replacement of PPh_3 by $P(cyclohexyl)_3$, the most basic ligand used, shows that steric effects as well as electronic must be of importance.

In these reactions the expected inversion of the carbon atom which undergoes nucleophilic attack has not yet been demonstrated.

The reaction of perfluoroethyl iodide may also have a similar mechanism but, with this electronegative alkyl, nucleophilic attack may occur at the iodine atom.

$$[(C_5H_5)Rh(CO)_2] + CF_3CF_2I \rightarrow [(C_5H_5)RhI(CF_2CF_3)(CO)] + CO$$

In this case no CO insertion occurs probably due to the strength of the CF_3CF_2—Rh bond[35]. With allylic halides on the other hand[34] the initially formed σ-allyl complexes could readily convert to π-allylic complexes with loss of CO and the products obtained can be rationalised by an S_N2 or S_N2' mechanism.

$$(C_5H_5)Co(CO)_2 + CH_2{=}CH{-}CH_2I \rightarrow [(C_5H_5)Co(CO)(\pi\text{-}CH_2CHCH_2)]^+ I^- + (C_5H_5)CoI(\pi\text{-}CH_2CHCH_2)$$

4.4.2.2 *Vitamin B_{12S} and its model compounds*

S_N2 mechanisms have also been established[36, 37] in the related reactions of alkyl halides with vitamin B_{12S} (i.e. the cobalt(I) form of vitamin B_{12}), cobaloxime(I) and other models of vitamin B_{12S}. (For general introduction to the chemistry of model compounds of vitamin B_{12} see Reference 38). Schrauzer has shown a very similar kinetic behaviour between these reactions and those reactions known to proceed by S_N2 mechanisms. Although the cobalt(I) nucleophiles used by Schrauzer are among the most powerful known, the relative changes in the kinetics on varying both R and X of the alkyl halide, RX, are very similar to those observed with other nucleophiles. For instance, the rates are in the

$$\underset{\substack{|\\ \mathrm{L}}}{(\mathrm{Co^I})^-} + \mathrm{R{-}X} \rightarrow \underset{\substack{|\\ \mathrm{L}}}{\overset{\substack{\mathrm{R}\\ |}}{(\mathrm{Co^{III}})}} + \mathrm{X^-}$$

order, X = I > Br > Cl but are far less sensitive to the change of halide than are free-radical reactions. The rate of the reaction also increases with an increase in the σ-basicity and a decrease in the π-acidity of the axial ligand L and this is consistent with the metal acting as a nucleophile. The $\Delta S^{\ddagger}$–values are in the range -20 to -30 cal $\deg^{-1}$ mol^{-1} and this can be related to an increased order in the transition state possibly due to an increase in coordination number of five to six. The small solvent-dependence of the kinetics is as expected for a reaction between an ion and a neutral molecule. The S_N2 nature of this reaction was finally confirmed by Jensen *et al.*[39] who showed that the reaction of substituted cyclohexyl halides with cobaloxime(I) led to an inversion at the carbon atom. Stereospecificity has been similarly demonstrated in the reaction of *sec*-octyl bromide with cobaloxime(I) complexes[40].

The reaction has also been reported between propargyl halide, $CH{\equiv}CCH_2X$, and cobaloxime(I) and shown to give an allenyl–Co^{III} complex[42, 43] and in this case the apparent reaction mechanism is S_N2'.

$$\underset{\substack{|\\ \mathrm{L}}}{(\mathrm{Co^I})^-} + \mathrm{CH{\equiv}C{-}CH_2X} \longrightarrow \underset{\substack{|\\ \mathrm{L}}}{\overset{\substack{\mathrm{CH{=}C{=}CH_2}\\ |}}{(\mathrm{Co^{III}})}} \qquad + \mathrm{X^-}$$

There are several examples of substitution by non-metallic nucleophiles where non-concerted processes occur and yet the reactions apparently have S_N2' mechanisms[41]. This may also be the case with metallic nucleophiles. Allenyl complexes are only obtained from the reaction with terminal acetylenes, whereas the disubstituted acetylene $MeC{\equiv}CCH_2Cl$ gives the $MeC{\equiv}CCH_2$–Co^{III} complex[42], probably by the normal S_N2 mechanism described above for saturated alkyl halides.

Nucleophilic substitution of vinylic halides by cobaloxime(I) has been shown to occur with retention of the configuration of the olefin[44, 46].

Two mechanisms have been proposed for the nucleophilic substitution of

$$(Co^{I})^{-}(Py) \xrightarrow{\text{cis-HBrC=CHPh}} (Co^{III})(Py)\text{-CH=CHPh (cis)} + Br^{-}$$

$$(Co^{I})^{-}(Py) \xrightarrow{\text{trans-HBrC=CHPh}} (Co^{III})(Py)\text{-CH=CHPh (trans)} + Br^{-}$$

vinylic halides: for nucleophiles B^- (e.g. OMe^-) which are strong bases there is an initial abstraction of HX from RCH=CHX by B^- to give an acetylene and BH, followed by *trans*-addition of BH across the triple bond; while for softer nucleophiles of lower basicity (e.g. SMe^-) an addition–elimination process occurs with retention of configuration of the olefin, and it is the second mechanism that appears to be operative with cobalt(I) nucleophiles, for example:

$$(Co^{I})^{-}(Py) + \text{H(Br)C=C(Ph)H} \longrightarrow \text{Br-CH}(Co^{III}Py)\text{-}\bar{C}\text{HPh} \longrightarrow \text{H}(Co^{III}Py)\text{C=C(Ph)H} + Br^{-}$$

The reactions of cobaloxime(I) with phenylacetylene are pH dependent[44–46]. In alkaline solution (pH 11) the product obtained is the β-styryl compound resulting from *trans*-addition across the triple bond, while *cis*-addition to the acetylene (demonstrated by deuterium studies[45]) occurs in neutral solution (pH 8) to give the α-styryl compound. It has been shown that in neutral medium a cobalt hydrido-complex is the reactive species[45].

Analogous rhodoxime(I) complexes react very similarly with alkyl halides but rather more slowly than the corresponding cobaloxime(I) complexes[47]; since their reaction profiles are very similar (i.e. the relative changes of rate with different alkyl halides) presumably the rhodium(I) complexes also react as nucleophiles by an S_N2 reaction mechanism. Once the alkyl rhodium(III) complexes are formed they are considerably more stable than the corresponding cobalt complexes, and this probably accounts for the metabolic inhibition by such rhodium complexes.

Pyridinecobaloxime (I) —($PhC{\equiv}CH$, pH 11)→ $(Co^{III})(Py)$–CH=CH–Ph (H and Ph on the two carbons, i.e. (Co^{III}) and Ph on opposite carbons)

Pyridinecobaloxime (I) —(pH 8)→ $(Co^{III})(Py)$–C(Ph)=CH_2

Major products

It has not been possible to make a further comparison with the reactivity of iridium complexes because of the difficulty in reducing iridium(III) complexes to the corresponding iridoxime(I) complexes[47]. Similar reactions have also been demonstrated[48] between alkyl halides and neutral 5-coordinate cobalt(I) complexes, for example [Co(I)(CR)Br] (CR = a neutral tetradentate nitrogen ligand coordinating in a planar manner) reacts with RX to give cationic complexes of the type $[RCo(CR)Br]^+$ and the mechanism is very likely the same as for anionic cobalt(I) complexes.

The reactions of cobalt(I) anions with alkyl halides are specific examples of the general reactions of metallate anions with alkyl halides. S_N2 mechanisms have been demonstrated by the inversion of configuration at the carbon atom in the substitution of alkyl halides by $[Mn(CO)_5]^-$ [50] or by $[Fe(CO)_2(\pi\text{-}C_5H_5)]^-$ [49, 50]. For instance, $[Fe(CO)_2(\pi\text{-}C_5H_5)]^-$ reacts to give the alkyl complex $[Fe(CHDCHDCMe_3)(CO)_2(C_5H_5)]$ with inversion of configuration[49]. It is possible that similar mechanisms occur in all such reactions when the anion is coordinatively saturated:

$$[L_nM]^- + R\text{—}X \rightarrow L_nM\text{—}R + X^-$$

A semi-quantitative order of nucleophilicities based on the rates of reaction with alkyl halides relative to $[Co(CO)_4]^-$ having a nucleophilicity of unity has been given[51], $[(C_5H_5)Fe(CO)_2]^-$, 7.0×10^7; $[(C_5H_5)Ru(CO)_2]^-$, 7.5×10^6; $[(C_5H_5)Ni(CO)]^-$, 5.5×10^6; $[Re(CO)_5]^-$, 2.5×10^4; $[(C_5H_5)W(CO)_3]^-$, ~500; $[Mn(CO)_5]^-$, 77; $[(C_5H_5)Mo(CO)_3]^-$ 67; $[(C_5H_5)Cr(CO)_3]^-$, 4; $[Co(CO)_4]^-$, 1.

4.4.3 Coordinatively unsaturated d^8-complexes

For the oxidative addition of alkyl halides to such complexes (notably of Rh^I or Ir^I) evidence has been provided in particular cases for transition states of type (A), while in others evidence points towards type (B) see p. 7. It may turn out that further work will resolve the conflicting evidence and all systems will show similar behaviour, but at the moment it appears that different mechanisms are operative in different cases. The chemical factors deciding which one occurs are far from clear.

4.4.3.1 cis- *or* trans-*Addition*

It has been shown that the addition of MeX to $[IrY(CO)L_2]$ (X and Y are halides, L = tertiary phosphines) to give $[IrMeXY(CO)L_2]$ (both in solution[24, 52] and for the reaction of gaseous MeX with crystalline complex[53]) is *trans*, and that these *trans*-additions are kinetically controlled. There is one report of *cis*-addition of MeI to $[IrCl(CO)(PPh_3)_2]$ [25] but later work[53] has suggested that the addition is also *trans* in this case. *Trans*-addition of acetyl halide to $[IrX(CO)(PMe_2Ph)_2]$ has also been reported[52].

However, the addition can be solvent-dependent. Whereas in benzene solution the addition of MeI to $[IrCl(CO)(PMe_2Ph)_2]$ is stereospecifically *trans*, in the better-ionising solvent methanol a mixture is obtained[52]:

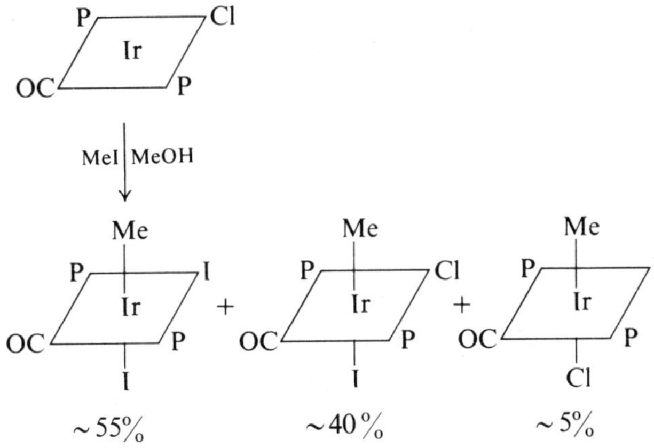

P, Cl, OC, P around Ir

MeI | MeOH

Me / P, I, OC, P around Ir / I — ~55% + Me / P, Cl, OC, P around Ir / I — ~40% + Me / P, I, OC, P around Ir / Cl — ~5%

The halide-exchange reaction must occur during the oxidative addition process since $[IrMeICl(CO)(PMe_2Ph)_2]$ (obtained by stereospecific *trans*-addition of MeI in benzene) can be recovered unchanged from refluxing methanol, whereas the addition reaction was carried out at room temperature. To explain this result it was suggested[52] that a reaction occurs via a transition state of type (B) to give a 5-coordinate or 6-coordinate solvent-containing cationic alkyl–iridium(III) complex and iodide ion, the liberated iodide rapidly substituting the chloride of unreacted starting material as would be expected[54]. This substitution would lead to the mixed products. In benzene solution either the ions recombine via an ion-pair before separation occurs, or possibly there is a different mechanism. There is evidence to suggest that such a combination of ions would give the *trans*-product as observed since attack by nucleophile on 5-coordinate iridium(III) complex $[IrHCl_2(PPh_3)_2]$ occurs at the most electropositive side of the complex, that is, *trans* to the ligand most strongly σ-bonded to the metal (that is, having greatest *trans*-influence), which is H in this example[55].

Thus, in the case of methyl halide addition the solvolysed halide would attack the iridium *trans* to the methyl to give overall *trans*-addition.

Trans-addition is not of itself evidence for a two-step process since as Pearson and Muir[53] have pointed out orbital symmetry considerations allow for either *cis*- or *trans*-addition to occur by one-step processes. Indeed

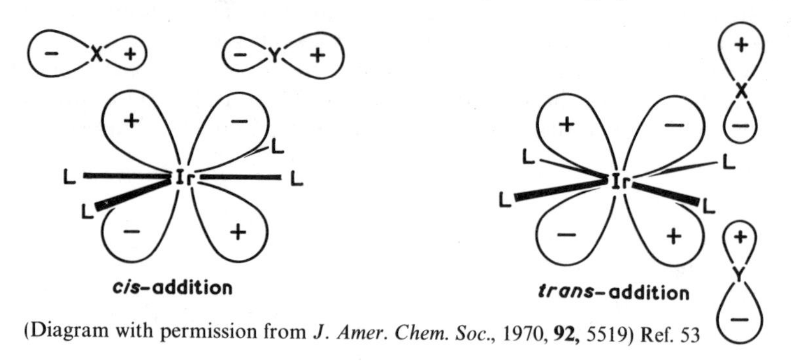

(Diagram with permission from *J. Amer. Chem. Soc.*, 1970, **92,** 5519) Ref. 53

the *trans*-addition of gaseous MeI to crystalline $[IrCl(CO)L_2]$ (L = PPh_3 or $PMePh_2$) is unlikely to go via an ionic mechanism[53]. Also there is one curious observation suggesting that halide ion is not liberated even in solution[57]. Addition of MeI to $[IrI(CO)(PPh_3)_2]$ in the presence of $^{137}I^-$ leads to negligible incorporation of labelled iodine into the adduct $[IrMeI_2(CO)(PPh_3)_2]$ even though rapid equilibration between free and coordinated iodide might be expected for the iridium(I) complex[54] before oxidation.

4.4.3.2 *Configuration changes at carbon*

Both retention[53] and inversion[60] of the alkyl group have been claimed for the addition of alkyl bromide to complexes of the type $[IrCl(CO)L_2]$ (L = tertiary phosphine) in dichloromethane. Optically active $CH_3CHBrCO_2Et$ has been added to $[IrCl(CO)(PMePh_2)_2]$ to give an adduct which retained optical activity. Cleavage of the Ir—C bond by Br_2 regenerated $CH_3CHBrCO_2Et$ with retained configuration[53]. Since it has been shown[58, 59] that alkyl–metal bonds are cleaved by Br_2 with retention, the configuration must also have been retained during the oxidative addition. However, an opposite result has been reported for the reaction of *trans*-2-fluorocyclohexyl bromide with $[IrCl(CO)(PMe_3)_2]$ for which inversion was demonstrated[60].

4.4.3.3 *Kinetic behaviour*

Chock and Halpern have shown that the addition of MeI to $[IrX(CO)(PPh_3)_2]$ (X = Cl, Br or I) is a second-order process, first-order in both complex and methyl iodide[4]. The dependence of the rate on the halide ligand is in the order $IrCl(CO)(PPh_3)_2$ > $IrBr(CO)(PPh_3)_2$ > $IrI(CO)(PPh_3)_2$ and only MeI (not MeBr or MeCl) can be added to these complexes. With more basic ligands (e.g. PMe_2Ph [52] or $PMePh_2$ [24]) the other two halides can also be added.

The reaction parameters obtained by Chock and Halpern are in the range $\Delta H^{\ddagger} = 5.6$ to 8.8 kcal mol^{-1} and $\Delta S^{\ddagger} = -43$ to -51 cal deg^{-1} mol^{-1} (in benzene). Such large negative $\Delta S^{\ddagger}$-values (see Table 4.2) are usually associated with a large increase in polarity or in considerable steric restriction on forming the transition state. Large negative $\Delta S^{\ddagger}$-values have also been observed in the

Table 4.2 Activation parameters for the oxidative addition of YZ to some rhodium(I) and iridium(I) complexes ($\Delta H^{\ddagger}_1$ and $\Delta S^{\ddagger}_1$) and for the reductive elimination ($\Delta S^{\ddagger}_{-1}$ and $\Delta S^{\ddagger}_{-1}$)

YZ	*Complex*	*Solvent*	$\Delta H^{\ddagger}_1$	$\Delta S^{\ddagger}_1$	$\Delta H^{\ddagger}_{-1}$	$\Delta S^{\ddagger}_{-1}$	*Reference*
MeI	$IrCl(CO)(PPh_3)_2$	Benzene	5.6	−51			4, 57
MeI	$IrBr(CO)(PPh_3)_2$	Benzene	7.6	−46			4
MeI	$IrI(CO)(PPh_3)_2$	Benzene	8.8	−43			4
MeI	$IrCl(CO)(PPh_3)_2$	C_6H_5Cl	8.0	−41			57
MeI	$IrCl(CO)(PPh_3)_2$	C_6H_5Br	7.9	−41			57
MeI	$IrCl(CO)(PPh_3)_2$	C_6H_5I	6.6	−45			57
MeI	$IrCl(CO)(PPh_3)_2$	Acetone	7.1	−42			57
MeI	$IrCl(CO)(PPh_3)_2$	DMF	16.4	−10			4, 57
MeI	$RhCl(CO)(PPh_3)_2$	MeI	10.2	−43	16.7	−20	61
MeI	$RhCl(CO)[P(MeOC_6H_4)_3]_2$	MeI	9.1	−44	18.4	−11	61
MeI	$RhCl(CO)[P(FC_6H_4)_3]_2$	MeI MeI	15.9	−29	17.6	−17	61
MeI	$RhCl(CO)_2(PPh_3)$	CH_2Cl_2	11.6±1.5	−33.4±5			63
Furoylazide	$IrCl(CO)(PPh_3)_2$	$CHCl_3$	7.6	−37			65
Phenylazide	$IrCl(CO)(PPh_3)_2$	$CHCl_3$	8.8	−34			65
Phenylazide	$RhCl(CO)(PPh_3)_2$	$CHCl_3$	6.7	−42			65
$HC{\equiv}CCH_2Cl$	$IrCl(CO)(PPh_3)_2$	Benzene	7.1	−49			42
$HC{\equiv}CCMe_2Cl$	$IrCl(CO)(PPh_3)_2$	Benzene	10.0	−40			42

addition of R_3SiH, a reaction unlikely to have an ionic mechanism and the second reason probably applies in this case (see Section 4.6.2). However, on studying the MeI-addition in a number of solvents considerable variation of the activation parameters was observed as shown in Table 4.2. The biggest change in $\Delta S^{\ddagger}$ was from -51 cal deg^{-1} mol^{-1} in benzene to -10 cal deg^{-1} mol^{-1} in dimethylformamide and the whole behaviour pattern on varying the solvent[4, 57] is very similar to that of the reaction of pyridine with ethyl iodide, a reaction known to have a highly polar transition state. Also, the changes in the rates of addition of a series of *para*-substituted benzyl bromides to $[IrCl(CO)(PPh_3)_2]$ parallel the changes in rate of reaction of pyridine with the same series of substituted benzyl bromides[57]. Thus, there appears to be strong evidence for a highly polar transition state. However, Wilkinson and Douek later added a word of warning in interpreting the data in dimethylformamide solution since they have shown that MeI slowly gives conducting solutions in this solvent and they suggest that a different mechanism might be operating in this case[61].

Other workers[62] have looked at the effect of varying the basicity of the neutral ligands L on the rate of addition of MeI or benzyl chloride to the iridium(I) complexes $[IrCl(CO)L_2]$, expecting that an increase in the basicity of the ligands might correlate with an increase in nucleophilicity of the complexes. They found little correlation between the rate of addition and the basicity of the ligands PEt_3, PEt_2Ph, $PEtPh_2$ and PPh_3 and regard steric effects as important. A more sensible correlation was obtained with the series $P(p\text{-}C_6H_4X)_3$ (where X = OMe, Me, H, Cl or F) where steric effects should be of little importance. The rate of addition increases in the same order as increasing basicity but nevertheless there is a poor relationship between the rates and the Hammett constants for these substituents.

The reactions of methyl iodide with *trans*-$[RhX(CO)L_2]$ or *trans*-$[RhX(CO)_2L]$ (X = halide; L = tertiary phosphine or tertiary arsine) show greater complexity than the iridium reactions described above, since there is an inductive period for the oxidative addition to *trans*-$[RhX(CO)L_2]$ which could not be fully explained, and in both cases the MeI adducts once formed undergo further reaction to give acetyl complexes[61,63]. The kinetics and changes in i.r. spectra of the solutions have been accounted for by the following reaction sequences ($L = PPh_3$):

$$\text{A. } [RhX(CO)L_2] + MeI \underset{k_{-1}}{\overset{k_1}{\rightleftharpoons}} [RhMeIX(CO)L_2] \xrightarrow{k_2} [Rh(COMe)IXL_2]$$

(Reference 61)

$$\text{B. } [RhX(CO)_2L] + MeI \xrightarrow{k_1} [RhMeIX(CO)_2L]$$

$$[RhMeIX(CO)_2L] \xrightarrow[MeI]{k_2} [Rh(COMe)IX(CO)L(MeI)]$$

$$[RhMeIX(CO)_2L] \xrightarrow[+S]{k_3} [Rh(COMe)IX(CO)L(S)]$$

$$[Rh(COMe)IX(CO)L(S)] \xrightarrow[\text{fast}]{+MeI} [Rh(COMe)IX(CO)L(MeI)]$$

(Reference 63)

The two reaction schemes are actually very similar except that the equilibrium set up in the initial oxidative addition step in scheme A was not observed in scheme B. Also in case B the methyl migration (carbonyl insertion) followed two paths operating simultaneously, one zero-order in [MeI] (probably a solvent-assisted reaction), the other first-order in [MeI]. Since for reaction A neat methyl iodide was used as solvent, it is reasonable that only one reaction path should be observed for the 'carbonyl insertion' reaction. Directly analogous complexes were isolated from solution in each case, i.e. $[Rh(COMe)IXL_2]$ [61] and $[Rh(COMe)IX(CO)L]$ [63] respectively and the MeI-containing species postulated in scheme B was not obtained.

For both reactions the oxidative addition was faster than the subsequent 'carbonyl insertion' step and its kinetics could be studied independently. The addition was shown to be second-order and the activation parameters are included in Table 4.2. These are of the same order as those for the addition of MeI to $[IrCl(CO)(PPh_3)_2]$ and again there is the problem of interpretation of the large negative activation entropies, but the dramatic solvent-dependence of the rate of reaction B (too fast to measure in dimethylformamide by normal spectroscopic methods and too slow in benzene) favours a highly polar transition state characteristic of S_N2 mechanisms with neutral reactants.

The participation of an ionic intermediate in the reactions is further suggested by the reductive elimination of iodobenzene from $[Pt^{IV}Ph_2I_2(PEt_3)_2]$, the first step of which is the fast reversible solvolysis of iodide ion[56]. The reaction is inhibited by iodide ion and has the rate law:

$$\text{Rate} = \frac{kK[PtPh_2I_2(PEt_3)_2]}{[I^-]}$$

This is consistent with the mechanism:

$$[PtPh_2I_2(PEt_3)_2] \underset{\text{fast}}{\overset{K}{\rightleftharpoons}} [PtPh_2I(PEt_3)_2(S)]^+I^- \xrightarrow{k} [PtPhI(PEt_3)_2] + PhI$$
$$(S = \text{solvent})$$

This mechanism is the reverse of the ionic mechanism of the oxidative addition to square-planar d^8-complexes it would be useful to establish the

Since there is conflicting evidence for the mechanism of alkyl halide addition to square-planar d^8-complexes it would be useful to establish the kinetics, the stereochemical course of the addition, and the conformational changes at the carbon atom for a reaction involving the same substrates and under the same reaction conditions. Until this has been done it is not possible to resolve Pearson and Muir's contention that the reactions occur in a one-step concerted process without the formation of intermediates[53] with the conclusion of Osborn and co-workers that these reactions follow an S_N2 mechanism[60]. (See note added in proof, p. 154.)

4.4.3.4 *Allylic halides*

Allylic halides have been shown to add in a rather different way from saturated alkyl halides giving *cis*-addition in benzene with the tertiary phosphine ligands which were originally mutually *trans* in the planar complex becoming mutually *cis* in the adduct[64]. It has been suggested[64] that the cationic π-allylic

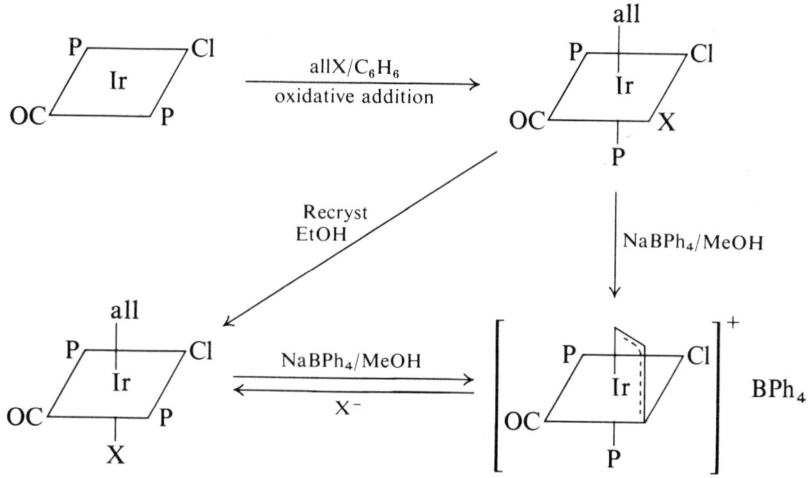

P Cl Ir OC P
$allX/C_6H_6$
oxidative addition
all P Cl Ir OC X P
Recryst EtOH
$NaBPh_4/MeOH$
all P Cl Ir OC P X
$NaBPh_4/MeOH$
X^-
$[\ldots]^+$ P Cl Ir OC P BPh_4

complex shown in the diagram is an intermediate in the isomerisation of the initial adduct that takes place in ethanol. Since there is a facile route for this isomerisation by solvolysis of halide it is likely that no such solvolysis occurs during the initial *cis*-addition and a one-step addition process is the most plausible. It is worth noting that an overall *trans*-addition of allyl halide has occurred via an initial *cis*-addition, and that the addition is solvent-dependent, a situation not always appreciated.

4.4.3.5 *Propargyl halides*

Propargyl halides add to iridium(I) complexes to give allenyl–iridium(III) complexes[42] by a reaction analogous to that of cobaloxime(I). It has been suggested that this second-order reaction has an S_N2' mechanism, but this mechanism may only be apparent. Enthalpies of activation are somewhat higher than for methyl iodide addition, although the entropies of activation are quite similar (see Table 4.2). Collman[42] believes that the substrate undergoes preliminary coordination to the metal through the triple bond and, indeed, when non-terminal acetylenes are used adducts of this type are obtained. A similar preliminary coordination of the C—C multiple bond could also occur in the addition of allylic halides.

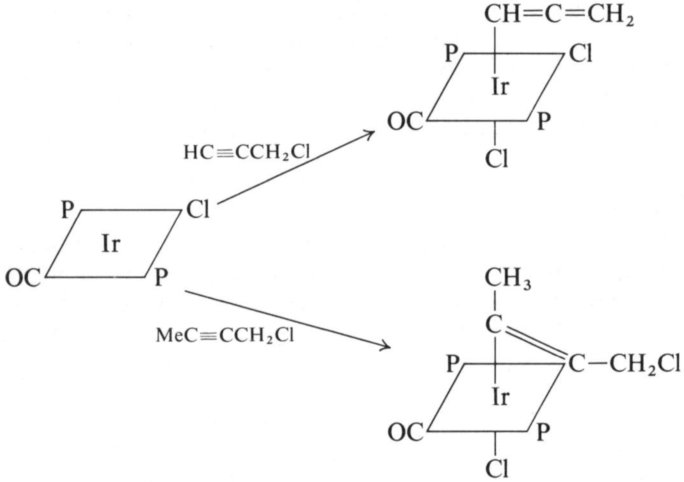

$$\text{Ir(CO)ClP}_2 \xrightarrow{\text{HC}\equiv\text{CCH}_2\text{Cl}} \text{Ir(CH=C=CH}_2\text{)(CO)Cl}_2\text{P}_2$$

$$\text{Ir(CO)ClP}_2 \xrightarrow{\text{MeC}\equiv\text{CCH}_2\text{Cl}} \text{Ir(CO)ClP}_2(\text{CH}_3\text{C}\equiv\text{C}-\text{CH}_2\text{Cl})$$

4.4.3.6 *Acyl halides*

Acyl halides have been shown to react more rapidly with iridium(I) complexes than do alkyl halides and Halpern[8] has speculated that this enhanced susceptibility to nucleophilic attack is a result of the mechanism:

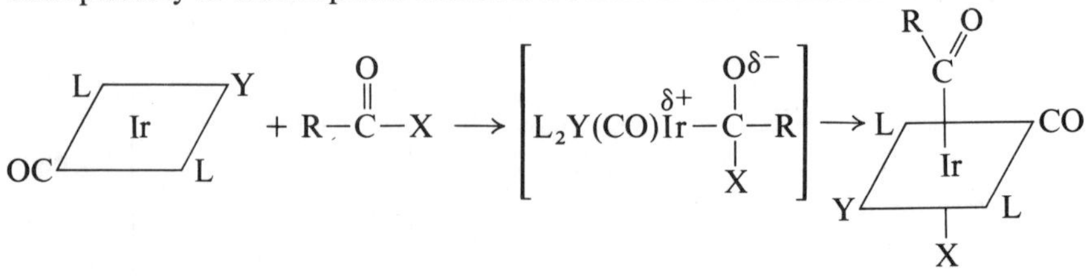

$$\text{Ir(CO)YL}_2 + \text{R}-\overset{\text{O}}{\overset{\|}{\text{C}}}-\text{X} \longrightarrow \left[\text{L}_2\text{Y(CO)}\overset{\delta+}{\text{Ir}}-\overset{\text{O}^{\delta-}}{\overset{|}{\underset{\text{X}}{\underset{|}{\text{C}}}}}-\text{R}\right] \longrightarrow \text{Ir(COR)(CO)XYL}_2$$

No detailed kinetic or other mechanistic studies have been reported, however, for simple acyl halides. Some related work is that of Collman and co-workers[65] who have looked at the kinetics of the reaction of $[MCl(CO)(PPh_3)_2]$ ($M = Rh$ or Ir) and analogous compounds with aryl- or acyl-azides to give the dinitrogen complexes $[MCl(N_2)(PPh_3)_2]$ (not isolated when $M = Rh$). The first step has been shown to be the oxidative addition of the organic azide to the d^8-metal complex to give an adduct possibly like that shown below, which gives the dinitrogen complex by a rapid interligand reaction and reductive elimination, although there is no evidence that the adduct is a true intermediate:

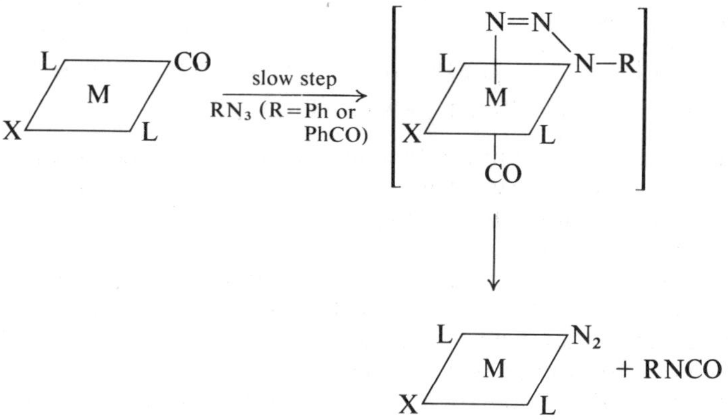

L, CO, M, X, L — slow step, RN_3 (R=Ph or PhCO) → [N=N, L, N—R, M, X, L, CO]

↓

L, N_2, M, X, L + RNCO

The reaction is second-order, first-order in complex and first-order in organic azide, and has activation parameters as included in Table 4.2, the $\Delta S^{\ddagger}$-values being characteristically large and negative. In agreement with Halpern's observation[8] the acyl halides are the most reactive with rates in the order $C_6H_5CON_3 > C_6H_5N_3 > p\text{-}CH_3C_6H_4N_3$ [65]. The variation of rates with ligands used follows the orders $X = I > Br > Cl > N_3, L = PPhEt_2 > PPh_2Me > PPh_3$ and also Ir > Rh; these orders being as expected for a reaction in which the complex acts as a nucleophile with the most basic ligands favouring faster reaction.

4.4.4 d^{10}-complexes

The complex $[IrCl(CO)(PPh_3)_2]$ can be reduced by sodium under carbon monoxide to the d^{10}-complex $Na[Ir(CO)_3(PPh_3)]$, which reacts oxidatively with methyl iodide in the presence of triphenylphosphine to give $[IrMe(CO)_2(PPh_3)_2]$[66]. The corresponding reaction with perfluoroalkyl iodides gives $[IrI(CO)_2(PPh_3)_2]$[66] probably because in such perfluoro-compounds the iodine atom is the most positive centre and subject to nucleophilic attack. Nothing is known of the mechanisms of these reactions and only a little more of the addition of alkyl halides to the d^{10}-complexes of the platinum triad. In the additions of MeI, benzyl bromide or CH_2ICH_2I (which gives ethylene and the di-iodo-complex) to $[Pt(PPh_3)_2(C_2H_4)]$ there is a preliminary dissociation of ethylene to give $[Pt(PPh_3)_2]$ which is the species which undergoes oxidative addition[17]. The kinetics follow a rate law consistent with the mechanism.

$$[Pt(PPh_3)_2(C_2H_4)] \overset{K}{\rightleftharpoons} [Pt(PPh_3)_2]+C_2H_4 \quad K = (3.0 \pm 1.5) \times 10^{-3} M$$

$$[Pt(PPh_3)_2]+XY \rightarrow [PtXY(PPh_3)_2]$$

where XY = MeI, $PhCH_2Br$ or CH_2ICH_2I

A similar mechanism is observed in the oxidative addition reactions of $Pt(PPh_3)_3$ or $Pt(PPh_3)_4$[67]. Solutions of these complexes both contain $Pt(PPh_3)_3$ as the predominant species in solution, and again there is a pre-equilibrium reaction to give the same reactive species as was obtained from the ethylene complex.

$$[Pt(PPh_3)_3] \overset{K}{\rightleftharpoons} [Pt(PPh_3)_2]+PPh_3 \quad K = (1.6 \pm 1) \times 10^{-4} M$$

$$[Pt(PPh_3)_2]+XY \rightarrow [PtXY(PPh_3)_2]$$

Thus, although the overall reaction scheme has been established for these compounds, nothing is yet known of the nature of the transition state for the alkyl halide addition to $Pt(PPh_3)_2$. Cook and Jauhal have speculated[68] that there is an intermediate or transition state of the type $(PPh_3)_2 Pt^{\delta+}$----R----$X^{\delta-}$ but only by analogy with the iridium(I) addition reactions[4].

That nickel(0) complexes react as nucleophiles with aryl halides was established by Fahey who looked at the reaction of $[Ni(PEt_3)_2(C_2H_4)]$ (prepared *in situ*) with 1,2,4-trichlorobenzene[69].

Treatment of the mixture of aryl complexes formed with dry HCl gave a mixture of ortho- (7%), meta- (6%) and para- (87%) dichlorobenzenes showing that nickel had reacted preferentially at the 2-position as does methoxide ion in its nucleophilic substitution (S_NAr) reactions. Fahey further showed that the nickel(0) complex reacts with 1,2-chlorobromobenzene with displacement of bromide preferentially. The ratio of rate constants $k_{Cl}/k_{Br} = 0.0030$, and is very similar to that observed by Schrauzer and Deutsch in the reactions of alkyl halides with cobalt(I) nucleophiles[36].

Chloro-olefins also react with d^{10}-complexes to give vinyl-complexes of d^8-metals, but since the mechanism of these reactions is believed to involve initial addition to give olefin complexes followed by isomerisation these reactions will be considered later in Section 4.10.2.

4.4.5 d^7-complexes

Halpern and co-workers have undertaken mechanistic studies on the reactions of alkyl halides with the cobalt(II) complexes $[Co(CN)_5]^{3-}$ [70] cobaloxime(II) complexes such as $[Co(glyoximato)_2B]$ (where B = coordinated base)[71] and Schiff's base complexes, for example [Co(salen)(MeIMD)] (where salen = bis(salicylidene)ethylenediamino and MeIMD = 1-methylimidazole)[72]. These studies[70] support the free-radical mechanism for the pentacyano-complex as shown below, for example:

$$\text{Total reaction:} \quad 2[Co(CN)_5]^{3-}+RX \rightarrow [Co(CN)_5R]^{3-}+[Co(CN)_5X]^{3-}$$

$$\text{R.D. step:} \quad [Co(CN)_5]^{3-}+RX \xrightarrow{k} [Co(CN)_5X]^{3-}+R^{\bullet}$$

$$[Co(CN)_5]^{3-}+R^{\bullet} \rightarrow [Co(CN)_5R]^{3-}$$

This scheme is in agreement with the stoichiometry of the reaction, the rate law $-d[Co(CN)_5^{3-}]/dt = 2k[Co(CN)_5^{3-}][RX]$, and also the variation of rate with variation of RX which follows a very similar pattern to that observed for the halogen atom abstraction from RX by the free radicals Na and Me. For example, the change in rate in the order X = I > Br > Cl is very much more dramatic than expected for nucleophilic substitution and is consistent with abstraction of $X^{\bullet}$ by the cobalt(II) complex. When RX is more complicated than methyl or benzyl halide the reaction may be more complex, for example isopropyl iodide gives propene and propane. These products may be explained in terms of cleavage of the alkyl to give olefin and hydrido-complex, but in all cases the rate-determining step is halogen atom abstraction so that the kinetic behaviour is always very similar.

Analogous reaction mechanisms are proposed for the complexes [Co(dimethylglyoximato)$_2$B][71] and the rate, although insensitive to variations in the oxime ligands, increases in the order B = nicotinamide < pyridine < γ-picoline, which is the order of increasing basicity of the ligand. When B = PR_3 the order of rate is R = cyclohexyl < Ph < Bu^n < Et < Me which shows no simple relationship to the basicity of the tertiary phosphine[8]. Although the Schiff base complexes show similar mechanisms in certain cases, with nitrobenzyl halides a different behaviour is observed and the rate-determining step is believed to be the electron-transfer reaction[72]:

$$[Co(salen)(MeIMD)_2] + RX \rightarrow [Co(salen)(MeIMD)_2]^+ + R^{\bullet} + X^-$$

and such reactions fall outside the classification of oxidative addition.

Similar free-radical mechanisms to those described above have also been proposed for various other addenda molecules and $[Co(CN)_5]^{3-}$ reacts with H_2O_2, NH_2OH and ICN similarly to alkyl halides[73], and the mechanism is not confined to d^7-complexes but also occurs in the reaction of alkyl halides with high-spin chromium(II) complexes[74].

4.5 HYDROHALIC ACIDS

4.5.1 Coordinatively saturated d^8-complexes

Additions of hydrohalic acids to d^8-complexes are generally very fast[4] and no kinetic study has yet been reported, while most information has come from qualitative comparisons of the reactivity of various complexes. The 5-coordinate d^8-complexes $[M(CO)_3(PPh_3)_2]$ (M = Ru or Os) react with hydrohalic acids HX to give $[MX_2(CO)_2(PPh_3)_2]$ and it was suggested[20] that the first step is protonation to give the cationic hydrides $[MH(CO)_3(PPh_3)_2]^+$, and such complexes, e.g. $[OsH(CO)_3(PPh_3)_2]ClO_4$, have since been isolated[75]. Although the parent carbonyl compounds are much less basic than the tertiary phosphine substituted species even these have been protonated:

$$M(CO)_5 + H^+ \rightleftharpoons [MH(CO)_5]^+ \quad (M = Fe^{76}, Ru^{77} \text{ or } Os^{77})$$

As expected[5] the basicity of such complexes is in the order Os > Ru > Fe and $M(CO)_3(PPh_3)_2 > M(CO)_5$ and, whereas $[OsH(CO)_5]PF_6$ can be isolated[77], $[FeH(CO)_5]^+$ can only be detected in strongly acidic media[76]. Reversible protonation of transition metals is now a well-established phenomenon and such behaviour has been reviewed[11]. The mechanism of these reactions could be very simple but as yet there has been no quantitative study either of the basicity or of the kinetics of protonation of 5-coordinate d^8-complexes.

Addition of HX to the isoelectronic compounds $[Ir(CO)_x(PR_3)_{5-x}]^+$ ($x = 1$, or 2) gives the compounds $[IrHX(CO)(PR_3)_3]^+$ with either CO or PR_3 being lost in the reaction[30,78]. However, it is not known whether the initial step in this reaction is the loss of the neutral ligand to give 4-coordinate species (as has been shown to be possible[78]) followed by HX addition, or whether it is protonation of the 5-coordinate complex as in the iron triad but in this case to give doubly charged cations. Doubly charged cationic hydrides of this type are feasible intermediates since similar species have been isolated[79] from the reaction: *trans*-$[Ir(CO)(dmpe)_2]^+ \xrightarrow{EtOH} [IrH(CO_2Et)(dmpe)_2] \underset{OEt^-}{\overset{+H^+}{\rightleftharpoons}} [IrH(CO)(dmpe)_2]^{2+}$ (dmpe = $Me_2PCH_2CH_2PMe_2$).

Substitution of CO from $[IrH(CO)_2(PR_3)_3]^{2+}$ or PR_3 from $[IrH(CO)(PR_3)_4]^{2+}$ by halides X^- would give the product $[IrHX(CO)(PR_3)_3]^+$. It is most probable the mechanisms of these reactions will prove to be very dependent upon the position of equilibrium between 4- and 5-coordinate species, and their relative nucleophilicities towards protons.

4.5.2 Coordinatively unsaturated d^8-complexes

The mechanism of oxidative addition of HX to 4-coordinate species is equally nebulous. The stereochemistry of addition has been established and shown to be kinetically controlled in a few cases. Crystals of $[IrY(CO)(PPh_3)_2]$ react with gaseous HX to give the adduct with X and H in mutually *cis*-positions (X and Y are halides)[80] and since rearrangements are unlikely to occur in the solid state this probably reflects the mechanism of addition with a one-step process most favoured. The addition of HBr in ether to a solution of $[IrCl(CO)L_2]$ (L = PPh_3[81] or $PMePh_2$[24]) in benzene has been reported to give the same equilibrium mixture of isomers as obtained by the addition of HCl to the bromo-complexes, but it was later shown that in completely anhydrous conditions (e.g. in dry benzene) there is a kinetically controlled *cis*-addition of HX[82]. The presence of ionising solvents was shown[82] to enable exchange of halide to occur either before, during, or after the oxidative addition to give the mixtures observed in earlier work[24,81]. There has also been a less-substantiated report of *trans*-addition of HCN to *trans*-$[IrCl(CO)(PPh_3)_2]$[83]. In ionising solvents the reaction most likely involves more than one step, probably addition of a proton and then of a halide, but whether in non-ionising solvents a one-step addition occurs has still to be established.

Relative basicities, as opposed to nucleophilicities, of the square-planar complexes have been determined by a study on the position of the equilibrium between d^8- and d^6-complexes in the reaction[84].

$$[Ir^IX(CO)L_2] + HY \overset{K}{\rightleftharpoons} [Ir^{III}HYX(CO)L_2]$$

where HY is a carboxylic acid. There is an approximately linear relationship between log $[Ir^{I}]/[Ir^{III}]$ and the pK_a-value of non-sterically hindered acids, and on varying X and L and using the same acid HY the equilibrium constant increases in the order I>Br>Cl, $PMe_3>PMe_2Ph>PMePh_2>PPh_3$, $AsMe_2Ph>PMe_2Ph$ and $AsPh_3>PPh_3$. These orders show that an increase in the basicity of the ligands favours large equilibrium constants. Tertiary arsines are exceptions to this and although they have much lower basicity than tertiary phosphines they have greater stabilising influence on the higher oxidation state of the metal. It was tentatively suggested[84] that determining factors may be (a) the greater size of the arsenic atom which reduces steric compression in the adduct by enabling the alkyl or aryl groups on the arsenic to be further from the metal and/or (b) the greater interaction between the soft arsenic and the soft iridium atoms which induces greater electron density on to the metal stabilising the higher oxidation state. This greater stabilisation by $AsPh_3$ relative to PPh_3 may be a general phenomenon since it has also been observed for the addition of O_2 [66] and H_2 [85].

4.5.3 d¹⁰-complexes

The reaction of the d^{10}-complex $Pt(PPh_3)_4$ with HX is reported to lead to the series of equilibria:

$$Pt(PPh_3)_4 \rightleftharpoons Pt(PPh_3)_3 \overset{H^+}{\rightleftharpoons} [PtH(PPh_3)_3]^+ \overset{X^-}{\rightleftharpoons} PtHX(PPh_3)_2 \rightleftharpoons PtH_2X_2(PPh_3)_2$$

and it was believed that the 3-coordinated species was protonated[86], but although this is the predominate species in solution the 2- or 4-coordinated species may be involved instead. Recently the complexes $[Ni\{P(OEt)_3\}_4]$ [87] and $[Ni(Ph_2PCH_2CH_2PPh_2)_2]$ [88] have been protonated to give 5-coordinate complexes of the type $[NiHL_4]^+$. The complex $[Ni\{P(OEt)_3\}_4]$ behaves differently from the platinum and palladium complexes in showing little tendency to dissociate in solution (the slow substitution reaction of this complex is probably a related phenomenon)[16], and Tolman[89] has shown that protonation occurs on the coordinatively saturated molecule to give $[NiH\{P(OEt)_3\}_4]^+$ by a fast reaction(followed by stopped-flow techniques) which is first-order in complex and first-order in acid. $\Delta H^\ddagger$ and $\Delta S^\ddagger$ for the forward reaction are 13 ± 1 kcal mol^{-1} and -2.3 ± 3 cal deg^{-1} mol^{-1} respectively and for the reverse reaction 8 ± 1 kcal mol^{-1} and -26.2 ± 3 cal deg^{-1} mol^{-1} respectively. It can be seen from these data that the protonation reaction is endothermic but there is a large compensating gain in entropy presumably due to the reduction in solvent-ordering by the proton on coordination. However, little solvent-ordering can be lost on formation of the transition state since the entropy of activation for the protonation reaction is close to zero.

Following the protonation reaction a slow decomposition reaction occurs by initial dissociation of the 5-coordinate species[89]:

$$NiP_4 \overset{H^+}{\rightleftharpoons} [NiHP_4]^+$$

$$[NiHP_4]^+ \overset{-H^+}{\rightleftharpoons} [NiHP_3]^+ + P$$

$$H^+ + [NiHP_3]^+ \rightarrow [Ni]^{2+} + H_2 + 3P$$

$$(P = P(OEt)_3)$$

A kinetic study of the reaction of $[NiH\{P(OEt)_3\}_4]^+$ with butadiene to give π-crotyl complexes also suggested a similar preliminary dissociation of $P(OEt)_3$ must occur[90].

4.6 HYDRIDES OF GROUP IV ELEMENTS

4.6.1 Carbon–hydrogen bond cleavage by oxidative addition

There is an increasing number of oxidative addition reactions reported in which cleavage of a carbon–hydrogen bond occurs to give an alkyl (or aryl) hydrido-complex. The earliest example[91] was the tautomerism between a ruthenium(0) and a ruthenium(II) complex in which the metal cleaved a hydrogen from an alkyl:

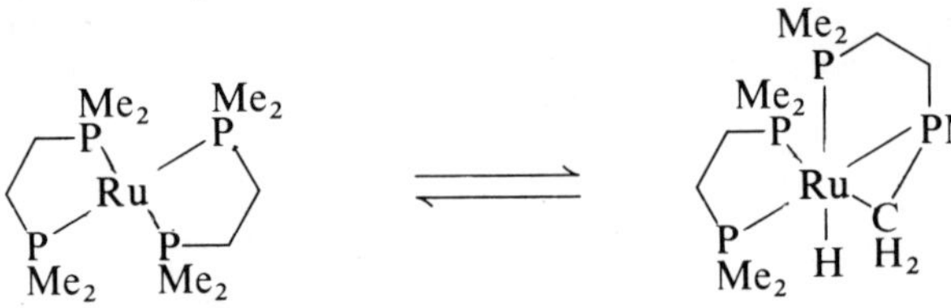

Me₂ Me₂ P P Ru P P Me₂ Me₂ ⇌ Me₂ P Me₂ P PMe Ru C P H H₂ Me₂

Further the d⁸-complex $[IrCl(PPh_3)_3]$ has proved to be one of the more reactive complexes known toward oxidative addition, and in refluxing solvents the metal undergoes an irreversible intramolecular oxidation[92].

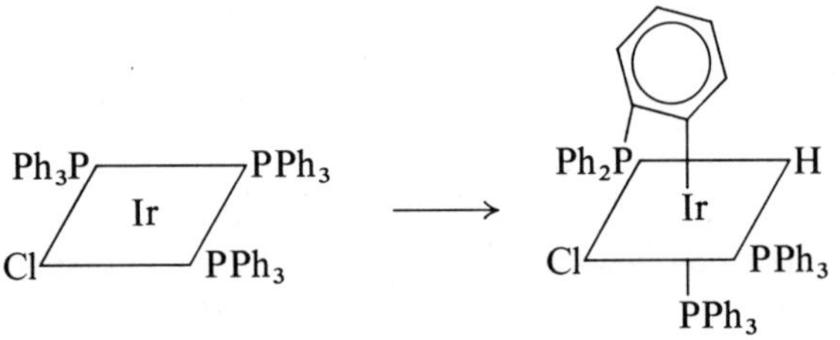

Ph₃P PPh₃ Ir Cl PPh₃ ⟶ Ph₂P H Ir Cl PPh₃ PPh₃

The formation of metal–aryl bonds by *ortho*-hydrogen abstraction has been widely reported and recently reviewed by Parshall[10]. Some recently reported examples are for cobalt[93], rhodium[94], platinum[95] and the formation of an aryl–hydrido-complex of tungsten from arene is believed to involve a similar reaction[96].

In the majority of cases aryl–hydrido-species are not isolated as they have been for the ruthenium and iridium complexes above, but a proton is effectively lost in the reaction with no resultant oxidation state change of the metal; for example the complex $[Pt^{II}Cl(PBu^t_2Ph)(PBu^t_2C_6H_4)]$ is obtained by refluxing a solution of $[Pt^{II}Cl_2(PBu^t_2Ph)_2]$ and HCl is lost in this reaction[95]. The mechanisms of these reactions are not well understood. Indeed, in some reactions the metal may act as an electrophile in a process akin to normal electrophilic aromatic substitution (the mechanism favoured for Pd^{II} complexes[10]), while in other systems the metal may act as a nucleophile, i.e. it undergoes oxidative addition which may or may not be followed by reductive elimination[10].

In a clear example of oxidative addition Bennett and Milner[92] have demonstrated that the first-order intramolecular oxidation of the iridium in $[IrCl\{P(aryl)_3\}_3]$ (see above) is promoted by electron-donating substituents in the *para*-position of the aryl rings with rates in the order F < H < MeO < Me for the *para*-substituent. The increased rate is probably a reflection of the increased nucleophilicity of the iridium in this order, and the small isotope effect (not accurately measured) on using *ortho*-deuterated PPh_3 is in agreement with a 3-centred transition state (A). The other possibility is

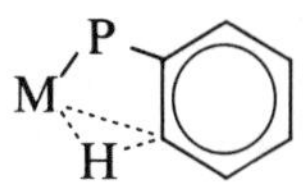

(A)

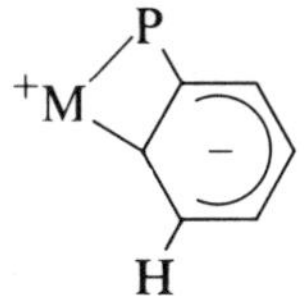

(B)

that the intermediate (B) is formed as in nucleophilic aromatic substitution (since the metal must surely act as a nucleophile); however, such a reaction would not be expected to show the substituent effects described above, but the rate should be enhanced by electron-withdrawing substituents. The driving force of these reactions is not clear, but in the iridium example above it could well be the enthalpy gain with the formation of strong Ir—C and Ir—H bonds. Shaw *et al.*[95] have also shown that steric crowding can favour carbon–hydrogen bond cleavage. Presumably there is a release of steric compression on forming the transition-state to enable the complex *trans*-$[PtCl_2(PBu^t_2Ph)_2]$ to undergo a ready reaction, since *trans*-$[PtCl_2(PMe_2Ph)_2]$ shows no such tendency. In certain complexes it has been shown that the *ortho*-hydrogens of aryl rings may be very close to the metal atom in the solid state[10], and presumably in a very sterically-favoured position for reaction.

As well as the cleavage of C—H bonds by oxidative addition there are various reactions reported in which the reverse reaction occurs, that is, the formation of C—H bonds by reductive elimination. Examples are the decarbonylation of aldehydes catalysed by $[RhCl(PPh_3)_3]$[97,98], the cleavage of Pt^{II}–carbon bonds by HCl[99,100] and the deuteration of C_6H_6 by D_2 catalysed by $(C_5H_5)_2TaH_3$ or $IrH_5(PPh_3)_2$[101]. The reaction pathway for the decarbonylation of aldehydes is proposed to be[97]:

$$RCH_2CH_2CHO + RhClP_3\ (P{=}PPh_3) \rightarrow RCH_2CH_2CO{-}RhHClP_2$$
$$\downarrow$$
$$RhCl(CO)P_2 + RCH_2CH_3 \leftarrow RCH_2CH_2{-}RhHCl(CO)P_2$$
$$RhCl(CO)P_2 + H_2 + RCH{:}CH_2 \leftarrow RCH_2CH_2{-}RhHCl(CO)P_2$$

The basic outline of this scheme has been justified by the reaction of RCDO to give RD with 100% retention of the deuterium which supports the intramolecular nature of the reaction[98]. Now the two steps which involve the carbon bonded to the aldehyde group are the decarbonylation and the reductive elimination steps. The total result of these reactions is stereospecific retention of configuration of this carbon, i.e. R_3C^*CHO gives R_3C^*H with retention[98], and since carbonylation and decarbonylation are known to occur with retention of configuration[49] so must the reductive elimination step. This is good evidence for a one-step 3-centred transition state analogous to that described for the oxidative addition reaction.

In the reactions of hydrochloric acid with $[PtPh_2(PEt_3)_2]$ to give $[PtPhCl(PEt_3)_2]$[99] and with $[PtMeCl(PEt_3)_2]$ to give $[PtCl_2(PEt_3)_2]$[100], Belluco *et al.* have shown that the first step of the reaction is the rapid reversible oxidative addition of HCl to the platinum(II) complexes. The reductive eliminations of benzene or methane respectively from the platinum(IV) adducts are the slow steps of these reactions. Some kinetic data are available but few mechanistic conclusions can be made about the reductive eliminations, although Belluco prefers a 3-centred transition state.

4.6.2 Addition of silanes, germanes and stannanes

4.6.2.1 Coordinatively unsaturated d^8-complexes

Trialkyl-, triaryl-, trialkoxy- or trihalo- silanes can be added to iridium(I) complexes to give octahedral iridium(III) silyl hydrido-complexes[102, 105], for example:

$$[IrCl(CO)(PPh_3)_2] + R_3SiH \rightleftharpoons [IrHCl(SiR_3)(CO)(PPh_3)_2]$$

The additions are readily reversible, but with electron-withdrawing substituents on the silanes, for example alkoxy-groups, the equilibrium lies sufficiently to the right for these compounds to be isolated and characterised in the usual way[105]. The hydrosilation of olefins catalysed by d^8 metals is believed to involve the oxidative addition of silanes, at least in certain cases, and this has made this reaction a significant one to study. Analogous adducts of R_3SnH have also been isolated[103], however, with R_3GeH the dihydrido-germyl-species $[IrH_2(GeR_3)(CO)(PPh_3)_2]$ and R_3GeCl were obtained rather than the simple addition compounds[104]. It was believed that these germyl compounds were formed by successive oxidative addition and reductive elimination processes as shown below:

$$Ir^{I}Cl \underset{-HMR_3}{\overset{+HMR_3}{\rightleftharpoons}} Ir^{III}HCl(MR_3) \xrightarrow{-ClMR_3} Ir^{I}H \underset{-HMR_3}{\overset{+HMR_3}{\rightleftharpoons}} Ir^{III}H_2(MR_3)$$

Later it was shown that similar reaction sequences were possible for iridium complexes derived from silanes[102, 104] and in this case iridium(I) hydrido-species could be isolated from the reaction[102]. It is possible that a similar sequence of reactions occurs in the $[IrCl(CO)(PPh_3)_2]$-catalysed exchange of silanes[107, 108]:

$$Si^{*}H + SiD \rightleftharpoons Si^{*}D + SiH$$

It is likely that addition and elimination of silanes to iridium(I) hydrides are involved in this exchange reaction[102]. An observation related to the mechanism of addition of silanes is that the above exchange occurred with retention of configuration at the silicon atoms, and although both addition and elimination steps could take place with inversion it is more likely that both occur with retention. In view of the comparison of iridium(I) complexes with carbenes it might be noted that dibromocarbene inserts into Si—H bonds with retention[109].

A kinetic study[110] has been made of the addition of R_3SiH to the

4-coordinate complex $[Ir(Ph_2PCH_2CH_2PPh_2)_2]BPh_4$, and the following rate laws were obtained:

$$\textit{Addition: } -d[Ir^I]/dt = k_1[Ir^I]R_3SiH$$
$$\textit{Elimination: } -d[Ir^{III}]/dt = k_{-1}[Ir^{III}]$$

Activation data included in Table 4.3 show large negative activation entropies for the addition reaction as observed for other addition reactions. Such values have been interpreted for the oxidative addition of MeI to signify highly polar transition states[4]. However, there is virtually no change in either $\Delta H_1^\ddagger$ or $\Delta S_1^\ddagger$ on changing the solvent from tetrahydrofuran ($\varepsilon \sim 4$) to acetonitrile ($\varepsilon \sim 38$) and this argues against a transition state in which the Si—H bond is highly polarised. It might be added that ΔS^0–values are also large and negative, while the values of $\Delta S^\ddagger_{-1}$ are quite close to zero, from which it would seem that the transition state has rather similar steric restrictions and polarity to those of the final adduct. Thus the data are consistent with the 3-centred transition state:

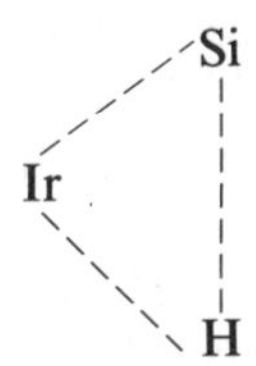

Table 4.3 **Activation parameters for reactions of $Me_n(EtO)_{3-n}SiH$ with $[(diphos)_2Ir][BPh_4]$ (ΔH-values in kcal mol^{-1} and ΔS-values in cal deg^{-1} mol^{-1})[110]**

n	*Solvent*	$\Delta H_1^\ddagger$	$\Delta S_1^\ddagger$	$\Delta H^\ddagger_{-1}$	$\Delta S^\ddagger_{-1}$	ΔH°	ΔS°
0	THF	5.57 ± 0.46	-48.0 ± 1.6	23.6 ± 0.80	3.5 ± 3.0	-18.05 ± 0.35	-51.6 ± 1.2
1		5.80 ± 0.15	-47.0 ± 0.5	21.8 ± 0.5	-0.3 ± 1.5	-16.00 ± 0.35	-46.7 ± 1.0
0	MeCN	5.56 ± 0.28	-47.3 ± 0.9	25.3 ± 0.75	8.2 ± 2.5	-19.70 ± 0.45	-55.5 ± 1.5
1		5.62 ± 0.45	-46.5 ± 1.5	22.7 ± 0.65	1.2 ± 2.2	-17.10 ± 0.20	-47.7 ± 0.7
2		5.59 ± 0.50	-48.2 ± 1.0	19.2 ± 0.89	0.1 ± 2.5	-13.63 ± 0.39	-48.3 ± 1.5

4.6.2.2 *Coordinatively saturated d^8-complexes*

The 5-coordinate complex $[IrH(CO)(PPh_3)_3]$ adds R_3SiH by *cis*-addition to give the same product as that obtained by prolonged treatment of $[IrCl(CO)(PPh_3)_2]$ with R_3SiH[111]. The kinetics of this reaction are consistent with a preliminary dissociation of triphenylphosphine with the silane adding oxidatively to the 4-coordinate species so produced[112].

$$[IrH(CO)(PPh_3)_3] \underset{k_{-1}}{\overset{k_1}{\rightleftharpoons}} [IrH(CO)(PPh_3)_2] + PPh_3$$

$$R_3SiH + [IrH(CO)(PPh_3)_2] \xrightarrow{k_2} [IrH_2(SiR_3)(CO)(PPh_3)_2]$$

The rate law for this reaction is

$$\frac{-d[Ir^I]}{dt} = \frac{k_1 k_2 [R_3SiH][Ir^I]}{k_{-1}[PPh_3] + k_2[R_3SiH]}$$

from which values of k_1 and k_{-1}/k_2 were obtained. This mechanism is different from that of the reaction of MeI or halogen with 5-coordinate d^8-complexes where ionic mechanisms have been established[33], and is consistent with the idea that R_3Si—H adds by 3-centred transition states with ionic mechanisms not feasible. One would expect that if this were the case dissociation of the 5-coordinate species would be the determining factor in the addition of silanes to complexes which do not readily dissociate, and indeed it was found[113] that iron pentacarbonyl only reacts with $HSiCl_3$ under conditions where CO dissociation can occur, i.e. under photochemical conditions to give *cis*-$[FeH(SiCl_3)(CO)_4]$, or above 140 °C when $Fe(SiCl_3)_2(CO)_4$, among other products, is obtained.

4.7 HYDROGEN

In the homogenous catalytic hydrogenation of olefins molecular hydrogen may be 'activated' in one of three possible ways as exemplified below and discussed by Halpern elsewhere[14, 114].

(a) $[Ir^{I}Cl(CO)(PPh_3)_2]+H_2 \rightarrow [Ir^{III}H_2Cl(CO)(PPh_3)_2]$ Metal oxidised by two
(b) $2[Co^{II}(CN)_5]^{3-}+H_2 \rightarrow 2[Co^{III}H(CN)_5]^{3-}$ Metal oxidised by one
(c) $[Ru^{III}Cl_6]^{3-}+H_2 \rightarrow [Ru^{III}HCl_5]^{3-}+H^{+}+Cl^{-}$ Metal not oxidised

The heterolytic cleavage of H_2 in reaction C does not involve any oxidation of the metal and will not be discussed further. Homolytic cleavage of the H—H bond which occurs in reactions (a) and (b) requires 104 kcal mol^{-1} and only by the exothermic formation of two M—H bonds (to the same metal in (a) or to different metal atoms in (b)) can a highly endothermic step be avoided.

4.7.1 Coordinatively unsaturated d^8-complexes

Both 4- and 5-coordinate d^8-complexes undergo reactions of type (a) and these will be discussed first. One of the early and important reactions reported by Vaska[115] for the compound $[IrCl(CO)(PPh_3)_2]$ was the reversible *cis*-addition of hydrogen. The *cis*-nature of the addition was initially established by i.r. studies[115] and later by some n.m.r. work[116, 52]. There is now a body of evidence to support the *cis*-addition of hydrogen to a variety of d^8-complexes[5], and such evidence supports a 3-centred transition state as has later evidence described below.

Rate constants and equilibrium constants have recently been determined for the reaction:

$$[IrX(CO)L_2]+H_2 \underset{k_{-1}}{\overset{k_2}{\rightleftharpoons}} [IrH_2X(CO)L_2]$$

The equilibrium constant K showed the following dependence on the ligands used:

(1) Ligands X. Constants K are in the order OCN < SCN < Cl < Br < I

for the reaction of H_2 with $[IrX(CO)(PPh_3)_2]$ in chlorobenzene at 40 °C[117], while in toluene at 80 °C the order is Cl < Br ~ I with the precise order of I and Br dependent upon the tertiary phosphine used[118].

(2) Ligand L. The order of equilibrium constants for X = Cl in toluene at 80 °C is $P(benzyl)_3 < P(cyclohexyl)_3 < PPh_3 < P(p\text{-tolyl})_3 < P(OPh)_3 < P(Pr^i)_3 < PBuPh_2$ but the order is somewhat different when the halide is changed[118].

(3) Other ligand. When the carbonyl is replaced by thiocarbonyl to give $[IrCl(CS)(PPh_3)_2]$ no uptake of hydrogen was observed (25 °C, 1 atm) which might reflect the greater π-acidity of the CS ligand[119]. A similar replacement of CO by PPh_3, which is a poorer π-acid and a better σ-base, results in an irreversible uptake of hydrogen[92].

Thus, (1) and (3) above suggest that the greater the electron density on the metal the higher its basicity towards hydrogen. However, the changes in K with change in tertiary phosphine show that other factors are important, and a more carefully controlled selection of phosphine ligands (e.g. tri-*p*-substituted-phenylphosphines) would be required to enable meaningful conclusions to be drawn from such data.

The kinetics of addition of hydrogen to the complexes $[IrX(CO)L_2]$ have been studied by several authors[4,7,120,118] who all report a reaction first-order in both complex and hydrogen concentrations with reverse (elimination) reaction being first-order in adduct concentration. Such behaviour is expected for a simple addition reaction. Activation parameters obtained are given in Table 4.4 together with thermodynamic data for the overall

Table 4.4 Activation parameters for the addition of hydrogen

	$\Delta H^\ddagger$ kcal mol^{-1}	$\Delta S^\ddagger$ cal deg^{-1} mol^{-1}	ΔH° kcal mol^{-1}	ΔS° cal deg^{-1} mol^{-1}	*Reference*
$H_2 + [IrCl(CO)(PPh_3)_2]/C_6H_5Cl$	11.5	−20	−15.0	−29	7
$H_2 + [IrCl(CO)(PPh_3)_2]/C_6H_6$	10.8	−23			4
$H_2 + [IrBr(CO)(PPh_3)_2]/C_6H_6$	12.0	−14			4

reaction. The isotope effect is not large ($k_{H_2}/k_{D_2} = 1.22$)[4], while moderately large negative activation entropies are obtained in this as in other oxidative addition reactions. It should be noticed that there is a much greater enthalpy than entropy difference on going from the transition state to the final state of the reaction, suggesting that the main difference between these two states is in the degree of Ir—H bond formation and H—H bond cleavage rather than in the polarity or structuring of the molecule. This is as observed in the addition of silanes[110] and is consistent with 3-centred transition state:

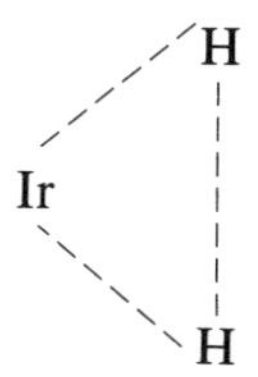

The reaction rate of $[IrX(CO)L_2]$ increases in the order X = I > Br > Cl but the dependence of k_1 on L is not simple[118]. Strohmeier[120] has compared changes in k_1 with the apparent electron density changes on the metal as estimated from $\nu(CO)$[121] and has found a maximum rate for P(*p*-tolyl)$_3$, as shown in Table 4.5, but a ready interpretation of this behaviour is difficult.

Table 4.5

	$P(C_6H_{11})_3$	$P(Pr^i)_3$	PBu_3	P(benzyl)$_3$	P(*p*-tol)$_3$	PPh_3	$P(OPh)_3$
k_1 (toluene/30 °C) $M^{-1}s^{-1}$	0.1	0.8	36	64	89	59	7.2
			Decreasing electron density on the metal. →				

4.7.2 Coordinatively saturated d^8-complexes

The addition of hydrogen to 5-coordinate d^8-complexes has not been studied quantitatively but several qualitative observations have some bearing on the mechanisms of these reactions. 5-coordinate d^8-complexes of the iron triad will react with H_2 but only under conditions where ligand-dissociation becomes likely (high temperature or photochemical conditions)[122–125]. Since the adducts formed are very stable toward dissociation and hydrogen is not displaced by carbon monoxide, kinetic rather than thermodynamic factors probably control the reaction with hydrogen.

$$H_2+[Os(CO)_3(PPh_3)_2] \xrightarrow{130\,°C/120\,atm} [OsH_2(CO)_2(PPh_3)_2]+CO \quad ^{122}$$
$$H_2+[Os(CO)_4(PPh_3)] \xrightarrow{130\,°C/120\,atm} [OsH_2(CO)_3(PPh_3)]+CO \quad ^{122}$$
$$H_2+[Os(CO)_5] \xrightarrow{100\,°C/80\,atm} [OsH_2(CO)_4]+CO \quad ^{123}$$
$$H_2+[Ru(CO)_3(PPh_3)_2] \xrightarrow{h\nu} [RuH_2(CO)_2(PPh_3)_2]+CO \quad ^{124}$$
$$H_2+[Fe(PF_3)_5] \xrightarrow{h\nu} [FeH_2(PF_3)_4]+PF_3 \quad ^{125}$$

Thus it seems that hydrogen can only react with coordinatively unsaturated molecules and then presumably by a one-step concerted process as described in Section 4.7.1.

This is supported by the reactivity of the isoelectronic complexes of the cobalt triad which can undergo reaction with hydrogen at room temperature and at atmospheric pressure, and this might be related to their more facile dissociation into 4-coordinate species. Unlike the complexes of the iron triad, these adducts can readily lose hydrogen by reacting with neutral ligands.

$$H_2+[Ir(CO)_3(PPh_3)_2]^+ \rightleftharpoons [IrH_2(CO)_2(PPh_3)_2]+CO \quad ^{78}$$
$$H_2+[Ir(CO)_2(PMePh_2)_3]^+ \rightleftharpoons [IrH_2(CO)(PMePh_2)_3]^+ +CO \quad ^{78}$$
$$H_2+[Ir(CO)(PMe_2Ph)_4]^+ \rightleftharpoons [IrH_2(CO)(PMe_2Ph)_3]^+ +PMe_2Ph \quad ^{30}$$

The lack of reactivity of vitamin B_{12S} and its analogues towards hydrogen, even though they are excellent nucleophiles, is most likely due in part to their lack of dissociation but more significantly to the planar chelate ligand system not allowing H_2 to enter two mutually *cis* coordination sites.

The reductive elimination of H_2 from $[IrH_2(CO)_2L_2]^+$ (L = tertiary phosphine) under the action of neutral ligand L′ to give $[Ir(CO)_2L_2L']^+$ has been studied mechanistically[127]. A unimolecular dissociation of hydrogen followed by a rapid reaction with L′ has been proposed.

$$[IrH_2(CO)_2L_2]^+ \underset{k_{-1}}{\overset{k_1}{\rightleftharpoons}} [Ir(CO)_2L_2]^+ + H_2 \xrightarrow[+L']{k_2} [Ir(CO)_2L_2L']^+.$$

The rate law expected for such a mechanism assuming a steady-state situation is $-d[IrH_2(CO)_2L_2{}^+]/dt = k_1k_2[IrH_2(CO)_2L_2{}^+][L']/\{k_{-1}[H_2]+k_2[L']\}$. Now if [L′] and k_2 are large then $k_2[L'] \gg k_{-1}[H_2]$ and the rate is given by $k_1[IrH_2(CO)_2L_2{}^+]$, which is in agreement with its observed independence of the ligand L′ (either its concentration or identity). Thus the rate of the reductive elimination step could be measured directly and gave the activation parameters $\Delta H^\ddagger = 20.3 \pm 0.6$ kcal mol^{-1} and $\Delta S^\ddagger = -5 \pm 2$ cal deg^{-1} mol^{-1} L = $PMePh_2$. The small change in entropy on forming the transition state and the isotope effect ($k_{H_2}/k_{D_2} = 2.1 \pm 0.3$) are consistent with a transition state similar to that for the oxidative addition reactions described earlier. That is, the controlling factor in the reaction is the enthalpy change (possibly dominated by that associated with the cleavage of the Ir—H bonds) rather than in re-structuring of the molecule or in charge separation on going from the dihydride to the transition state. The rate of reductive elimination of hydrogen decreases in the order L = $PPh_3 > AsPh_3 > PMePh_2 > PEtPh_2 > PEt_2Ph > PEt_3 > P(C_6H_{11})_3 \sim PPr^i_3$ which (apart from $AsPh_3$) is the reverse order of the basicity of the tertiary phosphine ligands.

4.7.3 *d⁷- complexes*

The reaction of the low-spin d^7-complex $[Co^{II}(CN)_5]^{3-}$ with molecular hydrogen (reaction type(b)) has been studied by a number of workers whose results have recently been confirmed by Halpern and co-worker[128]. The product is the d^6-complex $[Co^{III}(CN)_5H]^{3-}$ which is formed by a reaction with the rate law $-d[Co(CN)_5^{3-}]/dt = 2k_1[Co(CN)_5^{3-}]^2[H_2] - 2k_{-1}[Co(CN)_5H^{3-}]^2$ which is consistent with the equilibrium:

$$2[Co(CN)_5^{3-}] + H_2 \underset{k_{-1}}{\overset{k_1}{\rightleftharpoons}} 2[Co(CN)_5H^{3-}]$$

The energy required to cleave the hydrogen molecule is compensated for by the formation of two Co—H bonds by a concerted process, while a bimolecular reaction of $[Co(CN)_6]^{3-}$ with H_2 to give $[Co(CN)_5H]^{3-} + H\cdot$ would be prohibitively endothermic. The observed rate law, however, cannot be used to distinguish a termolecular process from one involving a preliminary fast equilibrium between $[Co(CN)_5]^{3-}$ and its dimer (in low concentration), which then undergoes a slow reaction with hydrogen. Other authors[129] have observed more complicated kinetics for this reaction which they interpret in terms of an initial disproportionation of the cobalt(II) complex to a cobalt(III) and a cobalt(I) complex, $[Co(CN)_4]^{3-}$, which undergoes oxidative addition of H_2 to give $[Co^{III}H_2(CN)_4]^{3-}$ which is finally converted to the monohydrido-product $[CoH(CN)_5]^{3-}$. These results are in direct conflict with those of Halpern and co-worker[128] who suggest that the more complicated kinetic behaviour is due to a side reaction which can be avoided by using higher hydrogen pressures.

4.7.4 d^{10}- complexes

Finally, no adducts of hydrogen with d^8- or d^{10}-complexes of the platinum triad have been isolated unambiguously, but whether for kinetic or thermodynamic reasons is uncertain. That the d^{10}-complex $[Pt(PPh_3)_2]$ has such low reactivity toward H_2 is worth comment since, in general, this d^{10}-complex forms more stable adducts (for instance, with O_2) than does Vaska's compound[13]. But, although no H_2-adducts have been isolated, a weak interaction is believed to occur as suggested by the exchange of the *ortho*-hydrogens of the PPh_3 ligands of $[Pt(PPh_3)_2]$ with D_2 [13].

4.8 OXYGEN AND OTHER SMALL MOLECULES

Some of the simple molecules that have been added to $[IrCl(CO)(PPh_3)_2]$ are oxygen, sulphur dioxide and carbon disulphide. Since there is good evidence in each case that the metal shows its typical nucleophilic behaviour these additions can be regarded as oxidative. The kinetics of addition of O_2 and H_2 to $[IrX(CO)(PPh_3)_2]$ follow a very similar pattern with rates in the order $I > Br > Cl$ and activation parameters of the same order (Tables 4.4 and 4.8) [4,7].

Table 4.8 Activation parameters for the addition of O_2 to $[IrX(CO)(PPh_3)_2]$

X	*Solvent*	$\Delta H_1^{\ddagger}$ kcal mol^{-1}	$\Delta S_1^{\ddagger}$ cal deg^{-1} mol^{-1}	ΔH° kcal mol^{-1}	ΔS° cal deg^{-1} mol^{-1}	*Reference*
Cl	C_6H_5Cl	9.3	−33	−17.1	−37	7
Cl	Benzene	13.1	−21			4
Br	Benzene	11.8	−24			4
I	Benzene	10.9	−24			4

Nevertheless, little can be said with any confidence about the transition state of this reaction or the meaning of the large negative entropies of activation.

The iridium complex above, the ruthenium(0) complex $[Ru(NO)(NCS)(PPh_3)_2]$ [147] and the complex $[Pt(PPh_3)_4]$ [148, 17] all form oxygen adducts and also catalyse the oxidation of triphenylphosphine to its oxide. Reports on the kinetics of the reaction of O_2 with the ruthenium complex have been particularly concerned with this catalytic process. Data were reported[147] for the cleavage of the coordinated O_2 but not for the addition of O_2 itself, which was a fast step in the catalytic cycle. On the other hand, similar studies on the platinum system[148] have established that oxygen adds to

$[Pt(PPh_3)_3]$ to give $[PtO_2(PPh_3)_2]$, while the alternative addition to $[Pt(PPh_3)_2]$, present in lower concentrations in these systems, cannot be detected by kinetic studies. This is in contrast to the addition of other molecules (for example, alkyl halides[17] and acetylenes[67]) which add exclusively to the dicoordinated species.

4.9 CLEAVAGE OF CARBON—CARBON BONDS BY OXIDATIVE ADDITION

Transition-metal complexes which are known to readily activate H—H bonds by oxidative addition show a considerably reduced reactivity towards C—H and C—C bonds. Such activations should be thermodynamically feasible since the energies of these bonds are comparable and also both alkyl and hydrido–metal bonds (for example, with iridium(III)) are known to have considerable stability. Carbon–hydrogen bond activation is well-established in a number of cases (see Section 4.5.1) but only recently have reports been made of C—C bond activation by oxidative addition. Reactions of d^8-complexes with strained three-membered carbon rings are known, for example, cyclopropane reacts with $[RhCl(CO)_2]_2$ to give the complex $[RhCl(C_4H_6O)(CO)]_2$[130] and triphenylpropenyl cation with $[IrCl(CO)(PMe_3)_2]$ to give $[Ir(C_3Ph_3)Cl(CO)(PMe_3)_2]^+$[131]. The apparent structure of the rhodium compound and the x-ray determined structure of the iridium compound are shown below:

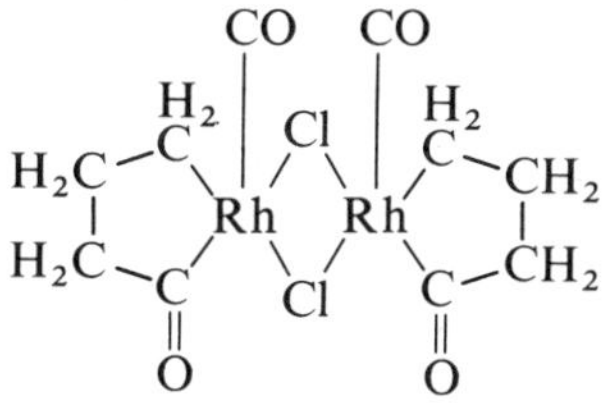

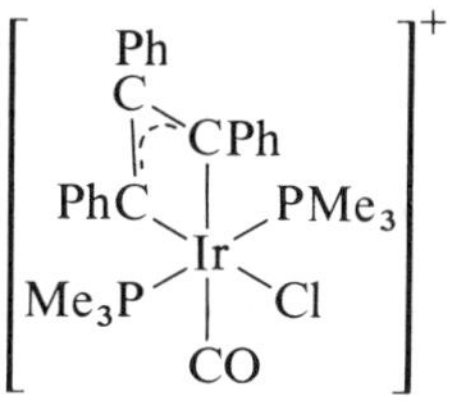

Now various isomerisations of hydrocarbons which are thermally 'forbidden' by the Woodward–Hoffmann orbital symmetry conservation rules can be catalysed by transition metal compounds (for example see Reference 132). It has been suggested that the 'forbidden' reactions become 'allowed' by a combination of the ligand and metal orbitals to give a new set of filled molecular orbitals which are now of suitable symmetry[132]. However, Halpern has proposed that in certain cases an alternative mechanism operates[134,135]. There is evidence that in the isomerisations of cubane to *syn*-tricyclo-octadiene[135], of quadricyclene to norbornadiene[134,133] and of hexamethylprismane to hexamethyl 'Dewar benzene'[134] catalysed by $[Rh(diene)Cl]_2$ (diene = cyclo-octa-1,5-diene or norbornadiene) an important step is the oxidative addition to the d^8-complex with cleavage of the C—C bond of the adding molecule. The evidence is based mainly on the isolation of complexes formed apparently by C—C bond cleavage and carbonyl insertion[134, 135] and illustrated below:

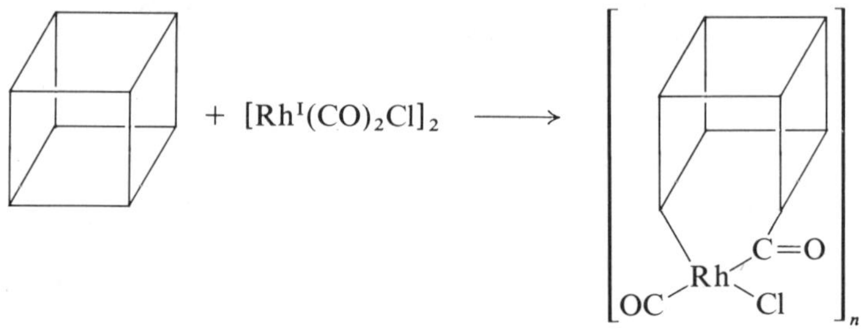

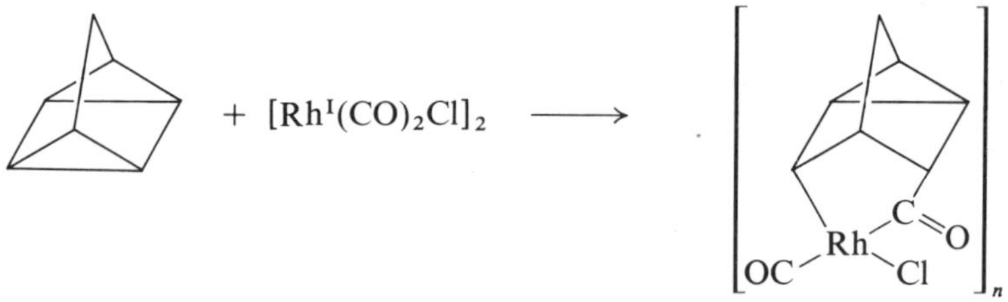

$$+ [Rh^{I}(CO)_2Cl]_2 \longrightarrow$$

OC Rh C=O Cl n

$$+ [Rh^{I}(CO)_2Cl]_2 \longrightarrow$$

OC Rh C=O Cl n

The kinetics of addition[135] of cubane to $[Rh(CO)_2Cl]_2$ follows the rate law $-d[\text{cubane}]/dt = k[\text{cubane}][Rh_2(CO)_4Cl_2]$ which is a very similar expression to that for the isomerisations catalysed by $[Rh(\text{diene})Cl]_2$ [133]. However, the mechanistic path of the addition itself is unknown, but a one-step concerted reaction with a 3-centred transition state is most immediately attractive.

4.10 ADDITION OF OLEFINS AND ACETYLENES

4.10.1 d^8-complexes

Complexes of the type $[IrCl(CO)(PR_3)_2]$ will react with olefins or acetylenes but the equilibrium constants for dissociation of the adducts formed is considerably dependent upon the nature of the olefin or acetylene. Strohmeier and co-worker have measured constants for the equilibrium[136].

$[IrCl(CO)L_2(ol)] \overset{K}{\rightleftharpoons} [IrCl(CO)L_2] + ol$ (ol = olefin) and for $[IrCl(CO)\{P(OPh)_3\}_2]$ found the values for K: maleic anhydride, 1.1×10^{-4}; dimethylmaleate, 1.0×10^{-3}; ethyl acrylate, 2.6×10^{-2}; dimethylfumarate, 2.2×10^{-1}; 1-heptene, 9.4; styrene, 37 (measured in toluene at 20 °C). It can be seen that olefins with electron-withdrawing substituents give adducts of the greatest stability. Similar conclusions were made by Vaska on studying similar equilibria but with a different range of olefins, and dissociation constants were found in the order $(CN)_2C{=}C(CN)_2 < C_2F_4 < (CN)HC{=}CH(CN) < H_2C{=}CH(CN) < C_2H_4$[7]. Vaska has also estimated that increased transfer of electron density from the metal to the olefin as judged by the change in $\nu(CO)$ correlates with an increase in stability of the adducts[7]. Adducts have been isolated only for the more electronegative of these olefins.

It has been a point of discussion as to whether such additions can be regarded as oxidative at all, i.e. whether the adducts should be regarded as (A) or (B)

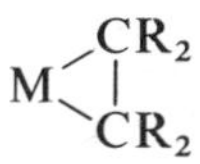

$$\mathrm{M}\langle\begin{smallmatrix}CR_2\\|\\CR_2\end{smallmatrix} \qquad (A) \qquad\qquad \mathrm{M}\leftarrow\begin{smallmatrix}CR_2\\\|\\CR_2\end{smallmatrix} \qquad (B)$$

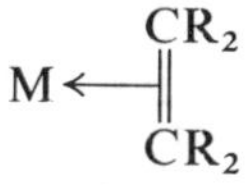

It appears that with highly electronegative substituents on the olefin there is a large degree of back-donation from the metal to the olefin using model (B), and under these conditions the two models approach and one can with some justification regard the metal as oxidised. It is in terms of model (A) that we will consider these complexes.

Green and co-workers[138] have looked at the reaction of $[IrCl(CO)(PPh_3)_2]$ with *trans*-$CF_3(CN)C{=}C(CF_3)CN$ which gives an adduct in which the stereochemistry of the olefin had been retained, and in which incidentally

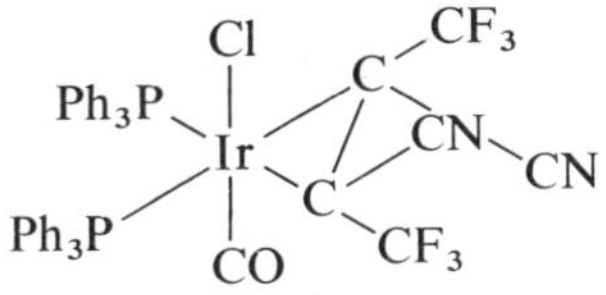

$$[IrCl(CO)(PPh_3)_2\{C(CF_3)(CN)C(CF_3)(CN)\}]$$

the tertiary phosphines had become mutually *cis*, as had been shown earlier in some similar examples[139, 140, 52]. The addition reaction showed the rate law $-d[Ir^I]/dt = k[Ir^I][\text{olefin}]$ and gave the activation parameters shown in Table 4.6.

Table 4.6 Activation parameters for the addition of *trans*-$CF_3(CN)C{=}C(CF_3)CN$ to *trans*-$[IrCl(CO)(PPh_3)_2]$ [138]

Solvent	$\Delta H^\ddagger$(kcal mol^{-1})	$\Delta S^\ddagger$(cal deg^{-1} mol^{-1})
Benzene	7.2	−38.7
Chlorobenzene	6.9	−40.6
Benzonitrile	10.9	−30.0

Two possible transition states have been suggested by these workers which are broadly equivalent to the two transition states in alkyl halide addition discussed in Section 4.4.

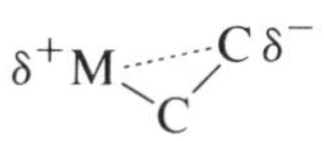

$$\delta^+\mathrm{M}\cdots C\,\delta^- \text{ (with M–C and C–C to second C)} \quad (A) \qquad\qquad \mathrm{M}\vdots\begin{smallmatrix}C\\|\\C\end{smallmatrix} \quad (B)$$

In transition state (A) one carbon interacts more strongly with the metal than the other and there is a considerable charge separation. The solvent effect on the activation parameters were not sufficiently great, however, to support transition state (A) unambiguously[138]. Transition state (A) could, but not necessarily, lead to the intermediate $^+M{-}C{-}C^-$ in which there is free rotation about the C—C bond before ring closure and there is evidence reported which may support such a rotation. It was shown[138] that a *cis/trans* mixture of olefins reacted without change in the composition of the mixture to give exclusively the adduct with coordinated *trans*-olefin. For this to be observed both isomers must react equally fast to give the same *trans*-olefin complex and this implies an ionic intermediate, unless there is an alternative mechanism for isomerisation after the initial addition.

The similarity is very apparent between this situation and that of the substitution of vinyl halides by cobalt(I) anions, where there is evidence for a similar ionic intermediate (see Section 4.4.22). In the addition to [IrCl(CO)$(PPh_3)_2$] there is a ring closure of the intermediate M—C C, while loss of halide ion from the cobalt intermediate gives the vinyl complex. In the example for cobalt, the halide ion must be lost from the intermediate sufficiently fast (or there remains a fairly large barrier to rotation about the apparently single C—C bond) to enable retention of the olefin configuration, whereas the iridium intermediate must be sufficiently long-lived to allow *cis/trans* isomerisation of the olefin. Such mechanistic ideas are at the moment only based on limited evidence and must be regarded with considerable caution.

Similar ionic intermediates may also occur in the addition of electronegative olefins[137, 141] or acetylenes[141] to the 5-coordinate d^8-complexes of iron, ruthenium and osmium. This reaction scheme it was proposed[141] could account very conveniently for the possible 'insertion' steps that were observed in these reactions. For example, for hexafluorodimethylacetylene there is either insertion by carbon monoxide or another acetylene molecule.

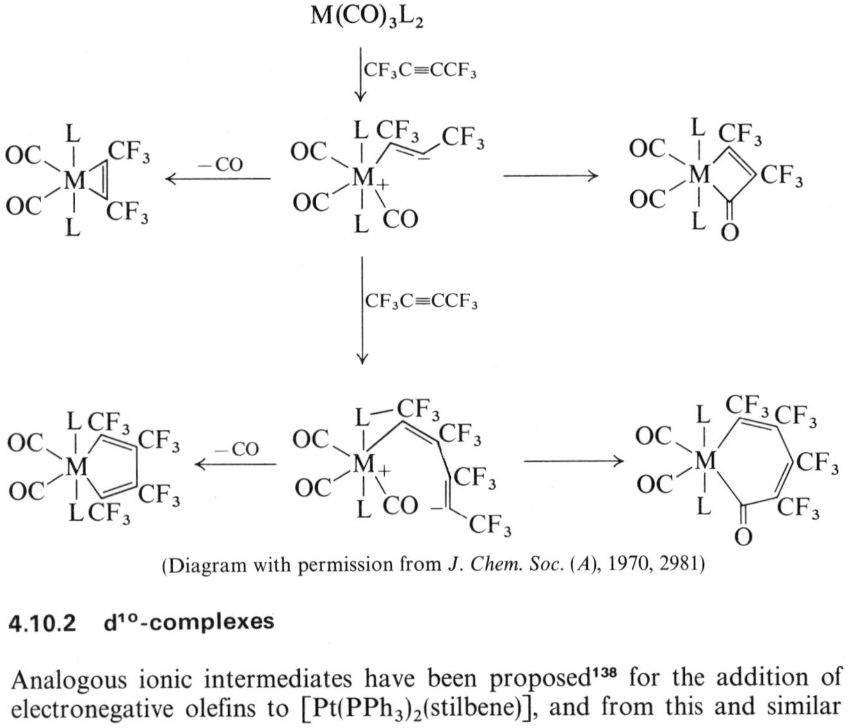

$M(CO)_3L_2$

$CF_3C{\equiv}CCF_3$

–CO

$CF_3C{\equiv}CCF_3$

–CO

(Diagram with permission from *J. Chem. Soc.* (*A*), 1970, 2981)

4.10.2 d^{10}-complexes

Analogous ionic intermediates have been proposed[138] for the addition of electronegative olefins to [$Pt(PPh_3)_2$(stilbene)], and from this and similar reactions a range of complexes of the type $[Pt(PPh_3)_2(olefin)]$ have been isolated. But with the addition of chloro-olefins to platinum(0) compounds the initially-formed olefin complex (moderately stable in benzene solution) may isomerise to the platinum(II) vinyl complexes in more polar solvents[143]. In the example below a chlorine atom has been transferred to the metal from the tetrachloroethylene ligand and similar isomerisations are reported for the cyano- and bromo- but not for the fluoro-olefins.

$$[Pt(PPh_3)_4] + C_2Cl_4 \xrightarrow{-2PPh_3} [Pt(PPh_3)_2(C_2Cl_4)]$$
$$\rightarrow trans\text{-}[PtCl(CCl{=}CCl_2)(PPh_3)_2]$$

The 'vinyl rearrangement' has been the subject of some mechanistic studies[144, 145] and it has been shown that the rate of rearrangement is first-order and dependent upon the solvent polarity. The activation data for the rearrangement of $[Pt(PPh_3)_2(C_2Cl_4)]$ are given in Table 4.7.

Table 4.7 Activation data for the 'vinyl rearrangement' reaction of $[Pt(C_2Cl_4)(PPh_3)_2]$ [144]

Solvent	$\Delta H^{\ddagger}$(kcal mol^{-1})	$\Delta S^{\ddagger}$(cal deg^{-1} mol^{-1})
MeOH	15.0 ± 1.0	-26 ± 4
n-PrOH	19.8 ± 1.0	-15 ± 4
EtOH	21.4 ± 1.0	-9 ± 4

The kinetic behaviour of this reaction (notably the negative value of $\Delta S^{\ddagger}$ and the solvent-dependence of the activation parameters) is consistent with an S_N1 (lim) mechanism, the rate-determining step being the solvolysis of chloride followed by rapid attack of chloride on the metal[144].

$$(Ph_3P)_2Pt(CCl_2{=}CCl_2) \longrightarrow (Ph_3P)_2Pt(CCl_2\text{–}\overset{+}{C}Cl) + Cl^- \longrightarrow trans\text{-}[PtCl(CCl{=}CCl_2)(PPh_3)_2]$$

In the corresponding reaction of C_2Cl_3H to give *cis*-$[PtCl(CH{=}CCl_2)(PPh_3)_2]$ there appears to be evidence that the S_N1 process is not of the limiting type (unassisted dissociation) but that an intramolecular process contributes to the kinetics[145]. The total picture of these reactions is of an oxidative addition of vinyl halide to a platinum(0) complex, the C—Cl bond being activated by preliminary coordination to the metal through the C—C bond. These isomerisations are more facile for nickel and palladium and only the vinyl complexes are generally observed for these metals.

Additions of acetylenic compounds to platinum(0) compounds to give the complexes $[PtL_2ac]$ (L = tertiary phosphine, ac = acetylene) are also very well established, and Chatt and co-workers[146] have shown by a qualitative study of the equilibrium: $PtL_2ac + ac' \rightleftharpoons PtL_2ac' + ac$ that stabilities depend on the acetylene (ac) in the order $C_2H_2 <$ alkyl–C≡CH < alkyl–C≡C–alkyl < PhC≡CH < PhC≡CPh < $C_2(p\text{-}NO_2C_6H_4)_2$, and it has been confirmed by many more examples that electron withdrawing groups stabilise the adducts. Halpern and co-workers have shown[67] that the addition of acetylenes to either $[Pt(PPh_3)_2(C_2H_4)]$ or $[Pt(PPh_3)_3]$ goes via preliminary dissociation of ethylene or triphenylphosphine respectively, and the second-order rate constants for the addition to the species so formed are in the order: diphenylacetylene < phenylacetylene < acetylene. This is an example of the rate of addition being totally unrelated to the stability of the adduct, and in this case the rates and the stabilities are in the reverse order.

4.11 SUMMARY

The mechanisms of oxidative addition reactions are dependent upon various factors: (a) The addendum molecules. These can be broadly divided into those which only add via 3-centred transition states, e.g. H_2 or silanes, and those which may add to the metal through only one atom generally giving ionic intermediates, e.g. halogens and alkyl halides. There is no apparent reason why the latter type should always add in this way, and for alkyl halides the mode of reaction with square-planar d^8-molecules is still in dispute. (b) The coordination number and electron configuration of the complexes. At the moment it appears that the first type of addendum molecule can only add to complexes which are coordinatively unsaturated (with the exception of d^7-complexes which behave differently to d^8- or d^{10}-complexes). (c) The ligands. Although the effect of different ligands on thermodynamic aspects of oxidative addition has been the subject of several qualitative studies[5], the effect of ligands on the mechanism of addition has received scant appraisal. At present it appears that they have only a secondary effect and that the mechanism is broadly determined by factors (a) and (b). It is in reactions with preliminary ligand-dissociation that the nature of the ligands might well become dominant. A true assessment of the above factors must, however, await further studies.

NOTE ADDED ON PROOF

Stereochemical changes at carbon on addition of alkyl halides

It has been shown that optically active *sec*-octyl bromide reacts with a cobaloxime(I) complex to give an alkyl-cobalt(III) complex which on subsequent reaction with bromine regenerates the *sec*-octyl bromide with retained configuration[40], (see page 126). Thus both the oxidative addition and the cleavage reaction must have taken place with retention or both with inversion. Similar observations had previously been made on the addition to an iridium(I) complex and Pearson and Muir concluded that the addition had occurred with retention on the assumption that the cleavage had also occurred with retention[53], (see page 130). However, a recent report that iron-carbon σ-bonds can be cleaved with inversion has meant that there are examples of both inversion and retention for the cleavage reaction. Therefore the assumption that the iridium–carbon bond had been cleaved with retention must be re-examined as must the conclusions of Pearson and Muir concerning the addition reaction.

References

1. Montelatici, S., van der Ent, A., Osborn, J. A. and Wilkinson, G. (1968). *J. Chem. Soc.(A)*., 1054 and references therein
2. Strohmeier, W. and Onodo, T. (1969). *Z. Naturforsch.*, **24B,** 1493
3. James B. R. and Memon, N. A. (1968). *Can. J. Chem.*, **46,** 217
4. Chock, P. B. and Halpern, J. (1966). *J. Amer. Chem. Soc.*, **88,** 3511
5. Collman, J. P. and Roper, W. R. (1968). *Adv. Organometal. Chem.*, **7,** 53
6. Collman, J. P. (1968). *Accounts Chem. Res.*, **1,** 136
7. Vaska, L. (1968). *Accounts Chem. Res.*, **1,** 335
8. Halpern, J. (1970). *Accounts Chem. Res.*, **3,** 386
9. Cramer, R. (1968). *Accounts Chem. Res.*, **1,** 186
10. Parshall, G. W. (1970). *Accounts Chem. Res.*, **3,** 139
11. Shriver, D. F. (1970). *Accounts Chem. Res.*, **3,** 231
12. Ugo, R. (1968). *Coord. Chem. Revs.*, **3,** 319
13. Ugo, R., La Monica, G., Cariati, F., Cenini, S. and Conti, F. (1970). *Inorg. Chim. Acta.*, **4,** 390 and references therein
14. Halpern, J. (1968). *Adv. Chem. Series*, **70,** 1
15. McClure, G. L. and Baddley, W. H. (1970). *J. Organometal. Chem.*, **25,** 261
16. Tolman, C. A. (1970). *J. Amer. Chem. Soc.*, **92,** 2956
17. Birk, J. P., Halpern, J. and Pickard, A. L. (1968). *J. Amer. Chem. Soc.*, **90,** 4491
18. Jonas, K. and Wilke, G. (1969). *Angew. Chem. (Int. Ed.)*, **8,** 519
19. Blake, D. M. and Nyman, C. T. (1970). *J. Amer. Chem. Soc.*, **92,** 5359
20. Collman, J. P. and Roper, W. R. (1966). *J. Amer. Chem. Soc.*, **88,** 3504
21. Noack, K. (1968). *J. Organometal. Chem.*, **13,** 411
22. Kruse, A. E. and Angelici, R. J. (1970). *J. Organometal. Chem.*, **24,** 231
23. Skinner, C. E. and Jones, M. M. (1969). *J. Amer. Chem. Soc.*, **91,** 4405
24. Collman, J. P. and Sears, C. T. (1968). *Inorg. Chem.*, **7,** 27
25. Bennett, M. A., Clark, R. J. H. and Milner, D. L. (1967). *Inorg. Chem.*, **6,** 1647
26. Bennett, M. A., Erskine, G. J. and Nyholm, R. S. (1967). *J. Chem. Soc.(A)*, 1260 and references therein
27. Deeming, A. J. and Shaw, B. L. (1971). *J. Chem. Soc.(A)*, 376
28. Adams, D. M., Cook, D. J. and Kemmitt, R. D. W. (1968). *J. Chem. Soc.(A)*, 1067 and references therein
29. Collman, J. P. and Roper, W. R. (1966). *Chem. Commun.*, 244
30. Deeming, A. J. and Shaw, B. L. (1970). *J. Chem. Soc. (A)*, 3356
31. Oliver, A. J. and Graham, W. A. G. (1970). *Inorg. Chem.*, **9,** 2653
32. Oliver, A. J. and Graham, W. A. G. (1970). *Inorg. Chem.*, **9,** 243
33. Hart-Davis, A. J. and Graham, W. A. G. (1970). *Inorg. Chem.*, **9,** 2658
34. Heck, R. F. (1963). *J. Org. Chem.*, **28,** 604
35. McCleverty, J. A. and Wilkinson, G. (1964). *J. Chem. Soc.*, 4200
36. Schrauzer, G. N. and Deutsch, E. (1969). *J. Amer. Chem. Soc.*, **91,** 3341
37. Schrauzer, G. N., Deutsch, E. and Windgassen, R. J. (1968). *J. Amer. Chem. Soc.*, **90,** 2441
38. Schrauzer, G. N. (1968). *Acc. Chem. Res.*, **1,** 97 and references therein
39. Jensen, F. R., Madan, V. and Buchanan, D. H. (1970). *J. Amer. Chem. Soc.*, **92,** 1414
40. Johnson, M. D. and Dodd, D. (1970). Unpublished work
41. Bordwell, F. G. (1970). *Accounts Chem. Res.*, **3,** 281
42. Collman, J. P., Cawse, J. N. and Kang, J. W. (1969). *Inorg. Chem.*, **8,** 2574
43. Johnson, M. D. and Mayle, C. (1970). *J. Chem. Soc. (D)*, 192
44. N-V-Duong, K. and Gaudemer, A. (1970). *J. Organometal. Chem.*, **22,** 473
45. Naumberg, M., N-V-Duong, K. and Gaudemer, A. (1970). *J. Organometal. Chem.*, **25,** 231
46. Johnson, M. D. and Meeks, B. S. (1970). *J. Chem. Soc. (D)*, 1027
47. Weber, J. H. and Schrauzer, G. N. (1970). *J. Amer. Chem. Soc.*, **92,** 726
48. Ochiai, E., Long, K. M., Sperati, C. R. and Busch, D. H. (1969). *J. Amer. Chem. Soc.*, **91,** 3201
49. Whitesides, G. W. and Borschetto, D. J. (1969). *J. Amer. Chem. Soc.*, **91,** 4313
50. Johnson, R. W. and Pearson, R. G. (1970). *J. Chem. Soc. (D)*, 986
51. Dessey, R. E., Pohl, R. L. and King, R. B. (1966). *J. Amer. Chem. Soc.*, **88,** 5121

52. Deeming, A. J. and Shaw, B. L. (1969). *J. Chem. Soc. (A)*, 1128
53. Pearson, R. G. and Muir, W. R. (1970). *J. Amer. Chem. Soc.*, **92**, 5519, and (1970), 160*th Amer. Chem. Soc. Nat. Meeting*, Inorg. 56, (Chicago)
54. Gray, H. B. and Wojcicki, A. (1960). *Proc. Chem. Soc.*, 358
55. Blake, D. M. and Kubota, M. (1970). *J. Amer. Chem. Soc.*, **92**, 2578
56. Ettorre, R. (1969). *Inorg. Nucl. Chem. Lett.*, **5**, 45
57. Halpern, J. and Chock, P. B. (1967). *Proc.* 10*th Int. Conf. Coord. Chem.*, 135, (Tokyo)
58. Jensen, F. R. and Gale, L. H. (1960). *J. Amer. Chem. Soc.*, **82**, 148
59. Jensen, F. R., Whipple, L. D., Wedegaertner, D. K. and Landgrebe, J. A. (1960). *J. Amer. Chem. Soc.*, **82**, 2466
60. Labinger, J. A., Braus, R. J., Dolphin, D. and Osborn, J. A. (1970). *J. Chem. Soc. (D)*, 612
61. Douek, I. C. and Wilkinson, G. (1969). *J. Chem. Soc. (A)*, 2604
62. Ugo, R., Pasini, S., Cenini, S., Fusi, A. and Conti, F. (1968). *Proc. 11th Int. Conf. Coord. Chem.*, 156, (Haifa)
63. Uguagliati, P., Palazzi, A., Deganello, G. and Belluco, U. (1970). *Inorg. Chem.*, **9**, 724
64. Deeming, A. J. and Shaw, B. L. (1968). *J. Chem. Soc. (D)*, 751 and (1969) *J. Chem. Soc. (A)*, 152
65. Collman, J. P., Kubota, M., Vastine, F. D., Sun, J. Y., and Kang, J. W. (1968). *J. Amer. Chem. Soc.*, **90**, 5430
66. Collman, J. P., Vastine, F. D. and Roper, W. R. (1968). *J. Amer. Chem. Soc.*, **90**, 2282
67. Birk, J. P., Halpern, J. and Pickard, A. L. (1968). *Inorg. Chem.*, **7**, 2672
68. Cook, C. D. and Jauhal, G. S. (1967). *Can. J. Chem.*, **45**, 301
69. Fahey, D. R. (1970). *J. Amer. Chem. Soc.*, **92**, 402
70. Chock, P. B. and Halpern, J. (1969). *J. Amer. Chem. Soc.*, **91**, 582 and references therein
71. Schneider, P. W., Phelan, P. F. and Halpern, J. (1969). *J. Amer. Chem. Soc.*, **91**, 77
72. Marzilli, L. G., Marzilli, P. A. and Halpern, J. (1970). *J. Amer. Chem. Soc.*, **92**, 5753
73. Chock, P. B., Dewar, R. B. K., Halpern, J. and Wong, L. Y. (1969). *J. Amer. Chem. Soc.*, **91**, 82
74. Kochi, J. K. and Powers, J. W. (1970). *J. Amer. Chem. Soc.*, **92**, 137
75. Laing, K. R. and Roper, W. R. (1969). *J. Chem. Soc. (A)*, 1889
76. Iqbal, Z. and Waddington, T. C. (1968). *J. Chem. Soc. (A)*, 2958 and references therein
77. Deeming, A. J., Johnson, B. F. G. and Lewis, J. (1970). *J. Chem. Soc. (A)*, 2967
78. Church, M. J., Mays, M. J., Simpson, R. N. F. and Stefanini, F. P. (1970). *J. Chem. Soc. (A)*, 2909
79. Ibekwe, S. D. and Taylor, K. A. (1970). *J. Chem. Soc. (A)*, 1
80. Vaska, L. (1966). *J. Amer. Chem. Soc.*, **88**, 5325
81. Vrieze, K., reference given by Baird, M. C., Mague, J. T., Osborn, J. A. and Wilkinson, G. (1967). *J. Chem. Soc. (A)*, 1347
82. Blake, D. M. and Kubota, M. (1970). *Inorg. Chem.*, **9**, 989
83. Benzoni, L., Zanzottera, C., Camia, M., Venturi, M. T. and De Innocentiis, M. (1968). *Chim. Ind. (Milan)*, **50**, 1227
84. Deeming, A. J. and Shaw, B. L. (1969). *J. Chem. Soc. (A)*, 1802
85. Mague, J. T. and Wilkinson, G. (1968). *J. Chem. Soc. (A)*, 1736
86. Cariati, F., Ugo, R. and Bonati, F. (1966). *Inorg. Chem.*, **5**, 1128
87. Drinkard, W. C., Eaton, R. D., Jesson, J. P. and Lindsay, R. V. (1970). *Inorg. Chem.*, **9**, 392
88. Schunn, R. A. (1970). *Inorg. Chem.*, **9**, 394
89. Tolman, C. A. (1970). *J. Amer. Chem. Soc.*, **92**, 4217
90. Tolman, C. A. (1970). *J. Amer. Chem. Soc.*, **92**, 6777
91. Chatt, J. and Davidson, J. M. (1965). *J. Chem. Soc.*, 843
92. Bennett, M. A. and Milner, D. L. (1969). *J. Amer. Chem. Soc.*, **91**, 6983
93. Parshall, G. W. (1968). *J. Amer. Chem. Soc.*, **90**, 1669
94. Keim, W. (1969). *J. Organometal. Chem.*, **19**, 161
95. Cheney, A. J., Mann, B. E., Shaw, B. L. and Slade, R. M. (1970). *J. Chem. Soc. (D)*, 176
96. Green, M. L. H. and Knowles, P. J. (1970). *J. Chem. Soc. (D)*, 1677
97. Tsuji, J. and Ohno, K. (1968). *J. Amer. Chem. Soc.*, **90**, 99 and references therein
98. Walborsky, H. M. and Allen, L. E. (1970). *Tetrahedron Lett.*, 823
99. Belluco, U., Croatto, U., Uguagliari, P. and Peitropaola, R. (1967). *Inorg. Chem.*, **6**, 718

100. Belluco, U., Gustiniana, M. and Grazini, M. (1967). *J. Amer. Chem. Soc.,* **89,** 6494
101. Barefield, E. K., Parshall, G. W. and Tebbe, F. N. (1970). *J. Amer. Chem. Soc.,* **92,** 5234
102. Chalk, A. J. (1969). *J. Chem. Soc. (D),* 1207 and references therein
103. Lappert, M. F. and Travers, N. F. (1968). *J. Chem. Soc. (D),* 1569, and (1970), *J. Chem. Soc. (A),* 3303
104. Glockling, F. and Wilbey, M. D. (1969). *J. Chem. Soc. (D),* 286, and (1970), *J. Chem. Soc. (A),* 1675 and references therein
105. de Charentenay, F., Osborn, J. A. and Wilkinson, G. (1968). *J. Chem. Soc. (A),* 787
106. Chalk, A. J. (1970). *J. Organometal. Chem.,* **21,** 207
107. Sommer, L. H. and Lyons, J. E. (1968). *J. Amer. Chem. Soc.,* **90,** 4197
108. Sommer, L. H., Lyons, J. E. and Fujimoto, H. (1969). *J. Amer. Chem. Soc.,* **91,** 7051
109. Brook, A. G., Duff, J. M. and Anderson, D. G. (1970). *J. Amer. Chem. Soc.,* **92,** 7567
110. Harrod, J. F. and Smith, C. A. (1970). *J. Amer. Chem. Soc.,* **92,** 2699
111. Harrod, J. F., Gilson, D. F. R. and Charles, R. (1969). *Can. J. Chem.,* **47,** 2205
112. Harrod, J. F. and Smith, C. A. (1970). *Can. J. Chem.,* **48,** 870
113. Jetz, W. and Graham, W. A. G. (1969). *J. Amer. Chem. Soc.,* **91,** 3375
114. Halpern, J. (1968). *Discuss. Faraday Soc.,* **46.** 7
115. Vaska, L. (1966). *J. Amer. Chem. Soc.,* **88,** 4100 and references therein
116. Craig Taylor, R., Young, J. F. and Wilkinson, G. (1966). *Inorg. Chem.,* **5,** 20
117. Vaska, L. (1968). 155*th Amer. Chem. Soc. Nat. Meeting,* M51, (San Fransisco)
118. Strohmeier, W. and Muller, F. J. (1969). *Z. Naturforsch.,* **24B,** 931
119. Yagupsky, M. P. and Wilkinson, G. (1968). *J. Chem. Soc. (A),* 2813
120. Strohmeier, W. and Onoda, T. (1969). *Z. Naturforsch.,* **24B,** 515 *0*
121. Strohmeier, W. and Onoda, T. (1968). *Z. Naturforsch.,* **23B,** 1377
122. L'Eplattenier, F. and Calderazzo, F. (1968). *Inorg. Chem.,* **7,** 1290
123. L'Eplattenier, F. and Calderazzo, F. (1967). *Inorg. Chem.,* **6,** 2092
124. Valentine, D., Sarver, B. and Collman, J. P. (1970), manuscript in preparation (referred to by Valentine, J. S. and Valentine, D. (1970). *J. Amer. Chem. Soc.,* **92,** 5795)
125. Kruck, von Th. and Prasch, A. (1969). *Z. anorg. Allg. Chem.,* **371,** 1
126. Sacco, A. and Rossi, M. (1967). *Chem. Commun.,* 316
127. Mays, M. J., Simpson, R. N. F. and Stefanini, F. P. (1970). *J. Chem. Soc. (A),* 3000
128. Halpern, J. and Pribanić, M. (1970). *Inorg. Chem.,* **9,** 2616
129. Banks, R. G., Das, P. K., Hill, H. A. O., Pratt, J. M. and Williams, R. J. P. (1968). *Discuss. Faraday Soc.,* **46,** 80
130. Roundhill, D. M., Lawson, D. N. and Wilkinson, G. (1968). *J. Chem. Soc. (A),* 845
131. Tuggle, R. M. and Weaver, D. L. (1970). *J. Amer. Chem. Soc.,* **92,** 5523
132. Pettit, R., Sugahara, H., Wristers, J. and Merk, W. (1969). *Discuss. Faraday Soc.,* **47,** 71 and references therein
133. Hogeveen, H. and Volger, H. C. (1967). *J. Amer. Chem. Soc.,* **89,** 2486
134. Cassar, L. and Halpern, J. (1970). *J. Chem. Soc. (D),* 1082
135. Cassar, L., Eaton, P. E. and Halpern, J. (1970). *J. Amer. Chem. Soc.,* **92,** 3515
136. Strohmeier, W. and Fleischmann, R. (1969). *Z. Naturforsch,* **24B,** 1217
137. Cooke, M. and Green, M. (1969). *J. Chem. Soc. (A),* 651
138. Ashley-Smith, J., Green, M. and Wood, D. C. (1970). *J. Chem. Soc. (A),* 1847
139. Ibers, J. A., McGinnety, J. and Kime, N. (1967). *Proc.* 10*th Int. Conf. Coord. Chem.,* 93, (Tokyo)
140. McGinnety, J. A. and Ibers, J. A. (1968). *Chem. Commun.,* 235
141. Burt, R., Cooke, M. and Green, M. (1970). *J. Chem. Soc. (A),* 2975 and 2981
142. Cooke, M. and Green, M. (1969). *J. Chem. Soc. (A),* 651
143. Bland, W. J. and Kemmitt, R. D. W. (1968). *J. Chem. Soc. (A),* 1278
144. Bland, W. J., Burgess, J. and Kemmitt, R. D. W. (1968). *J. Organometal. Chem.,* **14,** 201 and **15,** 217
145. Bland, W. J., Burgess, J. and Kemmitt, R. D. W. (1969). *J. Organometal. Chem.,* **18,** 199
146. Chatt, J., Rowe, G. A. and Williams, A. A. (1957). *Proc. Chem. Soc.,* 208
147. Graham, B. W., Laing, K. R., O'Connor, C. J. and Roper, W. R. (1970). *J. Chem. Soc. (D),* 1272
148. Halpern, J. and Pickard, A. L. (1970), *Inorg. Chem.,* **9,** 2798
149. Whitesides, G. W. and Boschetto, D. J. (1971). *J. Amer. Chem. Soc.,* **93,** 1529

5
Inorganic Photochemistry

R. B. BUCAT and D. W. WATTS
University of Western Australia

5.1 INTRODUCTION

Photochemical investigations are evident in the very early literature in inorganic chemistry. There has been a great expansion of interest in recent years in the study of main block elements as well as transition metal co-ordination complexes and organometallic compounds.

We have taken the view that, although this article is to be one of a series of biennial reviews, since it is the first there should be an attempt to present a reasonably balanced picture of the topics discussed, and thus references before the 1968–1970 period are often presented.

Some selection has been necessary in order to make the task manageable and the review coherent. We have chosen to omit solid-state photochemistry and the photochemistry of absorbed layers, and thus exclude the field of photography. We have also excluded the photochemistry of metal chelate compounds of biological origin such as porphyrins.

The photochemistries of the alkali and alkaline earth metals are omitted, although the increasing interest here, particularly in metal amine solutions, is recognised by a recent review[1]. There is also developing a vast literature in the applied area of the photochemical reactions of atmospheric pollution[2,3] and this is ignored.

The hydrated electron plays a vital role in the photochemical behaviour of aqueous anions and cations. Three recent reviews[4–6] have dealt with much of this chemistry, and thus no special Section is devoted to this topic. The very much smaller field of inorganic photochromism, although interesting, has not received special mention, but some examples are included where relevant.

We have considered for the most part only the primary photochemical processes and have omitted the subsequent reactions of radicals and other fragments. This decision alone has allowed us to give small coverage of the organometallic field in which two major reviews have recently appeared[7,8]. No Section is devoted to experimental procedures.

The review is written in two major Sections. The first is a discussion of the photochemistry of the metallic elements which include the transition metals, the lanthanides and actinides and certain of the heavy main block elements. The second Section contains a review of the recent work on aqueous solutions of anions of the non-metallic elements.

5.2 PHOTOCHEMISTRY OF COMPOUNDS OF THE METALLIC ELEMENTS

Balzani and Carassiti[7] have published a comprehensive treatise on just this field, and a number of more or less extensive summaries have appeared[7–24].

It is not always possible to find similarities in the photochemistries of metals of the same group of the Periodic system; however there remain advantages in comparing metals in the various triads of the transition series and in vertical groups in the main block of the Periodic Table. What photochemistry has been studied for the lanthanides and actinides will be discussed together. Metal carbonyl complexes and the related organometallic compounds with metals in low oxidation states are gathered together because of the similarity of their photochemical initiation.

5.2.1 Scandium, yttrium and lanthanum

Inclusion of this group is essentially *pro forma* in that virtually no observations have been reported. It is known that both Y^{3+} and La^{3+} ions in aqueous solution can photosensitise the polymerisation of vinyl compounds[25–28]. It appears that no photochemistry for scandium has been reported.

5.2.2 Titanium, zirconium and hafnium

No reports of photochemical activity in zirconium and hafnium compounds have been found. There are a number of reports, beyond the scope of this review, of the photochemical activity of TiO_2. Aqueous solutions of Ti^{III} have been shown to sensitise the polymerisation of acrylic monomers[29, 30], probably by photoelectron production[12]. Aqueous titanium(IV) lactate solutions are photo-reduced to titanium(III) in the presence of methylene blue. This reaction seems to follow the photo-reduction of the dye[31]. Ungurenasu and Cecal[32] report the photolysis of bis-(π-cyclopentadienyl)titanium(IV) cyclotetrasulphan, $(\pi\text{-}C_5H_5)_2TiS_4$. Finally, the photo-reduction of oxalatotitanium(IV) has been reported[33].

5.2.3 Vanadium, niobium and tantalum

No photochemistry has been reported for tantalum compounds. Vanadium(II), (III) and (IV), like titanium(III), photosensitises the polymerisation of acrylic monomers[29, 30] in aqueous solution. Vanadium(III) accelerates the reduction of thionine[34], while vanadium(V) photo-oxidises a number of organic carbonyl compounds[35].

Niobium halides, $NbCl_4$ and $NbCl_3$, but not $NbCl_5$, bleach following irradiation in aqueous hydrochloric acid solutions[36], and finally precipitate Nb_2O_5. Alcohols increase this photosensitivity.

Van Bronswyk[37] has shown that neither of the clusters $Nb_6Cl_{12}^{2+}$ nor $Ta_6Cl_{12}^{2+}$ shows photochemical activity.

Some carbonyls of these metals show photochemical activity (see Section 5.2.12).

5.2.4 Chromium, molybdenum and tungsten

There are significant photochemistries for all three of these elements; for chromium the (II), (III) and (VI) oxidation states have been investigated, while to date the interest in molybdenum and tungsten has centred around the $M(CN)_8^{4-}$ and $M(CN)_8^{3-}$ ions where M = Mo or W.

5.2.4.1 *Chromium*

The photochemical publications on chromium(III) coordination compounds contain some of the most precise information available in the field. This is in no small part due to the thermal stability of both the irradiated complexes and the products which result from simple acts of ligand-for-ligand replacement, and not from electron transfer. In those cases where the act of thermal substitution and photochemical substitution are distinguishable by product analysis it is clear that at least one, and probably two, modes of electronically-excited-state chemistry have been established, although it has been suggested that one of these modes may be the reactions of 'hot' ground-state molecules.

The results in this field are already extensively reviewed[7–16, 18, 38–40]. The 'hot' ground-state model, which was suggested in two forms[21, 41], has received little support, and presently two excited-state mechanisms are under debate.

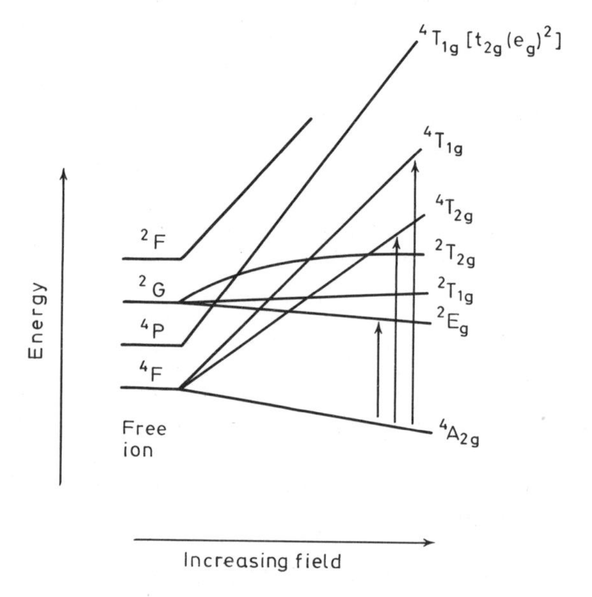

Figure 5.1 Qualitative energy level diagram for the ligand field states of octahedral chromium(III) complexes

The first mechanism[41–43], further developed by Schläfer[38, 44], considers the lowest doublet state ^2E_g (see Figure 5.1) to dominate the photochemical reactions, while Adamson[39] has proposed that the quartet state $^4T_{2g}$ at least is involved. In discussing these alternatives, it is preferable to consider the symmetrical (O_h) complexes (the CrA_6 type) separately from the less-symmetrically substituted complexes such as the CrA_5B and CrA_4B_2 types.

(a) O_h *Complexes*—The term diagram for the d^3 electron configuration in a symmetrical octahedral field is given in Figure 5.1. In addition to the quartet ground state $^4A_{2g}$, three doublet states arise, ^2E_g, $^2T_{1g}$ and $^2T_{2g}$, the first two of these having identical energies in a pure octahedral field and being referred to only as ^2E_g. This degeneracy is lost in distorted octahedral fields and is removed by spin-orbit coupling. The lowest energy configurations resulting from the excitation of a t_{2g} electron to an e_g orbital are the quartet states $^4T_{2g}$ and $^4T_{1g}$. The $(t_{2g})(e_g)^2$ configuration gives rise to a quartet $^4T_{1g}$ state.

This simple term diagram, although in general complicated by delocalisation of metal electrons on to the ligand, can account for the general features of the absorption, phosphorescence and fluorescence spectra, these last two being measured on complexes in rigid matrices at low temperatures. Valentine[18] and Balzani and Carassiti[7] have reviewed the main features of these spectra.

Firstly, it seems likely that both the ^2E_g and $^4T_{2g}$ states of chromium(III) will undergo substitution in their lifetime. This is suggested by comparison of these states with the labile vanadium(III) complexes, or by the Schläfer[38] calculation of crystal field activation energies following Basolo and Pearson[45].

The data so far accumulated for ligand exchange in chromium(III) complexes of O_h symmetry[7, 12, 14, 18] show that the quantum yield is independent of the irradiating wavelength, and thus give convincing evidence that at least one of these states, thermally equilibrated, is the photochemically-reactive intermediate. The argument from here rests on the likely lifetimes of these two excited states, and upon their relative energy.

The Schläfer[38] and Plane[41, 46, 47] doublet mechanism depends on all reaction occurring through the ^2E_g state irrespective of the wavelength of the irradiation, the wavelength-independence of the quantum yield being consistent with the justified longer lifetime and lower energy for this doublet compared with the quartet state, which leads to a near 100% conversion of excited quartet molecules to the doublet state. Such an interpretation is consistent with luminescence experiments, which admittedly were obtained at a much lower temperature.

Adamson[39] maintains that the thermally-equilibrated $^4T_{2g}$ state could lie below the thermally-equilibrated ^2E_g state leading to an efficient $^2E_g \rightarrow {}^4T_{2g}$ transition, and that the lifetime of the quartet state, being distorted, could be longer than calculated by a factor of up to 10^2. Valentine[18] considers it unlikely that thermal distortion could lead to the inversion of the ^2E_g and $^4T_{2g}$ states, whose relative energies have been clearly established by low-temperature luminescence experiments to be $^4T_{2g}$ above ^2E_g.

On the evidence so far for symmetrical O_h complexes, the doublet mechanism seems more likely.

(b) *Lower-symmetry octahedral chromium(III) complexes*—The main interest in the photolysis of these complexes (e.g. *cis*-$[Cr(OH)_2(en)_2]^+$ and $[CrX(NH_3)_5]^{2+}$ where X = Cl, Br, I and SCN) is the difference between acts of thermal and photolytic substitution, which for *cis*-$[Cr(OH)_2(en)_2]^+$ lead to different proportions of isomerisation and aquation, and for $[CrX(NH_3)_5]^{2+}$-type complexes lead to varying amounts of NH_3 and X aquation. In addition, for many of these complexes the photolytic products and quantum yields are found to be wavelength-dependent[7, 48, 49].

Adamson[12, 14, 22, 23, 39] has proposed a set of empirical rules to predict the results of photolysis in these systems:

Rule 1. Consider the six ligands to lie in pairs at the ends of three mutually-perpendicular axes. The axis having the smallest average crystal field will be the one labilised, and the total quantum yield will be about that of an O_h complex of the same average field.

Rule 2. If the labilised axis has two different ligands, then the ligand of greater field strength preferentially aquates.

Rule 3. The discrimination implied in Rules 1 and 2 will occur to a greater extent if the irradiation is in the first quartet band rather than the second.

Balzani and Carassiti[7] claim that the rules do not stand the test of experiment, and quote in particular the photolysis[48, 49] of $[CrCl(NH_3)_5]^{2+}$ leading predominantly to *cis*-$[CrCl(OH_2)(NH_3)_4]^{2+}$, which according to them points to the labilised axis being an NH_3—NH_3 axis and not the NH_3—Cl axis predicted by Rule 1. Balzani and Carassiti make the assumption that the water ligand enters *cis* to the departing group, for which there is no evidence. The *cis*-$[CrCl(OH_2)(NH_3)_4]^{2+}$ product can arise equally well from a labilisation of the NH_3 *trans* to the chloro ligand.

The simplicity of the O_h energy-level diagram is lost in these cases, but at least there seems good evidence that the doublet and quartet states will lie closer in energy, and that the photochemical activity could derive from the chemical activity of two excited states, either a doublet and quartet or two quartet states. Irradiation at much shorter wavelengths[7, 48–50] (~ 250 nm) of complexes such as $[CrX(NH_3)_5]^{2+}$ shows an increase of X^- labilisation. Schläfer thus identifies X^--release with a charge-transfer state, and prefers to identify the NH_3-release with the doublet state. Similarly, Schläfer[51] has identified a charge-transfer contribution in the photolysis of $[Cr(en)_3]^{3+}$. Balzani and Carassiti[7] warn that a charge-transfer contribution could be interfering with the interpretation of the results for irradiation in the ligand field bands, where spin-forbidden charge-transfer states could exist.

Chen and Porter[52] have recently identified photochemical activity of both a doublet and a quartet state in the reactions of the reineckate ion. Adamson, Martin and Camessei[53], in studying the photosensitised reactions of $[Cr(NCS)(NH_3)_5]^{2+}$, find support for their previously-expressed view[54] that here the doublet state leads to only thiocyanate aquation and the quartet to only ammonia release.

There is no doubt that there remain many conflicting results as well as theories in this area of photochemistry, and that there is a great need for carefully-determined quantum yields studied both as a function of temperature and of wavelength. In many systems a better understanding of the competing thermal reactions is also required. In addition, it is fair to say that at this stage little cognisance has been taken of the role of the solvation sphere[55, 56], and studies of anation reactions[57], and of photolyses in non-aqueous media[58–60], must prove fruitful. Such experiments must lead to information on the identity of the intermediate which is thought by Valentine[18] to be 7-coordinate, but is clearly expected by others to be produced by ligand dissociation[7].

Finally, on these complexes the recent work on photochemical resolution using circularly-polarised light points to the growing potential of specific photochemical preparative procedures in coordination chemistry[61–63].

Irradiation of labile chromium(II) (d^4, spin-free) species in aqueous solutions leads to the production of H_2 [64]. The primary step was postulated[65] to be electron transfer from metal to solvent water leading to hydrogen atom production. Results on similar systems suggest that photoelectron production may well be the mechanism[7,12]. Recently Hartmann and Filss[66,67] have measured the hydrogen isotope effect in these systems.

The chromium(VI) equilibrium system of CrO_4^{2-} and $Cr_2O_7^{2-}$ in aqueous solution in its photochemical oxidation of organic substrates has long been of interest, and regrettably most of the work appeared before there was an understanding of the pH-dependent CrO_4^{2-}–$Cr_2O_7^{2-}$ equilibrium. As a result much of this work should be re-studied. Such reactions are relevant to many printing processes and these reactions are reviewed[68,69]. Two more recent papers have appeared on the photo-oxidation of alcohols[70,71].

A flash photolysis study of chromyl chloride[72] stands out in Cr^{VI} photochemistry. Two transient absorptions were observed and the following chain postulated:

$$CrO_2Cl_2 \xrightarrow{h\nu} [CrO_2Cl_2]^* \rightarrow CrO_2Cl \xrightarrow{h\nu} [CrO_2Cl]^* \longrightarrow CrO_2.$$

The photochemistry of the carbonyls of chromium is important (see Section 5.2.12).

5.2.4.2 Molybdenum and tungsten

The important work in the case of these two metals involves the octacyano complexes of the general formula $M(CN)_8^{n-}$, where $n = 3$ or 4. The general chemistries of these ions have been reviewed recently[73–77]. It is generally agreed that the shape of these ions is either dodecahedral or antiprismatic, and in solution Kepert and Blight have postulated a fluctional structure[78]. The photochemical behaviour, the interpretation of which is controversial, has been thoroughly reviewed[7,12].

In the case of the $M(CN)_8^{4-}$ ions, irradiation leads to an immediate red compound with an increase in pH. Prolonged photolysis leads to a blue final product. The first stage is reversible in the dark and surely represents the mechanism of photochemical cyanide exchange[79,80], while the formation of the blue product is not reversed by a dark reaction.

There has been a long dispute between Jakób and co-workers[81–90] and primarily the schools of Carassiti[7,91–94], and of Adamson[12,95,97] and others[80,98]. We find the evidence at this stage compelling in its support of Adamson and Carassiti, and interpret the photolysis in ligand field bands as leading to the photo-aquation reaction:

$$\underset{}{M(CN)_8^{4-} + 2H_2O} \overset{h\nu}{\rightleftarrows} \underset{\text{(red intermediate)}}{[M(CN)_7(OH_2)]^{3-}} + HCN + OH^-$$

Jakób favoured the formulation $[M(CN)_8(OH_2)_2]^{4-}$ for the red intermediate. The recent isolation[99,100] of the red product as $Ag_3[Mo(CN)_7(OH_2)]$ seems convincing despite some attractive features in the $[M(CN)_8(OH_2)_2]^{4-}$ mechanism.

There is general agreement that the final blue product is a tetracyano complex and that it arises from a reaction of the type:

$$[M(CN)_7(OH_2)]^{3-} + OH^- \xrightarrow{h\nu} [M(CN)_4(O)(OH)]^{3-} + 2HCN + CN^-$$

This final product was originally thought to be 8-coordinate, but it now seems likely to be $[M(CN)_4(O)(OH)]^{3-}$ in solution and $[M(CN)_4(O)_2]^{4-}$ in the solid[101,102].

Irradiation of both these molybdenum and tungsten ions by flash photolysis has been shown to produce photoelectrons[103,104], with yields of about 0.3, as determined by N_2O scavenging.

Although the photochemistry of the $M(CN)_8^{3-}$ has been known and studied for 60 years, the picture has only become clear during the last 10 years following comprehensive studies by Carassiti and Balzani[105–108]. The following reaction scheme has been established:

$$M(CN)_8^{3-} + H_2O \xrightarrow{h\nu} M(CN)_8^{4-} + H^+ + OH$$

The quantum yield, which is in excess of unity, suggests the involvement of the OH radical in the chain:

$$M(CN)_8^{3-} + OH \rightarrow M(CN)_8^{4-} + H^+ + \tfrac{1}{2}O_2$$

The even higher quantum yield for the molybdenum leads to the postulation of the secondary chain reaction:

$$Mo(CN)_8^{3-} + H_2O + OH \rightarrow Mo(CN)_8^{4-} + 2OH + H^+$$

Although these equations happily explain the quantum yields there remains some inconsistency in detail[7], and the still unrepeated observation of Perumareddi[12,97] of the formation of cyanogen suggests that the possible formation of cyanide radicals should not be forgotten.

Finally, we have photochemical reduction of both Mo^{VI} and W^{VI} by alcohol[109,110] and hydroxy acids[111], and the photochromism exhibited by certain piperidine tungstates[112]. Aqueous colloidal MoO_3 photolyses with ultra-violet light in the presence of H_2 and CO to give molybdenum blue[113,114]. Both molybdenum and tungsten, like chromium, show an extensive photochemistry over a wide range of carbonyl complexes (Section 5.2.12).

5.2.5 Manganese, technetium and rhenium

The majority of the photochemical studies on this group involve either the cyano and oxalato complexes of manganese or manganese carbonyl derivatives which are discussed later (Section 5.2.12). Photochemical studies have been made of manganese(II) and manganese(III) complexes, mainly in the low-spin configurations t_{2g}^5 (for example $Mn(CN)_6^{4-}$, and $Mn(CN)_5(NO)^{2-}$) and t_{2g}^4 (for example $Mn(CN)_6^{3-}$). The photochemistry of the oxyanions of manganese; MnO_4^{3-}, MnO_4^{2-} and MnO_4^- has also been studied. Little new appears to have been published since two recent comprehensive reviews[7,12].

No photochemical effect was detected in irradiation of $Mn(CN)_6^{3-}$ and $Mn(CN)_6^{4-}$, but Balzani and Carassiti[7] suggest that this is due to the thermal lability of the systems. The corresponding pentacyanonitrosyl complexes,

$Mn(CN)_5(NO)^{2-}$ and $Mn(CN)_5(NO)^{3-}$, have been extensively studied[115–119]. In buffered solutions, and in the absence of added CN^- (which through subsequent thermal reactions gave $Mn(CN)_6^{4-}$), $Mn(CN)_5(NO)^{3-}$ reacted as follows:

$$Mn(CN)_5(NO)^{3-} \xrightarrow{h\nu} Mn^{2+} + 5CN^- + NO$$

$Mn(CN)_5(NO)^{2-}$ has been investigated at the same time by these workers, and has been interpreted as:

$$Mn(CN)_5(NO)^{2-} + H_2O \xrightarrow{h\nu} Mn^{2+} + 5CN^- + NO_2^- + 2H^+$$

with the subsequent thermal reaction:

$$Mn(CN)_5(NO)^{2-} + CN^- \rightarrow Mn(CN)_5(NO)^{3-} + \tfrac{1}{2}(CN)_2$$

and the subsequent photolysis of the $Mn(CN)_5(NO)^{3-}$. Balzani and Carassiti[7] emphasise inconsistencies between this mechanism and the pH measurements, and attempt to correlate the irradiation frequency-dependence of quantum yield with the molecular orbital assignments of Manoharan and Gray[120].

Porter *et al.*[121] have investigated the wavelength-dependence of the photochemical activity of $Mn(C_2O_4)_3^{3-}$, and propose the mechanism:

$$Mn(C_2O_4)_3^{3-} \xrightarrow{h\nu} Mn(C_2O_4)_2^{2-} + C_2O_4^-$$
$$Mn(C_2O_4)_3^{3-} + C_2O_4^- \rightarrow Mn(C_2O_4)_2^{2-} + C_2O_4^{2-} + 2CO_2$$

Adamson *et al.* emphasise the thermal instability of the complex, and that in this case one cannot exclude the possibility of photo-aquation as the initiation[12]. The thermally unstable diaqua complex, $Mn(C_2O_4)_2(OH_2)_2^-$, also decomposes photochemically[122, 123].

The photochemistry of the oxyanions of Mn^{V}, Mn^{VI} and Mn^{VII} have not been recently re-investigated and are reviewed[7, 12].

The photochemistry of the technetium(IV) ion, $TcCl_6^{2-}$, has been studied[124, 125]. The primary product is $TcCl_5(OH_2)^-$ through:

$$TcCl_6^{2-} + H_2O \xrightarrow{h\nu} TcCl_5(OH_2)^- + Cl^-$$

The eventual products are TcO_4^- and TcO_2.

Rhenium(VII) is photo-reduced in the presence of thiourea to give a green-yellow complex of use in photometric determination of rhenium[126].

5.2.6 Iron, ruthenium and osmium

There is an extensive photochemistry of both iron(II) and iron(III), a small amount of work has been done on ruthenium and osmium complexes, while the carbonyl compounds of all three have been investigated (Section 5.2.12).

5.2.6.1 Iron

The photochemistry of the coordination compounds of iron can be divided into that of iron(II) and iron(III), the d^6 and d^5 states. The work has been recently reviewed[7, 9–12, 15, 20, 21].

Iron(II)—These complexes fall into two classes; the weak-field $t_{2g}^4e_g^2$ complexes, and the strong-field t_{2g}^6 complexes. Most weak field complexes of iron(II) are unstable in aqueous solution, and thus weak-field photochemistry is substantially that of the $Fe(OH_2)_6^{2+}$ ion. Most of the strong-field complexes so far studied contain at least a majority of cyanide ligands.

The $t_{2g}^4e_g^2$ complexes have a quintet ($^5T_{2g}$) ground state, and there is only one spin-allowed ligand field band, that to the ^5E_g state. The strong-field ground state is the $^1A_{1g}$ state from which two spin-allowed ligand field bands originate; a $^1A_{1g} \rightarrow {}^1T_{1g}$ and a $^1A_{1g} \rightarrow {}^1T_{2g}$ transition. Charge-transfer bands also occur in the higher energy region which have been interpreted as ligand-to-metal, metal-to-ligand and charge transfer-to-solvent bands[127–131].

(a) *Strong-field complexes*—The most studied system in this category is the $Fe(CN)_6^{4-}$ ion which is very thermodynamically stable in water and shows basic properties[132], existing as $H[Fe(CN)_6]^{3-}$ at pH 3–4 and $H_2[Fe(CN)_6]^{2-}$ at pH 1. Two photo-reactions are found to occur; one producing the solvated electron, and the other resulting in photo-aquation. Photo-aquation appears to occur regardless of the wavelength of the irradiation, but appears to have a higher photochemical efficiency if the irradiation is in the ligand field bands[133]. Photoelectron production occurs with highest efficiency in the charge-transfer bands, but has been claimed in the ligand field bands[130], although some scepticism must be expressed here since there is appreciable band overlap. At this time it can be fairly said that no clear picture of the independence or interdependence of these two reaction modes has evolved.

The photo-aquation reactions have been studied by a number of workers[7] with very little agreement, almost certainly because of the photochemical and thermal instability of the primary product $[Fe(CN)_5OH_2]^{3-}$.

Photoelectron production from charge-transfer irradiation has been studied using a variety of scavengers, although nitrous oxide scavenging is perhaps the best understood. These results can be accounted for by the following primary reaction and four thermal processes[129–131, 134].

$$\begin{aligned}
Fe(CN)_6^{4-} &\xrightarrow{h\nu} Fe(CN)_6^{3-} + e^-(aq) \\
e^-(aq) + Fe(CN)_6^{3-} &\longrightarrow Fe(CN)_6^{4-} \\
e^-(aq) + N_2O &\longrightarrow N_2 + O^- \\
O^- + H^+ &\longrightarrow OH \\
OH + Fe(CN)_6^{4-} &\longrightarrow OH^- + Fe(CN)_6^{3-}
\end{aligned}$$

The primary quantum yields determined by different authors vary from 0.3–1.35, and are certainly a function of the thermal chemistry of scavenging. Quantum yields are also wavelength-dependent as must be expected where a number of bands overlap. Airey and Dainton[135] find that in acid solution irradiation of $H[Fe(CN)_6]^{3-}$ gives a lower yield than $Fe(CN)_6^{4-}$, and admit hydrogen atom production as a possibility. The production of hydrated electrons in these systems has been reviewed[136, 137].

Various photochemical substitution reactions of $Fe(CN)_6^{4-}$ have been studied using ligands such as 2,2′-dipyridyl and 1,10-phenanthroline[7]. All these reactions occur through ligand substitution into the products of primary photo-aquation.

The other comprehensively studied iron(II) complex is the $[Fe(CN)_5NO]^{2-}$ ion which can be thought to contain an NO^+ ligand. Here again there is much confusion in the literature[7], although a recent study[138] has established that both the following primary processes occur following radiation at > 300 nm:

$$Fe(CN)_5NO^{2-} \xrightarrow{h\nu} Fe(CN)_5^{2-} + NO$$

$$Fe(CN)_5NO^{2-} \xrightarrow{h\nu} Fe(CN)_5NO^{-} + e^{-}(aq)$$

Some other substituted cyano complexes of iron(II) have been studied and reviewed[7,12].

(b) *Weak-field complexes*—Here a great deal of work has been published on the photolysis of acid aqueous solutions of Fe^{2+}. The work prior to 1952 was reviewed[139], and later summaries have appeared[7,12].

It has been established that the primary process involves photoelectron production[131,140]

$$Fe^{2+}(aq) \xrightarrow{h\nu} Fe^{3+}(aq) + e^{-}(aq)$$

The subsequent thermal processes are extraordinarily complicated, involving pH-dependent processes, hydrogen atom production and, depending on pH, bridged complex species.

The photochemical oxidation of $Fe^{2+} \rightarrow Fe^{3+}$ in aqueous solution in the presence of both arsenic acid and thionine has been well studied and reviewed[7].

Iron(III)—As in the case of iron(II), the complexes of interest with iron(III) fall into two categories; the first involves the metal in saturation-coordination with ligands usually of strong-field character (e.g. $Fe(CN)_6^{3-}$, $Fe(phen)_3^{3+}$, etc.), and the second group involves weak-field ligands in one or two coordination sites with the rest of the sites occupied by the solvent, usually water.

In the weak-field case, the electronic configuration is $t_{2g}^3e_g^2$, with the result that complexes have no crystal field stabilisation energy. The ground state is $^6A_{1g}$, and all the ligand field excited states are of different multiplicity, and thus ligand field excitation is spin-forbidden. This, of course, does not preclude the population of these states following excitation into higher energy states, and phosphorescence has been observed for the isoelectronic manganese(II) system[141,142]. Such emission in iron(III) will occur in the near infra-red and will be technically difficult to find. When the spin-forbidden ligand field bands occur, they are on the edge of massive charge-transfer (ligand to metal) bands (e.g. $Fe(C_2O_4)_3^{3-}$)[143,144].

In the strong ligand case the ground state is t_{2g}^5 (one unpaired electron, low-spin) which leads to spin-allowed ligand field transitions. These ligand field bands normally appear as shoulders on charge-transfer (ligand to metal) bands.

(a) *Strong-field complexes*—The most thorough investigation of $Fe(CN)_6^{3-}$ was carried out by Moggi *et al.*[145], who clearly showed how complicated are the reactions of this system in that subsequent photochemical and thermal reactions abound. At the same time these results cast doubt on the quantitative photochemistry of Legros[146]. The present situation can be summarised as follows:

$$
\begin{array}{ccccc}
 & & & & Fe(OH)_3 \\
 & & & & \uparrow \\
 & & & & OH^- \\
 & & & & | \\
Fe(CN)_6^{3-} & \xrightarrow{h\nu} & Fe(CN)_5(H_2O)^{2-} & \xrightarrow{h\nu} \ldots\ldots \longrightarrow & Fe(OH_2)_6^{3+} \\
 & \searrow^{h\nu} & \downarrow h\nu & & \\
 & & Fe(CN)_5(H_2O)^{3-} & \xrightarrow{h\nu} \ldots\ldots \longrightarrow & \{Fe^{2+}\} \\
 & & & & \downarrow \\
 & & & & \text{PRUSSIAN BLUE}
\end{array}
$$

The path (dotted arrow) has not been authenticated, and many of the subsequent processes shown as photochemical certainly have thermal contributions.

An interesting photochemical substitution reaction of azide ion into $Fe(CN)_6^{3-}$ to give $Fe(CN)_5N_3^{3-}$ has been studied by Kenney *et al.*[147] using flash photolysis. The product retains photo-excitation, and further exchanges ligands contributing to 'thermal reversal'.

The visible spectra of $Fe(dipy)_3^{3+}$ and $Fe(phen)_3^{3+}$ are dominated by a ligand to metal charge-transfer absorption at 600 nm. Following excitation the excited states of these complexes oxidise water to OH radicals[148], or other potential reducing agents such as ethyl alcohol, peroxide, oxalic acid or formic acid, and may oxidise other ground-state complex ions leading to oxidised but coordinated ligands[149].

(b) *Weak-field complexes*—Aqueous solutions of Fe^{3+}, best shown as $Fe(OH_2)_6^{3+}$, and mono-substituted derivatives, $FeX(OH_2)_5^{2+}$ where X = OH, Cl, Br, SCN and N_3, have been extensively reviewed[7, 12, 139, 150]. Excitation into charge-transfer (ligand to metal) bands results in either oxidation of coordinated water (eventually to oxygen), oxidation of X, or oxidation of other solution molecules (often organic in nature). Such oxidations of solution molecules are achieved by radicals formed in the primary photochemical process, OH or X, which can also initiate the photo-induced polymerisation of vinyl monomers. Dainton *et al.*[151] have found evidence for the existence of a second excited state in the photolysis of aqueous Fe^{3+}.

Non-aqueous solutions of $FeCl_3$ have been the subject of quite a lot of work[7]. Many of the systems need re-study in that the exact nature of the absorbing species is not defined by solution equilibrium studies. Some of the recent work using the dipolar aprotic solvent *N,N*-dimethylformamide (DMF) is interesting[152–155]. Here the absorbing species is $FeCl_4^-$, and the primary process:

$$FeCl_4^- \xrightarrow{h\nu} FeCl_2(DMF)_x + Cl^- + Cl$$

In all these organic solvent systems the subsequent thermal fate of the radicals is complicated.

The vast photochemistry of aqueous ferric solutions containing carboxylic acid molecules (satisfactorily reviewed in Reference 7) warrants re-investigation in order to separate the roles of the thermal reactions of coordinated solvent-originated hydroxyl radicals and radicals formed from the oxidation of coordinated acid molecules in the primary process.

The photochemistry of the $Fe(C_2O_4)_3^{3-}$ is of special interest and of course still holds a special place in photochemistry because of its role in actinometry. Much significant work was carried out after Döbereiner[156] first reported photochemical activity in 1831, this work being dominated by the extensive studies of Hatchard and Parker[157]. They established that the system could be used for actinometry through the whole charge-transfer region, less than 550 nm, but found very small quantum yields through the spin-forbidden d–d band centred at 680 nm. Subsequent flash photolysis studies by the same workers[158] established the following mechanism:

$$Fe(C_2O_4)_3^{3-} \xrightarrow{h\nu} [Fe(C_2O_4)_3^{3-}]^*$$

$$[Fe(C_2O_4)_3^{3-}]^* \rightarrow [\text{Metastable State}]$$

$$[\text{Metastable State}] \rightarrow Fe(C_2O_4)_2^{2-} + C_2O_4^-$$

$$C_2O_4^- + Fe(C_2O_4)_3^{3-} \rightarrow [(C_2O_4)_2Fe \begin{smallmatrix} O\text{—}CO \\ | \\ \dot{O}\text{—}CO \end{smallmatrix}]^{2-} + C_2O_4^{2-}$$

$$[(C_2O_4)_2Fe \begin{smallmatrix} O\text{—}CO \\ | \\ \dot{O}\text{—}CO \end{smallmatrix}]^{2-} \rightarrow Fe(C_2O_4)_2^{2-} + 2CO_2$$

5.2.6.2 Ruthenium and osmium

The luminescent properties of both ruthenium(II) and osmium(II) complexes have been investigated and reviewed[7], but no photochemistry appears to have been reported for osmium(II). The aqueous photochemistry of $Ru(CN)_6^{4-}$ has been shown to be similar to that of $Fe(CN)_6^{4-}$ [104, 159].

Two significant studies of pentakis-(ammine)ruthenium(II) complexes containing molecular nitrogen[160], and pyridine, acetonitrile, water and ammonia[161] as the sixth ligand have been reported. In the first study, following irradiation at 221 nm only photo-oxidation of the ruthenium to Ru^{III} was confirmed, and the search for reduction products of nitrogen was fruitless. No photo-aquation was observed. In the second study, both photo-oxidation and aquation were observed, and the relative yields were found to be wavelength-dependent. Both the sixth ligand and ammonia are removed in the aquation reactions, and hydrogen is confirmed as one of the products of the photo-oxidation of $Ru(NH_3)_5(py)^{2+}$.

Ruthenium and osmium carbonyls have been investigated (see Section 5.2.12).

5.2.7 Cobalt, rhodium and iridium

Quite an extensive photochemistry is now evident for these metals in the d^6 configuration. As more work emerges on rhodium(III) similarities now appear which suggest that its photochemistry may be analogous to cobalt(III) and not chromium(III) as previously suggested[12].

5.2.7.1 *Cobalt(III)*

The photo-sensitivity of 'Werner-type' cobalt(III) complexes was clearly established in the 1900s, particularly in respect to the photo-redox process which usually leads to a cobalt(II) species and an oxidised form of one of the ligand molecules. In addition to being well covered by general reviews[7, 9–15, 18, 20–23], one review entirely on cobalt(III) photochemistry has appeared[16], and more recently Endicott has written a review which is almost exclusively on the charge-transfer photochemistry of cobalt(III) complexes[24].

The spectra of cobalt(III) complexes as they pertain to their photochemistry are well described[7, 12, 15, 16, 18, 21]. In O_h symmetry the ground state is $^1A_{1g}$, and the first ligand field excited states are shown in Figure 5.2. The two intense ligand field bands which normally appear in the visible or near ultraviolet regions are due to the spin-allowed transition $^1A_{1g} \rightarrow {}^1T_{1g}$ and $^1A_{1g} \rightarrow {}^1T_{2g}$. The spin-forbidden transition to the ^3T states are sometimes

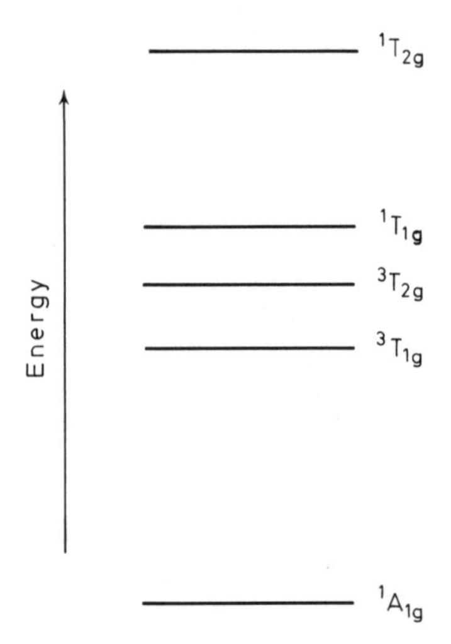

Figure 5.2 Qualitative energy level diagram for the ligand field states of octahedral cobalt (III) complexes

found on the low energy side of these major bands, but if present are of very low intensity, and irradiation into such bands is not photochemically significant.

The first two charge-transfer bands are assigned to the promotion of an electron in a ligand-centred molecular orbital to a metal-centred orbital (e_g, antibonding). The low-energy band corresponds to promotion from a π orbital, and the higher energy one to the promotion of a σ electron.

The frequency of the charge-transfer bands moves to longer wavelengths and the bands become more intense as the ease of oxidation of the ligand increases, and as a result in many cases the charge-transfer absorptions overlap the ligand field bands.

In contrast to chromium(III) complexes, luminescence results for cobalt(III) complexes are sparse, and are only recorded for cyano complexes[162–165]. These emissions are identified as $^3T_{1g} \rightarrow {}^1A_{1g}$, a transition which has been observed in absorption in some complexes[166].

There are two modes of photochemical reaction of cobalt(III); firstly, the poorly understood photo-substitution reactions, and secondly, the more extensively studied photo-redox reactions. To date, the only photo-substitution reaction for which much data is available is the photo-aquation reaction, and here two types of complexes have been studied, firstly, the cobalt(III) cyanide and oxalate complexes for which quite high quantum efficiencies are found, and secondly, the 'Werner-type' ammine and ethylenediamine complexes where low quantum yields are found.

(a) *Photo-aquation of ligand field excited states*—The quantum yields for photo-activation of ligand field bands of cobalt(III) ammine complexes are extraordinarily low compared to the analogous chromium(III) complexes[7,9,12,18]. Consideration of the ligand field activation energies for the excited states suggests that the $^1T_{1g}$ state of cobalt(III), like the ^2E_g and $^4T_{2g}$ states of chromium(III), should be highly reactive, and that the $^3T_{1g}$ state will be of lability comparable to the ground state $^1A_{1g}$ [38]. The low efficiency of the photochemical reactions correlates with the lack of luminescence data, and is associated with a low crossing probability between the excited states ($^1T_{1g} \rightarrow {}^3T_{1g}$) and high crossing probabilities between the excited states $^1T_{1g}$ and $^3T_{1g}$ and the ground state $^1A_{1g}$ [7,12,18]. In the d^6 systems with very strong ligands (e.g. $[Co(CN)_5X]^{3-}$), where high aquation quantum yields are found, the energy separation between the ground state and the ligand field excited states is high, and one must expect a decrease in the efficiency of the radiation-less deactivation[7]. Adamson, Chiang and Zinato have recently published an extensive study of this system[167].

The oxalato complexes of cobalt(III) have been extensively studied and reviewed[7,12]. They show a thermal redox process, as well as photochemical redox, in charge-transfer and ligand field bands. Some monodentate and multidentate oxalato ammine complexes have been recently studied[168].

(b) *Photo-activity of the charge transfer excited states*—The majority of the photochemistry of cobalt(III) complexes falls in this category, and it is here that the most significant advances have been made in the last 2 years. A naive approach to the complexity of the predominantly photo-redox reactions of excited states leads to conflicting results in the earlier literature. Balzani and Carassiti[7] and Endicott[24] have very recently reviewed this field, and the latter review is most valuable in that it includes unpublished results some of which are now appearing on flash photolysis experiments using chemical scavengers[169–171]. The work of Caspari, Hughes, Endicott and Hoffman[170] is an exhaustive study of the often investigated $[CoCl(NH_3)_5]^{2+}$ system, and to quote, shows that 'the prevailing prejudice that the photochemistry of cobalt(III) is simple, and that mechanistic principles were established long ago, has apparently inhibited the critical evaluation of these systems'.

This work shows that all results so far are consistent with the existence of two short-lived excited states of the complex, the first, the normal excited state, involving charge transfer from the chlorine ligand, crosses to a scavengeable metastable state. The pH-dependence of the quantum yield is accommodated if the second species is involved in proton transfer to solvent. This is consistent with the greater acidity of the product of a one-electron oxidation of ammonia. It is now certain that the simple radical-pair mechanism of Adamson[12, 14, 22, 23, 172] cannot account for the now established complexity of this system.

A recent paper by Schrauzer, Lee and Sibert[173] notes the similarity between the radical production in these complexes and the photo-dealkylation of alkylcobalamins and alkylcobaloximes, and attributes the light stability of some methylcobalt[174] derivatives to the high efficiency of the recombination $Co^{II} + CH_3\cdot \rightarrow Co—CH_3$.

Irradiation of the charge transfer bands, in addition to leading to redox processes, yields a small amount of photo-substitution.

There is little doubt that much of this data is questionable. The careful work of Vogler and Adamson[172, 175] on *trans*-$[CoCl(NCS)(en)_2]^+$ confirms that the charge-transfer aquation is of different mode to the ligand field excitation, which as shown previously differs from the thermal mode[176]. There is no doubt that before conclusions can be drawn about these reactions more care must be taken in accounting for the oxidation products which accompany cobalt(II) production in the predominant redox process. It must be agreed that whether the mechanism of redox is through the Adamson radical pair or the more complex Endicott complexes, an intermediate octahedral cobalt(II)-containing species exists, and it will of course be substitutionally labile.

Recently, three extensive publications have appeared by Shagisultanova *et al.* on a range of 'acido-ammine' complexes, but have not been viewed by the authors[177–179]. Norden has studied the oxidation of ethylenediamine tetra-acetate (EDTA) by irradiation of $[Co(en)_3]^{3+}$ in the presence of EDTA[180].

(c) *Photo-sensitised reaction of cobalt(III) complexes*—A number of workers have recently used organic sensitisers to induce photochemical activity in cobalt(III) complexes[181–185]. The mechanism involves energy transfer from a donor triplet state to a triplet state of the complex. Cobalt(III)–ammine complexes are activated to the redox mode, suggesting a low-lying charge transfer excitation of the complex, while Porter[184] in the case of $Co(CN)_6^{3-}$ finds that biacetyl transfers energy to the $^3T_{1g}$ ligand field triplet state. Adamson, Vogler and Lantzke have recently achieved intramolecular energy transfer following irradiation of the 'free-ligand' band of the coordinated *trans*-4-stilbenecarboxylate ion[186].

5.2.7.2 Rhodium and iridium

As emphasised in recent reviews[7, 10, 12, 16, 18], there are few quantitative photochemical results on the complexes of rhodium(III) and iridium(III), although a number of useful photochemical preparative procedures have been developed recently[187–193]. All the luminescence results have been summarised[7], except for a recent thorough study of a range of rhodium(III) complexes by Demas and Crosby[194].

Until very recently the only definitive quantitative results were those of Moggi[195], who irradiated the two d–d bands of $[RhCl(NH_3)_5]^{2+}$. A variation in quantum yield between the two bands was explained in terms of the excitation of a sigma antibonding electron into a d_{z^2} orbital in one case and a $d_{x^2-y^2}$ orbital in the other. These results led people to anticipate similarities between rhodium(III) and iridium(III) photochemistry, and that of chromium(III) rather than cobalt(III)[12,196].

However, Kelly and Endicott[197] have shown, by studying $[RhI(NH_3)_5]^{2+}$ which has $I^- \rightarrow Rh^{III}$ charge-transfer bands at relatively low energy, that excitation into these charge-transfer bands leads to a significant photo-redox process involving the production of iodine atoms. These results point to a new and interesting photochemistry analogous to that of cobalt(III) in the initial excitation and mechanism, although the final products will differ because of the differences in stability of cobalt(II) and rhodium(II) complexes.

5.2.8 Nickel, palladium and platinum

5.2.8.1 Nickel and palladium

Palladium(II), d^8-complexes are usually square-planar, while nickel(II) complexes, although often square-planar with strong ligands, can exhibit a range of structures. The complexes are labile, and as a result there is considerable doubt concerning the validity of the few results so far collected on the photochemistry of these systems.

The conclusion[198] that no photo-aquation occurred in either $Ni(CN)_4^{2-}$ or $Pd(CN)_4^{2-}$ may be in error because of the speed of the thermal anation reaction. The observation of transients in flash photolysis of $Ni(CN)_4^{2-}$ suggests this[12].

Trans-bis-glycinatopalladium(II), although reported as insensitive to radiation at 313 nm, decomposes to palladous oxide when radiated at 254 nm[199].

5.2.8.2 Platinum

Both platinum(II) and platinum(IV) complexes have been shown to possess photochemical activity. Although the work on these complexes is not as extensive as that on some other transition metals, the concentration of good work in the last few years is significant.

In platinum(II) complexes (d^8), the $d_{x^2-y^2}$ orbital of the d set is vacant, and ligand field bands are found and attributed to excitation of electrons from filled d orbitals to this empty orbital. Some discussion involving the relative energy of the d orbitals, now agreed as $d_{x^2-y^2} > d_{xy} > d_{xz}, d_{yz} > d_{z^2}$, and on whether the main ligand field bands are singlet–singlet or singlet-triplet has occurred[200–202], and been summarised[7,12]. The intense bands appearing at higher energy are normally assigned as ligand to metal ($d_{x^2-y^2}$) charge-transfer transitions.

Platinum(IV) has nominally the same d^6 configuration as cobalt(III), but is modified in a fashion consistent with it having a higher oxidation number and being of the third transition series. The most important photochemical studies on these systems are on the PtX_6^{2-} ions, where X is Cl, Br or I, and these spectra have been analysed as an overlap of ligand field bands and charge-transfer bands due to excitation of electrons from bonding ligand orbitals to antibonding sigma metal orbitals[203, 204].

Some emission studies on both platinum(II) and platinum(IV) complexes have been attempted, and are summarised by Balzani and Carassiti[7].

(a) *Platinum(II) complexes*—The results here fall into either the category of straight substitution reactions, often by solvent, or isomerisation, which can be through an intermolecular substitution or, in chelate complexes, by an intramolecular twist mechanism.

The reaction

$$H_2O + PtCl_4^{2-} \xrightarrow{h\nu} PtCl_3(OH_2)^- + Cl^-$$

has been most thoroughly investigated by Carassiti *et al.*[205, 206], who reasonably suggest that the constancy of quantum yield through the charge-transfer region is due to crossing to the $^1B_{1g}$ ligand field state which corresponds to a $(d_{z^2})^1(d_{xy}, d_{yz})^4(d_{xy})^2(d_{x^2-y^2})^1$ configuration, which would be vulnerable to nucleophilic attack in the axial position. The constant yield at 313, 404 and 472 nm is attributed to crossing to the lowest component of the ^3E_g triplet state, which would be similarly vulnerable to solvent attack. Such a mechanism is consistent with the recent biacetyl-sensitised photolysis of $PtCl_4^{2-}$, although Langford and Sastri suggest that not all the excited molecules reach the ^3E_g state, which they suggest could have a quantum yield close to unity[207].

Bromide ion exchange with $PtBr_4^{2-}$ is similarly light-sensitive, and involves aqua intermediates[208, 209]. Unpublished results of Adamson[12] show that both ammonia and chloride ion are released in the photolysis of $PtCl(NH_3)_3^+$, and in view of the interest in the double substitution modes in cobalt(III) and chromium(III) complexes this deserves further investigation.

Extensive studies of the $PtBr(dien)^+$ system[210, 211] have led to the following mechanism for the photolysis of the complex in aqueous nitrite solutions:

$$PtBr(dien)^+ \xrightarrow{h\nu} [PtBr(dien)^+]^*$$

$$[PtBr(dien)^+]^* \longrightarrow PtBr(dien)^+$$

$$[PtBr(dien)^+]^* \longrightarrow Pt(dien)(OH_2)^{2+}\cdot Br^-$$

$$Pt(dien)(OH_2)^{2+}\cdot Br^- \longrightarrow PtBr(dien)^+ + H_2O$$

$$Br^- + Pt(dien)(OH_2)^{2+}\cdot Br^- \longrightarrow PtBr(dien)^+ + Br^- + H_2O \quad (5.1)$$

$$NO_2^- + Pt(dien)(OH_2)^{2+}\cdot Br^- \longrightarrow Pt(NO_2)(dien)^+ + Br^- + H_2O \quad (5.2)$$

Such a mechanism is in keeping with our knowledge of the thermal chemistry[212, 213]. It is a pity that radioactive bromide ions were not competed with the nitrite ion to see if the following equilibrium was attained:

$$Pt(dien)(OH_2)^{2+}\cdot Br^- \rightleftharpoons Pt(dien)(OH_2)^{2+} + Br^-$$

or whether the anion-exchange processes (5.1) and (5.2) were single steps.

Recently, Bartocci and Scandola[214] have established the formation of N_3 radicals in the irradiation of $PtN_3(dien)^+$ as an intermediate in nitrogen formation.

Cis- and *trans*-$PtCl_2(NH_3)_2$ were reported to give chloride aquation with radiation at 313 nm[215], and previously to yield polymeric Pt^{II}–Pt^{IV} products at higher energy[216].

The light-induced photo-isomerisation of square-planar platinum(II) complexes involving bidentate ligands was first reported in 1910[217], but the recent results in the three system; *cis*- and *trans*-$PtCl_2(PEt_3)_2$ [218, 219], *cis*- and *trans*-$Pt(gly)_2$ (where gly = glycinate ion)[220–222] and *cis*- and *trans*-$PtCl_2(py)_2$ (where py = pyridine)[223], deserve special mention.

In the $Pt(gly)_2$ system it was shown quite clearly that the thermal isomerisation was an intermolecular process and that the photochemical process was intramolecular, the former requiring free glycinate ion in solution and the latter not.

Scandola *et al.*[222] have presented sound evidence for the destabilisation of square-planar geometry, in favour of a distorted tetrahedral structure, by promotion of a non-bonding d electron into the $d_{x^2-y^2}$ orbital. Such a structure has also been suggested[224] for an excited state of $Ni(CN)_4^{2-}$.

Such an excited state could deactivate by isomerisation or by returning to the original geometry, but would also be labile to substitution. It would appear that deactivation by substitution is more probable in complexes containing monodentate ligands, and that in chelate complexes intramolecular isomerisation is favoured. In $PtCl_2(py)_2$ both reactions are manifest[223]. The photochemical steady state is very solvent-dependent, more polar solvents favouring the polar *cis* isomers, as emphasised by the recent work of Mastin and Haake[225] on the $PtX_2(Ph_3P)_2$ isomers. Recent work on $Pt(C_2O_4)(Ph_3P)_2$ is mentioned in Section 5.2.12.

(b) *Platinum(IV) complexes*—Although platinum(IV) complexes, like cobalt(III) compounds, are non-labile, which is advantageous in photochemical study, they are very sensitive to trace amounts of Pt^{II} which is known to catalyse the thermal reactions[9].

Quantum yields for the reactions:

$$PtCl_6^{2-} + Cl^{-*} = (PtCl_5Cl^*)^{2-} + Cl^-$$

and

$$PtCl_6^{2-} + X^- = PtCl_5X^{2-} + Cl^-$$

where X = Br or I, are found to be of the order 10^3 and to increase with increasing temperature[226, 227], pointing to chain mechanisms. Taube and Rich[226] postulated the formation of a chorine atom and a platinum(III) species in the photo-initiation step of Cl^--exchange with $PtCl_6^{2-}$. In the absence of exchangeable anions photo-aquation occurs according to:

$$PtCl_6^{2-} + H_2O \xrightarrow{h\nu} PtCl_5(OH)^{2-} + H^+ + Cl^-$$

This process is reversed thermally by addition of chloride ions[228, 229].

More recently, similar behaviour has been confirmed for the $PtBr_6^{2-}$ system, which gives a wavelength-independent quantum yield for aquation in acid solution, leading to the conclusion that radiation-less transitions led to the population of a triplet ligand field state from which aquation occurred[220, 230, 231]. Photo-induction of Br^--exchange (quantum yield ~ 10^2) led to the postulation of a radical chain similar to that mentioned above for $PtCl_6^{2-}$ [208, 232, 233].

Irradiation of PtI_6^{2-} is complicated by the fast thermal processes. Aquation is however accelerated[230], and a long-lived transient is identified in flash photolysis[234].

Balzani and Carassiti[7] discuss the modes of induction of exchange and aquation, and conclude that aquation is independent of the redox processes which lead to exchange. The observation of the transient Br_2^- species in the aquation of *trans*-$PtBr_2(NH_3)_4^{2+}$ points to radical formation at least in this system[12].

Of the reactions of platinum(IV) complexes so far reported, the preliminary report of the linkage nitro–nitrito isomerisation in $Pt(NO_2)(NH_3)_5^{3+}$ is most interesting[235].

5.2.9 Copper, silver and gold

We have specifically excluded the photochemistry of photographic emulsions and solid-state photochemistry, which excludes most of the results in this field. Although qualitative results on the photochemical activity of copper(I), copper(II), silver(I) and gold(III) complexes are available, little progress has been made since recent reviews[7, 12].

Ziolo and Dori, following their recent preparation of the thiocyanato-bis-(triphenylphosphine)copper(I) complex[236], have observed photoluminescence in phosphine complexes of both Cu^I and Ag^I [237].

Sundararjan and Wehry[238] have photo-reduced bis-(2,9-dimethyl-1,10-phenanthroline)copper(II) to the copper(I) complex which does not photo-oxidise; OH radical production is postulated.

Van den Berg and Schneider have re-investigated the Fe^{II}-sensitised photolysis of $Cu(C_2O_4)_2^{2-}$, which yields carbon dioxide and copper[239]. A rate-determining step:

$$[Fe(C_2O_4)_2^{2-}]^* + CuC_2O_4 \rightarrow [CuC_2O_4]^* + Fe(C_2O_4)_2^{2-}$$

is postulated, which seems unlikely. Some 'ammine' complexes of copper(II) have also been re-investigated[240], in addition to alcoholic solutions of copper(II) chloride where the photolysed solutions have been studied by e.p.r. spectroscopy[241].

5.2.10 Zinc, cadmium and mercury

Since the two recent reviews[7, 12] in this field nothing major has been accomplished in the photochemistry of the complex ions of these divalent metal ions. De Waal and Van den Berg have investigated the photo-decomposition of mixtures of oxalate ion and mercuric ion[242]. There is very appreciable photochemistry of mercury alkyls and other organo-mercury compounds[7], the complexities of which arise from the thermal reactions of the radicals formed in the primary process.

5.2.11 Lanthanides and actinides

Publications on the photochemistry of these elements are dominated by the work on cerium, in the lanthanides, and the uranyl ion (UO_2^{2+}), in the actinides. In addition to recent general reviews[7, 12], uranyl ion photochemistry was reviewed in 1964[243].

5.2.11.1 Cerium

The recent literature features the extensive studies of ceric ammonium nitrate solutions by Martin and co-workers[244–247], Hayon and co-workers[248, 249], and Moorthy and Weiss[250]. Three distinct mechanisms for NO_3 radical production have been suggested. Firstly, Martin *et al.*[244] postulated intramolecular electron transfer:

$$\begin{aligned} &Ce^{IV}NO_3^- \xrightarrow{h\nu} [Ce^{IV}NO_3^-]^* \\ &[Ce^{IV}NO_3^-]^* \longrightarrow Ce^{III} + NO_3 \end{aligned} \qquad (5.3)$$

where the exact complexation of the metal is not specified. Hayon and Saito[248] favoured the photolysis of HNO_3 solvent molecules:

$$\begin{aligned} &HNO_3 \xrightarrow{h\nu} OH + NO_2 \\ &\text{and} \\ &OH + HNO_3 \rightarrow H_2O + NO_3 \end{aligned} \qquad (5.4)$$

Later, Hayon and Dogliotti[249], following flash photolytic studies, concluded that OH was first produced by

$$Ce^{IV}H_2O \xrightarrow{h\nu} Ce^{III} + OH + H^+ \qquad (5.5)$$

The recent papers of Martin *et al.*[245, 246] reinforce the conclusion that the photoactive molecule is not the solvent, but conclude that the original postulates[244] (Equations (5.3) and (5.4)) are completely consistent with the results. Martin points out that poor analysis of the NO_3-decay data could have resulted in the discrepancies in kinetic order established by the two groups. Martin further emphasises that the presence of water is not a necessary prerequisite of NO_3 radical formation, which is formed in anhydrous nitric acid, although they acknowledge that the primary process (5.5) could occur where NO_3^- is absent. Martin states[246] that a further publication is imminent.

In a further paper, Martin and Glass[247] have followed the rate of the reaction:

$$NO_3 + Ce^{III} \rightarrow Ce^{IV}NO_3^-$$

by flash photolysis. Dogliotti and Hayon[249] also detected HSO_4 radicals in the flash photolysis of ceric sulphate solutions.

The photolysis of aqueous solutions of cerium(III) and the potential of the Ce^{III}-Ce^{IV} system as a source of chemical energy from solar energy is discussed[7, 251].

Recently, the photochemical reduction of Ce^{IV} in aqueous solution has been studied[252], also the photo-oxidation of tertiary alcohols and carboxylic acids in the presence of Ce^{IV} [253].

5.2.11.2 *Uranyl ion (UO_2^{2+})*

Through photochemical reactions, the uranyl ion causes oxidation of a wide range of substrates, both organic and inorganic, and these reactions have been reviewed[7, 12, 243, 254]. There is a wide use of the uranyl oxalate system in actinometry. The photochemical yield is measured by oxalate loss, and the products, mainly CO_2, CO, and HCOOH, vary in their proportion with pH, but are independent of the wavelength and temperature. Low pH favours CO production. Heidt, Tregay and Middleton[255] have made a thorough study of the pH-dependence of CO yield, and correlate the increased CO production with the breaking of one of the carbon–oxygen bonds by protons in the energy-rich complex:

$$[O_2U^{VI}OOCH^+]^* + H^+ \rightarrow O_2U^{VI}OH^+ + HCO^+$$
$$HCO^+ \rightarrow H^+ + CO$$

The excited species derives from the reactions

$$O_2U^{VI}C_2O_4 \xrightarrow{h\nu} O_2U^{V+}OCO^- + CO_2$$
$$O_2U^{V+}OCO^- + H_2O \rightarrow [O_2U^{VI}OOCH^+]^* + OH^-$$

5.2.11.3 *Other complexes*

There is an extensive literature on the luminescence of lanthanide ions[7], where the screening effect of the outer 5s and 5p orbitals prevents a rapid radiation-less deactivation of the excited states which involve the 4f orbitals. These properties make lanthanide ions sensitive probes in organic photochemistry[256, 257]. The fluorescence and photochemistry of Eu^{3+} ions has recently been re-investigated[258].

The photo-acceleration of U^{IV}–U^{VI} exchange has been studied under Hg lamp illumination[259]. The photo-oxidation of solid PuF_4 to PuF_6 has been studied[260], also the oxidation of Pu^{III} in acid solutions[261].

5.2.12 Metal carbonyls and organo-metallics

The present interest in the thermal reactions in carbonyl systems is reflected in accelerated activity in the field of photochemically-induced processes. An extraordinarily comprehensive review has been published recently by Von Gustorf and Grevels[8], in addition to that of Balzani and Carassiti[7]. Others have reviewed some aspects of the field[12, 262–264]. It is not possible to review here the ramifications of the complicated thermal reaction consequences of photochemical activation (see Reference 8). We thus simply cite the very recent publications relevant to the primary process, and some interesting new preparations.

The ultraviolet spectra of metal carbonyl complexes have been reviewed[8]. Although the spectra can be complicated, the photochemical initiation is simple, and in the ultraviolet spectral region leads to carbon monoxide release following absorption into bands which are metal to ligand charge-transfer transitions. The whole process can be written

$$M(CO)_n \xrightarrow{h\nu} [M(CO)_n]^* \rightarrow M(CO)_{n-1} + CO$$

The consequences of irradiation at longer wavelengths, where ligand field excitation could occur, are variously interpreted[7,12].

Working in rare gas matrices at 15 K, infrared evidence substantiates the existence of $Cr(CO)_5$, $Mo(CO)_5$, $W(CO)_5$ and $Ni(CO)_3$ in the photolysis of $Cr(CO)_6$ $Mn(CO)_6$, $W(CO)_6$ and $Ni(CO)_4$ respectively[265–268]. McIntyre[269] has determined the rate constant for the recombination of $Cr(CO)_5$ with CO in cyclohexane and polystyrene. Recently, Rest *et al.*[270,271] have produced evidence for $Mn(CO)_3NO$ and $Mn(CO)_2NO$ in the photolysis of $Mn(CO)_4$ NO, and $HMn(CO)_4$ from $HMn(CO)_5$. These coordinatively-unsaturated complexes recapture carbon monoxide on warming in a process which can be photochemically accelerated[269,272]. Flash photolysis of $Fe(CO)_5$ produces $Fe(CO)_4$, which lasts long enough to extract a CO molecule from $Ni(CO)_4$ yielding $Ni(CO)_3$ [273]. $Os(CO)_4$ seems certain to be the intermediate in the recent photolytic condensation of $Os(CO)_5$ to give $Os_2(CO)_9$ and Os_3 $(CO)_{12}$ [274].

Such simple reversible processes are not always found, as shown by Ogilvie[275] in that acetylmanganese carbonyls can result from the photolysis of methylmanganese pentacarbonyl and trifluoromethylmanganese pentacarbonyl at 17 K.

The coordinatively-unsaturated $M(CO)_{n-1}$ species can be trapped by a vast variety of electron donor molecules in solvents of poor coordinating ability, or can be trapped by solvent species (e.g. ether[276] and THF[277]). The types of bases so far used are well summarised by Von Gustorf and Grevels[8], but the range is exemplified by the following very recent publications: phosphines[278–284], arsines[279,282] and phosphites[283,285,286], silanes[287], diaryldichalcogens[288], bis-(trialkylstannyl)selenide and telluride[289], stannous chloride[290], olefins and fluorolefins[291], difluoromethylene groups[296], imido groups[297] and hexa-alkylborazoles[297]. Perhaps the most remarkable of all is the formation of icosahedral metalloboranes[298]. The isomerisation of dienes by irradiation of solutions containing $W(CO)_6$ and the diene, has been interpreted as involving a triplet excited state of the complex $[W(CO)_5(\text{diene})]$[299].

Valentine, Sarver and Collman[300] have recently reported that CO expulsion initiates the addition of H_2 to $Ru(PPh_3)_2(CO)_3$ to give $Ru(PPh_3)_2$ $(CO)_2H_2$. It must be remembered in substituted carbonyls of the type $ML(CO)_{n-1}$ that the ligand 'L' is often expelled along with the carbon monoxide molecules[7,8,12]. The recent transient production of norbornadien-7-one by its expulsion from its stabilised coordination in (norbornadien-7-one)iron tricarbonyl is interesting[301].

Valentine and Valentine[302] accomplished a photo-induced oxidative addition of 9,10-phenanthrenequinone to $Ir(PPh_3)_2COCl$, but favour here a mechanism involving excited phenanthrenequinone molecules rather than excited complex.

A great deal of work has been carried out on cyclopentadienyl carbonyls[8], and in this field the formation of $[C_5H_5Mn(CO)_2]_2C_5H_6$ with a bridging cyclopentadiene molecule is worthy of mention[303]. There is an increasing literature on metallocene complexes[7, 8, 304–308] as well as the metal benzenes[7, 8, 309]. Recently, van Reil has produced a photo-substitution at the ferrocene nucleus[310].

The recent work of Nyman *et al.*[311–313] on oxalato-bis-(triphenylphosphine) platinum(II) complexes is interesting. Here the photochemically-produced unsaturated $Pt[P(C_6H_5)_3]_2$ is the initial product except in the presence of hydrogen gas, where $Pt[P(C_6H_5)_3]_2CO$ appears to initiate the formation of the cluster $Pt_4[P(C_6H_5)_3]_5(CO)_3$.

5.2.13 Other metallic elements

There is a growing interest in the photochemistry of thallium(III) solutions. The charge transfer to metal excitation is written:

$$Tl^{III}(aq) \xrightarrow{h\nu} H^+ + Tl^{II} + OH$$

in acid solution, but can be written as the photolysis of the ion-pair $Tl(OH)^{2+}$. Stranks and Yandell[314, 315] have investigated the photo-induced chain reaction between Tl^I and Tl^{III}, while others have studied the radical propagated oxidation of carboxylates[316] and alcohols[317].

There is a vast photochemistry of stannanes and lead alkyl compounds, which follow the other Group IV elements silicon and germanium, which have been extensively investigated as silanes and germanes. All this chemistry follows the pattern of organic photochemistry, particularly in respect to the thermal radical reactions. There is some recent interest in the halogen compounds of germanium[318] and tin(IV); in the latter, photolysis at 355 nm of SnI_4 in heptane leads to metallic tin and iodine molecules[319].

Essentially similar production of metastable atoms of phosphorus, arsenic and antimony is found in the flash photolysis of phosphine, arsine and stibine[320]. The halides, $AsCl_3$, PCl_3, $SbCl_3$ produce radicals AsCl, PCl and SbCl[321–323] arising from secondary photolysis of the dichloro-radicals. Callear and Oldham[324] have observed atomic arsenic in flash photolysis of $AsCl_3$.

5.3 PHOTOCHEMISTRY OF THE ANIONIC COMPOUNDS OF NON-METALLIC ELEMENTS

5.3.1 Anions in aqueous solution

Dramatic progress has been made in the elucidation of aqueous photochemical mechanisms in the last 5 years. This is in no small way due to the wider use of flash photolysis techniques which allow the identification of transient intermediates, the presence of which would remain pure speculation following the investigation of end-point analysis alone. Two recent reviews have appeared which summarise a great deal of this chemistry[5, 325].

Photochemical studies and absorption spectral analysis have complemented each other in the identification of the primary photochemical step. A broad survey of the absorption spectra assignments for dissolved ions, including charge-transfer to solvent (CTTS) spectral theories, has been published[326]. Hence this paper will emphasise the kinetic and mechanistic evidence provided by photochemical investigation.

When dissolved anions absorb energy in the ultraviolet region, reaction may be initiated by one of two processes:

(a) Charge-transfer to solvent may occur

$$Y^{n-}(aq) \xrightarrow{h\nu} \{Y^{n-}(aq)\}^* \rightarrow Y^{(n-1)-}(aq) + e^-(aq)$$

Such a reaction occurs with SO_4^{2-}, SO_3^{2-}, OH^- and the halide ions.

(b) An intramolecular transition often leads directly to bond rupture, as in the cases of NO_3, NO_2^-, CO_3^{2-} and the halate ions.

A number of anions (e.g. $S_2O_3^{2-}$, NCS^- and N_3^-) have overlapping absorptions bands, and thus even monochromatic light can effect both these primary processes, resulting, through secondary reactions, in vastly different decomposition mechanisms.

We recognise the need for the tabulation of rate constants for the reactions of a great variety of transient species (especially H, OH, O^- and the solvated electron). These rate constants have been accumulated mainly by competitive scavenging kinetics in both photochemistry and radiation chemistry. Many reactions of this type will be mentioned as they pertain to the aqueous photochemistry which follows, but no attempt is made to draw them together.

5.3.1.1 Oxyanions of sulphur

(a) *Sulphate, bisulphate and persulphate ions*—The ultraviolet spectrum of SO_4^{2-} below 200 nm has been assigned to a charge-transfer to solvent process[326]:

$$SO_4^{2-}(aq) \xrightarrow{h\nu} \{SO_4^{2-}(aq)\}^* \rightarrow SO_4^-(aq) + e^-(aq)$$

This has been supported by kinetic evidence in the presence of electron scavengers such as N_2O [327], lower alcohols[328] and H_3O^+ [328]. The quantum yield, determined under total scavenging conditions, has been recorded as 0.71 and 0.64 in separate investigations[327, 328].

In 1967, Hayon and McGarvey[329] left no further doubt about the mechanism when, in the flash photolysis of aerated aqueous SO_4^{2-} solutions, transient species with maximum absorptions at 450 nm and 240 nm were positively identified as SO_4^- and O_2^- radical ions. The latter was formed by the capture of the solvated electrons by molecular oxygen.

The absorption at 450 nm was identified with a band found in the flash photolysis of persulphate ion in aqueous solution where O—O bond scission occurs[330]:

$$(O_3S—O—O—SO_3)^{2-} \xrightarrow{h\nu} 2SO_4^-$$

In the acid region, pH changes and the presence of O_2 do not affect either the formation or bimolecular decay kinetics of this radical, which is more stable than previously reported[331]. However, at pH values > 8.5, the decay rate increases, until at pH 10.8 the absorption band is completely eliminated. In aerated solutions, it is replaced by a band centred at 430 nm due to the ozonide radical O_3^-. The following mechanism is suggested:

$$SO_4^- + H_2O \rightarrow OH + SO_4^{2-} + H^+$$
$$OH + O_2 \rightarrow O_3^- + H^+$$

Barrett *et al.*[328] isolated the contribution to hydrogen production by HSO_4^- ion when 1 M sulphuric acid solutions were irradiated at 185 nm in the presence of methanol. The kinetics indicated a charge-transfer to solvent process for this ion also:

$$HSO_4^-(aq) \xrightarrow{h\nu} \{HSO_4^-(aq)\}^* \rightarrow HSO_4(aq) + e^-(aq)$$

with a quantum yield for cage production of 0.55.

(b) *Sulphite and dithionate ions*—Dogliotti and Hayon[332] showed the existence of two transient radical ions in the photolysis of 10^{-2} M sulphite ion solutions. The first, at 720 nm, was assigned to HSO_3^{2-}, which was formed independently of pH(7–12), and in the presence of O_2 decayed by a first-order process, and was not formed in the presence of N_2O. The second, at 275 nm, decayed more slowly by a second-order process and was assigned to SO_3^-. This assignment was made partly on the basis of the e.s.r. spectra after photolysis at 77 K.

The following mechanism, initiated by a charge transfer to solvent transition was proposed:

$$SO_3^{2-}(aq) \xrightarrow{h\nu} \{SO_3^{2-}(aq)\}^* \rightarrow SO_3^-(aq) + e^-(aq)$$
$$e^-(aq) + SO_3^{2-}(aq) \rightarrow HSO_3^{2-} + OH^-$$

or in the presence of O_2,

$$e^- + O_2 \rightarrow O_2^-$$
$$O_2^- + SO_3^{2-} + H_2O \rightarrow HSO_3^{2-} + OH^- + O_2$$

The fate of these radicals and their mechanism of decay seems uncertain.

Fox[5] indicates that the spectrum of sulphite ion involves an internal transition overlapping with the major charge-transfer to solvent band.

Solutions of dithionate ion, analogously to sulphate and persulphate[329, 330], in flash photolysis produce a band identical to SO_3^- in SO_3^{2-} photodecomposition[333]. Dithionate probably ruptures an S—S bond in a manner comparable to the O—O bond scission in persulphate[330].

$$(O_3S—SO_3)^{2-} \xrightarrow{h\nu} 2SO_3^-$$

(c) *Thiosulphate ion*—The photochemistry of the thiosulphate ion is an extremely interesting case. What was thought to be a single band, λ_{max} $\sim$215 nm, in the absorption spectrum of $S_2O_3^{2-}$, has been resolved into three overlapping bands[334] characterised as follows:

Band *A* λ_{max} $\sim$240 nm, n$\rightarrow\pi^*$ (non-bonding electrons on S)
Band *B* λ_{max} $\sim$215 nm, CTTS
Band *C* λ_{max} < 200 nm, n$\rightarrow\pi^*$ (non-bonding electrons on O)

Dogliotti and Hayon[332] find support for these spectral assignments in observing two independent primary photochemical processes in the flash photolysis of 10^{-3} M $S_2O_3^{2-}$ solutions: one due to a charge-transfer to solvent transition:

$$S_2O_3^{2-}(aq) \xrightarrow{h\nu} \{S_2O_3^{2-}(aq)\}^* \rightarrow S_2O_3^-(aq) + e^-(aq)$$

and the other due to an intramolecular transition which results in S—O bond-breaking:

$$S_2O_3^{2-}(aq) \xrightarrow{h\nu} \{S_2O_3^{2-}(aq)\}^* \rightarrow S_2O_2^- + OH + OH^-$$

The two processes occur to an almost equal extent.

The radical ions $S_2O_3^-$ and $S_2O_2^-$ were identified as transient absorptions with maxima at 380 nm and 280 nm respectively. The intensities of these transient absorptions are affected differently by the presence of scavengers, and by the use of light filters with cut-off wavelengths at $\sim$237 nm and $\sim$215 nm. On this basis, $S_2O_3^-$ production is attributed to excitation within band B and $S_2O_2^-$ production to absorption in band C.

As is the case with a number of these photolysis systems, further reactions of the observed transients, and the mechanism of formation of the final products (in this case, H_2S, SO_3^- and S) are at present quite speculative[332]. The self-scavenging of electrons by $S_2O_3^{2-}$, evidence of which was found by Sperling and Treinin[334], could play a part in final product formation.

5.3.1.2 *Oxyanions of nitrogen*

(a) *Nitrate ion*—The nitrate ion in aqueous solution exhibits a weak absorption with peak at $\sim$300 nm and a strong absorption at $\sim$200 nm. These bands have been assigned to n$\rightarrow\pi^*$ and $\pi\rightarrow\pi^*$ intramolecular transitions[335].

Under the action of ultraviolet light, nitrates are reduced to nitrite with the liberation of O_2. The stoichiometry of this reaction is assumed to be

$$NO_3^- \rightarrow NO_2^- + \tfrac{1}{2}O_2$$

This system was first studied in 1912[336], but astonishingly, little quantitative information had been obtained before 1966. The reaction was known to have a low quantum yield[337] ($\phi_{NO_2^-} = 0.25$ at $\lambda = 207$ nm) which decreased very rapidly at $\lambda > 250$ nm. In alkaline solutions the rate of nitrite formation is about tenfold the rate in acid solutions[338].

Petriconi and Papee[339] found the yield of nitrite to increase as the pH was raised from 5 to 12, but then to decrease again at higher pH levels. These workers have also detected the formation at $\lambda < 280$ nm of a species thought to be the pernitrite ion ($OONO^-$).

By irradiation in the 300 nm band of neutral NO_3^- solutions, Daniels *et al.*[340] observed a decreasing rate of production of NO_2^-, which seemed consistent with the following processes:

$$\begin{aligned} &NO_3^- \overset{h\nu}{\rightleftarrows} \{NO_3^-\}^* \rightarrow NO_2^- + O \\ &O + NO_3^- \rightarrow O_2 + NO_2^- \\ &O + NO_2^- \rightarrow NO_3^- \end{aligned} \tag{5.6}$$

To account for the enhancement of yield in the presence of hydroxyl radical scavengers, a second concurrent primary process is postulated, becoming dominant only when nitrate recombination within a 'cage' is prevented:

$$\begin{aligned} &NO_3^- \overset{h\nu}{\rightleftarrows} \{NO_3^-\}^* \rightarrow (NO_2 + O^-)_{cage} \\ &(NO_2 + O^-) + H^+ \rightarrow (NO_2 + OH)_{cage} \\ &(NO_2 + O^-) + H^+ \rightarrow NO_3^- + H^+ \end{aligned}$$

The sharp increase of nitrite yield at pH > 6 is attributed to a pH-dependence of reaction (5.6) which is considered to be a step-wise process:

$$O + NO_3^- \rightleftarrows (O{-}O{-}N{\lesssim}^{O}_{O})^-$$

(peroxynitrate)

In acid solution:

$$NO_4^- + H^+ \rightleftarrows H{-}O{-}O{-}N{\lesssim}^{O}_{O}$$

(peroxynitric acid)

The latter product re-arranges to eliminate oxygen

$$H{-}O{-}O{-}N{\lesssim}^{O}_{O} \rightarrow O_2 + N{\lesssim}^{O-H}_{O}$$

A more facile elimination of oxygen from the peroxynitrate ion species in alkaline solutions leads to a higher yield under these conditions.

The decrease of NO_2^- yield at very high pH was explained by ionisation of the hydroxyl radical:

$$OH \xrightleftharpoons{pK\,=\,12} O^- + H^+$$

However in a flash photolysis study, Barat *et al.*[341, 342] detected a transient absorption with peak at 300 nm attributed to the peroxynitrite ion, which was thus considered to be an intermediate in the nitrite-forming secondary reactions. It was shown by lack of scavenging effects that neither of the alternative primary products O atoms or O^- radicals were precursors of the peroxynitrite. The peroxynitrite absorption increased with first order kinetics to a maximum after 100 μs.

These workers were also unable to find any evidence of the hydrated electron in photolysed nitrate solution, but a weak absorption due to the

NO_2 radical was detected. On this basis they assumed the reactions

$$NO_3^- \xrightarrow{h\nu} (NO_3^-)^*$$

$$(NO_3^-)^* \xrightarrow{H_2O} NO_2 + OH + OH^-$$

$$2NO_2 + H_2O \rightarrow NO_2^- + NO_3^- + 2H^+$$

The origin of molecular O_2 was uncertain, but it was not a product of peroxynitrite decomposition.

Very recently, Treinin and Hayon[343] provided strong evidence for the formation of NO_2 as a transient in NO_3^- solutions containing NO_2^-.

Shuali *et al.*[344] have reported work on photolysis in the high energy (~195 nm) band of NO_3^-. The hydroxyl radical was indirectly identified as an intermediate, thus supporting the primary process proposed by Barat. Transient absorptions due to both peroxynitrous acid and the peroxynitrite ion were detected, and the pH dependence of their intensities allowed calculation of the pK of the acid, 6.0 ± 0.3, in good agreement with values obtained for the acid produced by other means.

The mechanism of peroxynitrite formation and its role in the photolytic process, the role of hydroxyl radicals and the mechanism of oxygen formation were not explained further. To account for the pH-dependence, a species such as $\left(\begin{matrix}O\\ |\\ O\end{matrix}\!\!>\!N{-}O\right)^-$ was assumed which reacted in alkaline solution as follows:

$$\left(\begin{matrix}O\\ |\\ O\end{matrix}\!\!>\!N{-}O\right)^- + NO_3^- \rightarrow 2NO_2^- + O_2$$

but whose conjugate acid took part in a back process

$$\left(\begin{matrix}O\\ |\\ O\end{matrix}\!\!>\!N{-}OH\right) + NO_2^- \rightarrow HNO_2 + NO_3^-$$

Differences in the effect of radical scavengers on product yields exist in the reports of the recent work[340–342, 344].

It is obvious that the system is not a simple one, but the correlation between the pH-dependence of peroxynitrite ion concentration and the quantum yields of both products, O_2 and NO_2^-, seems unmistakable.

(b) *Nitrite ion*—There is no net reaction when aqueous nitrite ion absorbs ultraviolet light[345], and as a result the system has been sparsely studied.

In a classic example of the value of the flash photolysis technique, Treinin and Hayon[343] very recently showed that the zero net reaction is obtained despite an efficient primary decomposition. The observation of transient absorptions due to Br_2^-, CO_3^- and O_3^- in the presence of Br^-, CO_3^{2-} and O_2 respectively, indicates the formation of OH radicals, and evidence was produced to show that transient absorption bands with peaks at 400 nm, 340 nm and < 250 nm were due to NO_2, N_2O_4 and N_2O_3 respectively.

Extrapolation of transient absorbances to zero time showed a 1:1 ratio in the initial yields of NO_2 and O^-, and the elimination of the transient absorptions by hydroxyl scavengers pointed to O^- or OH as precursors.

The data are explained by the mechanism:

$$NO_2^- \xrightarrow{h\nu} NO + O^-$$
$$O^- + H_2O = OH + OH^-$$
$$OH + NO_2^- \rightarrow NO_2 + OH^-$$
$$O^- + NO_2^- \xrightarrow{+H_2O} NO_2 + 2OH^-$$
$$NO_2 + NO_2 = N_2O_4$$
$$NO + NO_2 = N_2O_3$$

Nitrite is completely re-generated by hydrolysis of the nitrogen oxides. Because both NO_2 and N_2O_4 decay with pH-independent first-order kinetics, and because $k_{N_2O_4} = k_{N_2O_3} = 2 \times k_{NO_2}$, the authors suggested that decay can occur by a process

$$NO_2 + H_2O \xrightarrow{\text{slow}} H_2NO_3$$

$$H_2NO_3 + NO_2 \xrightarrow{\text{fast}} NO_3^- + NO_2^- + 2H^+$$

as well as by the previously-accepted mechanism

$$N_2O_4 + H_2O \rightarrow NO_3^- + NO_2^- + 2H^+$$

The N_2O_3 decay rate increases as the pH increases. This oxide is thought to react through either

$$N_2O_3 + H_2O \rightarrow 2NO_2^- + 2H^+$$
$$N_2O_3 + OH^- \rightarrow 2NO_2^{\ -} + H^+$$

Although this explanation is qualitatively satisfactory, a mathematical treatment of the system containing N_2O_3 does not fully agree with experimental observations on the decay kinetics[343].

5.3.1.3 *Oxyanions of the halogens*

(a) *Oxybromate ions*—The photo-decomposition of the aqueous BrO_3^- ion produces BrO^- and O_2, whilst BrO^- breaks down to Br^- and O_2 [346]. The one-step elimination of O_2 suggested by Farkas and Klein seems unlikely[346]. This photo-reduction was considered[347] to be initiated by a CTTS excitation, but a study of the solvent environmental effect on the spectra of the halates led Treinin and Yaacobi[348] in 1964 to assign the primary absorption process above 200 nm to intramolecular electronic transitions. The spectra of hypohalites have also been assigned to intramolecular transitions[326].

Recent flash photolysis studies by Treinin and co-workers[349–351] indicate that at least two primary processes are important in oxybromate ion photolysis. These are the dissociation reactions

$$BrO_n^- \xrightarrow{h\nu} (BrO_n^-)^* \rightarrow BrO_{n-1} + O^- \tag{5.7}$$

and

$$BrO_n^- \xrightarrow{h\nu} (BrO_n^-)^* \rightarrow BrO_{n-1}^- + O(^3P) \tag{5.8}$$

In the flash photolysis of BrO_3^- solutions, a transient absorption (λ_{max} ~ 475 nm) was assigned to the BrO_2 radical, and an absorption at ~ 445 nm in strongly alkaline solutions in the presence of oxygen was attributed to O_3^- [349]. That these intermediate products arise from the primary process (5.7) with 1:1 production of BrO_2 and O^- is supported by a doubling of the BrO_2 yield in air-free solutions in the presence of H_2 or BrO_2^-. This effect is explained by the reactions:

$$O^- + H_2 \rightarrow OH^- + H$$
$$H + OH^- \rightarrow H_2O + e^-(aq)$$
$$e^-(aq) + BrO_3^- + H_2O \rightarrow BrO_2 + 2OH^-$$

or

$$O^- + BrO_2^- + H_2O \rightarrow BrO_2 + 2OH^-$$

In air-free bromite solutions BrO_2 is again detected but not the BrO radical because of the very fast reaction

$$BrO + BrO_2^- \rightarrow BrO_2 + BrO^-$$

The fast reaction

$$O^- + BrO_2{}^-(aq) \rightarrow BrO_2 + 2OH^-$$

also occurs, and explains the absence of an effect due to hydrogen on BrO_2 yields in these solutions. Also because of this competing reaction, O_3^- is not formed in aerated solutions at high ($> 8 \times 10^{-4}$ M) concentrations of BrO_2^-.

In aerated BrO^- solutions, O_3^- is again detected with a yield very dependent upon BrO^- concentration.

Further evidence for the primary process (5.7) is obtained in the 254 nm photolysis of BrO_3^- in a boric acid glass[350]. Under these conditions transient species may be stabilised. Absorptions with peaks near 320 nm, 350 nm and 470 nm are attributed to the BrO_3, BrO and BrO_2 radicals respectively. These radicals are formed by the sequence of reactions:

$$BrO_3^- \xrightarrow{h\nu} BrO_2 + O^- \text{ (or OH)}$$
$$BrO_3^- + OH \rightarrow BrO_3 + OH^-$$
$$BrO_3 \rightarrow BrO + O_2$$

Although this last step explains the first-order decay of BrO_3, and that BrO_3 appears to be precursor to BrO, its validity is doubted. It is found on warming that the BrO_3 decomposes most readily, the second-order decay of BrO_2 and BrO depending on the diffusion of the species. Decay of the radicals leads to the formation of an absorption band due to Br_2 at ~ 410 nm.

Flash photolysis of aerated solutions, of each of the ions BrO_3^-, BrO_2^- and BrO^-, over a wide pH range, gives rise to an absorption at 260 nm[351]. This band is suppressed in the absence of O_2 and also with increasing concentrations of various additives. The absorption is attributed to O_3, the variable yield of which is due to competition reactions of oxygen atoms in the ground state, formed in primary photolytic act (5.8).

$$O + O_2 \rightarrow O_3$$
$$O + S \rightarrow \text{products}$$

S represents a wide range of chemical reagents whose relative second-order rate constants for reaction with O atoms are tabulated.

Amichai and Treinin[351] have estimated relative quantum yields of O and O^- production in solutions of BrO_3^-, BrO_2^- and BrO^- to be 1.6, 1.3 and 0.8 respectively.

(b) *Oxyiodate ions*—Flash photolyses of IO_3^- and IO^- solutions[352, 353] show a chemistry which indicates that these ions are photo-reduced as a result of a primary process analogous to reaction (5.7).

$$IO_3^- \xrightarrow{h\nu} IO_2 + O^-$$

$$IO^- \xrightarrow{h\nu} I + O^-$$

However, no evidence is found of the 260 nm band of O_3 in aerated solutions. It is concluded that at wavelengths longer than 200 nm the oxyiodate anions do not photo-dissociate via a primary process corresponding to (5.8) to yield $O(^3P)$ atoms.

(c) *Oxychlorate ions*—Amichai and Treinin[351] found no evidence for a radical type of dissociation in ClO_3^- photolysis. However, Barat *et al.*[354] detected absorptions attributable to IO_2, BrO_2 and ClO_2 in the flash photolysis of IO_3^-, BrO_3^- and ClO_3^- ions, respectively. A primary process corresponding to (5.7) is suggested in each case.

5.3.1.4 Carbonate ions

The absorption by aqueous carbonates in the region below 200 nm has been characterised as a CTTS excitation[355]

$$CO_3^{2-}(aq) \underset{}{\overset{h\nu}{\rightleftarrows}} \{CO_3^{2-}(aq)\}^* \rightleftarrows CO_3^-(aq) + e^-(aq)$$

No work on the photochemistry or flash photolysis of aqueous carbonate was reported before 1967, probably because, as in the case of nitrite ion, no net reaction had been observed.

In 1967, Hayon and McGarvey[329] detected transient absorptions at 600 nm and 240 nm in the flash photolysis of aerated, alkaline CO_3^{2-} solutions. These absorptions are due to CO_3^- and O_2^- formed from the CTTS process followed by

$$e^-(aq) + O_2 \rightarrow O_2^-$$

Both intermediates decayed by second-order kinetics with equal rate constants, suggesting the decay reaction:

$$CO_3^- + O_2^- \rightarrow \text{products}$$

Behar *et al.*[356] recently reported decay of the 600 nm band to be bimolecular, and that the 260 nm absorption decayed initially ($\sim$ ms) by second-order kinetics followed by slow first-order kinetics ($t_{\frac{1}{2}} \sim 2$ s). The long-lived intermediate was not O_2^-. The formation of a new transient species, CO_5^{2-}, was proposed, this ion decaying by a first-order process, probably to the initial solution components:

$$CO_3^- + O_2^- \rightarrow CO_5^{2-}$$
$$CO_5^{2-} \rightarrow CO_3^{2-} + O_2$$

5.3.1.5 Phosphate ions

The absorption spectra below 200 nm of aqueous solutions of various phosphate anions have been attributed to CTTS transitions[357]. The same conclusion was reached[358] as a result of the kinetic effect of electron scavengers in the 185 nm photolysis of aqueous solutions of Na_2HPO_4.

In a flash photolysis study[359], the detection of transient intermediates by their absorption spectra confirmed the above indications for each of the ions $H_2PO_4^-$, HPO_4^{2-} and $P_2O_7^{4-}$. In addition, the spectra between 300 and 180 nm have been measured for $H_2PO_2^-$, HPO_3^{2-}, $H_2PO_3^-$ and $H_2P_2O_6^{2-}$, and assigned as CTTS[360].

5.3.1.6 Halide and pseudo-halide ions

(a) *Halide ions*—A review by Fox concentrates largely on aqueous halide photolysis[5]. Jortner *et al.*[361] investigated the photolysis of aqueous solutions of Cl^-, Br^-, and I^- at 253.7, 228.8, and 184.9 nm and concluded that a CTTS process occurred. The evidence further indicated that solvated electron scavengers could compete with radical pair re-combination within a 'cage'. Subsequently it was pointed out[362–364] that at concentrations far less than 'cage' theories predict, competition by scavengers can be efficient.

Czapski *et al.*[365] have demonstrated, at N_2O scavenger concentrations up to 1.6×10^{-2} M, that the dependence of quantum yield on scavenger concentration can be explained by competition between the scavenger and the final photolysis products for reaction with the solvated electrons. These solvated electrons are formed homogeneously with quantum yields which are independent of scavenger concentration, and there is no evidence for cage scavenging. However, in I^- photolysis, at N_2O concentrations up to 1.2 M [366], the marked dependence of I_3^- yield on N_2O concentration is consistent with competition between cage-scavenging and geminate re-combination.

$$I^-(aq) \rightarrow \{I^-(aq)\}^* \rightleftarrows \{I + e^-(aq)\}_{cage}$$
$$\{I + e^-(aq)\}_{cage} + N_2O \rightarrow I + N_2O^-$$
$$N_2O^- \rightarrow N_2 + O^-$$

There was no evidence[366] for geminate re-combination in aqueous chloride photolysis, where radicals achieve homogeneous distribution with unit efficiency. The importance of geminate re-combination in Br^-(aq) photolysis is uncertain, as are the reasons for the differences in the Cl^-, Br^- and I^- systems.

(b) *Thiocyanate ion*—The aqueous thiocyanate absorption band with a maximum at ~220 nm has been assigned CTTS[367,368], despite contrary suggestions[369]. The presence of solvated electrons has been indicated in the flash photolysis of this ion[370]. Products detected in NCS^- photolysis are sulphur, H_2S, CN^- and in aerated solutions, a small yield of SO_4^{2-} [332,371].

Flash photolysis showed a transient absorption at 485 nm[332]. Comparison with a transient found in pulse radiolysis studies identified this intermediate as the $(NCS)_2^-$ radical ion, formed as a result of an electron-transfer primary process:

$$NCS^-(aq) \overset{h\nu}{\rightleftharpoons} \{NCS^-(aq)\}^* \rightleftharpoons NCS + e^-(aq)$$
$$NCS + NCS^- \rightleftharpoons (NCS)_2^-$$

The dimer radical, over a wide pH range, is found to decay by first-order kinetics in the absence of O_2, and by second-order kinetics in the presence of O_2. This latter phenomenon is possibly due to the reactions:

$$e^-(aq) + O_2 \rightarrow O_2^-$$
$$(NCS)_2^- + O_2 \rightarrow \text{products}$$

Differences exist in the decay kinetics of the transient absorptions assigned to the same radical ion in pulse radiolysis and photolysis studies.

The use of filters indicates that the CTTS band of NCS^- lies below ~230 nm, and it is thus no surprise that Luvia and Treinin[371] detected only a small yield of solvated electrons in photolysis at 254 nm. Under these conditions, the following features are found. Firstly the decrease of yield with time, and suppression of photolysis by added CN^-, indicate an efficient back reaction. At constant pH, yield increases with increasing NCS^- concentration up to a limiting value, and as pH increases, $\phi(CN^-)$ decreases sharply by a factor of ~1.5 at pH 7.5. Fitting of the data to diffusion-controlled scavenging rate laws indicates that this effect is the result of a pH-dependence in the yield of the primary dissociative process. Finally the products are found to slowly disappear by a post-radiation process.

To satisfy the above observations, another primary process is proposed involving a competitive self-scavenging by NCS^- of S atoms formed in the primary process.

$$NCS^- \overset{h\nu}{\rightleftharpoons} (NCS^-)^* \rightleftarrows (S + CN^-)_{cage} \quad (5.9)$$
$$H^+ \downarrow\uparrow \; pK_a = 7.5$$
$$HNCS^{-*} \rightleftarrows (S + HCN)_{cage} \quad (5.10)$$
$$NCS^- + (S + CN^-)_{cage} \rightarrow 2CN^- + S_2$$
$$CN^- + (S + CN^-)_{cage} \rightarrow NCS^- + CN^-$$
$$nS_2 \rightarrow S_{2n}$$
$$CN^- + S_{2n} \xrightarrow{\text{slow}} NCS^- \text{ re-generation}$$

Dissociation (5.9) is considered less efficient than dissociation (5.10).

At this stage, this mechanism is not well-substantiated.

(c) *Azide ion* – The photolysis of aqueous azide solutions is complicated by the overlap of a CTTS absorption band at 200 nm and a band due to an $n \rightarrow \pi^*$ intramolecular transition (λ_{max} ~240 nm)[372]. This interpretation has been questioned[368].

Photolysis of dilute solutions at 254 nm into the $n \rightarrow \pi^*$ band gave rise to N_2, NH_2OH and OH^- products without any evidence for photo-ionisation of the azide ion[374]. However, more recently[341, 375], flash photolysis into the CTTS band at wavelengths less than 230 nm has identified both the azide radical and the solvated electron formed by the primary process:

$$N_3^-(aq) \overset{h\nu}{\rightleftharpoons} \{N_3^-(aq)\}^* \rightleftarrows N_3 + e^-(aq)$$

A steady-state photolysis study by Treinin and co-workers[376] has shown 229 nm photolysis to be essentially identical with photolysis at 254 nm. However, 214 nm excitation initiates a different chemistry of decomposition. At pH 7.7 the products N_2, NH_2OH and NH_3 are formed with a quantum yield for azide loss of 0.31. NH_3 is considered to result from reaction of NH_2OH and the solvated electron because of the following observations. Firstly, as irradiation proceeds, $\phi(NH_2OH)$ decreases to a limiting value while $\phi(NH_3)$ increases to a limiting value. The addition of $\geqslant 3 \times 10^{-4}$ M NH_2OH prior to irradiation causes the limiting yields to be reached immediately. Finally, addition of scavengers (e.g. acetone) to the system can result in elimination of the NH_3 product.

Thus, the following mechanism is proposed for the system when limiting yield conditions have been reached.

$$N_3^- \overset{h\nu}{\rightleftharpoons} \{N_3^-\}^* \text{ (quantum yield, } \phi)$$

A fraction $(1-\alpha)$ of the excited azide ions reacts similarly to those excited by 254 nm light[374], and may in fact be the identical excited species

$$\{N_3^-)^* + H_2O \rightarrow NH + OH^- + N_2$$
$$NH + H_2O \rightarrow NH_2OH$$

The remaining fraction, α, ionises and gives rise to secondary reactions

$$N_3^{-*} \rightarrow N_3 + e^-(aq)$$
$$e^-(aq) + NH_2OH \rightarrow NH_2 + OH^- \qquad (5.11)$$
$$NH_2 + NH_2OH \rightarrow NH_3 + NHOH$$
$$N_3 + NH_2OH \rightarrow N_3^- + H^+ + NHOH \qquad (5.12)$$
$$2NHOH \rightarrow N_2 + 2H_2O$$

Reaction (5.12) accounts for evidence of a reaction of NH_2OH which is not as pH-dependent (due to NH_3OH^+ formation) as (5.11).

Analysis leads to the values $\phi = 0.40$, $\alpha = 23\%$, so that only $\sim 9\%$ of azide ions excited at this wavelength are photo-ionised.

5.3.1.7 Hydroxide ion

Aqueous OH^- ions are generally recognised to decompose photolytically via an initial CTTS act[377]. Recently, Basco *et al.*[378] have identified a species postulated to be an hydrated dielectron, formed by bimolecular combination of solvated electrons in hydroxide photolysis in H_2-saturated solutions.

$$OH^-(aq) \xrightarrow{h\nu} OH + e^-(aq)$$
$$OH + H_2 + OH^- \rightarrow 2H_2O + e^-(aq)$$
$$2e^-(aq) \rightarrow (e_2^{2-})_{aq}$$

References

1. Gaathon, A. and Ottolenghi, M. (1970). *Israel J. Chem.*, **8,** 165
2. Katz, M. (1970). *Can. J. Chem. Eng.*, **48,** 3
3. Altshuller, A. P. (1969). *Bull. WHO,* **40,** 616
4. Stein, G. (1969). *Actions Chim. Biol. Radiat.*, **13,** 119
5. Fox, M. F. (1970). *Quart. Rev. Chem. Soc.*, **24,** 565
6. Anbar, M. (1968). *ibid.*, **22,** 578
7. Balzani, V. and Carassiti, V. (1970). *Photochemistry of Coordination Compounds.* (London and New York: Academic Press)
8. Von Gustorf, E. K. and Grevels, F. W. (1969). *Fortschr. Chem. Forsch.*, **13,** 366.
9. Basolo, F. and Pearson, R. G. (1967). *Mechanisms of Inorganic Reactions.* (New York: N.Y., John Wiley and Sons, 2nd Ed.)
10. Wehry, E. L. (1967). *Quart. Rev. Chem. Soc.*, **21,** 213
11. Szychlinsky, J. (1962). *Wiad. Chem.*, **16,** 607
12. Adamson, A. W., Waltz, W. L., Zinato, E., Watts, D. W., Fleischauer, P. D. and Lindholm, R. D. (1968). *Chem. Rev.*, **68,** 541
13. Adamson, A. W. (1968). In: *Proc. 11th Int. Conf. Coord. Chem.*, p. 63. (Haifa-Jerusalem, Ed. M. Cais. Amsterdam: Elsevier)
14. Adamson, A. W. (1968). *Rec. Chem. Progr.*, **29,** 191
15. Balzani, V., Moggi, L. and Carassiti, V. (1968). *Ber. Bunsenges Physik. Chem.*, **72,** 288
16. Balzani, V., Moggi, L., Scandola, F. and Carassiti, V. (1967). *Inorg. Chim. Acta Rev.*, **1,** 7
17. Strohmeier, W. (1964). *Angew. Chem.*, **76,** 873
18. Valentine, D., Jr. (1968). *Advan. Photochem.*, **6,** 123
19. Ohnesorge, W. E. (1966). Fluorescence of Metal Chelate Compounds, *Fluorescence and Phosphorescence Analysis.* (Ed. D. M. Hercules, New York: Interscience)
20. Valentine, D., Jr. (1969). *Ann. Survey Photochem.*, **1,** 459
21. Balzani, V., Carassiti, V. and Scandola, F. (1966). *Gazz. Chim. Ital.*, **96,** 1213
22. Adamson, A. W. (1968). *Coord. Chem. Rev.*, **3,** 169
23. Adamson, A. W. (1969). *Pure Appl. Chem.*, **20,** 25
24. Endicott, J. F. (1970). *Israel J. Chem.*, **8,** 209
25. Simionescu, C., Crusos, A. and Feldman, D. (1964). *Chem. Abstr.*, **60,** 14610
26. Simionescu, C., Feldman, D. and Oprea, S. (1964). *J. Appl. Polymer Sci.*, **8,** 447
27. Hrihorov, M., Feldman, D. and Simionescu, C. (1965). *Rev. Roumaine Chim.*, **10,** 77
28. Hrihorov, M., Feldman, D. and Simionescu, C. (1965). *Chem. Abstr.*, **63,** 7109
29. Dainton, F. S. and James, D. G. L. (1951). *J. Chim. Phys.*, **48,** C17.
30. Dainton, F. S. and James, D. G. L. (1958). *Trans. Faraday Soc.*, **54,** 649
31. Oster, G. K. and Oster, G. (1959). *J. Amer. Chem. Soc.*, **81,** 5543
32. Ungurenasu, C. and Cecal, A. (1969). *J. Inorg. Nucl. Chem.*, **31,** 1735
33. Matsumoto, A., Tokunaga, S. and Shikawa, J. (1970). *Kogyo Kagaku Zasshi,* **73,** 1243; *Chem. Abstr.*, **73,** 93573i
34. Havemann, R., Pietsch, H. and Sachse, E. (1960). *Z. Wiss. Phot.*, **54,** 185
35. Panwar, K. S. and Gaur, J. N. (1967). *Talanta,* **14,** 127
36. Wojtczak, J. (1967). *Poznan. Towarz. Przyjaciol. Nauk, Wydzial Mat. Przyrod. Pr. Kom. Mat. Przyr.*, **12,** 25; *Chem. Abstr.*, **67,** 77860
37. Van Bronswyk, W. (1968). *J. Chem. Soc., A,* 692
38. Schläfer, H. L. (1965). *J. Phys. Chem.*, **69,** 2201
39. Adamson, A. W. (1967). ibid., **71,** 798
40. Schläfer, H. L. (1970). *Z. Chem.*, **10,** 9
41. Plane, R. A. and Hunt, J. P. (1957). *J. Amer. Chem. Soc.,* **79,** 3343
42. Schläfer, H. L. (1957). *Z. Physik. Chem. Frankfurt,* **11,** 65
43. Schläfer, H. L. (1959). *Acta Chim. Acad. Sci. Hung.*, **18,** 376
44. Schläfer, H. L. (1960). *Z. Elektrochem.*, **64,** 887
45. Basolo, F. and Pearson, R. G. (1958). *Mechanisms of Inorganic Reactions.* (New York, N.Y.: John Wiley and Sons)
46. Edelson, M. R. and Plane, R. A. (1959). *J. Phys. Chem.*, **63,** 327
47. Edelson, M. R. and Plane, R. A. (1964). *Inorg. Chem.*, **3,** 231
48. Wasgestian, H. F. and Schläfer, H. L. (1968). *Z. Physik. Chem. Frankfurt,* **62,** 127
49. Wasgestian, H. F. and Schläfer, H. L. (1968). ibid., **57,** 282

50. Riccieri, P. and Schläfer, H. L. (1970). *Inorg. Chem.*, **9,** 727
51. Schläfer, H. L. (1969). *Z. Physik. Chem.*, **65,** 107
52. Chen, S. N. and Porter, G. B. (1970). *Chem. Phys. Lett.*, **6,** 41
53. Adamson, A. W., Martin, J. E. and Camessei, F. D. (1969). *J. Amer. Chem. Soc.*, **91,** 7530
54. Zinato, E., Lindholm, R. and Adamson, A. W. (1969). ibid., **91,** 1076
55. Adamson, A. W. (1966). *Alcuni Aspetti Della Chimica Dei Complessi Ottaedrici.* (Rome: National Research Council Publication)
56. Adamson, A. W. (1965). Mechanism of Inorganic Reactions, *Advances in Chemistry* Series No. 49, p. 237. (Washington, D.C.: American Chemical Society)
57. Adamson, A. W. (1960). *J. Inorg. Nucl. Chem.*, **13,** 275
58. Behrendt, S., Langford, C. H. and Frankel, L. S. (1969). *J. Amer. Chem. Soc.*, **91,** 2236
59. Watts, D. W. (in press). 'Kinetics and Mechanism in Organic Solvent Systems', *Physical Chemistry of Organic Solvent Systems.* (London: Plenum Press)
60. Sastri, V. S. and Langford, C. H. (1970). *J. Phys. Chem.*, **74,** 3945
61. Adamson, A. W., Chen, A. and Waltz, W. L. (1967). *Proc. X., I.C.C.C.*, p. 305. (Tokyo)
62. Stevenson, K. L. and Verdieck, J. F. (1968). *J. Amer. Chem. Soc.*, **90,** 2974
63. Stevenson, K. L. and Verdieck, J. (1969). *Mol. Photochem.*, **1,** 271
64. Dainton, F. S. and James, D. G. L. (1958). *Trans. Faraday Soc.*, **54,** 649
65. Collinson, E., Dainton, F. S. and Malati, M. A. (1959). *ibid.*, **55,** 2096
66. Hartmann, H. and Filss, P. (1969). *Z. Phys. Chem. Frankfurt*, **66,** 48
67. Hartmann, H. and Filss, P. (1969). ibid., **66,** 60
68. Kosar, J. (1965). *Light Sensitive Systems.* (New York, N.Y.: John Wiley and Sons Inc.)
69. Suzuki, S., Matsumoto, K., Harada, K. and Tsubura, E. (1968). *Photogr. Sci. Eng.*, **12,** 2
70. Sasaki, M. and Kikuchi, S. (1967). *Chem. Abstr.*, **68,** 118468
71. von Hornuff, G. and Kaeppler, E. (1964). *J. Prakt. Chem.*, **23,** 54
72. Halonbrenner, R., Huber, J. R., Wild, U. and Günthard, H. H. (1968). *J. Phys. Chem.*, **72,** 3929
73. Chadwick, B. M. and Sharpe, A. G. (1966). *Advan. Inorg. Chem. Radiochem.*, **8,** 83
74. Parish, R. V. (1966). ibid., **9,** 315
75. Parish, R. V. (1966). *Coord. Chem. Rev.*, **1,** 439
76. Muetterties, E. L. and Wright, C. M. (1967). *Quart. Rev. Chem. Soc.*, **21,** 109
77. Lippard, S. J. (1967). *Prog. Inorg. Chem.*, **8,** 109
78. Blight, D. G. and Kepert, D. L. (1968). *Theoret. Chim. Acta*, **11,** 51
79. Adamson, A. W., Welker, J. P. and Volpe, M. (1950). *J. Amer. Chem. Soc.*, **72,** 4030
80. Goodenow, E. L. and Garner, C. S. (1955). ibid., **77,** 5268
81. Jakób, W., Samotus, A., Stasicka, Z. and Golebiewsi, A. (1966). *Z. Naturforsch*, **21b,** 819
82. Stasicka, Z., Samotus, A. and Jakób, W. (1966). *Roczniki Chem.*, **40,** 967
83. Jakób, W., Kosinska-Samotus, A. and Stasicka, Z. (1962). *ibid.*, **36,** 611
84. Jakób, W. and Jakób, Z. (1952). ibid., **26,** 492
85. Jakób, Z. and Jakób, W. (1956). *Zeszyty Nauk. Uniw. Jagiel. Ser. Nauk. Mat. Przyrod. Mat. Fiz. Chem.*, **2,** 49
86. Jakób, W., Kosinska-Samotus, A. and Stasicka, Z. (1962). *Proc. 7th I.C.C.C.*, p. 238. (Stockholm)
87. Jakób, W., Kosinska-Samotus, A. and Stasicka, Z. (1962). *Roczniki Chem.*, **36,** 165
88. Jakób, W. and Jakób, Z. (1962). ibid., **36,** 593
89. Jakób, W., Samotus, A. and Stasicka, Z. (1964). *Theory and Structure of Complex Compounds, Proceedings of the Symposium, Wroclaw, Poland,* p. 211. (London: Pergamon Press, Ed. B. Jezowska-Trzebiatowska)
90. Jakób, W., Samotus, A. and Stasicka, Z. (1966). *Roczniki Chem.*, **40,** 1383
91. Carassiti, V. and Balzani, V. (1960). *Ann. Chim. Rome*, **50,** 630
92. Balzani, V., Manfrin, M. F. and Moggi, L. (1969). *Inorg. Chem.*, **8,** 47
93. Carassiti, V. and Claudi, M. (1959). *Ann. Chim. Rome*, **49,** 1976
94. Bertoluzza, A., Carassiti, V. and Marinangeli, A. M. (1960). *ibid.*, **50,** 645
95. Adamson, A. W. and Perumareddi, J. R. (1965). *Inorg. Chem.*, **4,** 247
96. Adamson, A. W. and Sporer, A. H. (1958). *J. Amer. Chem. Soc.*, **80,** 3865
97. Perumareddi, J. R. (1962). *Ph.D. Dissertation,* University of Southern California, U.S.A.
98. MacDiarmid, A. G. and Hall, N. F. (1953). *J. Amer. Chem. Soc.*, **75,** 5204
99. Mitra, R. P. and Sharma, B. K. (1969). *Canad. J. Chem.*, **47,** 2317
100. Mitra, R. P., Sharma, B. K. and Mohan, H. (1969). *Indian J. Chem.*, **7,** 1162
101. Lippard, S. J. and Russ, B. J. (1967). *Inorg. Chem.*, **6,** 1943

102. Lippard, S. J., Nozaki, H. and Russ, B. J. (1967). *Chem. Commun.*, 118
103. Waltz, W. L., Adamson, A. W. and Fleischauer, P. D. (1967). *J. Amer. Chem. Soc.*, **89**, 3923
104. Waltz, W. L. and Adamson, A. W. (1969). *J. Phys. Chem.*, **73**, 4250
105. Carassiti, V. and Balzani, V. (1961). *Ann. Chim. Rome*, **51**, 518
106. Balzani, V. and Carassiti, V. (1961). ibid., **51**, 533
107. Balzani, V. and Carassiti, V. (1962). ibid., **52**, 432
108. Balzani, V., Bertoluzza, A. and Carassiti, V. (1962). *Bull. Soc. Chim. Belges*, **71**, 821
109. Nemodruk, A. A. and Bezrogova, E. V. (1968). *Zh. Anal. Khim.*, **23**, 884
110. Nemodruk, A. A. and Bezrogova, E. V. (1969). ibid., **24**, 408
111. Korobova, Z. P. and Kharlamov, I. P. (1970). *Chem. Abstr.*, **72**, 8953r
112. Neu, F. and Schwing-Weill, M. J. (1968). *Bull. Soc. Chim. France*, 4821
113. Dodonova, N.Ya. (1965, 1966). *Zh. Fiz. Khim.*, **39**, 3033; (1965). *Chem. Abstr.*, **64**, 9131a (1966)
114. Dodonova, N. Ya. (1966). *Zh. Fiz. Khim.*, **40**, 1543. *Chem. Abstr.*, **65**, 11591h
115. Jakób, W., Golebiewski, A. and Senkowski, T. (1964). *Theory and Structure of Complex Compounds, Proceedings of the Symposium, Wroclaw, Poland.* (Ed. by B. Jezowska-Trzebiatowska), 231. (London: Pergamon Press)
116. Jakób, W. and Senkowski, T. (1964). *Roczniki Chem.*, **38**, 1751
117. Jakób, W. and Senkowski, T. (1966). ibid., **40**, 1601
118. Senkowski, T. (1968). ibid., **42**, 2007
119. Jakób, W., Czaja, J., Rudowska, D. and Senkowski, T. private communication reported in reference 7.
120. Manoharan, P. T. and Gray, H. B. (1966). *Inorg. Chem.*, **5**, 823
121. Porter, G. B., Doering, J. G. W. and Karanka, S. (1962). *J. Amer. Chem. Soc.*, **84**, 4027
122. Rackow, B. (1966). *Z. Physik. Chem. Leipzig*, **233**, 103, 110, 113
123. Rackow, B. (1965). *Z. Chem.*, **5**, 397
124. Fujinaga, T., Koyama, M. and Kanchiku, Y. (1967). *Bull. Chem. Soc. Japan*, **40**, 2970
125. Koyama, M., Kanchiku, Y. and Fujinaga, T. (1968). *Coord. Chem. Rev.*, **3**, 285
126. Nemodruk, A. A. and Bezrogova, E. V. (1969). *Zh. Anal. Khim.*, **24**, 1534
127. Alexander, J. J. and Gray, H. B. (1968). *J. Amer. Chem. Soc.*, **90**, 4260
128. Fyfe, W. S. (1962). *J. Chem. Phys.*, **37**, 1894
129. Shiron, M. and Stein, G. (1964). *Nature (London)*, **204**, 778
130. Stein, G. (1965). *Advan. Chem.*, **50**, 238
131. Airey, P. L. and Dainton, F. S. (1966). *Proc. Roy. Soc. Lond.*, **A291**, 340
132. Jordan, J. and Ewing, G. J. (1962). *Inorg. Chem.*, **1**, 587
133. Emschwiller, G. and Legros, J. (1965). *Compt. Rend.*, **261**, 1535
134. Ohno, S. (1967). *Bull. Chem. Soc. Japan*, **40**, 1770
135. Airey, P. L. and Dainton, F. S. (1966). *Proc. Roy. Soc. Lond.*, **A291**, 478
136. Ohno, S. (1970). *Nippon Kagaku Zasshi*, **91**, 91, *Chem. Abstr.*, **72**, 125, 504k
137. Ohno, S. (1970). *Nippon Kagaku Zasshi*, **91**, 103, *Chem. Abstr.*, **72**, 138, 265h
138. Buxton, G. V., Dainton, F. S. and Kalecinski, J. (1969). *Int. J. Radiat. Phys. Chem.*, **1**, 87
139. Uri, N. (1952). *Chem. Rev.*, **50**, 374
140. Dainton, F. S. and Jones, F. T. (1965). *Trans. Faraday Soc.*, **61**, 1681
141. Orgel, L. E. (1955). *J. Chem. Phys.*, **23**, 1958
142. Porter, G. B. and Schläfer, H. L. (1964). *Ber. Bunsenges Physik, Chem.*, **68**, 316
143. Parker, C. A. (1954). *Trans. Faraday Soc.*, **50**, 1213
144. Parker, C. A. (1953). *Proc. Roy. Soc. Lond.*, **220A**, 104
145. Moggi, L., Bolletta, F., Balzani, V. and Scandola, F. (1966). *J. Inorg. Nucl. Chem.*, **28**, 2589
146. Legros, J. (1967). *Compt. Rend.*, **265**, 225
147. Kenney, D. L., Clinckemaillie, G. G. and Michiels, C. L. (1968). *J. Phys. Chem.*, **72**, 410
148. Glikman, T. S. and Podlinyaeva, M. E. (1955). *Ukrain. Khim. Zhur.*, **21**, 211
149. Baxendale, J. H. and Bridge, N. K. (1955). *J. Phys. Chem.*, **59**, 783
150. Dainton, F. S. (1954). *J. Chem. Soc. (Spec. Publ.)*, **1**, 18
151. Adamson, M. G., Baulch, D. L. and Dainton, F. S. (1962). *Trans. Faraday Soc.*, **58**, 1388
152. Dainton, F. S. and Jones, R. G. (1967). ibid., **63**, 1512
153. Bamford, C. H., Jenkins, A. D. and Johnston, R. (1958). *J. Polymer Sci.*, **29**, 355
154. Bengough, W. I., MacIntosh, S. A. and Ross, I. C. (1963). *Nature (London)*, **200**, 567
155. Bengough, W. I. and Ross, I. C. (1966). *Trans. Faraday Soc.*, **62**, 2251
156. Döbereiner, J. (1831). *Schw. J.*, **62**, 90

157. Hatchard, C. G. and Parker, C. A. (1956). *Proc. Roy. Soc. Lond.*, **A235**, 518
158. Hatchard, C. G. and Parker, C. A. (1959). *J. Phys. Chem.*, **63**, 22
159. Emschwiller, G. (1959). *Compt. Rend.*, **248**, 959
160. Sigwart, C. and Spence, J. T. (1969). *J. Amer. Chem. Soc.*, **91**, 3991
161. Ford, P. C., Stuermer, D. H. and McDonald, D. P. (1969). ibid., **91**, 6209
162. Zuloaga, F. and Kasha, M. (1968). *Photochem. Photobiol.*, **7**, 549
163. Mingardi, M. and Porter, G. B., Reported in Reference 7
164. Crosby, G. A. (1967). *J. Chim. Phys.*, **64**, 160
165. Mingardi, M. and Porter, G. B. (1966). *J. Chem. Phys.*, **44**, 4354
166. Jorgensen, C. K. (1954). *Acta Chem. Scand.*, **8**, 1502
167. Adamson, A. W., Chiang, A. and Zinato, E. (1969). *J. Amer. Chem. Soc.*, **91**, 5467
168. Way, H. and Filipescu, N. (1969). *Inorg. Chem.*, **8**, 1609
169. Kantrowitz, E. R., Endicott, J. F. and Hoffman, M. Z. (1970). *J. Amer. Chem. Soc.*, **92**, 1776
170. Caspari, G., Hughes, R. G., Endicott, J. F. and Hoffmann, M. Z. (1970). ibid., **92**, 6801
171. Endicott, J. F., Hoffman, M. Z. and Beres, L. S. (1970). *J. Phys. Chem.*, **74**, 1021
172. Adamson, A. W. and Vogler, A. (1970). ibid., **74**, 67
173. Schrauzer, G. W., Lee, L. P. and Sibert, J. W. (1970). *J. Amer. Chem. Soc.*, **92**, 2997
174. Yamada, R., Shimizu, S. and Fukui, S. (1966). *Biochim. Biophys. Acta*, **124**, 197
175. Vogler, A. and Adamson, A. W. (1967). Abstracts of Pacific Conference on Chemistry and Spectroscopy, A.C.S. Western Regional Meeting, Anaheim, California, October 30th
176. Lindholm, R. D. and Adamson, A. W. (1967). *J. Phys. Chem.*, **71**, 3713
177. Shagisultanova, G. A. and Orisheva, R. M. (1969). *Khim. Vys. Energ.*, **3**, 265; *Chem. Abstr.*, **71**, 96826p
178. Shagisultanova, G. A. and Orisheva, R. M. (1969). *Khim. Vys. Energ.*, **3**, 266; *Chem. Abstr.*, **71**, 96827q
179. Shagisultanova, G. A., Poznyak, A. L. and Budkevich, B. A. (1970). *Khim. Vys. Energ.*, **4**, 286; *Chem. Abstr.*, **73**, 50696j
180. Norden, B. (1970). *Acta Chem. Scand.*, **24**, 1703
181. Vogler, A. and Adamson, A. W. (1968). *J. Amer. Chem. Soc.*, **90**, 5943
182. Valentine, D., Jr., Unpublished results quoted in reference 18
183. Scandola, M. A. and Scandola, F. (1970). *J. Amer. Chem. Soc.*, **92**, 7278
184. Porter, G. B. (1969). *ibid.*, **91**, 3980
185. Scandola, M. A., Scandola, F. and Carassiti, V. (1969). *Mol. Photochem.*, **1**, 403
186. Adamson, A. W., Vogler, A. and Lantzke, I. R. (1969). *J. Phys. Chem.*, **73**, 4183
187. Bauer, R. A. and Basolo, F. (1968). *J. Amer. Chem. Soc.*, **90**, 2437
188. Broomhead, J. A. and Crumley, W. (1968). *Chem. Commun.*, 1211
189. Sleight, T. P. and Hare, C. R. (1968). *J. Phys. Chem.*, **72**, 2207
190. Sleight, T. P. and Hare, C. R. (1968). *Inorg. Nucl. Chem. Lett.*, **4**, 165
191. Moggi, L., Varani, G., Manfrin, N. F. and Balzani, V. (1970). *Inorg. Chim. Acta*, **4**, 335
192. Berka, L. H. and Philippon, G. E. (1970). *J. Inorg. Nucl. Chem.*, **32**, 3355
193. Baranovskii, I. B., Kovalenko, G. S. and Babaeva, A. V. (1969). *Zh. Neorg. Khim.*, **14**, 2806
194. Demas, J. N. and Crosby, G. A. (1970). *J. Amer. Chem. Soc.*, **92**, 7262
195. Moggi, L. (1967). *Gazz. Chim. Ital.*, **97**, 1089
196. Adamson, A. W. (1967). *J. Phys. Chem.*, **71**, 798
197. Kelly, T. and Endicott, J. F. (1970). *J. Amer. Chem. Soc.*, **92**, 5733
198. Moggi, L., Bolletta, F., Balzani, V. and Scandola, F. (1966). *J. Inorg. Nucl. Chem.*, **28**, 2589
199. Balzani, V., Carassiti, V., Moggi, L. and Scandola, F. (1965). *Inorg. Chem.*, **4**, 1243
200. Martin, D. S., Tucker, M. A. and Kassman, A. J. (1965). ibid., **4**, 1682
201. Martin, D. S., Tucker, M. A. and Kassman, A. J. (1966). ibid., **5**, 1298
202. Basch, H. and Gray, H. B. (1967). ibid., **6**, 365
203. Jorgensen, C. K. (1959). *Advan. Chem. Phys.*, **5**, 33
204. Jorgensen, C. K. (1959). *Mol. Phys.*, **2**, 309
205. Balzani, V. and Carassiti, V. (1968). *J. Phys. Chem.*, **72**, 383
206. Scandola, F., Traverso, O. and Carassiti, V. (1969). *Mol. Photochem.*, **1**, 11
207. Sastri, V. S. and Langford, C. H. (1969). *J. Amer. Chem. Soc.*, **91**, 7533
208. Grinberg, A. A. and Shagisultanova, G. A. (1955). *Izvest. Akad. Nauk, SSSR Otdel. Khim. Nauk.*, 981
209. Grinberg, A. A., Nikoleskaya, L. E. and Shagisultanova, G. A. (1955). *Dokl. Akad. Nauk. SSSR*, **101**, 1059

210. Bartocci, C., Scandola, F., Traverso, O. and Balzani, V. Annual Meeting Chimica Inorganica (Padova) 23, Quoted in Reference 7
211. Bartocci, C., Scandola, F. and Balzani, V. (1969). *J. Amer. Chem. Soc.*, **91,** 6948
212. Gray, H. B. and Olcott, R. J. (1962). *Inorg. Chem.*, **1,** 481
213. Langford, C. H. and Gray, H. B. (1966). *Ligand Substitution Processes.* (New York: Benjamin)
214. Bartocci, C. and Scandola, F. (1970). *Chem. Commun.*, 531
215. Perumareddi, J. R. and Adamson, A. W. (1968). *J. Phys. Chem.*, **72,** 414
216. Babaeva, A. V. and Mosyagina, M. A. (1953). *Izvest. Akad. Nauk. SSSR, Otdel. Khim. Nauk.*, 227
217. Ramberg, L. (1910). *Chem. Ber.*, **43,** 580
218. Haake, P. and Hylton, T. A. (1962). *J. Amer. Chem. Soc.*, **84,** 3774
219. Shaw, B. L. and Brookes, P. R. (1968). *Chem. Commun.*, 919
220. Balzani, V. and Carassiti, V. (1968). *J. Phys. Chem.*, **72,** 383
221. Balzani, V., Carassiti, V. Moggi, L. and Scandola, F. (1965). *Inorg. Chem.*, **4,** 1243
222. Scandola, F., Traverso, O., Balzani, V., Zucchini, G. L. and Carassiti, V. (1967). *Inorg. Chem. Acta*, **1,** 76
223. Sabbatini, N., Varani, G., Balzani, V., Moggi, L. and Manfrin, M. F., reported in Reference 7
224. Ballhausen, C. J., Bjerrum, N., Dingle, R., Eriks, K. and Hare, C. R. (1965). *Inorg. Chem.*, **4,** 514
225. Mastin, S. H. and Haake, P. (1970). *Chem. Commun.*, 202
226. Rich, R. L. and Taube, H. (1954). *J. Amer. Chem. Soc.*, **76,** 2608
227. Dreyer, R. and König, K. (1966). *Z. Chem.*, **6,** 271
228. Blasius, E., Preetz, W. and Schmitt, R. (1961). *J. Inorg. Nucl. Chem.*, **19,** 115
229. Dreyer, R., Dreyer, I. and Rettig, D. (1963). *Z. Physik. Chem. Leipzig*, **224,** 199
230. Balzani, V., Manfrin, M. F. and Moggi, L. (1967). *Inorg. Chem.*, **6,** 354
231. Sastri, V. S., reported in Reference 7
232. Schmidt, G. and Herr, W. Z. (1961). *Z. Naturforsch.*, **16a,** 748
233. Adamson, A. W. and Sporer, A. H. (1958). *J. Amer. Chem. Soc.*, **80,** 3865
234. Penkett, S. A. and Adamson, A. W. (1965). ibid., **87,** 2514
235. Balzani, V., Sabbatini, N. and Carassiti, V. (1968). *Progress in Co-ordination Chemistry*, (Ed. by M. Cais), 80. (Amsterdam: Elsevier)
236. Ziolo, F. R. and Dori, Z. (1968). *J. Amer. Chem. Soc.*, **90,** 6560
237. Ziolo, R. F., Lipton, S. and Dori, Z. (1970). *Chem. Commun.*, 1124
238. Sundararajan, S. and Wehry, E. L. (1970). *Chem. Commun.*, 267
239. Van den Berg, J. A. and Schneider, P. (1968). *J. S. Afr. Chem. Inst.*, **21,** 113
240. Il'Vukevich, L. A. and Shagisultanova, G. A. (1969). *Khim. Vys. Energ.*, **3,** 207
241. Poznyak, A. L. (1969). ibid., **3,** 380
242. De Waal, D. J. A. and Van den Berg, J. A. (1969). *Tydskr. Natuurwetensk.*, **9,** 1; *Chem. Abstr.*, **71,** 96806g
243. Rabinowitch, E. and Belford, R. L., Eds. (1964). 'Spectroscopy and Photochemistry of Uranyl Compounds', Vol. 1, *Monographs on Nuclear Energy, Chemistry Division.* (New York, N.Y.: MacMillan Co.)
244. Martin, T. W., Rummel, R. E. and Gross, R. C. (1964). *J. Amer. Chem. Soc.*, **86,** 2595
245. Glass, R. W. and Martin, T. W. (1970). ibid., **92,** 5084
246. Martin, T. W., Swift, L. L. and Venables, J. H., Jr. (1970). *J. Chem. Phys.*, **52,** 2138
247. Martin, T. W. and Glass, R. W. (1970). *J. Amer. Chem. Soc.*, **92,** 5075
248. Hayon, E. and Saito, E. (1965). *J. Chem. Phys.*, **43,** 4314
249. Dogliotti, L. and Hayon, E. (1967). *J. Phys. Chem.*, **71,** 3802
250. Moorthy, P. N. and Weiss, J. J. (1965). *J. Chem. Phys.*, **42,** 3127
251. Marcus, R. J. (1956). *Science*, **123,** 399
252. Shchegoleva, I. S. and Glikman, T. S. (1969). *Ukr. Khim. Zh.*, **35,** 157
253. Greatorex, D. and Kemp, T. J. (1969). *Chem. Commun.*, 383
254. Sakuraba, S. and Matsushima, R. (1970). *Bull. Chem. Soc. Jap.*, **43,** 2359
255. Heidt, L. J., Tregay, G. W. and Middleton, F. A., Jr. (1970). *J. Phys. Chem.*, **74,** 1876
256. Filipescu, N. and Mushruch, G. W. (1968). ibid., **72,** 3516
257. Filipescu, N. and Mushrush, G. W. (1968). ibid., **72,** 3522
258. Haas, Y., Stein, G. and Tomkiewicz, M. (1970). ibid., **74,** 2558
259. Kato, Y. and Fukutomi, H. (1969). *Bull. Tokyo Inst. Technol.*, 1

260. Trevorrow, L. E., Gerding, T. J. and Steindler, M. J. (1969). *Inorg. Nucl. Chem. Lett.*, **5**, 837
261. Palei, P. N., Nemodruk, A. A. and Bezgova, E. V. (1969). *Radiokhimiya*, **11**, 300
262. Chini. P. (1968). *Inorg. Chim. Acta Rev.*, **2**, 31
263. Dobson, G. R., Stoltz, I. W. and Sheline, R. K. (1966). *Advances in Inorganic and Radiochemistry*, Ed. by J. H. Emeléus and A. G. Sharpe, Vol. 8, 1–82. (London: Academic Press Inc.)
264. Strohmeier, W. (1964). *Angew. Chem. Int. Ed. Engl.*, **3**, 370
265. Dobson, G. R., El-Sayed, M. F. A., Stoltz, I. W. and Sheline, R. K. (1962). *Inorg. Chem.*, **1**, 526.
266. Stoltz, I. W., Dobson, G. R. and Sheline, R. K. (1963). *J. Amer. Chem. Soc.*, **85**, 1013
267. Stoltz, I. W., Dobson, G. R. and Sheline, R. K. (1962). ibid., **84**, 3589
268. Rest, A. J. and Turner, J. J. (1969). *Chem. Commun.*, 1026
269. McIntyre, J. A. (1970). *J. Phys. Chem.*, **74**, 2403
270. Rest, A. J. (1970). *Chem. Commun.*, 345
271. Rest, A. J. and Turner, J. J. (1970). *ibid.*, 375
272. Graham, M. A., Rest, A. J. and Turner, J. J. (1970). *J. Organometal. Chem.*, **24**, C54
273. von Gustorf, E. K. and Langmuir, M. E., unpublished results quoted in Reference 8
274. Moss, J. R. and Graham, W. A. G. (1970). *Chem. Commun.*, 835
275. Ogilvie, J. F. (1970). ibid., 323
276. Dobson, G. R. (1965). *J. Phys. Chem.*, **69**, 677
277. Strohmeier, W. and Mueller, F. J. (1969). *Chem. Ber.*, **102**, 3608
278. Carty, A. J., Efraty, A., Ng, T. W. and Birchall, T. (1970). *Inorg. Chem.*, **9**, 1263
279. Strohmeier, W. and Mueller, F. J. (1969). *Chem. Ber.*, **102**, 3613
280. Anisimov, K. N., Kolobova, N. E. and Pasynskii, A. A. (1969). *Izvest. Akad. Nauk. SSSR, Ser. Khim.*, 2238
281. Fischer, E. O. and Eckhart, E. (1969). *J. Organometal. Chem.*, **18**, 26
282. Haines, R. J. and Du Preez, A. L. (1970). *ibid.*, **21**, 181
283. Mesmeyanov, E. N., Makarova, L. G. and Polonyanyuk, I. V. (1970). ibid., **22**, 707
284. Silverstein, H. T., Beer, D. C. and Todd, L. J. (1970). ibid., **21**, 139
285. Benedict, J. J. (1969). ibid., **17**, 37
286. King, R. B. and Kapoor, R. N. (1969). *J. Inorg. Nucl. Chem.*, **31**, 2169
287. Jetz, W. and Graham, W. A. G. (1969). *J. Amer. Chem. Soc.*, **91**, 3375
288. Hieber, W. and Kaiser, K. (1969). *Z. Naturforsch.*, **B24**, 778
289. Schumann, H. and Weis, R. (1970). *Angew. Chem. Int. Ed. Engl.*, **9**, 246
290. Nesmeyanov, A. N., Anisimov, K. N., Kolobova, N. E. and Khandozhko, V. N. (1969). *Izv. Akad. Nauk. SSSR, Ser. Khim.*, 1950
291. Booth, B. L., Haszeldine, R. N. and Mitchell, P. R. (1970). *J. Organometal. Chem.*, **21**, 203
292. Fields, R., Germain, M. M., Haszeldine, R. N. and Wiggans, P. W. (1970). *J. Chem. Soc. A*, 1964
293. Nesmeyanov, A. N., Anisimov, K. N., Kolobova, N. and Pasynskii, A. A. (1970). *Izv. Akad. Nauk. SSSR, Ser. Khim.*, 727
294. Nesmeyanov, A. N., Kolobova, N. E., Anisimov, K. N. and Skripkin, V. V. (1969). ibid., 2859
295. Seel, F. and Roeschenthaler, G. V. (1970). *Angew. Chem. Int. Ed. Eng.*, **9**, 166
296. Beck, W. and Schier, E. (1970). *Z. Naturforsch.*, **B25**, 221
297. Deckelmann, E. and Werner, H. (1970). *Helv. Chim. Acta*, **53**, 139
298. Wegner, P. A., Guggenberger, L. J. and Muetterties, E. L. (1970). *J. Amer. Chem. Soc.*, **92**, 3473
299. Wrightson, M., Hammond, G. S. and Gray, H. B. (1970). ibid., **92**, 6068
300. Valentine, D., Sarver, B. and Collman, J. P., unpublished results quoted in Reference 302
301. Landesberg, J. M. and Sieczkowski, J. (1969). *J. Amer. Chem. Soc.*, **91**, 2120
302. Valentine, J. S. and Valentine, D. (1970). ibid., **92**, 5795
303. Bathelt, W., Herberhold, M. and Fischer, E. O. (1970). *J. Organometal. Chem.*, **21**, 395
304. Tarr, A. M. and Wiles, D. M. (1968). *Can. J. Chem.*, **46**, 2725
305. Richards, J. H. and Pisker-Trifunac, N. (1969). *J. Paint Technol.*, **41**, 363
306. Mesmeyanov, A. N., Vol'Kenau, N. A. and Shilovtseva, L. S. (1970). *Dokl. Akad. Nauk. SSSR*, **190**, 857
307. Hoshi, Y., Akiyama, T. and Sugimori, A. (1970). *Tetrahedron Lett.*, 1485

308. Sato, T., Shimada, S. and Hata, K. (1969). *Bull. Chem. Soc. Jap.*, **42,** 2731
309. Traverso, O., Scandola, F., Balzani, V. and Valcher, S. (1969). *Mol. Photochem.*, **1,** 289
310. van Riel, H. C. H. A., Fischer, F. C., Lugtenburg, J. and Havinga, E. (1969). *Tetrahedron Lett.*, 3085
311. Daniel, D. M. and Nyman, C. J. (1969). *Chem. Commun.*, 483
312. Hayward, P. J., Blake, D. M., Nyman, C. J. and Wilkinson, G. (1969). *Chem. Commun.*, 987
313. Blake, D. M. and Nyman, C. J. (1970). *J. Amer. Chem. Soc.*, **92,** 5359
314. Stranks, D. R. and Yandell, J. R. (1965). *Exchange Reactions, Proc. Symp.*, p. 83. (Upton, N.Y.)
315. Stranks, D. R. and Yandell, J. K. (1969). *J. Phys. Chem.*, **73,** 840
316. Kochi, J. K. and Bethea, T. W. (1968). *J. Org. Chem.*, **33,** 75
317. Burchill, C. E. and Wolodarsky, W. H. (1970). *Can. J. Chem.*, **48,** 2955
318. Guillory, W. A. and Smith, C. E. (1970). *J. Chem. Phys.*, **53,** 1661
319. Bell, T. N., Boonstra, M. and Dobud, P. A. (1970). *J. Amer. Chem. Soc.*, **92,** 4521
320. Basco, N. and Yee, K. K. (1967). *Nature (London)*, **216,** 998
321. Basco, N. and Yee, K. K. (1967). *Chem. Commun*, 1146
322. Basco, N. and Yee, K. K. (1967). ibid., 1255
323. Basco, N. and Yee, K. K. (1968). *Spectroscopy Letters*, **1,** 13, 19
324. Callear, A. B. and Oldham, R. J. (1968). *Trans. Faraday Soc.*, **64,** 840
325. Treinin, A. (1970). *Israel. J. Chem.*, **8,** 103
326. Blandamer, M. J. and Fox, M. F. (1970). *Chem. Rev.*, **70,** 59.
327. Dainton, F. S. and Fowles, P. (1965). *Proc. Roy. Soc. Lond.*, **A287,** 312
328. Barrett, J., Fox, M. F. and Mansell, A. L. (1967). *J. Chem. Soc. A.*, 483
329. Hayon, E. and McGarvey, J. J. (1967). *J. Phys. Chem.*, **71,** 1472
330. Dogliotti, L. and Hayon, E. (1967). ibid., **71,** 2511
331. Woods, R., Kolthoff, I. M. and Meehan, E. J. (1963). *J. Amer. Chem. Soc.*, **85,** 2385
332. Dogliotti, L. and Hayon, E. (1968). *J. Phys. Chem.*, **72,** 1800
333. Dogliotti, L. and Hayon, E. (1968). *Nature (London)*, **218,** 949
334. Sperling, R. and Treinin, A. (1964). *J. Phys. Chem.*, **68,** 897
335. Meyerstein, D. and Treinin, A. (1961). *Trans. Faraday Soc.*, **57,** 2104
336. Baudisch, O. (1911). *Chem. Ber.*, **44,** 1009
337. Warburg, E. (1919). *Z. Elektrochem.*, **25,** 334
338. Villars, D. S. (1927). *J. Amer. Chem. Soc.*, **49,** 326
339. Petriconi, G. L. and Papee, H. M. (1968). *J. Inorg. Nucl. Chem.*, **30,** 1525
340. Daniels, M., Meyers, R. V. and Belardo, E. V. (1968). *J. Phys. Chem.*, **72,** 389
341. Barat, F., Hickel, B. and Sutton, J. (1969). *Chem. Commun.*, 125
342. Barat, F., Gilles, L., Hickel, B. and Sutton, J. (1970). ibid., 1982
343. Treinin, A. and Hayon, E. (1970). *J. Amer. Chem. Soc.*, **92,** 5821
344. Shuali, U., Ottolenghi, M., Rabani, J. and Yelin, Z. (1969). *J. Phys. Chem.*, **73,** 3445
345. Holmes, M. (1926). *J. Chem. Soc.*, 1898
346. Farkas, L. and Klein, F. S. (1948). *J. Chem. Phys.*, **16,** 886
347. Bridge, N. K. and Matheson, M. S. (1960). *J. Phys. Chem.*, **64,** 1280
348. Treinin, A. and Yaacobi, M. (1964). *ibid.*, **68,** 2487
349. Amichai, O., Czapski, G. and Treinin, A. (1969). *Israel J. Chem.*, **7,** 351
350. Amichai, O. and Treinin, A. (1970). *J. Phys. Chem.*, **74,** 3670
351. Amichai, O. and Treinin, A. (1969). *Chem. Phys. Lett.*, **3,** 611
352. Amichai, O. and Treinin, A. (1970). *J. Phys. Chem.*, **74,** 830
353. Amichai, O. and Treinin, A. (1970). *J. Chem. Phys.*, **53,** 444
354. Barat, F., Gilles, L., Hickel, B. and Sutton, J. (1969). *Chem. Commun.*, 1485
355. Weeks, J. L., Meaburn, G. M. A. and Gordon, S. (1963). *Radiation Res.*, **19,** 559
356. Behar, D., Czapski, G. and Duchovny, I. (1970). *J. Phys. Chem.*, **74,** 2206
357. Halmann, M. and Platzner, I. (1965). *J. Chem. Soc.*, 1440
358. Halmann, M. and Platzner, I. (1966). *J. Phys. Chem.*, **70,** 2281
359. Huber, J. R. and Hayon, E. (1968). ibid., **72,** 3820
360. Benderly, H. and Halmann, M. (1967). ibid., **71,** 1053
361. Jortner, J., Ottolenghi, M. and Stein, G. (1964). ibid., **68,** 247
362. Dainton, F. S. and Fowles, P. (1965). *Proc. Roy. Soc. Lond.*, **A287,** 312
363. Stein, G. (1967). *The Chemistry of Ionisation and Excitation.* (London: Taylor and Francis)
364. Matheson, M. S., Mulac, W. A., Weeks, J. L. and Rabani, J. (1966). *J. Phys. Chem.*, **70,** 2092

365. Czapski, G. and Ottolenghi, M. (1968). *Israel J. Chem.*, **6,** 75
366. Czapski, G., Ogdan, J. and Ottolenghi, M. (1969). *Chem. Phys. Lett.*, **3,** 383
367. Gusersky, E. and Treinin, A. (1965). *J. Phys. Chem.*, **69,** 3176
368. Leopold, J., Shapira, D. and Treinin, A. (1970). ibid., **74,** 4585
369. McDonald, J. R., Scherr, V. M. and McGlynn, S. P. (1969). *J. Chem. Phys.*, **51,** 1723
370. Matheson, M. S., Mulac, W. A. and Rabani, J. (1963). *J. Phys. Chem.*, **67,** 2613
371. Luria, M. and Treinin, A. (1968). ibid., **72,** 305
372. Burak, I. and Treinin, A. (1963). *J. Chem. Phys.*, **39,** 189
373. McDonald, J. R., Rabalais, T. W. and McGlynn, S. P. (1970). ibid., **52,** 1332
374. Burak, I. and Treinin, A. (1965). *J. Amer. Chem. Soc.*, **87,** 4031
375. Treinin, A. and Hayon, E. (1969). *J. Chem. Phys.*, **50,** 538
376. Burak, I., Shapira, D. and Treinin, A. (1970). *J. Phys. Chem.*, **74,** 568
377. Devonshire, R. and Weiss, J. J. (1968). ibid., **72,** 3815
378. Basco, N., Kenney, G. A. and Walker, D. C. (1969). *J. Chem. Soc. D*, 917

6
Mechanism and Steric Course of Octahedral Substitution

C. H. LANGFORD and V. S. SASTRI
University of Carleton, Ottawa

List of Abbreviations
en = ethylenediamine
PR_3 = trialkylphosphine (triphenylphosphine)
DMG = dimethylglyoxime

cyclam = 1,4,8,11-tetra-azacyclotetradecane
DMSO = dimethylsulphoxide
S,S-trien = 1,4,7,10-tetra-azadecane, (triethylenetetramine)
2,3,2-tet = 1,4,8,11-tetra-azaundecane. RR and RS designations refer to the configurations of the secondary nitrogens when coordinated.
acac = acetylacetone
dipy = dipyridyl
N_4 = $(NH_3)_4$
py = pyridine
tn = trimethylenediamine
EDTA = ethylenediaminetetra-acetate
terpy = terpyridyl
phen = phenanthroline
NTA = nitrilotriacetic acid
S_6 = six solvent molecules
OAc = acetate
ox = oxalate
biguan H = protonated biguanide
mal = malonate
pn = propylenediamine
DMF = dimethylformamide
TMPA = trimethylphosphoramide
dien = diethylenetriamine

6.1 INTRODUCTION

6.1.1 Scope of the chapter

Not too many years ago a review with the present title would have been considered pretentious. It would have been, in fact, principally a review of the ligand substitution chemistry of the substituted amine complexes of Co^{III}, with perhaps some consideration of Cr^{III} systems and a nod in the direction of the Ni^{II} and Rh^{III} systems, and esoteric complexes of a few other metals. This is not at all the present situation. Good data exist for complexes of almost all of the very many metal ions whose coordination numbers are characteristically six. (It is in this loose sense, not in a group theoretic sense, that mechanists tend to speak of octahedral complexes.)

The early theorists of 'octahedral substitution' were not, however, truly arrogant, but more nearly *idiots savants*. The categories of analysis and intellectual tools developed for the discussion of mechanistic problems of Werner complexes of Co^{III} have proved their worth over the whole field. Cobalt(III) complexes have been, and continue to be, extremely fruitful model systems. This has surprised the pessimistic school of inorganic kineticists whose motto is 'every little reaction has a mechanism all its own'. But, such surprise may reflect an epistemological error. As information on other systems has accumulated, the way of asking questions about the model Co^{III} systems has undergone a subtle evolution. We were lucky in the early stages, but if it were really true that every reaction had a unique mechanism, there would be little point in mechanistic study. A mechanism is a theoretical construction. If it is not powerful (general), it tells the theoretician that he is asking the wrong questions. The successful ability of the model for the behaviour of Co^{III} complexes to explain the behaviour of a wide range of complexes of other metals indicates that many correct questions have been put to test.

The study of octahedral substitution is now a mature field. More or less general agreement exists on the right questions to ask and the main ways of asking them. Smugness may be the greatest current danger. Suppose that current thinking about 'octahedral substitution' is 'right'. The next challenge will be to integrate the theory of octahedral substitution with theories of the photochemical, the oxidation–reduction, and the (neglected) addition–elimination reactions of metal complexes. If wider generality is the aim, it will be especially important to be clear about the precise meaning of the questions we now ask about substitution. After a brief summary of some other sources of review information and a statement on the scope of this review, the nature of the questions asked will be analysed as a starting point for the consideration of particular results.

There is little point in undertaking a comprehensive review back to a pre-1967 date. The second edition of Basolo and Pearson's important book appeared that year[2]. The mechanistic literature up to 1967 was covered in The Annual Reviews of Physical Chemistry[3] and the steric course was reviewed by Archer[4] in an account covering the literature up to 1968. Other reviews related to this subject have appeared since 1967[5–7]. In consequence, we have taken the period 1968–1970 for review, with earlier results quoted only to provide clarity and continuity to the argument. Carbonyl and organometallic compounds are commonly treated mechanistically in the context of their structural chemistry. Although there is no reason to suppose that they require a fundamentally different attitude toward mechanisms, we have omitted them in order to achieve a fairly complete treatment of the complexes of 'hard' ligands.

6.1.2 The tools of mechanistic analysis

The solution kineticist has available three fundamental types of experiment: *rate law*, *stereochemistry*, and *variation of rate constant with structure or environment*. Each permits the analysis of slightly different aspects of a reaction, and it is useful to develop a scheme of reaction classification which clarifies the behaviour of a reaction in each experimental situation. It is worth while to give a brief account of the role of each of the three types of experimental investigation in the study of octahedral substitution.

6.1.2.1 Rate laws

As a consequence of the 'solvent-cage' effect, any slow reaction occurring between two or more species in solution may be sub-divided into two steps;

the first a diffusional encounter, and the second an 'isomerisation' of the encounter complex into the product[8,9]. If the overall rate is more than an order of magnitude slower than the diffusion-controlled limit, the first step may be treated as a pre-equilibrium. The mechanism from which the rate law must be derived is:

$$\begin{array}{c} MX+Y \underset{}{\overset{K_E}{\rightleftharpoons}} MX,Y \\ MX,Y \xrightarrow{k} MY,X \end{array} \tag{6.1}$$

where K_E is the encounter equilibrium constant, MX,Y is the encounter complex (or in the present context an 'outer-sphere complex'), and k is the *first-order* rate constant for the slow step of 'isomerisation'. Octahedral substitution pathways represented by equation (6.1) have been called *interchange* (I) pathways[10]. It is clear that an understanding of outer-sphere complex formation (which is frequently a matter of ion-pairing) is important to interchange pathways. The subject has been reviewed[11,12].

There is another pathway which has been important in discussions of octahedral substitution and which is distinct from interchange. It is one in which encounter with an entering ligand does not preceed but follows the slow step. This is written:

$$\begin{array}{c} MX \underset{k_{-1}}{\overset{k_1}{\rightleftharpoons}} M+X \\ \quad\quad k_2 \downarrow +Y \\ \quad\quad MY \end{array} \tag{6.2}$$

The implication of encounter following the slow step is that M is an *intermediate* with a sufficiently long lifetime that competition between X and Y for a presence in encounter complexes with M is possible. In octahedral substitution, the intermediate M has a coordination number of five. The pathway (6.2) is called *dissociative* D [10].

Both paths (6.1) and (6.2) frequently lead to second-order rate expressions; first-order in complex and first-order in attacking ligand. In dilute solution, the observed second-order rate constant for path (6.1) is $k_{obs} = K_E k$. The observed constant may also be interpreted in terms of the [Y]-dependence of the competition between the k_{-1} and k_2 steps in pathway (6.2). Only when the rates have been carefully analysed over a wide concentration range is it possible to distinguish between paths (6.1) and (6.2) [13,14]. Of course, it must be remembered that encounter is perpetual when Y is the solvent, and in this case there may not be an operational distinction between pathways (6.1) and (6.2).

6.1.2.2 Stereochemistry

The steric course of the reaction is also closely related to the type of encounter which occurs with the entering ligand. The steric course of an interchange is likely to be determined by the encounter geometry, and may give very little information as to the nature of the transition-state geometry unless this can be explicitly related to the encounter geometry. Stereochemistry may be more revealing when a D-pathway is involved. One important point is that stereochemical rearrangements past the transition state may sometimes identify an intermediate of the D-pathway type whose lifetime is too short to be detected by X versus Y competition experiments.

6.1.2.3 Variation of rate constant

Variation of the rate constant for substitution as a function of structure, environment, temperature, or pressure gives, in the context of the transition-state theory, information about the relationship between *only two points* on the reaction hypersurface. These points are the ground state and the transition state. The variation of rate constants may be due to differences of free energy, entropy, enthalpy, or the volume of activation.

Variations in these quantities may be used to analyse bonding in the transition state but do not, by themselves, indicate whether the encounter with an entering ligand comes before or after the transition state. The *single* most important factor in octahedral substitution is the degree of specific bonding of the entering group in the transition state. If bonding to the entering group occurs, the energy of the transition state is lowered, and this lowering will vary as a function of the choice of entering group. If the entering

	Stoichiometric mechanism		
Mode of activation	*Intermediate of increased coordination number*	*One-step process*	*Intermediate of reduced coordination number*
Associative activation	A	I_a	
Dissociative activation		I_d	D

Figure 6.1 Classification of ligand substitution mechanisms

group does not specifically bond to the metal centre in the transition state, there can be no entering-group lowering of the activation energy. Under these circumstances, activation energy must be determined from the energetics of bond breaking. The activation energy must accumulate in the original complex by thermal fluctuation.

It is clear that the D-pathway implies *no* entering-group assistance, since the entering group is *not* a stoichiometric component of the transition state. Therefore, it is convenient to call this unassisted activation *dissociative* (labelled *d*). The mode of approach to the transition state which is specifically assisted by the entering group is conveniently called *associative* (*a*). The key test which distinguishes between *d* and *a* modes of activation is the sensitivity of the rate constant to the nature of the entering group. It should be clear that an *interchange* pathway (I) may use either the *a* or *d* modes of activation so that we expect I_a and I_d processes. I_d processes are those in which the 'dissociated' transition state does not drop to an intermediate of reduced coordination number with lifetime longer than an average encounter-complex. Figure 6.1 collects the mechanistic labels used in this review.

6.2 ENCOUNTERS AND OUTER-SPHERE COMPLEXING

The value of the equilibrium constant for encounter is clearly central to a discussion of interchange pathways. This was first suggested, in fact, by

Werner in 1912 [15]. Most work on ligand substitution has been based on the assumption that encounter equilibria could be calculated from the ion-pairing equation of Fuoss[16], which was derived in turn from a consideration of diffusion-controlled reactions by Eigen[17]. At zero ionic strength, the equation is:

$$K_E = \frac{4\pi a^3 N_0}{300} \exp\left\{\frac{-z_1 z_2 e^2}{\varepsilon a k T}\right\} \tag{6.3}$$

where K_E is the encounter equilibrium constant, a is the distance of closest approach of the two species, ε is the dielectric constant of the medium, and the other symbols have their usual meaning[16]. Equation (6.3) may be corrected to finite ionic strength through the use of extended Debye–Hückel expressions. In practice, a is an adjustable parameter.

Some substitution work, especially that on Co^{III} complexes in both water and non-aqueous media, has relied on the experimental study of encounter equilibria[14, 18–23]. This is important, because at least three lines of experimental evidence reveal the weaknesses of the simple electrostatic model of outer-sphere complexation expressed by equation (6.3). These lines are:

(a) Formation of anionic species by overcompensation of the original positive charge of a metal complex,

(b) lack of evidence of outer-sphere complexation in situations where it is to be expected on electrostatic grounds, e.g. $Coen_3^{3+}$ with $Fe(CN)_6^{3-}$ [14], and,

(c) the fact that outer-sphere complexes are preferred over inner-sphere complexes in several systems.

In attempts to understand the deviations of encounter (outer-sphere) equilibria from the predictions of the simple electrostatic theory, several authors have drawn attention to the possible role of hydrogen bonding between anionic ligands and water or amine ligands of the inner coordination sphere[18–20, 25]. Others have suggested a role for dispersion forces[26, 27].

Whatever weaknesses may be found in equation (6.3), there can be no doubt that it has served well so far in the correlation of kinetic data. Its success is especially apparent for fast complex-formation reactions in water. Equation (6.3) will be used below as a reference point in the discussion of reactions of especially labile complexes, although its generality deserves careful experimental attention. Probably, the most promising and general method for testing equation (6.3) is the study of outer-sphere associations by ultrasonic absorption, a method which has provided support for the equation as a description of common aqueous electrolyte solutions[28]. Specific ultrasonic data on $Co(NH_3)_6^{3+}$ and $Coen_3^{3+}$ with SO_4^{2-} were reported by Elder and Petrucci[29].

One important point that has a bearing on the use of 'constant ionic strength media' is the capacity of ClO_4^- to form outer-sphere complexes. There is no reason to believe that this ion is immune from such behaviour. The recent observation[30] that *trans*-$Coen_2(H_2O)_2^{3+}$ is more favourable (versus *cis*) in nitrate media than in perchlorate media probably implies greater outer-sphere complexing by the ClO_4^- ion than by NO_3^- ion.

6.3 COBALT(III)—THE PARADIGM

6.3.1 Introduction

The systematic analysis of the mechanism and the steric course of octahedral substitution began with, and is still best developed in, studies of amine complexes of Co^{III} and their derivatives because of the substantial experimental advantages that these complexes provide. Thus, these complexes are easily synthesised, are inert, and are easily distinguished. The rich variety of relatively inert structures available has enabled the subtle and detailed testing of mechanistic hypotheses through studies of rate laws over wide ranges of concentration, studies of steric course, studies of the effect of structure-variation, and studies of the effects of solvent, temperature, isotope effects and pressure-dependence. More recently, studies of the influence of polydentate chain and macrocyclic ligands have been undertaken.

The subject can be divided into two parts. The first is the study of solvolytic pathways (called acid hydrolysis in aqueous media). Solvolytic pathways are central because of the predominantly important result that the rates for these reactions are insensitive to most entering ligands. In this situation, any coordinating solvent becomes the most important reactant because of its advantages of continual encounter with the complex. In fact, the only effective way to study the role of the entering group is to make the solvent the leaving group, so that solvent-entry leads to no net reaction. The only entering ligand which profoundly affects substitution rates is the -lyate ion of the solvent. Thus, the second part of the subject is base hydrolysis.

The various tools of mechanistic analysis will be introduced as they are applied to these two sub-divisions in the study of Co^{III} complexes. Our discussion of other metal ions will follow the format of the discussion of the Co^{III} paradigm.

6.3.2 Variation of rate constants in solvolysis and related reactions

6.3.2.1 Leaving-group effects

Given that entering-group effects are small, the mode of activation must be assumed to be *d* with, at most, weak binding in the transition state, or it must be assumed that the solvent as an entering group plays a *specific* role. (If the role were non-specific, other ligands would also be capable of such behaviour, and if none were capable of behaving in a significantly different manner the interaction would *not* be what is usually understood as bond formation.) The *d* hypothesis is more immediately attractive and has an important immediate consequence; reaction rates should be sensitive to the nature of the leaving ligand.

The expected sensitivity is particularly clearly demonstrated by the simple system $Co(NH_3)_5X^{2+}$, where X is a monovalent anionic ligand. First, there is a linear correlation between $\Delta G^{\ddagger}$ for the aquation reactions of $Co(NH_3)_5X^{2+}$ and the thermodynamic ΔG^0 values for these reactions, this relationship having, a slope of *unity*[31,32]. Unit slope means that changes in the overall ΔG values with variations of X, consisting mainly of variation in the Co—X bond strength and in the solvation of X^-, are quantitatively reflected in changes in the free energy of activation. The implication is clear. The X group in the transition state is very similar to the X^- product. The correlation holds over five decades in log k. Points have recently been added for NO_3^- [33] and N_3^- [34]. F^- has been shown to fit poorly[35]. A correlation has also been observed between $\Delta S^{\ddagger}$ for these aquations and the entropy of hydration of the halide ions which has a similar import[36]. A very striking correlation has been observed between the volumes of activation, $\Delta V^{\ddagger}$ and ΔV^0 values, for the overall reaction as reported by Jones and Swaddle[37]. This implies full electrostriction by the anion in the transition state.

The above correlations confirm that the leaving group X^- has 'left' in the transition state, but do not settle the issue of a specific role for the entering water ligand. Langford[31] has argued from Hammond's postulate that a transition state reached highly 'endothermically' from $Co(NH_3)_5H_2O^{3+}$ should also exhibit, at most, weak Co—OH_2 bonding. An interesting point is that the transition state for the water-exchange reaction must be symmetrical with respect to leaving and entering groups. Haim[38] has used the above rate and equilibrium data to calculate 'formal equilibrium constants' for the conversion of the transition states for water-exchange into the transition state for X^--substitution. These are found to be near unity and to depend only on the charge of X^-. This emphasises the similar role played by X^- and H_2O in these transition states, and confirms the weakness of binding of both the leaving and entering groups as required by the *d* description.

Leaving-group dependence qualitatively similar to that displayed by $Co(NH_3)_5X^{2+}$ is familiar in the chemistry of the more complicated Co^{III} derivatives. A successful approximate extension of the quantitative linear free energy relationship was described by Kernohan and Endicott in a discussion of the *trans*-di-X^- macrocyclic amine complexes of Co^{III} [39]. It is important to note that leaving-group orders can change when there is a major change in the nature of coordination about the metal (see, for example, the dimethylglyoximates[40]). The kinetic behaviour of amino acids as leaving groups was recently reported[41]. The theoretical framework for understanding leaving-group orders in terms of Co—X bonds and solvation factors will be discussed below.

6.3.2.2 *Steric effects*

Another factor which should help in discriminating between *d* and *a* transition states is steric crowding of the reaction centre. Dissociative reactions should be accelerated by steric crowding of the complex, which can be relieved by passing to a transition state of reduced coordination number. The effect on associative reactions is expected to be opposite. The classic example is the difference in reactivity between the *meso*- and *d*,l-dichloro-bis-

(butylenediamine)-cobalt(III) ions[42]. Basolo and Pearson have observed that the only difference between these isomers is that the methyl groups of the *meso* form are on the same side of the chelate ring with maximal steric interference. Aquation of the *meso* form is 30-times faster than aquation of its racemic ligand isomer.

Work on the preparation of 5-coordinate complexes supports the idea that moderate steric effects are sufficient to shift the energetic relationships between 5- and 6-coordination without the aid of esoteric electronic effects. Alkylation of a coordinated nitrogen is especially effective[43]. Alkyl amine complexes have come under extensive study recently by Chan[44,45], Staples[46], Parris[47], and their respective collaborators, and by Hay and Cropp[48]. The general pattern of results seems consistent with the predictions of the *d* mechanism. There is the inevitable difficulty in these studies of separating steric from electronic effects, and the authors have considered electronic effects. That electronic effects are not large is indicated by studies of substituted pyridine and aniline complexes[49,50]. (It is interesting that benzylamine complexes behave like aliphatic amine complexes[51].)

Perhaps the most important outstanding question related to the effects normally characterised as steric is that of solvation. It would be valuable to have results for solvolysis of some hindered complexes in dipolar aprotic solvents.

6.3.2.3 *Charge on the complex*

There is reason to suppose that dissociation of an anionic ligand will be retarded by increasing the overall charge on a complex, and that attack by an anionic ligand will be facilitated. Except for the role that charge plays in encounter, there is no way to study this factor that is not hopelessly entangled with variation in covalent binding effects[52]. Averaging over the various ligands used to change charge, the net electrostatic effect in Co^{III} does seem to be consistent with the *d* account[53]. This question has received little attention from recent commentators on the chemistry of Co^{III}, (their caution may have been wise!), but it is still an issue of interest in the chemistry of other metal ions, and the lesson of caution taught by Co^{III} chemistry needs to be applied.

6.3.2.4 *Effects of non-labile ligands*

The classic system for the study of non-labile ligand effects is aquation in the series *cis*- and *trans*-$Coen_2AX^{n+}$, where X is a relatively labile ligand (Cl^- or Br^-) and A is a variable ligand not undergoing substitution in the reaction. An attempt to interpret the results of studies of this system was first made by Ingold, Nyholm and Tobe[53]. It is clear that the relation of non-labile ligand effects to the question of the nature of the transition state for substitution depends upon a theory of electronic effects. Early workers (probably wisely) borrowed the theory from the successful account of substituent effects in reactions of aromatic organic systems. As more information has accumulated, it has become clear that the analogy between the effect of a substituent removed from a carbon (no d-orbital participation) reaction centre and a substituent directly bound to a Co^{III} reaction centre is tenuous and needs to be applied with caution[54, 55].

Tobe and his collaborators have completed an extensive survey of the $Coen_2AX^+$ system which was recently reviewed[56]. A few further A groups have been added by recent work[57]. With Cl^- as the leaving group, the order of labilisation by anionic A ligands is:

$$\textit{trans } A: OH^- > NO_2^- > N_3^- > CN^- > Br^- > Cl^- > SO_4^{2-} > NCS^-$$
$$\textit{cis } A: OH^- > Cl^- > Br^- > NO_2^- > SO_4^{2-} > NCS^-$$

The reactions of the *cis* complexes are a little faster than the reaction of the *trans* complexes, except for the NO_2^- and N_3^- cases. (Would CN^- also fit into this small family?). The range of rates in question is about two powers of ten except for the extremely low lability of *trans*-NCS^-. The development of a coherent electronic theory is difficult for solvation effects most decidedly play an important part in these comparatively small variations, and for this reason it is not possible to use these rate-variations alone (see Stereochemistry Section 6.3.4 below) to confirm or contest the assignment of the *d* mechanism. One important fact does however clearly emerge—there are no *profound* stereospecific (*trans* or *cis*) non-labile ligand effects in this system. This is most easily understood by assuming that Co 3d orbitals are more important to the σ-bonding framework of these complexes than are Co 4p orbitals, (which specifically connect *trans* related ligands).

Perhaps the most useful mechanistic conclusion from the study of the effect of A ligands on rate in the $Coen_2AX^+$ systems is that the mechanisms are most probably very similar. Since there are other reasons for assigning *d* pathways to some of these complexes, the small A effects can be used to argue that the *d* pathway is a fair description for all of these reactions.

Now, 3d orbitals lie below the orbitals of the fourth shell in high-valent cationic species. If the charge on cobalt is reduced, participation in σ-bonding by the 4p orbitals should increase and *trans* effects should appear[58]. Cobalt complexes of macrocyclic ligands, like dimethylglyoxime and the corrin system, have a chemistry which is sometimes considered indicative of Co^I behaviour. Specific *trans* effects might be expected. Moreover, such a change in metal charge would be expected to correlate with a higher lability via the *d* mechanism and a tendency to the formation of stable 5-coordinate intermediate species. Ablov[59, 60, 61] and his co-workers have used the products of competitive reactions to establish a *trans*-effect order in *cis*-dimethylglyoxime complexes. This procedure (which by-passes actual kinetic measurements) is dangerous but does suggest *trans*-labilising in the order; $SO_3^{2-} > OH^- > NCSe^- \sim NCS^- > SC(NH_2)_2 > I^- > NO_2^- > Br^- > Cl^- > H_2O$. Actual *kinetic* studies[59] have revealed only small *trans* effects. Alkyl groups do have a large labilising effect in dimethylglyoxime complexes. The order $(CH_3)_2CH— > CH_3CH_2— > CH_3—$, corresponding to increasing electron release to Co^{III}, is observed[62]. The labilising effect of PR_3 has been discussed in terms of σ-donor properties of the phosphine.

Large, but as yet poorly defined, labilising effects arise in corrin and porphyrin complexes[63, 64]. These last are really quite spectacular non-labile ligand effects. One tentative analysis assigns a factor of 10^{10} for the increase in reactivity of a corrin complex compared to that for an analogous cyclam complex[56]. This particular area is however poorly defined. Most Co^{III} complexes reported have four or five amine or alkylamine nitrogen donor ligands. Complexes with fewer than four nitrogen donor ligands are not well understood. The dimethylglyoxime complexes are an exception exhibiting no spectacular effect. Spectacular reactivity differences seem to arise in the corrins and also in complexes derived from $Co(H_2O)_6^{3+}$.

The only ligand for which simple and conclusive evidence of large *trans* labilising power in amine complexes of Co^{III} is available is SO_3^{2-}. A recent comprehensive account is given by Stranks and Yandell[66].

6.3.2.5 Solvent effects

Much of the recent work on solvent effects on reactions of Co^{III} complexes has compared reactions in water to reactions in alcohols and dipolar aprotic solvents. As was observed some years ago[67], the known effects of solvent are not as spectacular as those observed for ionic organic reactions. One of the most important features of the studies is that anions are less well solvated in most of the non-aqueous solvents than they are in water[68]. This has the effect of promoting ion association, and increasing encounter equilibrium constants, which is useful in the detailed analysis of rate laws to be considered below[69].

A pattern of solvent-dependence of rate constants that is typical is the one for the solvolysis of *cis*-dibromo (triaminotriethylamine)Co^{III} reported by Madan and Peone[70]. Even the solvent-dependence of reactions involving dimethylglyoxime complexes seems more or less similar[127, 164]. This is shown in Table 6.1.

Table 6.1 Solvent-dependence of solvolysis rates for *cis*-Co(triaminotriethylamine)Br_2^+ at 25 °C

Solvent	k (s^{-1}) 25 °C	$\Delta H^{\ddagger}$ (kcal mol^{-1})	$\Delta S^{\ddagger}$ (cal deg^{-1} mol^{-1})
Formamide	2.7×10^{-2}	10.0	−32
Dimethyl-formamide	4.3×10^{-4}	22	0
Dimethyl-sulphoxide	7.7×10^{-4}	18	−12
N-methyl-formamide	2.5×10^{-1}	—	—
Water	2.8×10^{-2}	15	−15

The rate constants span less than a factor of 10^3. Clearly, there is no simple relationship to solvent dielectric constant, and the variation in activation parameters suggests a complex situation.

Fitzgerald, Parker and Watts[71] have made a valuable start to unfolding the complexities by the application of Parker's solvent activity coefficients[68], but the results are not as yet unambiguous. The essence of the approach is to evaluate the solubilities of complexes and their ion-pairing constants in various solvents to get values for ground-state, solvent activity coefficients. Reaction rates are then measured which give values of $\Delta G^{\ddagger}$, leading to solvent activity coefficients for transition states. The difficulty arises in the interpretation of these values. It is easy to construct a model for a dissociative transition state in which a halide ligand is leaving by suggesting that halide is present in the free ion form. The contribution to the transition state activity coefficient for the halide can then be determined from free halide ion data. Appropriate analogues for the metal complex part of the transition state are harder to come by, and so far, fairly crude pictures have had to be used. The approach is, however, well worth developing. It might yield not only a deeper insight into the reactivity of Co^{III} complexes, but also allow a development of a generally applicable structural understanding of the role of the solvent (outer-sphere ligands) in reactions of complexes in solution.

A different approach to solvent effects is the attempt to find a linear correlation of reaction rates with parameters designed to measure microscopic ion solvation such as Winstein's Y parameter[72] or Kosower's Z parameter[73]. Such a correlation failed in the first case in which it was attempted[34]. Recently, a satisfactory correlation involving Y has been found by Burgess for reactions of *cis*-$Coen_2Cl_2^+$ in aqueous mixtures containing methanol, ethanol, acetone, dioxane and formic acid[74]. (The significance of the linear correlation may be questioned on the grounds that the range of rates correlated spans less than one power of ten.) An important point in common with the earlier effort[67] is that the slope of the correlation with Y (called m) is 0.35, which is much less than the value of 1.00 characteristic of dissociative solvolysis of alkyl halides. Such a low m value is just another way of expressing the modest solvent sensitivity of reactions of Co^{III} complexes. Langford[67] has suggested that the low sensitivity to solvent might reasonably be attributed to 'free ion-like' solvation of the leaving halides in the ground-state halo complex, which follows from the 'ionic' nature of the metal halide bond. (So far, the solubility measurements of Fitzgerald, Parker, and Watts[71] cannot separate this aspect of ground-state solvation from others.)

Jones, Edmondson and Taylor[75] have introduced the approach of correlating rates in a mixed solvent (aqueous ethanol) with the equilibrium constant for the *same reaction* in the mixed solvent. They report a linear free energy relationship between rate and equilibrium for the solvolysis of $Co(NH_3)_5Cl^{2+}$ in 0–55% (volume) ethanol with a slope of 0.44 at 29.9 °C. The interpretation must be similar to the explanations of low m values. This interesting approach could be extended.

Other workers reporting useful results on solvolytic reactions in non-aqueous solvents include Ašperger and co-workers[76,77] who have examined reactions in formamide and alcohols, and Chan and Leh[45,79] who have considered alcohols. Numerous papers by Watts and his collaborators should be consulted by all workers interested in dipolar aprotic solvents. (Reference 78 is important in this respect). Staples and his group are developing a programme involving the examination of reactions of Co^{III} complexes in strongly acid media[80,81]. Solvolysis of $Coen_2Cl_2^+$ has even been suggested as a structural probe for mixed solvents[82].

In general, it may be said that data on solvent effects are consistent with the notion that Co^{III}–amine complexes react mainly via *d* pathways, but the solvent-effect data is sufficiently complex to eliminate them as central links in the chain of argument leading to a preference for the *d* mechanism.

6.3.2.6 *Pressure effects*

A model pressure-dependence study is the recent one of Swaddle and Jones described above[37]. It involves explicit comparison of activation volumes with dilatometric results for the overall equilibrium of the reaction in question. As will be seen below, activation volumes are extremely difficult to interpret mechanistically unless a good model system has been developed.

6.3.2.7 *Isotope effects*

The effect of exchange of D for H in water as a solvent has been known for some time, and explanations have relied heavily on changes in ground-state solvation[83] determined from solubility measurements.

6.3.2.8 *Entering-group effects*

Entering-group effects in Co^{III} complexes are hard to study in aqueous solution because equilibria for so many ligands favour solvolysis under convenient conditions. A small range of entering-group effects is, of course, implied in the correlation of leaving-group effects and equilibrium effects discussed above. In a few cases, complex formation reactions may be analysed directly when equilibrium is favourable as it is with the azide[34]. Solvents of lower polarity also favour direct study of complex formation[78]. Chelating ligands also favour complex formation, and much of interest has appeared on reactions of oxalate[84–86] especially from Harris' laboratory. Oxalate entry is complicated by the number of steps involved.

If a chelating ligand can be studied in the step of closing the ring from the mono-coordinated state, a particularly simple analysis of the energetics of entering-ligand attack is possible. A puzzling example, which does not seem to fit easily into the overall pattern of reactions of Co^{III} amines, is the comparison of ring-closure reactions for $Coen_2(OH)CO_3$ [87] and $Coen_2(OH)PO_4$ [88]. The phosphate reaction is 500-times faster. This may be an example of a Co^{III}-promoted reaction between ligands of the type discussed by Buckingham, Sargeson and their collaborators (see, *inter alia*, Reference 89).

6.3.2.9 *The* $Co(OH_2)_6^{3+}$ *ion*

It is worth a special paragraph to note that the 'simplest' Co^{III} species in aqueous solution, $Co(OH_2)_6^{3+}$, has been largely omitted from the above discussion. This is a consequence of both the relative redox instability of the complex and its relative lability. McAuley, Malik, and Hill[65] have presented one of the few reports on this topic only recently. For the reactions:

$$Co(OH_2)_6^{3+} + Cl^- \xrightarrow{k_1} Co(OH_2)_5Cl^{2+} + H_2O \tag{6.4}$$

and

$$Co(OH_2)_5OH^{2+} + Cl^- \xrightarrow{k_2} Co(OH_2)_4OHCl^+ + H_2O \tag{6.5}$$

they report $k_1 = 2.5\ M^{-1}s^{-1}$ ($E_a = 26$ kcal mol^{-1}) and $k_2 = 200\ M^{-1}s^{-1}$ ($E_a = 12$ kcal mol^{-1}) at 8 °C. They suggest that the *d* mechanism is operative. Another unusual Co^{III} system is $CoCl_6^{3-}$. Its solvolysis in acetic acid was recently reported[90].

6.3.3 Rate laws over extended ranges

If a study of the effect of various structural and environmental parameters on rate constants suggests the dissociative mode of activation, there are two pathways to be anticipated, the D through an intermediate of 5-coordination, and the I where inner-sphere–outer-sphere ligand interchange occurs. Relevent studies have used the solvent as the leaving group. These two cases are characterised by the reaction schemes and overall rate laws:

$$Co{-}OH_2 \underset{k_{-1}}{\overset{k_1}{\rightleftharpoons}} Co + OH_2 \tag{6.6}$$

$$Co + Y \xrightarrow{k_2} CoY$$

$$\frac{d[CoY]}{dt} = \frac{k_1k_2[CoOH_2][Y]}{k_{-1} + k_2[Y]} \tag{6.7}$$

and:

$$Co{-}OH_2 + Y \overset{K_E}{\rightleftharpoons} Co{-}OH_2{,}Y \tag{6.8}$$

$$CoOH_2{,}Y \xrightarrow{k} CoY + OH_2$$

$$\frac{d[CoY]}{dt} = \frac{kK_E[Co{-}OH_2][Y]}{1 + K_E[Y]} \tag{6.9}$$

The limiting behaviour of both forms is similar. At low [Y], both give overall second-order reactions, and at high [Y] both give limiting rates. Once the complete rate laws are known, however, there are some indicative differences. On the D pathway, the limiting rate must equal the independently-measurable water-exchange rate, whereas the limiting rate on the I_d pathway should be smaller. This is because the Y group occupies only one outer-sphere coordination position in the encounter complex, the rest being occupied by water. On dissociation of the leaving water, water-exchange remains the more probable reaction. Also, the Y-dependence of the I_d path must match the independently-measurable encounter equilibria, which is not especially sensitive to any characteristic of Y other than its charge. On the D path, change of Y changes the ligand attacking a relatively stable intermediate. Differences may be large and do not parallel 'ion-pairing' tendencies. The classic example of the D path is in the reactions of $Co(CN)_5OH_2^{2-}$ studied by Wilmarth and his associates[91]. Evidence for $Co(NH_3)_4SO_3^+$

followed[92]. An interesting point is that the intermediates show considerable selectivity in their recombination reactions.

Reactions of $Co(NH_3)_5OH_2^{3+}$ were shown by Moore and Pearson[91] to be inconsistent with D path requirements. Langford and Muir[93] established the consistency with the I_d path in this system, including the expected statistical relation between the limiting rate of complex formation and the rate of water exchange. (An exceptionally slow reaction in the ion-pair with HPO_4^{2-} was reported by Lincoln and Stranks[94], although they favour a dissociative mechanism.) In Langford and Muir's work, only one encounter species was considered, and simple conditional 'equilibrium' constants were derived. Recently, Burnett[95, 96] has given a more sophisticated treatment of ion-pairing, and shown that a number of species including ion-pairs with the 'innocent' ClO_4^- ion are important. The analysis confirms earlier mechanistic assignment of the I_d path.

At one point, it seemed that the D path might be rare, but, evidence for it has been recently presented in reports on all of the following systems: *cis*- and *trans-bis*-unidentato-cyclam complexes[97], *trans*-$Coen_2(SO_3)_2^-$ and $Coen_2SO_3OH_2^+$ and its conjugate base[98], $(Co(DMG)_2SO_3OH_2)^-$ [98], and Co^{III}-hematoporphyrin[100]. The alternative I_d path is strongly suggested by many studies and clearly indicated in examples such as; reactions of *cis*-$Coen_2NO_2DMSO^{2+}$ in DMSO [101], alkyl-bis-(dimethylglyoximato)-aquo-Co^{III} (in striking contrast to SO_3^{2-}-containing complexes, which react via D)[62], and the reaction of oxalate with $Coen_2(OH_2)_2^{3+}$ [84]. Even the solid-state reaction of $Co(NH_3)_5OH_2^{3+}$ with halides seems interpretable in terms of the I_d framework[102].

Given the established instability of $Co(NH_3)_5^{3+}$, it is interesting that Williams and Hunt[103] now have reason (from nitrogen- and oxygen-exchange studies on $Co(NH_3)_5OH^{2+}$) to suggest the existence of the intermediate $Co(NH_3)_4OH^{2+}$ on a D pathway.

6.3.4 Stereochemistry

Stereochemical change can be an important criterion for distinguishing D from I pathways, and a valuable tool for the analysis of geometries (as opposed to energetics) of activated complexes. Moreover, any theory of reactivity that fails to account effectively for stereochemistry cannot be considered adequate. However, it is worth remembering that stereochemistry can be determined in octahedral substitution by two extremely subtle factors; the orientation in outer-sphere complexes, and the probably slight difference in energetics between trigonal-bipyramidal and square-pyramidal 5-coordinate geometry.

Werner[104] suggested that the stereochemical consequences of octahedral substitution result from specific orientation in the second coordination sphere. According to this view, geometrical isomerisation results from the placement of the entering group *trans* to the leaving group, and the retention of geometry arises from the placement of the entering group in the outer sphere adjacent (*cis*) to the leaving group. The existence of variable stereochemistry when the solvent (occupying all outer-sphere positions) is entering reminds us however that the 'geometry' about Co^{III} in the transition state is also quite important.

The consequences of the formation of a 5-coordinate Co^{III} species (which may act as a model for either the activated complex of an I_d path or the intermediate of a D path) are illustrated in Figures 6.2 and 6.3 for the typical cases of *cis*- and *trans*-CoA_4XY and D*-*cis*-Co(A—A)$_2$XY. The Figures

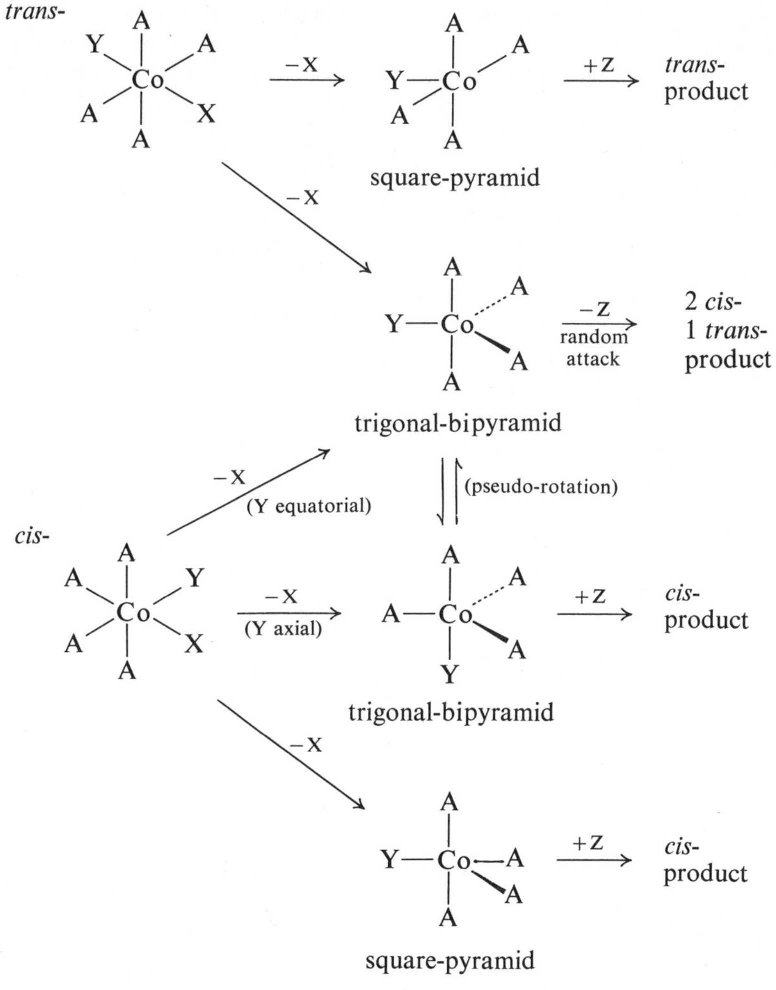

trans-
−X
+Z
trans-product
square-pyramid
−X
−Z
random attack
2 *cis*-
1 *trans*-
product
trigonal-bipyramid
−X
(Y equatorial)
(pseudo-rotation)
cis-
−X
(Y axial)
+Z
cis-product
trigonal-bipyramid
−X
+Z
cis-product
square-pyramid

Figure 6.2 Stereochemistry of dissociation of CoA_4YX isomers

show that a trigonal-bipyramid permits rearrangement, whereas there is no straightforward way to derive rearrangement from the square-pyramid. A reaction through a trigonal-bipyramid will yield more *cis* product starting with the *cis* reactant than with the *trans* reactant, *unless* pseudo-rotation[105, 106] can equilibrate the trigonal-bipyramids. (A test for the long-lived intermediate of the D path?) Optical isomerisation is not visualised as occurring through the square-pyramid. One of the two possible trigonal-bipyramids gives $\frac{2}{3}$ retention events and $\frac{1}{3}$ formation of *trans*-product. The other gives racemisation. Either nucleophilic attack (*a* mechanism) or dissociation of one end of a chelated ligand, followed by several steps, seems to be required to explain the occasional examples of inversion. Despite some elegant recent work[107, 108] inversions remain poorly understood.

Table 6.2 summarises the results available on the steric course of 'spontaneous' substitution reactions of *trans*-CoLAX systems, where L is a quadridentate ligand, a pair of bidentates, or four monodentate non-labile ligands. A is the stereochemical 'marker' ligand and X is the leaving group. A corresponding table of *cis* isomers was presented by Tobe[108]. In the 23

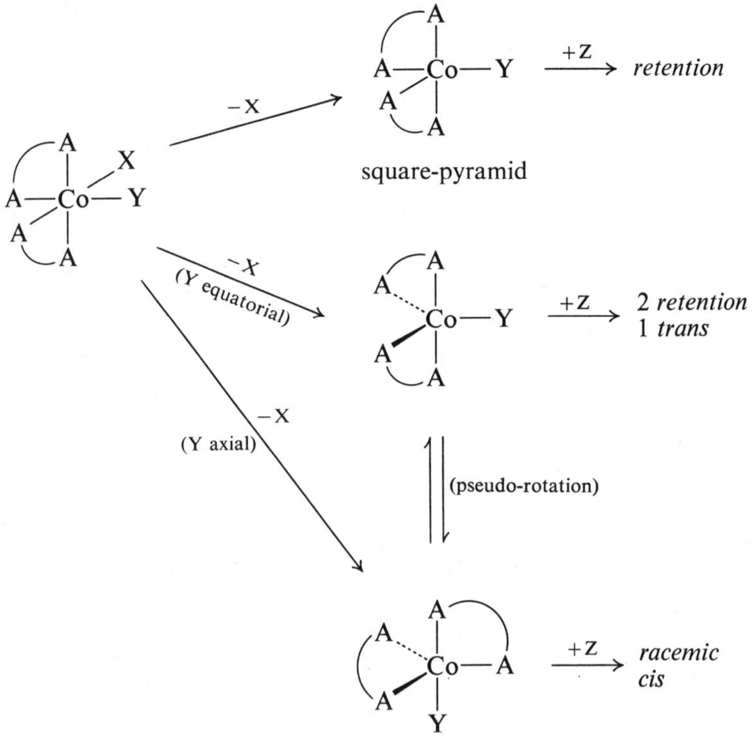

Figure 6.3 Stereochemistry of dissociation of D*-*cis*-Co(A—A)$_2$YX

examples quoted, no substitution of a *cis* isomer is accompanied by steric change. Tobe's interpretation suggests that all substitutions are *d*, but that the geometry at the transition state changes. All substitutions involving steric change have more *positive* values of $\Delta S^{\ddagger}$, and are in complexes with a π-donor ligand *trans* to the leaving group. These are considered to react via a trigonal-bipyramid, whereas the retentive (lower $\Delta S^{\ddagger}$) cases employ the square-pyramid. Tobe emphasises that a delicate balance of steric and electronic factors determines the geometry of the transition state. (Kernohan and Endicott have discussed the limitations of Tobe's correlation[39].) Some differences between ordinary 'spontaneous' aquation stereochemistry and the stereochemistry of the rapid Hg^{2+} 'induced' aquations has been a source of concern[118]. Tobe[111] has observed however that identity between the two paths would be expected only if an intermediate fully equilibrated with its solvation shell were formed.

The electronic factors which gives rise to trigonal-bipyramid and stereochemical changes have been considered. Basolo and Pearson have emphasised the π-bonding effect of the *trans* non-labile ligand[2]. Archer[4] has calculated

Table 6.2 Spontaneous aquation stereochemistries and activation parameters for Co^{III} complexes of the type *trans*-CoLAX (data from collections by Tobe[109] and Archer[110] except as noted). $\Delta H^{\ddagger}$ in kcal mol^{-1}, $\Delta S^{\ddagger}$ in cal deg^{-1} mol^{-1}

L	A	X	$\Delta H^{\ddagger}$	$\Delta S^{\ddagger}$	% steric change
en_2	OH	Cl	25.9	+20	75
en_2	Cl	Cl	26.2	+14	35
en_2	Cl	Cl	24.9	+ 3	50
en_2	NCS	Cl	30.2	+ 9	60±10
S,S,-trien	Cl	Cl	25.5	+16	100
RR,SS,-2,3,2-tet	Cl [111]	Cl	25.9	+12	50±20
en_2	N_3 [112]	Cl	20.7	− 6	<3
3,2,3-tet	Cl [113]	Cl	24.5	+ 4	0
R,S-2,3,2-tet	Cl [113]	Cl	24.6	+ 1	0
en_2	NH_3	Cl	23.2	−11	0
en_2	CN	Cl	22.5	− 2	0
en_2	NO_2	Cl	20.9	− 2	0
cyclam	OH	Cl	18.1	− 7	0
cyclam	Cl	Cl	24.6	− 3	0
RS-2,3,2-tet	Cl	Cl	23.7	− 1	0
$(NH_3)_4$	NH_3	Br	23.3	− 4	0
$(NH_3)_4$	NH_3 [114]	Cl	23.5	− 6	0
$(NH_3)_4$	Cl [114]	Cl	23.6	+8.6	55±10
$(acac)_2$	NO_2	amine[115]	—	—	0
$(acac)_2$	NO_2	NO_2 [116, 117]	—	—	0

non-bonding repulsions using the ligand field theory to predict the nature of the preferred trigonal-bipyramid, and has also considered whether overall low ligand field strength may be related to stereo-mobility. No 'rules' stating which ligands should prefer axial or equatorial positions in Co^{III} trigonal-bipyramids have been proposed in an analogous fashion to the rules developed for P^{V} [105] or Pt^{II} [10]. High degrees of steric change, such as those observed in the reactions of *trans*-$Coen_2OHCl^+$, may indicate something like pseudo-rotation, or they could merely imply effects of OH^- on the solvation shell (second sphere). In fact, outer-sphere orientation effects are observable in non-aqueous solvents where ion-pairs and triplets may be conveniently studied[119].

Recent evidence has finally produced likely examples of isomerisation not accompanied by substitution in Co^{III} complexes. This must always be considered. Some systems include *cis*-Co(cyclam)$OH(OH_2)^{2+}$ [120], $Coen_2(OH)_2^+$ [121] and the benzoylacetonate complex of Co^{III} [122].

Brady[123] (elaborating earlier suggestions[125]) has discussed the theory of intramolecular-twist isomerisations. An interesting intramolecular isomerisation of another sort has been considered by Chan[124]. This is substitutional isomerisation with unidentate glycine nitrile ligands flipping from a coordination involving the NH_2 end of the ligand to that involving the C≡N end now attached to an adjacent position. A new sort of stereochemical problem in Co^{III} substitutions has been raised by Smith and Haines[128] who have reported the stereospecific substitution of active propanediol into racemic *cis*-$Coen_2Cl_2^+$.

6.3.5 Base hydrolysis

The only attacking ligands which profoundly alter the rate of substitution at Co^{III} centres are the -lyate ions of the solvent; hydroxide in water, methoxide in methanol, etc. This reaction pattern is usually called base hydrolysis in contrast to the acid hydrolysis type solvolysis discussed above. After a lengthy debate, it is now generally agreed that a mechanism due to Garrick, and systematically developed for this case by Basolo and Pearson[2], is correct in its main outline. This is the *conjugate base* mechanism shown for $Co(NH_3)_5Cl^{2+}$ in equation (6.10).

$$\begin{aligned} Co(NH_3)_5Cl^{2+} + OH^- &\overset{K}{\rightleftharpoons} Co(NH_3)_4(NH_2)Cl^+ + H_2O \\ Co(NH_3)_4(NH_2)Cl^+ + H_2O &\overset{k}{\rightarrow} Co(NH_3)_5OH^{2+} + Cl^- \end{aligned} \tag{6.10}$$

The role of hydroxide is to deprotonate an ammine ligand, leaving NH_2^- which is seen to be the most powerful labilising group known. The amido complex undergoes rapid substitution in a process which is most probably *d*, and even D, in many cases. The main lines of evidence supporting the conjugate base hypothesis are: (1) the requirement for an acidic proton for base hydrolysis to occur, (2) evidence that other anions may function as the entering groups in hydroxide-catalysed reactions, (3) evidence from ^{18}O-exchange experiments that the normal entering ligand in base hydrolysis is water and not hydroxide, and (4) the general parallel in changes of reactivity of complexes in acid and base hydrolysis as the structure is varied. These points have been reviewed by Basolo and Pearson[2], and by Langford and Stengle[3]. The first requirement implies a role for an acid group, the second and third show that hydroxide is not essentially involved in the product-determining step, and the fourth suggests dissociative activation (by analogy to reactions occurring in acid solution) which would preclude hydroxide nucleophilic attack.

Current interest centres on the nature of the labilising effect of the amide ligand, the stereochemistry of the process at both Co and N, and the evidence for the 5-coordinate intermediate of the D pathway. These base hydrolysis studies, with the exception of the discussion of the acidity of the coordinated amine, fall into the same categories used above in the discussion of acid solution studies.

The unique problem of amine acidity is a convenient starting point. The specificity of hydroxide catalysis in aqueous solution implies that the proton-transfer steps in equation (6.10) are rapid in comparison to the substitution step, and may be considered a pre-equilibrium. Knowledge of K_a, the acid dissociation constant of the coordinated amine, would be equivalent to a knowledge of the desired proton-transfer equilibrium constant, K. Most efforts to measure K have been equivocal at best. The absence of departure from simple second-order kinetics in the base hydrolysis of $Co(NH_3)_5Cl^{2+}$ up to $[OH^-] = 1.00$ M [129] suggests that $K < 0.05$, but it could be much less. Chan[130] reported deviation from second-order kinetics in one case which he attributed to an alternative to the simple conjugate base mechanism, in which hydroxide acted as an ion-pair with the substrate complex. Such ion-pairs could be important at high $[OH^-]$, but Buckingham *et al.* were unable to confirm Chan's observation[131]. In any event, ion-pairing effects may yet prove to interfere with a clear kinetic identification of the conjugate base.

In contrast to the difficulties in the study of K, the kinetics of proton exchange have proved to be quite amenable to study. Recent work has mainly used ^{1}H n.m.r. spectroscopy. The measured rates are sensitive to the other ligands present, for example, protons *trans* to the chloride in *cis*- and *trans*-$Coen_2NH_3Cl^{2+}$ exchange 100–300 times faster than others whether they come from NH_3 or en[132]. *Cis* effects are also important, as illustrated by the more than 50-fold enhancement of the proton exchange of *trans*-$Coen_2(NO_2)_2^+$ compared to that of *trans*-$Coen_2F_2^+$ [133]. The mechanism of proton exchange has been studied by the investigation of the rate of exchange in comparison to the rate of racemisation at an optically active coordinated secondary amine[134]. Subtle effects in proton exchange are indicated by the difference in rates of *d*- and *l*-$Coen_3^{3+}$ exchange in the presence of *d*-tartrate[135]. In the base hydrolysis (very fast) of *trans*-Co(cyclam)Cl_2^+, it was shown that proton exchange in the reactant was not faster than hydrolysis, but that exchange was more extensive in the product[136]. This observation establishes the closest connection yet between deprotonation and the substitution step. A new case of a complex lacking an acidic proton and not subject to base hydrolysis is $Co(dipy)_2(COOCH_3)_2^+$ [137].

It should be remembered that the conjugate base hypothesis for amine complexes is similar to the theory put forward to account for the differences in reactivity of M—OH_2 and M—OH, where the study of the equilibria is straightforward enough to create no special problems. An interesting 'intermediate' case, where rates and equilibria for deprotonation can be examined, is the reaction of *trans*-bis-(dimethylglyoximato)cobalt(III) complexes[138].

Leaving-group effects in base hydrolysis suggest dissociative activation. There is a linear free energy relationship between rates and equilibria in the $Co(NH_3)_5X$ system[139]. (Davis and Lalor note that differences among halide leaving groups is primarily in $\Delta S^\ddagger$ values[140].) Similarity of mechanism among several different systems is suggested by Lalor's isokinetic relationships[141]. (Unusual activation parameters are observed when PO_4^{3-} is the leaving group[142].) Po and Jordan have reported on sulphato and sulphamato pentamine systems[143]. Steric effects of the nature expected in dissociative reactions are observed in the base hydrolysis of penta-kis-(alkyl amine)Co^{III} complexes[144] and Co(*m*-toluidine)en_2Cl^{2+} complexes[145].

With respect to non-labile ligand effects, it is difficult to say much that is unambiguous except that the amido group is strongly labilising and that a general parallel exists with labilising effects of other ligands in acid and base hydrolysis. Data on various substituent groups is still being accumulated. This promises a clearer picture in future. For example, cycloalkylamines were recently discussed[146]. The reference system, $Co(NH_3)_6^{3+}$ has been re-examined[147].

The role of the spectacularly labilising amido ligand has been extensively discussed. At first it was suggested that π-donation occurred in the transition state[2]. This was related to the weaker labilising effects of OH^- and Cl^-, for which steric change is observed, and this pattern of thinking was supported by the observation that steric change is common in base hydrolysis. In fact, π-stabilisation should be most effective if the amido group occupies an equatorial position in a trigonal-bipyramidal transition state structure. These ideas were questioned following Tobe's $\Delta S^‡$ correlations[109]. It was observed that stereoretentive multidentate nitrogen donors do prevent stereochemical change, and that this is accompanied by a decrease in the $\Delta S^‡$ value, which was thought to imply the square-pyramidal geometry. However, the amido ligand seems to *remain effectively* labilising in these complexes. Buckingham *et al.*[132] provided a direct approach to this problem with an elegant choice of ligand. In $Co(trenen)Cl^{2+}$ (trenen = 1,8-diamino-3-(2-ethylamino)-3,6-diazaoctane), an asymmetric nitrogen exists *trans* to the leaving Cl, which allows resolution into enantiomeric complexes. N.M.R. spectroscopy indicated that this proton exchanged much more rapidly than others in the complex. *No racemisation could be detected however for this nitrogen centre as a direct consequence of the act of base hydrolysis.* This suggests that the amido nitrogen does not adopt the planar configuration implied by the π-donor hypothesis. A contrary conclusion is suggested by the less complete work of Niththyananthan on $Co(2,3,2\text{-tet})Cl_2^+$, which suggests that nitrogen inversion is synchronous with the first act of base hydrolysis[148]. The two systems differ in that the inverted nitrogen in the second could be *cis* to the leaving group. Of course, the problem of amido π-bonding would assume much less importance if Archer's suggestion[4] of a correlation of stereochemical change with high-spin Co^{III} states were adopted. There is one significant piece of experimental evidence in support of this idea in the report by Watt and Knifton[149] of paramagnetic conjugate base species derived from (amino acid)Co^{III} complexes[149].

Two kinds of systems show unusualy reactivity[150] in base hydrolysis: *trans*-$Co(NH_3)_4X_2^+$(X = Cl, Br) and *cis*-$Coen_2pyCl^{2+}$. The reactivity of the former seems to be associated with the lability of the conjugate base[136]. It has been suggested that the reactivity of the latter is due to π-donation from the pyridine nitrogen following addition of hydroxide at the carbon of the ring[105], but little evidence is available to support or refute this suggestion.

Very little systematic work on base hydrolysis in non-aqueous solvents has been reported. One of the most interesting cases is the recent work in which Chester[151–154] describes the reaction of Co^{III} complexes in acetic acid with the -lyate ion of the solvent, acetate. Retardation by acetate was reported, which does not seem consistent with the conjugate base mechanism. In this connection, retardation of base hydrolysis of halogeno penta-amine cobalt(III) complexes by anions other than OH^- in aqueous solution should be noted[155].

Now, if base hydrolysis is a dissociative reaction of the conjugate base, we may ask if a 5-coordinate intermediate of the D pathway is detectable. Studies of competition between solvent and other ligands as entering groups should be the key. Two such types of studies were cited in the summary of pre-1968 studies in Section 6.3.5 above. The studies strongly support the conjugate base mechanism by showing that the catalytic hydroxide is not the entering group, but the question of the 5-coordinate intermediate is still not settled. Interchange reactions of the conjugate base are possible and consistent. If the intermediate is formed, two conditions should be met in competition reactions: (1) competition should be independent of the mode of formation of the entering group, and (2) some entering groups should be distinctly preferred over solvent.

The first condition is necessary but obviously not sufficient, the second is sufficient but not always necessary.

The issue of 5-coordinate intermediates has been the subject of elegant studies by Buckingham, Sargeson, and their collaborators[132, 156–159]. The steric course of the base hydrolysis of $Coen_2AX^{n+}$ is independent of X. When $^{15}NH_3$ is used to stereochemically signpost the simple complex $Co(NH_3)_5Cl^{2+}$, the product is 40% *cis*- and 60% *trans*-$Co(NH_3)_4(^{15}NH_3)(OH)^{2+}$, which is not the *statistical* distribution expected for a trigonal-bipyramid intermediate with the amido group derived from the *trans*-NH_3. Probably the most important observation is that base hydrolysis of $Co(NH_3)_5X^{2+}$ (X = Cl, Br, I, NO_3), in the presence of N_3^- or NCS^-, shows the rate of entry of these anions to be coincident with the rate of loss of X^- [158]! From the second condition above, this seems to require the formation of the intermediate.

If the intermediate exists, its shape should determine the steric course of base hydrolysis. The $^{15}NH_3$ label experiment seems to exclude the *a priori* attractive notion that the intermediate is a trigonal-bipyramid derived from the loss of a proton on the labile position *trans* to the leaving group, and retaining the amido ligand in the trigonal plane. A distorted square-pyramid has been proposed[156], but Nordmeyer[160] has pointed out that many stereochemical results may be explained if it is assumed that the effective amido group is formed *cis* to the leaving group. (Data on the steric course of base hydrolysis were recently collected and reviewed by Tobe[150].)

6.3.6 Bridged complexes

The study of binuclear Co^{III} systems has awaited the development of a reasonably clear picture of substitution in mononuclear complexes. This subject is, however, quite important for such binuclear species are useful in developing an understanding of oxygen coordination. Bridge closing of double-bridged species provides a fixed structure for studying interchange reactions[161, 162], and bridged intermediates have also been suggested in diamine-exchange reactions[163]. Sykes and his associates have initiated a considerable research programme in this particular area. In ring-closure reactions, leaving-group reactivity orders seem similar to those for mono-nuclear complexes[161,162], but the lability of water is greater than in $Co(NH_3)_5OH_2^{3+}$. The reactivity of Cl^- and SCN^- ions in their attack on hydroxo bridges is similar[165,166], and the rate of decomposition of the superoxido complex $(NH_3)_5Co{\cdot}O_2{\cdot}Co(NH_3)_5^{5+}$ is similar to that of the decomposition of $(NH_3)_5CoNH_2Co(NH_3)_5^{5+}$ [167]. A puzzling report of discrepancies in the solvent effects of dioxane, alcohols, and dimethylformamide on chloride loss from an NH_2-bridged complex has appeared[168].

6.3.7 Acid catalysis

Most leaving ligands discussed above are too weakly basic to attach a proton at ordinary pH values so that $[H^+]$ does not enter the relevant rate law expressions. There are, however, exceptions such as NO_2^- [169–171], NCO^- [172], and ReO_4^- [173]. The role of H^+ may be simply the labilisation of the leaving ligand (NO_2^-, F, CN^-) but with NCO^- and ReO_4^- the hydrogen ion is directly involved in reaction in the ligand without fission of a bond to Co^{III}. A curious H^+-catalysis effect is apparently involved in a bis-ditertiary-arsine complex which seems to isomerise 'intramolecularly' through an ill-defined acid-sensitive intermediate[174].

Another type of acid catalysis involved is the process frequently called 'Hg^{2+}-induced acid hydrolysis'. It has occasionally been suggested that such reactions should be called electrophilic substitutions, but the simplest interpretation in most cases (recent examples are cited in References 175 and 176) appears to be the rapid formation of a halide-bridged Co—X—Hg species in which the X—Hg group may be considered a good leaving group. Bifano and Linck[177] have shown that there is a close relationship (indicated by a linear free energy plot) between the transition state for Hg^{2+}-induced and ordinary spontaneous acid hydrolysis. These reactions are *not* however similarly related to the rate of Fe^{II} reduction of the Co^{III} complexes.

The Ag^+ ion should function similarly, but studies of its reactions in homogeneous solution are few. Lalor and Rusted[178] have reported Ag^+-catalysis of the solvolysis of $Co(NH_3)_5I^{2+}$, whose rate equation has a term *second-order* in Ag^+ concentration. Tanner and Higginson[179] have shown that a wide variety of metal ions can catalyse Cl^- loss from $Co(EDTA)Cl^-$ as well as amine-Co^{III} and -Cr^{III} complexes with non-labile inner spheres.

Electrophilic substitutions are illustrated by Hg^{2+}-dealkylation reactions of alkylcobaltoxime[180] and related complexes[181].

Morawetz and Vogel[182] have reported that polysulphonic acid solutions are spectacularly effective as mediators in the reactions between $Co(NH_3)_5Cl^{2+}$, $Co(NH_3)_5Br^{2+}$, and Hg^{2+}. Such behaviour should, perhaps, be more appropriately considered as transitional to heterogeneous catalysis. Heterogeneous catalysis by HgS, AgX (X = Cl, Br, I), Hg_2Cl_2, HgBr, Pt and Pd of the loss of halide from halogenopentamines of Co^{III} has been discussed by Archer and Spiro[183]. Platinum catalysis is reported for the solvolysis of halosulphito-bis-(dimethylglyoximato)Co^{III} complexes by Syrtsova and Korletyana[184].

6.3.8 Oxalato and carbonato complexes

Harris and co-workers have made an extensive study of the rates of isotopic exchange of $Co(N_4)CO_3^+$ ($N_4 = (NH_3)_4$, en_2, tn_2, $(NH_3)_2en$ etc.) with $^{14}CO_3^{2-}$ [185] and the acid-catalysed aquation for these complexes [186]. The isotopic exchange reaction exhibits duality of mechanism, and two associative and two dissociative paths have been postulated. Aquation reactions of these complexes proceed by a rate-determining 'carbonate' chelate ring-opening catalysed by water and the hydronium ion. This is supported by solvent isotope effects[187] observed in these systems. Decreased rates of aquation of carbonate complexes containing macrocyclic rings is attributed to the protection of the Co—O bond by the methyl groups of the non-labile amine ligand[188]. ^{18}O-tracer studies[87] of the base hydrolysis of $Co(en)_2CO_3^+$ show that carbonate ring-opening by the OH^- ion proceeds entirely by Co—O bond-breaking, and that water catalyses ring-opening by C—O bond-breaking. In order to provide a better understanding of the chemistry of carbonate complexes it would appear to be worth initiating some work on anation reactions.

Acid-catalysed aquation of $Co(C_2O_4)_3^{3-}$ and $Co(C_2O_4)_2(H_2O)_2^-$ has been reported[189, 190, 86]. The mechanism involves a fast proton pre-equilibration of the complex, followed by a rate-determining aquation step. Unlike Cr^{III}, Co^{III}-oxalate complexes undergo redox decomposition in addition to aquation.

The present studies reveal no evidence for the presence of oxalate radical intermediates, in disagreement with earlier work on these systems. Anation of $Co(en)_2(H_2O)_2^{3+}$ by oxalate ion seems to proceed by a dissociative interchange mechanism[84]. It is not possible to compare the anation results with those for the corresponding aquation of $Co(en)_2(C_2O_4)^+$ since that complex is stable even in highly acidic solutions.

6.3.9 Theoretical treatments

The development of an electronic theory of reaction pathways is difficult at best, and perhaps foolhardy for molecules as large as Co^{III} complexes. Yet it remains a fascinating possibility, and the availability of spectroscopic data in intimate detail, coupled with the high symmetry of the systems, provides a continuous temptation.

Theoretical efforts have borne fruit at least in the sense that they have provided a stimulus to the systematisation of observations in this field. Two approaches have dominated; ligand field and molecular orbital[2]. At present most of the theoretical studies continue to involve the ligand field approach with a particular focus on antibonding orbitals. Archer has considered possibilities for spin change in 5-coordinate intermediates using a model developed by Yatsimirskii[191]. Complete crystal field stabilisation energy calculations, enabling the prediction of intermediate geometries and activation energies in octahedral substitutions, have been made by Spees, Perumareddi and Adamson[192]. A model transition state involving C_{4v} symmetry and reduced coordination number apparently provides the best fit for Co^{III} data. In contrast, a model involving either C_{2v} or D_{5h} symmetry, with an enhanced coordination number, fits the Cr^{III} data best. An effort to use empirical σ-bonding parameters obtained from the spectra of C_{4v} complexes has been made by Langford[193]. In this model it has been suggested that the σ-bond to the leaving-group involving metal 3d orbitals is the crucial variable. Additionally, the role of non-labile ligands has been associated with their σ-donor properties. This approach apparently explains the over-estimates of the variation of reactivity obtained from simple crystal field treatments, and suggests a parallel between the Co^{III} and Cr^{III} cases.

6.4 CHROMIUM(III)

6.4.1 Introduction

Similarities between the chemistries of Cr^{III} and Co^{III} compounds have been recognised for a long time, and a comparison between the mechanisms of substitutions of octahedral complexes of Cr^{III} with those of Co^{III} seems a logical development. In addition to the similarity of many mechanistic approaches which unites all studies with those of Co^{III} is the numerical similarity between the parameter values for Co^{III} and Cr^{III}.

6.4.2 Leaving-group, steric, and non-labile ligand effects

Unlike the case of Co^{III}, with Cr^{III} it is possible to compare the aquation rates of the halogeno-pentammine series of complexes with those for the corresponding halogeno-penta-aquo complexes, although the preparation of the former presents some difficulties.

The aquation rates of halopentammine complexes decrease in the sequence $I^- > Br^- > Cl^- > F^- > NCS^-$ [194–196], which is in keeping with the strength of the corresponding Cr—X bond. The rates of aquation also show a much greater dependence on the nature of the leaving group than in the analogous Co^{III} complexes[197, 198]. The activation energies for the iodo, bromo and chloro complexes are essentially the same, and these complexes have been thought to undergo aquation by the *d* mechanism. Fluoro and thiocyanato complexes have activation energies greater than those for other halogeno complexes by about 3 kcal mol^{-1}, which prevents one from immediately assigning a *d* mechanism for the aquation of these complexes. Although it is tempting to suggest an *a* mechanism for these complexes, the activation volume data[199] on the aquation of $Cr(NH_3)_5SCN^{2+}$ seem to preclude such an assignment, and is more in keeping with a *d* pathway.

A lack of data on equilibrium constants for the aquation of halogeno-pentammine Cr^{III} complexes prevents one from attempting the construction of the familiar linear free-energy relationship which has yielded useful information on the role of the leaving group in the rate-determining step in aquation of the corresponding Co^{III} complexes. Jones and Phillips[195] have, however, established a linear relation between $\Delta S^{\ddagger}$ and S^0 and ligand solvation for the Cr^{III} series of complexes, with deviations in the case of the fluoride and the thiocyanate. This shows that solvation of the halide is important in the rate-determining step. A linear relationship between log k and ligand pk_a for this series of complexes has also been observed[195].

Anion-catalysed aquation of these ammine complexes has been reported[195]. Pyrophosphate was found to increase the rate of aquation of $Cr(NH_3)_5Br^{2+}$ by 5-times while the rate of aquation of $Cr(NH_3)_5F^{2+}$ increased by only 50%. This result was interpreted by suggesting a d pathway for the bromo complex and the importance of bond-making for the fluoro complex.

Much of the secondary evidence suggesting a d reaction pathway in the aquation of Co^{III} ammines is paralleled by similar evidence for Cr^{III} amines[2]. A new result that tends to reveal a divergence is the report of Parris and Wallace[200] on the rates of aquation of a series of complexes of the type $Cr(RNH_2)_5Cl^{2+}$ ($R = CH_3$, C_2H_5, n-C_3H_7, and n-C_4H_9). The methylamine derivative was found to aquate 30-times slower than the pentammine complex, and the n-butylamine derivative only 4-times slower than the pentammine complex. These results cannot be interpreted simply in terms of steric effects expected in d reaction pathways, for there is also a possible solvent-dependence[34].

For the aquo complexes, Swaddle and Guastalla[201] have established a linear free energy relationship in the aquation of acidopenta-aquo Cr^{III} complexes. Aquation rates follow the decreasing sequence $NO_3^- \sim I^- > Br^- > Cl^- > NCS^- > F^-$, and the slope of the plot of log k versus log K is 0.59. This suggests that the transition state of the aquation reaction resembles the products to a certain extent. It could probably be argued that the greater crystal-field contribution in the Co^{III} (d^6) series ($4Dq$) to the activation enthalpies is compensated by greater contributions from the enthalpies of hydration of the departing groups X^-, since the departing groups are essentially fully-separated and hydrated in the transition state (slope ≈ 1.0 in the linear free energy plot for $Co(NH_3)_5X^{2+}$ complexes), as opposed to partially-separated and hydrated in the $Cr(H_2O)_5X^{2+}$ complexes (CFSE = $2Dq$). Hence similar activation enthalpies should be observed for the aquation of both Cr^{III} and Co^{III} complexes. It should be noted that the earlier position of the Cr^{III} transition state on the reaction coordinate in comparison to that of Co^{III} is consistent with either d or a activation for Cr^{III}.

6.4.3 Entering groups

Interest in $M(OH_2)_6$ systems has tended to shift experimental studies, especially for labile complexes, from comparisons of leaving groups to comparisons of entering groups in the 'microscopic reversal' of the acid solution hydrolysis. $Cr(OH_2)_6^{3+}$ is, of course, stable and its 'anations' are attractive reactions to study. Anation rate constants[202] for $Cr(OH)(H_2O)_5^{2+}$ with Cl^-, Br^-, I^-, NO_3^-, SCN^-, and NCS^- span a range of $0.26–8.5 \times 10^{-5}$ $M^{-1}s^{-1}$, and for $Cr(H_2O)_6^{3+}$, with the aforementioned anions, cover a range of $0.08–73 \times 10^{-8} M^{-1}s^{-1}$. This result has been interpreted as implying that bond-making plays a role in these reactions as evidenced by the small but definite dependence of the formation rates on the nature of the ligand. The difference in the range of rates spanned by $Cr(H_2O)_5OH^{2+}$

is smaller in comparison to the range spanned by $Cr(H_2O)_6^{3+}$ with the same anions, and this may be explained on the basis that coordinated OH^- facilitates a dissociative mechanism through its π-bonding ability[2]. Oxy-anions such as SO_4^{2-} and H_3PO_2 appear to be special cases as evidenced by the large values for anation rates in the presence of these ions[203,204]. The abnormally high rate in the case of SO_4^{2-} may be solely a reflection of its higher charge, but with H_3PO_2 complex formation may be influenced by the conversion of $HOPH_2O$ to $HP(OH)_2$. If so, attack via an *a* pathway is implied.

Anation of $Cr(H_2O)_6^{3+}$ by glycine has been reported[205] to proceed 10-times faster than the water-exchange rate, and on this basis the authors have advanced an I_a mechanism.

Finally, turning to a different aspect of the same problem, the binding of the carboxyl groups in wool to Cr^{III} salts may be considered to be an anation reaction of $Cr(H_2O)_6^{3+}$ by $-COO^-$, and a dissociative mechanism has been suggested for this process[206].

6.4.4 Effect of pressure on the solvolysis rates

Recently studies on the effect of pressure on the rates of substitution reactions of complexes have begun to appear, and, in principle, these should yield useful information regarding the mechanisms of these processes through the interpretation of activation volumes. For such an interpretation it is necessary that information on three important factors, viz.; (1) intrinsic volumes of the ions, (2) electrostriction of the solvent, and (3) the compressibility of the ions, should be available. Since data on the third factor are lacking it has become common practice to substitute the volume changes for a series of related reactions in place of this factor.

Gray and Nalepa[199,207] have studied the effect of pressure on the aquation rates of $Cr(NH_3)_5NCS^{2+}$, $Cr(NCS)_6^{3-}$, and *trans*-$Cr(NH_3)_2(NCS)_4^-$, and the activation volumes have been found to be -8.6 ml mol^{-1}, 16 ml mol^{-1} and -24 ml mol^{-1} respectively.

Swaddle[208] reports a $\Delta V^\ddagger$ values of -9.3 cm^3 mol^{-1} for the $Cr(H_2O)_6^{3+}$–H_2O exchange system, and a value of -12.7 cm^3 mol^{-1} for the aquation of $Cr(H_2O)_5NO_3^{2+}$. He has suggested that both bond-making and bond-breaking are approximately of equal importance in these systems.

It should be pointed out at this stage that the lack of compressibility data on ions, coupled with the fact that data on the activation volumes of reactions is available for only a very few systems, is a considerable disadvantage in making any definitive interpretation of the activation volumes in substitution reactions, but important correlations may be expected soon since a substantial body of data of this description is now becoming available.

6.4.5 Non-aqueous solvents

Although kinetic results in mixed solvent media are difficult to interpret, they provide some clues regarding reaction mechanisms. In the case of Cr^{III} the use of the n.m.r. spectroscopic method suggested by Frankel, Stengle and Langford[209] has been of considerable advantage in allowing an estimate of the probability of encounter of the substrate with each of the solvent components. The n.m.r. spectroscopic method has been useful in assigning an I_d mechanism for the aquation of $Cr(NCS)_6^{3-}$ [210] and of *trans*-$Cr(NH_3)_2(NCS)_4^-$ [211,212]. Thomas and Holba[213] have reported the aquation of $Cr(en)(NCS)_4^-$ in water–methanol mixtures, and have attributed the decrease in rate observed to the decreased entropy of activation. (The encounter factor?). The first step in the reaction is the release of thiocyanate, which is independent of pH, and the release of amine occurs as a pH-dependent second step (in contrast to the case of *trans*-$Cr(NH_3)_2(NCS)_4^-$). Aquation rates of analogues of Reineckate ion, *trans*-$Cr(Am)_2(NCS)_4^-$ [214–217] (Am = *p*-toluidine, *p*-phenetidine) in ethanol–water and methanol–water mixtures show that two simultaneous reactions involving the release of the thiocyanate ligand and the corresponding amine occur, and as expected the step involving the release of amine is pH-dependent. The step in which water replaces the thiocyanate ligand can probably be explained on the basis of *d* processes.

6.4.6 Acid catalysed reactions

In the discussion of aquation reactions considered above, those reactions which are not sensitive to the H^+ ion concentration in solution have been discussed, and it has been seen that such behaviour is observed as long as the solutions are acidic enough to prevent base hydrolysis. With complexes in which groups like F^-, NO_2^-, N_3^-, SO_4^{2-}, $H_2PO_2^-$, CN^-, CH_3COO^-, are coordinated to the metal ion, however, a real increase in rate as a function of the hydronium ion concentration is observed and the rate law is composed of both acid-independent and acid-dependent terms. Presumably acid-dependent terms arise due to the reactions:

$$\begin{aligned} Cr(L)_5X^{2+} + H^+ &\longrightarrow Cr(L)_5HX^{3+} \text{ fast} \\ Cr(L)_5HX^{3+} + H_2O &\xrightarrow{k} Cr(L)_5(H_2O)^{3+} \end{aligned} \tag{6.11}$$

Acid-catalysed aquation of $Cr(NH_3)_5ONO^{2+}$ [219], and of $Cr(NH_3)_5N_3^{2+}$ [220] has been reported. One feature of the acid-catalysed path that appears to be general is that the gain in heat from the addition of the proton to the metal-bound ligand is reflected in lower enthalpies of activation for the acid-dependent reaction. Acid catalysed aquation of $Cr(H_2O)_5en^{2+}$ [221] and $Cr(H_2O)_5ONO^{2+}$ [222] has been reported, an interesting feature of the reaction of the latter complex being that terms first- and second-order in H^+ concentration have been observed. Aquation of $Cr(CN)_5NO^{3-}$ is unusual in that although all the cyanide groups are substituted by water the nitrosyl is highly resistant to substitution[223,224].

The acetato-penta-aquo Cr^{III} complex exhibits[225] both acid-dependent and acid-independent aquation paths. Two questions arise regarding this reaction; (1) the position of attachment of the proton, i.e. whether this attachment occurs to the carbonyl oxygen or to the metal-bound oxygen, and (2) the position of bond-scission. From the analogous acid-assisted reactions of the acetato-pentammine complexes of Co^{III}, Rh^{III}, Ir^{III}, and from similar values in enthalpies of activation between the acid hydrolysis of the methyl acetate (16.4 kcal mol^{-1}) and aquation of the acetato complexes ($\Delta H^{\ddagger}$ = 19 kcal mol^{-1}), the authors conclude that C—O bond-cleavage occurs in the aquation reaction. Arguments have also been presented favouring the attachment of the proton to the metal oxygen rather than the carbonyl oxygen, although it is not obvious that the carbonyl oxygen is less basic than the metal-bound oxygen atom.

Two particularly interesting acid catalysis studies are those of Matts and Moore[226] on the aquation of $Cr(H_2O)_5N_3^{2+}$ in the presence of scavenging anions like Cl^-, Br^-, NCS^-, HSO_4^-, and the metal ion induced solvolysis of Cr^{III}, which has attracted researchers in recent years. Examples of the latter reaction include the Tl^{III}-, Hg^{II}- and Ag^{I}-induced solvolysis of $Cr(H_2O)_5I^{2+}$ in aqueous methanol[231], and the displacement of Cr^{III} from $Cr(H_2O)_5(CH_2—C_5H_4NH)$ by Tl^{III} [232]. These studies are discussed further below as providing evidence of a D path.

The Cr^{II}-catalysed aquation of $Cr(H_2O)_5N_3^{2+}$ [229] and of $Cr(H_2O)_5CN^{2+}$ [230] has been reported. An interesting aspect is that the rate is independent of Hg^{2+} concentration in the latter system, although the authors suggest a transition state involving the entity $Cr(CN)Hg^{4+}$.

Mercury(II)-catalysed aquation and isomerisation of the sulphur-bonded monothiocyanate complex of Cr^{III} has also been reported[231], together with the linkage isomerisation of $Cr(CN)_6^{3-}$ in the presence of Fe^{II} [232]. Birk[233] has initiated a study of the Hg^{II}-induced aquation of the geometric isomers of the $Cr(OH_2)_4Cl_2^+$ ion, using HgX^+ (where X = Cl, Br, I).

The constant value of the enthalpy of activation (for all mercury halides) suggests that Cr—Cl bond-breaking is more important than Hg—Cl bond-making.

6.4.7 Stereochemistry

Garner and co-workers have made an extensive study of the aquation kinetics of the *cis* and *trans* isomers of halogeno-tetrammine-[234], dichloro-triaquo-ammine-[235], tetra-aquo-ethylenediamine-[236], dibromodiaquo-ethylenediamine-[237], 1,2,3- and 1,2,6-triaquo-diethylenetriamine[238], and 1,6-dibromo-2-aquo-3-ammine-ethylenediamine-Cr^{III} [239] complexes. Other studies include the aquation of *trans*-$Cr(en)_2Cl_2^+$ [240], $Cr(NH_3)_2(OH_2)_2Br_2^+$ [241], *cis*-$Cr(CN)_2(H_2O)_4^+$ [242] and the geometric isomerisation of *cis*-$Cr(en)_2Cl_2^+$ in anhydrous dimethylsulphoxide[243].

In such reactions, Cr^{III} complexes are slightly more reactive than their Co^{III} counterparts. Retention of configuration appears to be the general rule for the aquation reactions of Cr^{III} complexes. (In the aquation of *trans*-$Cr(en)_2Cl_2^+$ the products contain 14% *cis*, but at the same time some ethylenediamine is released.) Some data on the extent of the stereochemical change in Cr^{III} complexes have been tabulated by Archer[4]. In contrast to Co^{III} (d^6), with Cr^{III} we are dealing with a d^3 system in which the t_{2g} metal orbitals have fewer electrons capable of participating in π-bonding, and this may promote isomerisation through a trigonal-bipyramidal transition state. If the activation process involved an *a* pathway this would, of course, favour stereospecificity.

6.4.8 EDTA, oxalate and malonate complexes

An important characteristic of polycarboxylate complexes is that both their rates of formation and dissociation exhibit acid-dependent and acid-independent paths. In this respect the formation of the Cr^{III}-EDTA complex from $Cr(H_2O)_6^{3+}$ and EDTA [244–246], the stepwise protonation of EDTA, and the participation of various Cr^{III} hydroxo species in the reaction has been discussed[244–246]. The results can be explained in terms of *d* mechanisms. Since various acid–base equilibria are involved in these reactions, one would expect catalysis by anions like CO_3^{2-}, SO_3^{2-} and NO_2^- to be due to their effect on the acid–base equilibria, and a study to this effect has been made by Phatak, Bhat and Shanker[247].

Bannerjea and Chaudhuri[248–250] have studied the formation of the Cr^{III}-EDTA complex by reacting $Cr(en)_3^{3+}$, $Cr(ox)_3^{3-}$, $Cr(acac)_3$ and $Cr(biguanide_3^{3+}$ with EDTA. The pH-dependence of these reactions may be attributed to the existence of a pH-controlled chelate ring-opening process which occurs during the reaction. Although such studies with multidentate ligands are undoubtedly important, their kinetics are very difficult to unravel.

In aquation and anation reactions there is a great similarity between the oxalato and malonato complexes. Considerable work has been undertaken on the aquation of $Cr(C_2O_4)_3^{3-}$ [251] and the anation of *cis*-$Cr(C_2O_4)_2(H_2O)_2^-$ [252] with oxalate and bioxalate anions. The kinetics of the aquation of $Cr(mal)_3^{3-}$ has been the subject of detailed investigation[253–255]. The results obtained in the study of the aquation of $Cr(ox)_3^{3-}$ and of $Cr(mal)_3^{3-}$ may be explained on the basis of a rapid proton pre-equilibration of the complex followed by a rate-determining solvent replacement of the oxalate or malonate ligand. The lower thermodynamic stability of the malonato complex is reflected in the higher rate of its aquation in comparison to that of the oxalato analogue. Observations on the anation of *cis*-$Cr(C_2O_4)_2(OH_2)_2^-$ and on the water-exchange rates of the complex[256] may be correlated on the basis of outer-sphere ion association, followed by rate-determining substitution of water by oxalate.

Kinetic studies of the aquation of *cis*-$Cr(mal)_2(H_2O)_2^-$ and $Cr(ox)(H_2O)_2^+$ ions have been reported[257, 258]. In the latter system there is a distinct change in kinetics in going from 1.5–3.0 M $HClO_4$ solutions to higher acid concentrations (3.5–9.9 M $HClO_4$). Results on the formation of mono-oxalato and mono-malonato-Cr^{III} complexes from $Cr(H_2O)_6^{3+}$ can be accommodated in the framework of a dissociative pathway[259, 260].

Racemisation and ^{18}O-tracer experiments[261–263] yield totally unrelated rate expressions for both $Cr(bipy)(C_2O_4)_2^-$ and $Cr(phen)(C_2O_4)_2^-$. The interpretation of the results requires an interchange mechanism for ^{18}O exchange due to attack at the carbonyl carbon, in contrast to a total dechelation of the oxalate rings as an explanation for racemisation. Very extensive work has been reported[264–266] on the geometrical isomerisation of *trans*-$Cr(C_2O_4)_2(H_2O)_2^-$ and the analogous malonato complex[271]. The activation parameters of the malonato complex ($\Delta H^{\ddagger} = 30.5$ kcal mol^{-1}, $\Delta S^{\ddagger} = 17.6$ cal deg^{-1} mol^{-1}) and of the oxalato analogue ($\Delta H^{\ddagger} = 17.9$ kcal mol^{-1}, $\Delta S^{\ddagger} = -13$ cal deg^{-1} mol^{-1}) have been attributed to mechanisms involving Cr—O bond-breaking and a twisting process (without breaking) of the Cr—O bond respectively. Mention should also be made of the substitution of trien in the α-Cr(trien)$(C_2O_4)^+$ complex[268]. The racemisation of d-Cr$(ox)_3^{3-}$ in DMSO–water mixtures has been examined with the aid of n.m.r. spectroscopic solvation studies[269].

6.4.9 Binuclear complexes

In spite of the fact that binuclear species of tripositive ions are encountered in solution, and that bridged binuclear activated complexes are formed in many redox reactions, studies on the chemical properties of these complexes have been sparse. The early studies of Grant and Hamm[270], and of Connick and Thompson[271] on the acid cleavage of Cr^{III} complexes with two hydroxo bridges provided evidence for the intermediate formation of a single OH-bridged species. Wolcott and Hunt[272] have reported studies on the acid cleavage of di-μ-hydroxo-tetrakis-(*o*-phen)-dichromium(III). The experimental rate law fits two mechanisms in which the respective opening of the first bridge and of the second bridge have been considered as rate-determining. The acid cleavage of single hydroxy-bridged Cr^{III} complexes has also been reported[273]. One interesting observation is that the rate of these reactions is apparently increased by the presence of chloride ion. The kinetics of the formation of a cyanide-bridged adduct have also been reported[274]. More studies in this area, and on the interesting ring-closure reaction in binuclear complexes, would be most interesting.

6.4.10 Base hydrolysis

Presumably the role of OH^- in the base hydrolysis of Cr^{III} complexes is similar to its role in the corresponding reactions of Co^{III} complexes, with reactions proceeding through the conjugate base. From the smaller effects of π-donor ligands on aquation rates, and from a consideration of the equilibrium constant for the abstraction of protons from ammines by the hydroxide ion, it has been pointed out that the base hydrolysis reaction will be less pronounced for Cr^{III} complexes in comparison with their Co^{III} analogues. This problem has been much discussed in earlier reviews, and here the results of recent studies, which do not change the overall picture, are summarised.

Bannerjea and Chatterjee[275] have reported studies of the base hydrolysis of $Cr(NH_3)_5SCN^{2+}$ and $Cr(NH_3)_5ONO^{2+}$, and have suggested a normal I_d mechanism for the former, and a 'pseudo-substitution' of NO_2^- by OH^- without the rupture of the Cr—O bond. The base hydrolysis of trifluoroacetato-pentammine Cr^{III} [276] shows both first- and second-order dependence on $[OH^-]$, which has been explained as a concerted attack of two hydroxyl ions, one abstracting a proton and the other attacking the acyl carbon, which results in C—O bond-breaking. Steric effects[200] in the base hydrolysis of $Cr(RNH_2)_5Cl^{2+}$ (R = H, CH_3, C_2H_5, n-C_3H_7, n-C_4H_9) show a systematic increase in rates (from 1.92×10^{-3} to 9.91 $M^{-1}s^{-1}$) and the results have been interpreted in terms of a dissociative mechanism.

Solvent effects on the base hydrolysis of $Cr(NH_3)_5X^{2+}$ (X = Cl, F) have been reported[277]. Increased rates in alcohol–water mixtures in comparison to those in aqueous solutions may be explained both on the basis of increased ion-pair formation as well as on the favourable solvent effect on conjugate base formation. Despite this fact Chan has preferred to accommodate the results in the framework of an ion-pair mechanism. A similar argument in the case of Co^{III} has been considered above and rejected.

6.4.11 The question of D intermediates

The parallel between Cr^{III} and Co^{III} behaviour, often interpreted in terms of *d* activation, has naturally led to attempts to distinguish I_d pathways from their *D* counterparts. This involves detailed rate law analysis which has so far been most successful in proving the *absence* of 5-coordinate intermediates. Duffy and Earley[278] have found that the anation of $Cr(NH_3)_5OH_2^{3+}$ with Cl^- and SCN^- has limiting rates below the rate of solvent exchange. Baltisberger and Hanson[279] have reported the effects of CH_3OH substitution on solvent exchange and anation by Cl^- and SCN^- which are inconsistent with a common intermediate. The anation of $Cr(DMSO)_6^{3+}$ by SCN^- also appears to have a limiting rate below the rate of solvent exchange[280]. All these results have been discussed in terms of an I_d model.

Evidence in favour of a D intermediate, $Cr(OH_2)_5^{3+}$ has been suggested from competition studies. Matts and Moore[226] have reported the formation of $Cr(OH_2)_5X^{2+}$, as well as $Cr(OH_2)_6^{3+}$, in the solvolysis of $Cr(OH_2)_5N_3^{2+}$ in the presence of added X (X = Cl^-, Br^-, NCS^-, HSO_4^-). Such results must be interpreted with caution; Moore, Basolo, and Pearson[281] have pointed out that one water molecule of $Cr(OH_2)_5I^{2+}$ is labile, and that a product, $Cr(OH_2)_5X^{2+}$, may be formed by the path

$$\begin{aligned} Cr(OH_2)_5I^{2+} + X^- &\rightarrow Cr(OH_2)_4XI^+ + H_2O \\ Cr(OH_2)_4XI^+ + H_2O &\rightarrow Cr(OH_2)_5X^{2+} + I^- \end{aligned} \tag{6.12}$$

which gives no information about intermediates. Competition studies may be more valuable where the loss of the leaving X^- group is catalysed by an 'acid'[227–230]. In the study[227] of the solvolysis of $Cr(OH_2)_5X^{2+}$, induced by Tl^{3+}, Hg^{2+}, Ag^+ (or HONO if X = N_3^-), where X^- = I^-, Br^-, Cl^-, N_3^-, the same product ratio of $Cr(OH_2)_5CH_3OH^{3+}/Cr(OH_2)_6^{3+}$ was obtained suggesting a common intermediate. (The alternative is a common solvation-shell composition for all of these diverse reacting systems.)

Bannerjea and Sarkar[282] have observed an acetate-independent term in the kinetics of the anation of $Cr(NH_3)_5OH_2^{3+}$ which might suggest a D path. However, they do not report whether or not the reaction described is an approach to equilibrium where the term would represent aquation.

6.5 RHODIUM(III), IRIDIUM(III), PLATINUM(IV), AND RUTHENIUM (III)

6.5.1 Introduction

In this section, the discussion will deal with d^6 systems other than Co^{III}, the inclusion of Ru^{III} being justified because of the similar reaction rates of its complexes despite the fact that it is a d^5 system. The rate data which have accumulated on these systems during the last 3 years are collected in Tables 6.3–6.10. It will be observed that these larger ions of higher atomic number (and hence more extended valence orbitals) exhibit a behaviour suggesting the possibility of nucleophilic attack by the entering ligand, although bond-breaking probably remains the most important process.

6.5.2 Leaving-group effects

The aquation of halogenopentammine Rh^{III} complexes has recently been studied[283, 284] and the relative rates follow the decreasing order $Cl^- > Br^- > I^-$. A comparison of activation parameters for the aquation of $Rh(NH_3)_5Cl^{2+}$ with those for $Co(NH_3)_5Cl^{2+}$ show very little difference, indicating that the greater bond strength of Rh—Cl does not influence the process. This may be due to participation of the solvent in the reaction, a suggestion which is supported by the lower frequency factor for aquation of the Rh^{III} complex in comparison with its Co^{III} analogue. Acid-catalysed aquation of $Rh(NH_3)_5N_3^{2+}$ [285, 286] shows the existence of the familiar acid-independent and acid-dependent paths in this reaction. Studies in the presence of nitrous acid are interesting since in this case the rates follow the sequence $N_3^- > N_3H > Co(NH_3)_5N_3^{2+} > Rh(NH_3)_5N_3^{2+}$, which shows that the electronic environment at the terminal nitrogen of the azide group is more important than the overall charge. The aquation rates of $Ir(NH_3)_5X^{2+}$ are smaller than those of the corresponding Rh^{III} complex, an observation in agreement with the considerations of the crystal field theory. The existence of a linear free energy relationship for the aquation of $Ir(NH_3)_5X^{2+}$ complexes has been pointed out by Basolo and Pearson[2]. The aquation of $IrCl_6^{2-}$ [287, 288] and $IrBr_6^{2-}$ [289] has also been reported, and the mechanisms appear to be dissociative in nature. Aquation rates for $Ru^{II}(NH_3)_5X$ [290] (X = carboxylate anion), and for the geometrical isomers of Ru^{II} [291] have been obtained. The aquation of $Ru(NH_3)_5(pyX)^{2+}$ [292], with various groups in the X position, shows that the reaction rate may be correlated with the inductive effects of X as well as with the sign and magnitude of the Hammett σ parameter.

In contrast to other known systems, the aquation of $Ru(NH_3)_5X^{2+}$, (where X = Br, Cl, I), shows a decreasing sequence $Br^- > Cl^- > I^-$ which may be attributed to the greater importance of bond-making in this reaction[293]. The acid-catalysed reactions of the azido complexes of Ru^{III} [294, 295] provide

Table 6.3 The rates of ligand replacement reactions involving Rh^{III} complexes

Complex	*Leaving group*	*Entering group*	*Temperature* (°C)	μ	$k(s^{-1})$	$\Delta H^{\ddagger}$ (kcal mol^{-1})	$\Delta S^{\ddagger}$ (cal deg^{-1}mol^{-1})	*Reference*
$Rh(NH_3)_5OH_2^{3+}$	H_2O	Cl^-	65		4.2×10^{-3}			296
$Rh(NH_3)_5OH_2^{3+}$	H_2O	Br^-	65		7.9×10^{-3}			296
$Rh(NH_3)_5OH_2^{3+}$	H_2O	SO_4^{2-}	65		1.7×10^{-3}			296
$Rh(NH_3)_5OH_2^{3+}$	H_2O	Cl^-	35		0.89×10^{-4}			297
$Rh(NH_3)_5OH_2^{3+}$	H_2O	Cl^-	65		31×10^{-4}			297
$Rh(NH_3)_5OH_2^{3+}$	H_2O	Br^-	35		1.1×10^{-4}			297
trans-$Rh(en)_2(OH_2)_2^{3+}$	H_2O	Br^-	70		18.5×10^{-4}			297
trans-$Rh(en_2)Br(OH_2)^{2+}$	H_2O	Cl^-	45		39×10^{-4}			297
trans-$RhA_4L(OH_2)^{n+}$	H_2O							
$A = en_{\frac{1}{2}}$ $L = NH_3$	H_2O	Cl^-	50	0.2	3.41×10^{-4}*	25.15	3.31	317
$A = NH_3$ $L = NH_3$	H_2O	Cl^-	50	0.2	2.01×10^{-4}*	25.5	3.20	317
$A = NH_3$ $L = NH_3$	H_2O	Br^-	50	0.2	1.44×10^{-4}*	25.3	1.9	317
$A = en_{\frac{1}{2}}$ $L = H_2O$	H_2O	Br^-	50	0.2	1.52×10^{-4}*	31.99	22.86	317
$A = en_{\frac{1}{2}}$ $L = Cl$	H_2O	Cl^-	50	0.2	7.2×10^{-4}*	25.3	5.2	317
$A = en_{\frac{1}{2}}$ $L = OH$	H_2O	I^-	50	0.2	72.6×10^{-4}*	23.81	5.21	317
$RhCl_5(H_2O)^{2-}$	H_2O	Br^- 0.6 M	35	4.0	6.8×10^{-4}			299
$RhCl_5(H_2O)^{2-}$	H_2O	4.0 M			24.62×10^{-4}			299
$RhCl_5(H_2O)^{2-}$	H_2O	I^- 0.25 M	35	4.0	1.70×10^{-4}			299
$RhCl_5(H_2O)^{2-}$	H_2O	2.5 M			14.7×10^{-4}			299
$RhCl_5(H_2O)^{2-}$	H_2O	SCN^- 0.25 M	35	4.0	3.8×10^{-4}			299
$RhCl_5(H_2O)^{2-}$	H_2O	4.0 M			51.0×10^{-4}			299
$RhCl_5(H_2O)^{2-}$	H_2O	NO_2^- 0.25 M	35	4.0	2.74×10^{-4}			299
$RhCl_5(H_2O)^{2-}$	H_2O	4.0 M			8.40×10^{-4}			299
$RhCl_5(H_2O)^{2-}$	H_2O	N_3^- 0.25 M	35	4.0	11.5×10^{-4}			299
$RhCl_5(H_2O)^{2-}$	H_2O	4.0 M			1.0×10^{-4}			299

*Second-order reaction. k is in $mol^{-1}s^{-1}$

Table 6.4 Rates of aquation and base-hydrolysis reactions of Rh^{III} and Ir^{III} complexes

Complex	*Temperature* (°C)	μ	*Other*	$k(s^{-1})$	$E_a(kcal\ mol^{-1})$	log A	*Reference*
$RhCl_6^{3-}$	25.0			1.8×10^{-3}	25.0	$+10\dagger(\Delta S^{\ddagger})$	328
$IrCl_6^{3-}$	25	0.13		3.4×10^{-5}	28.11	16.13	287
$IrCl_5(H_2O)^{2-}$	25	0.13	0.1 M $HClO_4$	1.19×10^{-6}	—	—	287
$IrBr_6^{3-}$	25		0.1 M H^+	2.3×10^{-4}	—	—	289
cis-$Ir(py)_2Cl_4^-$	110		1–2.5 H^+	1.3×10^{-4}	34.0	15.5	319
trans-$Ir(py)_2Cl_4^-$	110	3.7	0.5 F H^+	8.0×10^{-5}	—	—	319
$Rh(NH_3)_5Cl^{2+}$	25			4.06×10^{-4}*	28.49	20.18	324
$Rh(NH_3)_5Br^{2+}$	25			3.37×10^{-4}*	30.55	21.40	324
$Rh(NH_3)_5I^{2+}$	25			7.26×10^{-5}*	32.85	22.60	324
$Rh(NH_3)_5N_3^{2+}$	87.5			2.94×10^{-7}*	34.60	20.80	285
$Ir(NH_3)_5I^{2+}$	25		OH^- (0.028 M)	9.4×10^{-10}*	36.32	18.54	325
trans-$Pt(CN)_4Br_2^{2-}$	25	1 M	pH 7–9	6.9×10^{-3}*	26.4$(\Delta H^{\ddagger})$	38.3†	326
$Rh(NH_3)_5Cl^{2+}$	77.06			2.27×10^{-5}	24.42	10.453	283

*Second-order reaction. k is $mol^{-1}s^{-1}$

† cal $deg^{-1}mol^{-1}$

Table 6.5 The rates of ligand replacement reactions involving Ir^{III} complexes

Complex	*Leaving group*	*Temperature* (°C)	$k(s^{-1})$	$\Delta H^\ddagger$ (kcal mol^{-1})	$\Delta S^\ddagger$ (cal deg^{-1} mol^{-1})	*Reference*
trans-$Ir(en)_2LX$						
L = Cl^-	Cl^-	105	5.9×10^{-6}	29.2	−6	318
Br^-	Cl^-	105	9.0×10^{-6}	27.1	−9	318
I^-	Cl^-	105	159×10^{-6}	25.3	−4	318
NO_2^-	Cl^-	105	107×10^{-6}	—	—	318
Br^-	Br^-	105	14.5×10^{-6}	27.2	−9	318
I^-	Br^-	105	104×10^{-6}	27.1	−5	318
I^-	I^-	105	58×10^{-6}	28.7	−3	318
trans-$Ir(py)_2Cl_3(OH_2)^-$	OH_2	110	4.4×10^{-4}*	—	—	319
$IrCl_5(H_2O)^{2-}$	OH_2	25	1.33×10^{-6}	(28.24 (E_a))	(14.81 (log *PZ*))	287

*Second-order reaction. k is $mol^{-1}s^{-1}$

Table 6.6 The rates of aquation reactions of Ru^{III} complexes

Complex	*Temperature*	μ	$k(s^{-1})$	$\Delta H^\ddagger$ (kcal mol^{-1})	$\Delta S^\ddagger$ (cal. deg^{-1} mol^{-1})	*Reference*
$Ru(NH_3)_5L$						
L = $HCOO^-$	25	0.1–0.5	1.28			290
HCOOH	25	0.1–0.5	20.10			290
CH_3COO^-	25	0.1–0.5	5.0			290
CH_3COOH	25	0.1–0.5	17.5			290
CF_3COOH	25	0.1–0.5	5.3	14	8	290
$Ru(NH_3)_5pyX$						
X = *p*-CH_3	25	2 M HCl	5.9×10^{-4}			292
m-CH_3	25	2 M HCl	3.12×10^{-4}			292
H	25	2 M HCl	2.88×10^{-4}			292
m-$COOCH_3$	25	2 M HCl	0.34×10^{-4}			292
m-Cl	25	2 M HCl	0.38×10^{-4}			292
trans-$Ru(NH_3)_4Cl_2$	25	0.1 M *p*-toluene-sulphonic acid	<0.1			291
trans-$Ru(NH_3)_4H_2OCl^+$	25	0.1 M	<0.1			291
cis-$Ru(NH_3)_4Cl_2^+$	25	0.1 M	16			291
cis-$Ru(NH_3)_4OH_2Cl^+$	25	0.1 M	4			291

Table 6.7 Data relating to the base hydrolysis reactions of Ru^{III} complexes

Complex	*Temperature* (°C)	μ	k_{OH} ($M^{-1}s^{-1}$)	E_a(kcal mol^{-1})	log *A* (*A in* s^{-1})	*Reference*
$Ru(NH_3)_5Cl^{2+}$	24.6	0.002	4.95	27.3	20.7	293
$Ru(NH_3)_5Br^{2+}$	24.6	0.001	11.3	23.1	18.0	293
$Ru(NH_3)_5I^{2+}$	24.6	0.002	5.11	24.3	18.5	293
cis-$Ru(en)_2OHCl^+$	25.0	0.04	0.5	—	—	293
cis-$Ru(en)_2OHBr^+$	25.0	0.04	1.42	20.3	15.0	293

evidence for the presence of molecular nitrogen and nitrene complexes. Although leaving-group effects in these systems still show the importance of bond fission, there is some suggestion of associative behaviour.

6.5.3 Entering-group effects

The anation of $Rh(NH_3)_5(H_2O)^{3+}$ with Br^-, Cl^- and SO_4^{2-} has been reported[296]. The order of reactivity is $SO_4^{2-} < Cl^- < Br^-$, which is consistent with the importance of increasing softness towards the relatively soft Rh^{III} centre. The anation rates are much higher than the water-exchange rates and are

Table 6.8 Data relating to the anation rates of Ru^{III} complexes

Complex	*Anating ion* (M)	*Temperature* (°C)	μ	$[H^+]$	10^3k ($M^{-1}s^{-1}$)	E_a(kcal mol^{-1})	*Reference*
$Ru(NH_3)_5H_2O^{3+}$	Cl^-(0.1)	54.7	0.26	0.1	2.10	—	307
	Br^-(0.1)	54.7	0.26	0.1	1.32	—	307
	I^-(0.15)	54.7	0.26	0.1	0.74	—	307
cis-$Ru(en)_2(H_2O)_2^{3+}$	Cl^-(0.25)	35.5	0.26	0.25	7.5	19.4	307
	Br^-(0.25)	35.5	0.26	0.25	5.3	20.2	307
cis-$Ru(en)_2Cl\ H_2O^{2+}$	Cl^-(0.1)	35.5	0.1	0.1	1.64	24.0	307
	Cl^-(0.5)	35.5	0.5	0.5	1.04	22.8	307
cis-$Ru(en)_2Br\ H_2O^{2+}$	Br^-(0.5)	35.5	0.5	0.5	0.65	22.0	307

Table 6.9 Data relating to the aquation rates of Ru^{III} complexes

Complex	*Temperature* (°C)	$k(s^{-1})$	$\Delta H^{\ddagger}$ (*or* E_a)	log $A(s^{-1})$	*Reference*
$Ru(NH_3)_5Cl^{2+}$	80.1	3.28×10^{-4}	23.2	10.8	293
$Ru(NH_3)_5Br^{2+}$	80.1	3.99×10^{-4}	23.2	11.0	293
$Ru(NH_3)_5I^{2+}$	80.1	1.64×10^{-4}	24.5	11.4	293
cis-$Ru(en)_2Cl_2^+$	54.5	8.68×10^{-4}	20.5	10.6	293
cis-$Ru(en)_2Br_2^+$	54.5	7.98×10^{-4}	21.1	11.0	293
cis-$Ru(en)_2I_2^+$	54.5	3.24×10^{-4}	22.7	11.6	293
cis-$Ru(en)_2(OH_2)Cl^{2+}$	54.5	3.86×10^{-4}	20.7	10.4	293
cis-$Ru(en)_2(OH_2)Br^{2+}$	54.5	4.08×10^{-4}	22.4	11.6	293
cis-$Ru(en)_2(OH_2)I^{2+}$	54.5	1.22×10^{-4}	25.0	12.7	293
$RuCl_6^{3-}$	25.0	~ 1.0	—	—	329
$RuCl_5(H_2O)^{2-}$	25.0	3.0×10^{-4}	—	—	330
$Ru(H_2O)_3Cl_3$	25.0	2.1×10^{-6}	—	—	329
$Ru(H_2O)_5Cl^{2+}$	25.0	$\sim 10^{-8}$	—	—	329
cis-$Ru(NH_3)_4Cl_2^+$	25.0	0.88×10^{-4}	22.4	11.3	293
cis-$Ru(en)_2Cl_2^+$	25.0	3.7×10^{-4}	20.5	10.6	293
cis-$Ru(trien)Cl_2^+$	25.0	42.0×10^{-4}	18.7	10.5	293
cis-$Ru(-)py_2Cl_2^+$	25.0	2.82×10^{-4}	18.8	9.3	293

suggestive of an *a* mechanism. Robb, Devstyn and Krüger[299] have reported results for the anation of $RhCl_5(H_2O)^{2-}$ and have established the reactivity order in this case to be $Br^- > I^- > SCN^- > NO_2^- > N_3^-$. These results suggest a dissociative mechanism in this case and bear some resemblance to the anation of $Co(CN)_5(H_2O)^{2-}$. (Possibly D?) The chloride anation of $Ir(NH_3)_5OH_2^{3+}$ is reported[300] to be 3-times faster than the water-exchange rate, which is attributed to an interchange mechanism with some associative character.

Anation studies[301] on $Ru(H_2O)_6^{2+}$ include evidence of substitution even by the ClO_4^- ion, and the reactions appear to proceed by a dissociative interchange mechanism. Davies and Mullins[302] have studied the anations of $Ru(terpy)(bipy)(H_2O)^{2+}$ and $Ru(bipy)_2(H_2O)_2^{2+}$ by N_3^-, NO_2^-, SCN^- and pyridine, and they have proposed an associative mechanism, although no comparisons with the water-exchange rates were made. Other studies include the formation rates of $Ru(NH_3)_5I^{2+}$ [303] from $Ru(NH_3)_5OH_2^{2+}$, of the dimerisation reactions of *cis*-$Ru(NH_3)_4(H_2O)_2^{2+}$ [304], and of the formation

Table 6.10 Data relating to the anation rates of Ru^{II} complexes at 25 °C

Complex	*Anating species*	$k(M^{-1}s^{-1})$	ΔH kcal mol^{-1}	E_a kcal mol^{-1}	$\Delta S^{\ddagger}$ cal deg^{-1} mol^{-1}	log A	*References*
$Ru(H_2O)_6^{2+}$	Cl^-	8.5×10^{-3}	20.2		−0.4		301
	Br^-	9.7×10^{-3}	19.8		−1.2		301
	I^-	9.0×10^{-3}	19.5		−2.4		301
	ClO_4^-	3.2×10^{-3}	19.4		−5.0		301
$Ru(H_2O)(bipy)(terpy)^{2+}$	N_3^-	5.13×10^{-1}		20.2		12.6	302
	NO_2^-	3.16×10^{-1}		32.3		21.2	302
	SCN^-	1.36×10^{-2}		12.7		5.6	302
	py	5.5×10^{-3}		6.2		0.45	302
$Ru(bipy)_2(H_2O)_2^{2+}$	N_3^-	2.55		17.3		11.2	302
	NO_2^-	1.38		17.5		10.1	302
	SCN^-	4.15×10^{-1}		16.4		9.8	302
	py	6.57×10^{-2}		7.8		2.7	302
$Ru(NH_3)_5H_2O^{2+}$	NO_2	7.21×10^{-2}					303
	CO	1.2×10^{-1}					303
	py	1.18×10^{-1}					303
	Isonicotinamide	6×10^{-2}					303
	N_2	8.0×10^{-2}		22.0	+10	15	304
	$Ru(NH_3)_5N_2^{2+}$	3.6×10^{-2}		19.9	+13.1	19.9	304
	cis-$Ru(NH_3)_4(H_2O)N_2^{2+}$	2.7×10^{-2}		20.4	+3	13.4	304
cis-$Ru(NH_3)_4(H_2O)_2^{2+}$	N_2	1.04×10^{-2}		20.4	+6	14	304
	$Ru(NH_3)_5N_2^{2+}$	6.8×10^{-2}		18.2	−3	12.2	304
	cis-$Ru(NH_3)_4(H_2O)N_2^{2+}$	7.2×10^{-2}		17.1	−7	11.4	304

of the $Ru(NH_3)_5N_2^{2+}$ complex[305]. The exchange of labelled nitrogen with a Ru^{II}–nitrogen complex appears to follow a dissociative pathway[306].

Anation rate studies of $Ru(NH_3)_5OH_2^{3+}$, *cis*-$Ru(en)_2(H_2O)_2^{3+}$ and *cis*-$Ru(en)_2ClH_2O^{2+}$ have been reported recently[307], although the results do not allow a choice between an associative and a dissociative mechanism. Results on the formation of $Ru(NH_3)_5NO^{3+}$ from $Ru(NH_3)_6^{3+}$ show[308] that the substitution of ammine is faster in the presence of NO than in its absence, which implies that bond-making by the entering group is important in this substitution reaction. Isotopic exchange of $RuCl_6^{3-}$ with $H^{36}Cl$ shows that the exchange follows the familiar aquation–anation pathway[309], and the decrease in exchange rate at high concentrations of HCl has been correlated with the decrease in the water activity.

The substitution reactions of Pt^{IV} complexes are known to proceed by Pt^{II}-catalysed paths[2]. Poë and co-workers[310, 311] have studied the interchange reactions of *trans*-$Pt(ox)_2L_2^{2-}$ and of *trans*-$Pt(en)_2L_2^{2-}$ ($L = Cl^-$ or Br^-), and shown that the reactions proceed through the intermediate formation of a Pt^{II} complex. In the case of the oxalato derivative little bond-breaking occurs in the transition state, while Pt—X and I—X bond-breaking appear to play important roles in the case of the cyano complex. An interesting aspect of the oxalate system is that the results indicate no intervention by a Pt^{III} intermediate. This casts doubt on the intervention of Pt^{III} intermediates in the reactions of $PtCl_6^{2-}$ and $PtBr_6^{2-}$, and it would be interesting to explore this discrepancy by a.c. polarography if this were feasible. Mason's studies[312, 313] on the substitution of $PtL_4Br_2^{n+}$ ($L = NH_3$, NO_2^-, CN^-), in which the observed reactivity order is $NH_3 > CN^- > NO_2^-$, may reflect only the unfavourable activation entropy for cyano and nitro complexes. The substitution of Cl in *trans*-$Pt(NH_3)_4Cl_2^{2+}$ by NH_3, pyridine and Br^-, which is catalysed by Pt^{II} complexes of different charge types and geometry, has been investigated[314, 315]. The charge and geometry of the Pt^{II} complexes appear to be of minor significance in this reaction, and the rates parallel those found in the oxidation of the Pt^{II} complex and in the reduction of the Pt^{IV} derivative. This result indicates that the reaction is an example of the well-established intervention of an inner-sphere redox reaction as a mediator in a substitution reaction. Although the rates of substitution of *trans*-$Pt(en)(NO_2)_2Cl_2$ by bromide are found to be solvent-sensitive, no intermediates have been detected in the substitution process. In general, it is difficult to substantiate that an example of a thermal Pt^{IV} substitution process free from redox catalysis, has as yet been observed. If they exist, such reactions must be extremely slow.

6.5.4 Stereochemistry

Unlike the case of Co^{III} complexes, studies on the stereochemistry of Rh^{III}, Ir^{III}, Ru^{II} and Ru^{III} complexes are not very extensive, but the results at present available indicate that retention generally occurs, as has been pointed out by Archer[4]. Poë and Shaw[317] pointed out the kinetic *trans* effect in the anation of *trans*-$RhA_4L(OH_2)^{n+}$ ($A = eny_{\frac{1}{2}}$, NH_3; $L = NH_3$, H_2O, Cl, OH) by halides such as Cl^-, Br^- and I^- proceeds according to the relative order $H_2O > NH_3 > Cl^- > OH^-$. The *trans* effect of the OH^- ion in Rh^{III} complexes is smaller than in the corresponding Co^{III} complexes, and this can be explained in terms of a more effective OH–metal π-bonding in the latter system. An increasing *trans* effect in the order $NCS < Cl < Br < NO_2 < I$ has been reported in the reactions of *trans*-$Ir(en)_2LX^+$. The aquation of *cis*-$Ir(py)_2Cl_4^-$ has been reported[319] to give 5% *trans* product, and a trigonal-bipyramidal intermediate has been suggested. Synthetic studies[320] on $Ru(NH_3)_4(py)_2^{2+}$ and a study of the substitution reaction of *cis*-$Ru(bipy)_2X_2$ [321] (X = F, Cl, Br) confirm the general retention of configuration in Ru^{II} complexes. Intramolecular linkage isomerisation has been reported for the $Ru(NH_3)_5N_2^{2+}$ complex ion[322], and complete retention has been observed[293] in the acid hydrolysis of *cis*-$RuL_2X_2^+$ (L = en, pn, trien$_{\frac{1}{2}}$; X = Cl, Br, or I). It should be noted that the emergence of *trans* effects with simple ligands (in contrast to the situation with Co^{III}) suggests an increased role for the *trans* specific metal p-orbitals in σ-bonding in these complexes.

6.5.5 Base hydrolysis

The lower sensitivity of Rh^{III}- and Ir^{III}-complexes (in comparison to that of Co^{III} complexes) towards base hydrolysis has been considered[2], although the similarities between these complexes and those of Co^{III} have been less emphasised. Lalor and collaborators have reported extensive work on the base hydrolysis of the halogenopentammine complexes of Rh^{III} [323, 141, 324] and Ir^{III} [325]. The difference in activation energies between the Rh^{III} series and the corresponding Co^{III} series amounts to 1–4 kcal mol^{-1}, which is in fair agreement with the value of 5 kcal mol^{-1} predicted on the basis of crystal field effects. Iso-kinetic relationships (E_a versus log A) for both Rh^{III} and Co^{III} complexes, and a linear relationship between the activation energy and the frequency of the first ligand field bond, have been established. These indicate a similarity in the reaction mechanism. Deviations for the azido group in the isokinetic plots suggests that the solvent plays an increased role in the transition state. (It is important to point out, in this respect, that a strict comparison between the azido group and the halides may not be completely justified.) A linear relationship between the activation energies and the ligand field parameter Δ (or $10Dq$) for the base hydrolysis of $M(NH_3)_5I^{2+}$-type complexes (M = Cr, Co, Rh and Ir) has been reported[325], and it is interesting to note that a common mechanism seems to be operating in all these systems. This mechanism probably involves the labilisation by an amido ligand in all these systems, a process which is mechanistically similar in each case although its efficiency must differ from one system to another.

Results[326] on the base hydrolysis of *cis*-$Ru(en)_2X_2$ and $Ru(NH_3)_5X^{2+}$ (X = Cl, Br, I) fit the familiar conjugate-base mechanism, and the observed *trans* effect may be explained in terms of an effective amido–Ru $p\pi$–$d\pi$ interaction as well as in terms of p-effects. The base hydrolysis of *trans*-$Pt(CN)_4Br_2^{2-}$ is interesting in that it is not possible to invoke the familiar conjugate-base hypothesis due to the absence of amino group in the complex. The observed rate law shows two terms, one dependent on the OH^- concentration, and the other on the Br^- concentration. The large positive entropy for this reaction enables the hydroxide-dependent term to be attributed to a dissociative interchange process. The bromide-dependent term may be explained as due to bromide association with the complex, forming a species similar to a tribromide, which is consistent with the negative entropy of activation.

In conclusion, there can be little doubt that the larger crystal field splittings for the heavier metals must be related to the same factor as the low lability of complexes of these metals, and, similarly, there can be little doubt that bond-breaking is the most important single factor in determining the nature of the transition state in the reactions of these complexes. There are however, sufficient differences between the behaviours of the complexes of these metals and those of Co^{III} to imply some associative character in their substitution reactions, as well as a change in the character of the orbitals (to metal p, from metal d) which dominate the σ-bonding framework in these complexes.

6.6 NICKEL (II)

6.6.1 Introduction

Although the octahedral complexes of Ni^{II} are labile, they are not excessively so. Consequently most fast-reaction methods, from stopped flow to ultrasonic absorption techniques, are applicable, and information regarding their behaviour is now abundant, the literature survey undertaken for this review yielding about 25 citations per year. The result of this activity is the development of a situation in which subtle mechanistic questions are now being asked in conjunction with experimental work on all aspects of these mechanisms. This state of affairs is in sharp contrast to the state of knowledge of other labile systems, where the bulk of the data available is solely confined to that on the rates of mono-complex formation and of solvent exchange. Complex formation by Ni^{II} complexes has not, however, been overlooked. Thus, Hewkin and Prince[331] have tabulated more than 130 complex-formation and solvent-exchange rate constants for Ni^{II} systems, and one of the early and major contributors to the study of octahedral Ni^{II} systems, R. G. Wilkins, has contributed a recent review of mechanistic conclusions[323]. In this section, we have not attempted to provide a complete collection of data, but rather, to show (within the context of that described above) how recent work continues to support Wilkins' generalisation, i.e., . . . 'reactions . . . are dissociative in character, and . . . bond-breaking takes precedence over bond-making. However, there is strong support for further refinements'[323].

6.6.2 Leaving-group effects

The general observation[332] that rates of complex formation at Ni^{II} centres depend on the charge of the entering ligand, (although not very strongly), implies that a parallel exists between the rate of ligand dissociation and the overall equilibrium constant for complex formation, in a very similar fashion to the existence of linear free energy relationships in Co^{III} chemistry. Thus, leaving-group effects are those expected for dissociative activation, a partial leaving-group lability order being[2]: $SO_4^{2-} > H_2O > CH_3PO_4^{2-} > CH_3COO^- > F^- > NCS^- > C_5H_5N > NH_3$. (Puzzling cases are more conveniently examined from other points of view as is attempted below.) There are a few special features of leaving-group behaviour which have been identified in Ni^{II} chemistry and may be of a general nature. Recent examples include the observation that the hydroxyl side-chain of serine seems to be labilising[333], the observation that the increased stability of 5- over 6-membered rings is based on a slower rate of ring opening[334], and the observation that amino polycarboxylate ligands containing propionate, rather than acetate, 'claws' dissociate less rapidly[335].

6.6.3 Entering-group effects. Normal versus sterically-controlled substitution

The subtle differences between entering-group effects assume considerable importance in the mechanistic discussion of labile systems, since in these systems complex formation, generally involving the aquo group, invariably occurs. For such systems a detailed stoichiometric mechanism of the I type has usually been assumed, in most cases without the desired proof from a full concentration range analysis of the rate laws. In most cases, however, this procedure has provided remarkably successful kinetic schemes for the elucidation of their mechanisms. Such a scheme is depicted in equation (6.13).

$$\begin{aligned} Ni(OH_2)_6^{2+} + L &\overset{K_E}{\rightleftharpoons} Ni(OH_2)_6^{2+},L \\ Ni(OH_2)_6^{2+},L &\underset{k_{-1}}{\overset{k_1}{\rightleftharpoons}} Ni(OH_2)_5L + H_2O \end{aligned} \qquad (6.13)$$

In equation (6.13), K_E is the encounter equilibrium constant discussed in Section 6.2, which is usually calculated theoretically assuming a distance of closest approach of about 5 Å. At low reactant concentrations, the rate law is:

$$d[NiL]/dt = K_E k_1 [Ni^{2+}][L] - k_{-1}[NiL] \qquad (6.14)$$

and the experimentally determined constant is $K_E k_1$. The success of the analysis is indicated by the small variation in k_1 using calculated values of K_E. Table 6.11 contains representative examples.

Table 6.11 Selected rate constants for the formation of Ni^{II} complexes with unidentate ligands at 25 °C

L	$10^{-3}k(=K_E k_1)$ $(M^{-1}s^{-1})$	K_0 (*calc.*) M^{-1}	$10^{-4}k_1$ (s^{-1})	*Reference*
H_2O*	—	—	3.0*	336
$CH_3PO_4^{2-}$	290	40†	0.7	337
CH_3COO^-	100	3	3	334
SCN^-	6	1†	0.6	338
F^-	8	1‡	0.8	339
HF	3	0.15	2	339
NH_3	5	0.15	3	340
C_5H_5N (py)	~4	0.15	~3	341

*First-order rate constant for the exchange of a single water molecule obtained from n.m.r. relaxation-time measurements.
†Experimental value
‡Lower than K_E for CH_3COO^-, because experiments were conducted at higher ionic strengths.

A more complex kinetic scheme must be written if a chelating ligand is used, because more steps are involved before the final product is reached.

For a bidentate ligand, equation (6.13) may be extended to equation (6.15).

$$Ni(OH_2)_6^{2+} + L—L \overset{K_E}{\rightleftharpoons} Ni(OH_2)_6^{2+},L—L$$

$$Ni(OH_2)_6^{2+},L—L \underset{k_{-1}}{\overset{k_1}{\rightleftharpoons}} Ni(OH_2)_5—L—L + H_2O \qquad (6.15)$$

$$Ni(OH_2)_5—L—L \underset{k_{-2}}{\overset{k_2}{\rightleftharpoons}} (OH_2)_4Ni\langle L—L\rangle + H_2O$$

There is no reason to assume that the first two steps are different from those in equation (6.13), and if $k_2 > k_{-1}$, the observed formation rate constant will still be $K_E k_1$. This is called *normal* substitution, and has been recognised for a number of *strongly binding* bidentate ligands for some time[341–343]. The reason for the correlation of *normal* substitution with strong binding is clear. If k_{-1}/k_1 is small, and k_2 is similar to k_1, then $k_2 \gg k_{-1}$.

We can anticipate different kinetics when the first-formed bond is as labile as the *cis* water molecule. The steady state assumption for intermediates yields the result $K_E k_1 k_2/(k_{-1} + k_2)$ for the observed rate constant for complex formation. When k_2 is no longer much larger than k_{-1}, the observed rate constant will be *less than* $K_E k_1$. This appears to be the case for the reactions of $Ni(OH_2)_6^{2+}$ with malonate and other dicarboxylate ligands[334], with $(HOCH_2CH_2)_2N(CH_2)_2N^+H(CH_2CH_2OH)_2$ [344], $CH_3COCO{=}C(OH)CH_3$ [345], and perhaps with 1,4,8,11-tetra-azacyclotetradecane[346]. The behaviour is also observed when water-molecule labilities have been substantially enhanced by the coordination of a number of amine ligands[332, 348, 349]. (This labilisation will be considered below.) Lw formation rates for chelates because of frequent first-bond formation and rupture before the closure of the first ring is called *sterically-controlled* substitution.

Despite the wide range of successful interpretations of complex formation rates through the assumption of *d* activation and using the I pathways of equations (6.13) and (6.15), there remain some anomalous cases. Aliphatic diamines react too rapidly[347], as does tetramethyl-ethylenediamine[332]. This has been explained by suggesting hydrogen-bonding interactions corresponding to an *internal conjugate-base* mechanism[340]. $Ni(terpy)(H_2O)_3^{2+}$ has a water-exchange rate constant of 5×10^4 s^{-1} [350]. Its reactions with phen, (rate constant $6.0 \times 10^3 M^{-1}s^{-1}$), bipy, ($2.0 \times 10^4 M^{-1}s^{-1}$), and terpy, ($2.2 \times 10^5 M^{-1}s^{-1}$) [341], suggest either a certain amount of nucleophilicity or quite anomalous variations in K_E. Both D and I_a interpretations of the order are possible (see below).

Studies of more complex chelation processes requiring more elaborate mechanisms than those depicted in equations (6.13) or (6.15), can probably be interpreted without the introduction of new mechanistic concepts for the elementary processes, although it is still too early to be quite sure of this point. Examples of such studies now include: the reaction of $Ni(NTA)^-$ with cyclohexane-diaminetetra-acetate[351], $Ni(EDTA)^{2-}$ with calmagite[352], $Ni(NTA)^-$ and $Ni(EDTA)^{2-}$ with Eriochrome Black T [353], mono- and bis-diethylenetriamine-nickel(II) with EDTA [354], and Ni^{II} in glacial acetic acid with hematoporphyrin IX [355]. Amino acid ligands can chelate through the involvement of either one of two sequences. Application of equation (6.15) leads to the conclusion that attack on the nitrogen atom occurs first, since substitution is normal. An elegant example of the disentanglement of many steps in a basic mechanism is given in the work of Margerum *et al.*[356–358], where the kinetic role of three to four cyanides in the formation of $Ni(CN)_4^{2-}$ has been explored.

6.6.4 Non-labile ligand effects

At the outset it can be stated that the overall charge on the complex does not affect the rate of dissociation of a ligand, so that ligand effects may be discussed simply in terms of electronic and steric factors. The simplest theory of labilising power is one which suggests that the more strongly σ-donating a ligand the more stabilisation it confers to the essentially 5-coordinate d transition state. This picture is reasonably consistent with the labilising order $N_3^- > OH^- > CH_3COO^- > H_2O > F^-$ [359], and with some observations regarding ethylenediamine-diacetate- and NTA-complexes[360]. It also accounts for the labilising effects of an increase in the number of coordinated nitrogens in ammine and polyamine complexes[349, 361–363]. (The results of studies of labilising effects show consistency between complex formation[344, 363] and n.m.r. spectroscopic water-exchange studies[361, 362].) The small labilising effect of coordinated amino acids probably arises from the coordinated nitrogen atom[364]. (*N*-methylated sarcosine does not exhibit the effect[365].) The σ-donor theory should also be capable of accounting for rate acceleration by Cl^- and SCN^- ions[366]. An important limitation of the theoretical interpretation, however, is that most of the effects interpreted are rather *small*.

An exception to the simple picture of labilisation is encountered in complexes containing bipy, terpy, or phen[348, 350, 367]. These are 'strong' ligands but they have no obvious labilising effect. Perhaps electron donor properties are neutralised by π-back donation. Another non-simple case is reported by Hoffman and Yeager[368]. They find that malonate is labilised by py, 2-aminoethylpyridine, en, 3,3′-iminopropylamine, and acetylacetonate. Only small effects are produced by bipyridyl, phen, and pyridine-2-aldoxime. Unfortunately, however, no correlation between these exceptions and the corresponding spectral shifts has, as yet, been reported. An interesting, if somewhat esoteric system has been studied by Kluiber, Horrocks, and their collaborators[369, 370]. The exchange of bis-(4-picoline) and bis-(4-picoline-1-oxide) adducts of bis-(β-diketonato)-Ni^{II} and -Co^{II} complexes was studied as a function of alkyl substituents on the diketone. $\Delta H^{\ddagger}$ and log k values correlate with Taft σ^* parameters, as would be expected from the simple model.

6.6.5 Non-aqueous solvents

Table 6.12 contains a list of values of solvent-exchange rate constants obtained by n.m.r. spectroscopic methods. This list contains the basic data for the examination of any mechanism, since it provides a means of comparing complex formation rates with a view to seeing if the pathway suggested in equation (6.13) is consistent with d activation. Furthermore, the Table also contains results for exchange studies involving several solvents, and in fact

Table 6.12 Solvent-exchange rate constants at 25 °C, (k in s^{-1}, $\Delta H^{\ddagger}$ in kcal mol^{-1})

*Complex**	*Solvent*	*Nucleus*	log k	$\Delta H^{\ddagger}$	*Reference*
$Ni(OH_2)_6^{2+}$	H_2O	^{17}O	4.5,4.4	12.1	367, 336
$NiCl(OH_2)_5^{+}$	H_2O	^{17}O	5.1	8	366
$Ni(NCS)_4(OH_2)_2^{2-}$	H_2O	^{17}O	6.0	6	371
$NiNH_3(OH_2)_5^{2+}$	H_2O	^{17}O	5.4	11	361
$Ni(NH_3)_2(OH_2)_4^{2+}$	H_2O	^{17}O	5.8	8	361
$Ni(NH_3)_3(OH_2)_3^{2+}$	H_2O	^{17}O	6.4	10	361
$Ni(en)(OH_2)_4^{2+}$	H_2O	^{17}O	5.4	10	362
$Ni(en)_2(OH_2)_2^{2+}$	H_2O	^{17}O	6.5	9	362
$Ni(terpy)(OH_2)_3^{2+}$	H_2O	^{17}O	4.5	12	350
$Ni(bipy)(OH_2)_4^{2+}$	H_2O	^{17}O	4.7	12.6	367
$Ni(bipy)_2(OH_2)_2^{2+}$	H_2O	^{17}O	4.8	13.7	367
$Ni(dien)(OH_2)_3^{2+}$	H_2O	^{17}O	6.1	5.5	383
$Ni(trien)(OH_2)_2^{2+}$	H_2O	^{17}O	6.5	7	383
$Ni(tren)(OH_2)_2^{2+}$	H_2O	^{17}O	5.8	8	383
$Ni(EDTA)(OH_2)^{2-}$	H_2O	^{17}O	5.8	8.0	384
$Ni(HEDTA)(OH_2)^{-}$	H_2O	^{17}O	5.3	10	384
$Ni(NH_3)_6^{2+}$	NH_3	^{14}N	5.0,5.2	11,10	372
	NH_3	^{1}H	4.8,5.1	11,10	373,374
$Ni(CH_3OH)_6^{2+}$	CH_3OH	^{1}H	3.0	16	375
$Ni(CH_3CN)_6^{2+}$	CH_3CN	^{1}H	3.4,3.6	12,11	376,377
$Ni(DMF)_6^{2+}$	DMF	^{1}H	3.5	15	378
$Ni(DMF)_6^{2+}$	DMF	^{17}O	3.8	—	379
$Ni(DMSO)_6^{2+}$	DMSO	^{1}H	3.8,3.5	8,13	380,381
$Ni(TMPA)_6^{2+}$	TMPA	^{1}H			382
$Ni(C_2H_5OH)_6^{2+}$	C_2H_5OH	^{1}H	4.0	11	385
$Co(OH_2)_6^{2+}$	H_2O	^{1}H	6.4	9	387
$Co(NCS)(OH_2)_5^{2+}$	H_2O	^{17}O	7.0	11	388
$CoCl(OH_2)_5^{+}$	H_2O	^{1}H	7.2	14	387
$CoNH_3(OH_2)_5^{2+}$	H_2O	^{17}O	7.2	13	389
$Co(NH_3)_2(OH_2)_4^{2+}$	H_2O	^{17}O	7.8	9	389
$Co(mal)(OH_2)_4$	H_2O	^{17}O	7.3	13	389
$Co(mal)_2(OH_2)_2^{2-}$	H_2O	^{17}O	>8	—	389
$Co(CH_3OH)_6^{2+}$	CH_3OH	^{1}H	4.3	14	375
$CoCl(CH_3OH)_5^{+}$	CH_3OH	^{1}H	7.0	12	390
$Co(DMF)_6^{2+}$	DMF	^{17}O	5.3	—	379
	DMF	^{1}H	5.5	14	378
$Co(CH_3CN)_6^{2+}$	CH_3CN	^{1}H	5.2	8	377
$Co(DMSO)_6^{2+}$	DMSO	^{1}H	5.2	10	381
$Co(NH_3)_6^{2+}$	NH_3	^{14}N	6.9	11	391
$Mn(OH_2)_6^{2+}$	H_2O	^{17}O	6.8	9	392
$Mn(phen)(OH_2)_4^{2+}$	H_2O	^{17}O	7.1	9	392
$Mn(phen)_2(OH_2)_2^{2+}$	H_2O	^{17}O	7.5	9	392
$Mn(NH_3)_6^{2+}$	NH_3	^{14}N	7.6	8	393
$Mn(DMF)_6^{2+}$	DMF	^{17}O	6.6	—	394
$Mn(CH_3OH)_6^{2+}$	CH_3OH	e.s.r.	6.0	7	395
$Mn(CH_3CN)_6^{2+}$	CH_3CN	^{1}H	7.1	7	396
$Fe(OH_2)_6^{2+}$	H_2O	^{17}O	6.5	8	336
$Fe(DMF)_6^{2+}$	DMF	^{17}O	4.3		394
$Fe(OH_2)_6^{3+}$	H_2O	^{17}O	4.3		336
$Fe(C_2H_5OH)_6^{3+}$	C_2H_5OH	^{1}H	4.3	6	385
$Fe(DMF)_6^{3+}$	DMF	^{1}H	1.5	13	385
$Fe(DMSO)_6^{3+}$	DMSO	^{1}H	1.7	10	397
$Fe(CH_3CN)_6^{3+}$	CH_3CN	^{1}H	<1.6		385
$Fe(CH_3OH)_6^{3+}$	CH_3OH	^{1}H	4.3		385
$Fe(OH_2)_5OH^{2+}$	H_2O	^{17}O	~6		336

Table 6.12—Continued

*Complex**	*Solvent*	*Nucleus*	log *k*	$\Delta H^{\ddagger}$	*Reference*
$FeCl_3(OH_2)_3$	H_2O	^{17}O	5.3	10	398
$Mg(OH_2)_6^{2+}$	H_2O	^{17}O	5.7	10	400
$Mg(OH_2)_6^{2+}$	H_2O/acetone		5.1	8	399
$Mg(CH_3OH)_6^{2+}$	CH_3OH/ acetone	^{17}O	4.4	12	399
$Al(OH_2)_6^{3+}$	H_2O	^{17}O	−0.8	27	401
$Al(DMSO)_6^{3+}$	DMSO	^{1}H	−1.3	20	402
$Al(DMF)_6^{3+}$	DMF	^{1}H	−0.7	18	403
$Ga(OH_2)_6^{3+}$	H_2O	^{17}O	3.3	6	401
$Ga(DMF)_6^{3+}$	DMF	^{1}H	0.2	15	401

*In some cases, 6-coordination and octahedral geometry is assumed without any definite evidence as to the species in solution.

the solvent-exchange reactions of NiS_6^{2+} with eight solvents are recorded. A certain amount of disappointment has been generated by the fact that the rates in these solvents are so similar. The situation is difficult to unravel because changing from one medium to another simultaneously changes all of the following factors; (1) the leaving group, (2) the entering group, (3) non-labile ligands and (4) the solvent medium. Early workers hoped that leaving-group effects would be dominant in these processes, and provide support for a dissociative mechanism in *all* solvents. This is not the case. Langford and Stengle[3] have recently suggested that in the *d* mechanism a balance occurs between factors (1) and (3) listed above. Thus, a good σ-donor ligand would be *both* a good labilising group and a poor leaving group, so that an increase in lability in the presence of poorer donor solvents would be compensated by replacing five good labilising ligands with five poor labilising ligands. The recent studies by Hunt, Dodgen, *et al.* on the water-exchange of mixed aquo–ammine complexes *in water* show the application of this suggestion. Using the correlation presented by Hunt[386, 361] it is possible to infer that five NH_3 groups labilise one water molecule by a factor of 10^3 in comparison with the effect of five OH_2 groups. Thus, the NH_3-exchange reaction rate of 10^5 s^{-1} for $Ni(NH_3)_6^{2+}$ in ammonia should correspond to a rate of 10^2 s^{-1} were it not for the labilisation produced by the five remaining NH_3 ligands, (using the doubtful assumption that NH_3 and H_2O, when involved as leaving groups, respond in an equivalent manner to substituent effects). The data of Margerum *et al.*[349] on the loss of NH_3 (i.e., NH_3 acting as a leaving group) from ammine complexes in water suggest that the factor is 10^4 not 10^3, although a complete correlation cannot be achieved. If this is accepted which means that NH_3 is now more sensitive to substituent effects than H_2O) the rate of ammonia exchange should be corrected for substituent effects to yield a rate constant of 10 s^{-1}. This may then be compared to the rate of ammonia loss from $(Ni(OH_2)_5NH_3^{2+}$ [349] in water which has a value of ~ 3 s^{-1}. Thus, on this basis the role of factors in these reactions is small.

A different view has been put forward by Bennetto and Caldin[406] who have reported the ratios of the rates of complex formation with bipyridine to solvent exchange in several solvents. They have suggested that solvent 'fluidity' is an important variable. That this is not important in the solvent-exchange process itself is shown by Frankel's[381] data on the reaction of $Ni(DMSO)_6^{2+}$ in mixtures of DMSO with CH_3NO_2. It is interesting to note that there are differences in the solvent-dependence of the complex formation reactions of Ni^{II} with 2,2′-bipyridine and 1,10-phenanthroline[409] in methanol–water mixtures.

6.6.6 The question of the D intermediate

Although the I_d pathway was assumed for the discussion in Section 6.5.3, there are few systems for which kinetics over a sufficiently wide range of conditions have been obtained to provide direct support for the pathway. What is needed are values for the rate constant k_1 of equation (6.13), either obtained directly or inferred from experimental values of K_E. These could then be compared with the solvent-exchange rates recorded in Table 6.12. The direct evaluation of k_1 is possible through the use of ultrasonic absorption kinetic experiments. This has been done for M^{II} sulphates by Eigen and Tamm[404]. The corresponding value for $Ni(OH_2)_6^{2+}$ was found to be 1.5×10^4 s^{-1}. Brintzinger and Hammes[337] were able to fit a value of K_E to the curvature in the plot for the ligand concentration versus the observed complex formation rate constant for $CH_3OPO_3^{2-}$ as a ligand, from which a k_1 value of 0.7×10^4 s^{-1} was obtained. These values are well below the solvent-exchange rate constant value of 2.7×10^4 s^{-1} and are consistent with an I_d pathway for $Ni(OH_2)_6^{2+}$. (See Section 6.3.3.) A recent careful analysis of pressure-jump data for complex formation in methanol specifically supports the I_d pathway[405].

A few systems exhibit anomalies that may reflect a change in the stoichiometric mechanism. Thus, Bennetto and Caldin[406] report a reduction in the apparent ratio of complex-formation rates to solvent-exchange rates in polar aprotic solvents which might be explained by the assumption of a 5-coordinate intermediate which reacted preferentially with the solvent. This idea is supported by Frankel's observation[381] that the solvent-exchange rates in DMSO—CH_3NO_2 mixtures are constant. He proposed $Ni(DMSO)_5^{2+}$ as an intermediate in this case. Frankel's proposal has been supported by a study of the simple complex-formation reaction of $Ni(DMSO)_6^{2+}$ with SCN^- in DMSO and DMSO–CH_3NO_2 mixtures[407]. The observed kinetic patterns were consistent with the existence of an intermediate which recombined faster with DMSO than SCN^-, but not with CH_3NO_2. It is interesting to note that the anomalies that have led to a discussion of a D intermediate in non-aqueous media are similar to the anomalies reported for the reactions of $Ni(terpy)(OH_2)_3^{2+}$ in water[350].

6.7 OTHER DIVALENT METALS

6.7.1 Introduction

The remaining divalent metals for which significant data exist include Co^{II}, Fe^{II}, Mn^{II}, V^{II}, Cr^{II}, Cu^{II}, Zn^{II}, Cd^{II}, Mg^{II}, Ca^{II}, Sr^{II} and Pt^{II}. For the same reason that the non-labile Pt^{II} square-planar complexes fall outside the scope of the review, the extremely labile Cu^{II} and Cr^{II} systems require only brief note. They are both subject to a strong tetragonal distortion so that they are not octahedral. (This is predicted from the Jahn–Teller theorem for d^4 and d^9 configurations.) Their extreme lability is assumed to be a consequence of the rapid exchange of the axial ligands[7]. One interesting point is that, for example, all the waters of ligands $Cu(OH_2)_6^{2+}$ exchange at the same rate[410]. This must mean that an axial–equatorial interchange occurs faster than the nearly diffusion-limited loss of axial water. Such a picture suggests that changes in lability occur when restrictive chelating ligands occupy equatorial positions[411, 412].

Amino acid complexes of Cu^{II} have been fully studied; with these complexes steric-controlled substitution is sometimes observed[364, 409]. Internal hydrolysis is also apparent in Cu^{II} chemistry as is evidenced by rapid reactions with en ligands[413]. Thus, it appears that Cu^{II} is extremely labile as a result of its non-octahedral geometry, but has reactivity patterns similar to those of Ni^{II}, and presumably the same will hold true for Cr^{II}.

The geometry of Zn^{II} and Cd^{II} complexes in solution is not really known although they are normally tetrahedral in the solid state. The limited kinetic studies available indicate that their reactions may be parallel to but faster than Ni^{II} reactions[7]. Ca^{II}- and Sr^{II}-complexes are extremely labile, although the rate of complex formation is not easily distinguishable from the diffusion limit[7]. Data on V^{II} complexes are quite limited[414–416], but those that exist are consistent with the Ni^{II}-like behaviour of these complexes with a water-exchange rate between 10^1 and 10^2 s^{-1}. The complexes of the remaining metals, namely Co^{II}, Fe^{II}, Mn^{II}, and Mg^{II} are better understood and require detailed treatment.

6.7.2 Cobalt(II)

In Table 6.13 a selection of older rate constants for the complex-formation reactions of $Co(OH_2)_6^{2+}$ are quoted (see References 7 and 422 for others) together with a number of recent values. This Table when compared with Tables 6.11 and 6.12 shows that the behaviour of Co^{II} is similar to that of Ni^{II} but the reaction rates are somewhat faster. Most of the rate constants in Table 6.13 can be related to the solvent-exchange rate constants in Table 6.12 assuming the I_d path of equation 6.13, but there are some 'slow' reactions exemplifying 'sterically controlled substitution'[355]. Tables 6.12 and 6.13 show that labilising effects of ligands on Co^{II} are qualitatively similar to those on Ni^{II}. In the cases of $M(OH_2)_6^{2+}$, $M(OH_2)_5Cl^+$, $M(OH_2)_5NH_3^{2+}$, and $M(OH_2)_4(NH_3)_2^{2+}$ the rate constant ratios are also quantitatively similar[386], i.e., $\delta\Delta G^\ddagger$ for substitution is approximately constant. This is a little surprising, for one might have expected a decrease in $\delta\Delta G^\ddagger$ as the system becomes more labile ($\Delta G^\ddagger$ decreases).

Some examples of behaviour more or less unique to Co^{II} systems include the clear demonstration of the internal conjugate-base effect in reactions with polyamino-alcohols[423], pyridine-exchange in $Co(py)_4Cl_2$ through tetrahedral $Co(py)_2Cl_2$ [424] (5-coordinate $Co(py)_3Cl_2$ is postulated as an intermediate), and suggestive evidence for a 7-coordinate species in CN^-

attack on the 1,2-diaminocyclohexane-tetra-acetato-Co^{II} ion[425]. An example of the value of substitution-rate studies is the application in O_2-oxidation studies[418, 426, 427]. A 5-coordinate species $Co(CN)_5^{3-}$ is suggested as participating in the decomposition pathway of $Co(CN)_5OH_2^{3-}$ [428].

Table 6.13 Complex-formation rate constants for Co k at 25 °C, (^{II} in $M^{-1}s^{-1}$), $\Delta H^{\ddagger}$ (in kcal mol^{-1})

Solvo complex	*Ligand*	log k	$\Delta H^{\ddagger}$	*Reference*
$Co(OH_2)_6^{2+}$	SO_4^{2-}	6.5	—	422
$Co(OH_2)_6^{2+}$	malonate^{2-}	7.0	—	422
$Co(OH_2)_6^{2+}$	NH_3	5.0	6	340
$Co(OH_2)_6^{2+}$	imidazole	5.0	—	422
$Co(OH_2)_6^{2+}$	picolinic acid	4.4	—	417
$Co(OH_2)_4$picolinate^{+}	picolinic acid	5.1	—	417
$Co(NH_3)_5OH_2^{2+}$	O_2	4.4	4	418
$Co(OH_2)_6^{2+}$	gly^{-}	6.2	—	364
$Co(OH_2)_4$gly^{+}	gly^{-}	6.3	—	364
$Co(OH_2)_6^{2+}$	ser^{-}	6.3	—	364
$Co(OH_2)_6^{2+}$	sarcosine^{-}	6.0	—	364
$Co(OH_2)_6^{2+}$	leucine^{-}	6.1	—	364
$Co(OH_2)_6^{2+}$	argenine^{0}	5.2	—	419
$Co(OH_2)_4$argenine^{2+}	argenine^{0}	5.9	—	419
$Co(OH_2)_6^{2+}$	carnosine^{0}	5.6	—	420
$Co(OH_2)_6^{2+}$	α-aminobutyric acid	5.4	—	345
$Co(OH_2)_6^{2+}$	β-aminobutyric acid	4.3	—	345
$Co(OH_2)_6^{2+}$	α-alanine	5.8	—	345
$Co(OH_2)_6^{2+}$	β-alanine	4.9	—	345
$Co(OH_2)_6^{2+}$	glycylsarcosine	5.7	—	420

6.7.3 Iron(II)

There are two types of d^6 iron(II) chemistry. Low-spin complexes are related to the low-spin d^6 Co^{III} systems and show a somewhat similar low lability. Detailed kinetic analysis is scarce but there has been some suggestion that these reactions are dissociative in character. High-spin complexes are labile and resemble Ni^{II}.

To complement the two solvent-exchange rate constants in Table 6.14, the rate constants for complex formation between $Fe(OH_2)_6^{2+}$ (high-spin) and NO [429], bipy [341], and 2,4,6-tripyridyl-s-triazine [430] have the values 5×10^5, 1.5×10^5, and 1.3×10^5 $M^{-1}s^{-1}$ respectively. These values are consistent with the I_d model. The labilising effects of substituents cannot be assessed.

A typical low-spin Fe^{II} complex is the tris-(1,10-phenanthroline)Fe^{II} ion. Burgess has examined the dissociation of this and related complexes in aqueous acid and mixed alcoholic solvents[431–433]. The results are consistent with dissociative mechanisms, except that an $[OH^-]$-dependent term is reported[431]. Solvent effects in mixed solvents seem to be mainly the result of a variation in ground-state solvation[432]. Jones and Twigg[434] have described substitution in the axial position of ferrous phthalocyanine. Imidazole appears to have a powerful *trans* effect.

6.7.4 Manganese(II)

The d^5 Mn^{II} system is normally high-spin and has no crystal field stabilisation energy. It is more labile than any other transition metal ion with the exception of the distorted Cu^{II} and Cr^{II} systems. The sparse data recorded in Table 6.12 suggest that this system is less sensitive to the labilising effects of substituents than either the Co^{II} or Ni^{II} systems. This is supported by complex formation studies[435]. Representative complex-formation rate constants at 25 °C are as follows: $Mn(OH_2)_6^{2+}$ with the 8-hydroxyquinolinate anion 1.1×10^8 $M^{-1}s^{-1}$ [435], with β-alanine 5×10^4 $M^{-1}s^{-1}$ (cf. Table 6.13)[436], with Cl^- 6×10^6 $M^{-1}s^{-1}$ [437] and with nitrilotriacetate 5×10^8 $M^{-1}s^{-1}$ [438].

More complicated reactions such as those of *trans*-1,2-cyclohexane-diaminetetra-acetate-Mn^{II} with Co^{II} and Cu^{II} have been examined[439]. The reaction of Mn^{II} with hematoporphyrin has been studied[440]. This involves substitution and oxidation to Mn^{III}.

6.7.5 Magnesium(II)

Magnesium(II) is the least labile of the divalent octahedral non-transition metal ions. (It has only recently been clearly shown to be 6-coordinate in solution[399].) This low lability is presumably attributable to the large value of the charge to radius ratio for the ion. (The difference in lability of Mg^{II} and Ca^{II} has a probable biochemical significance[441].) The water-exchange rate (Table 6.12) of 5.3×10^5 s^{-1} at 25 °C compares favourably with the value estimated some time ago from several complex formation rates, which were interpreted assuming the I_d pathway[404]. A recent result is the measurement of the rate of formation of Mg^{II} oxalate (6×10^6 $M^{-1}s^{-1}$, $\Delta H^\ddagger = 9$ kcal mol^{-1}) [442]. All of these results are consistent with dissociative pathways. Snellgrove and Plane have reported the acid-displacement of Mg^{2+} from porphyrin in ethanol and methanol–water mixtures[443]. The rate of this process has a complex acid-dependence and parallels the coordinating power of the solvent.

6.7.6 Lability order among divalent metal ions

Outside the transition metal series, the order of labilities of metal ions may be explained on a qualitative basis in terms of the charge to radius ratio for the metal ion which would be the obvious parameter in an electrostatic theory of dissociative substitution[6]. Again, qualitatively, the electrostatic model may be extended to transition metal ions if ligand field corrections are considered[2]. However, calculations of relative lability on the basis of the crystal field theory have tended either to over-emphasise differences amongst metals or to insist on more bond-formation with the entering ligand than the limited experimental sensitivity to entering group variation seems to warrant. The emergence of data on labilising effects of substituents suggests a resolution of this minor paradox.

Crystal field calculations of lability require the assignment of a Dq value to both the ground state and the transition state. This value can only be obtained empirically from ground-state spectra, and it is assigned to both ground state and transition state in simple (tractable) models. Such models lead to either an over-estimation of the activation energies or to transition state structures with too little bond-breaking to be consistent with experiment. If Dq were larger in the transition state than in the ground state, agreement between theory and experiment would be improved. Table 6.12 suggests that ligands with stronger field strengths are good labilising groups. If this is because the remaining ligands bind more strongly in the transition state after one ligand has departed, then larger Dq values should be chosen for the transition state than for the ground state, and the discrepancy between crystal field models and experimental results could be substantially reduced. This is, of course, an explanation of a discrepancy and not a suggestion for a new crystal field calculation. The factor which changes Dq is principally increased σ-bonding by the remaining ligands and for this reason a molecular orbital approach may be more fruitful than a crystal field calculation.

6.8 LABILE TRIVALENT IONS

6.8.1 Introduction

Trivalent ions have already been discussed above whenever their rates of substitution fell into the range amenable to classical rate studies. Several remain, however, which are too labile for classical rate techniques, although even here some cases of fairly 'slow' fast reactions are discussed. The system which has been best studied is that of octahedral Fe^{III}. V^{III} is also an interesting case because it has an empty t_{2g} orbital. Data has also been collected on species such as Al^{III}, Ga^{III}, Sc^{III}, and the lanthanides.

6.8.2 Iron (III)

The main difficulty in the interpretation of complex formation rates of octahedral aquo Fe^{III} complexes is the 'proton ambiguity' introduced by the acidity of the H_2O ligands coordinated to Fe^{III}, and the basicity of many other ligands. The two pathways shown in equations (6.16) and (6.17) are indistinguishable by rate law analysis because they have equivalent transition-state stoichiometries.

$$\begin{aligned} Fe(OH_2)_6^{3+} + X^- &\rightleftharpoons Fe(OH_2)_6^{3+},X^- \\ Fe(OH_2)_6^{3+},X^- &\rightarrow Fe(OH_2)_5X + H_2O \end{aligned} \tag{6.16}$$

$$\begin{aligned} Fe(OH_2)_6^{3+} + X^- &\rightleftharpoons Fe(OH)(OH_2)_5^{2+},HX \\ Fe(OH)(OH_2)_5,HX &\rightarrow Fe(OH_2)_5X + H_2O \end{aligned} \tag{6.17}$$

Equation (6.16) represents the conventional interchange pathway. Equation (6.17) represents an interchange reaction between the $Fe(OH)(OH_2)_5^{2+}$ complex (labilised by the presence of a hydroxyl group) and the conjugate acid of X^-. Now, it is found that the rates of reactions of $Fe(OH_2)_6^{3+}$ with anions of strong acids increase with decreasing $[H^+]$ confirming the labilising effect of OH^-, but some rates of reaction with anions of weak acids *decrease* with decreasing $[H^+]$ in the acid region[445] suggesting a pathway involving HX. Thus, a choice must be made between pathways (6.16) and (6.17) for the acid-independent terms in the rate equation, for the acid-dependent behaviour discussed above shows that pathway (6.17) cannot be ignored. For anions of moderate to weak acids, the observed acid-independent second-order rate constant increases with increasing basicity of the anion. It appears that a more or less constant rate of complex formation consistent with the common I_d mechanism is obtained if pathway (6.17) is assumed. The rate so obtained is near 10^4 $M^{-1}s^{-1}$ [445,446]. Reactions with anions of strong acids have rate constants near 10 $M^{-1}s^{-1}$, and hence the most plausible account seems to be that I_d reactions occur in all cases with the reactant changing from Fe $(OH_2)_6^{3+}$ to $Fe(OH_2)_5OH^{2+}$ as the ligand basicity increases.

Table 6.14 Rates of complex formation in Fe^{III} systems at 25 °C (*k* in M^{-1} s^{-1}, Δ ‡ in kcal mol^{-1})

Complex	*Ligand*	log *k*	$\Delta H^{\ddagger}$	*Reference*
$Fe(OH_2)_6^3$	Br^-	−0.5*	—	450
$Fe(OH_2)_6^{3+}$	HSO_4^-	1.6	—	451
$Fe(OH_2)_5OH^{2+}$	SO_4^{2-}	5.0	16	451,452
$Fe(OH_2)_6^{3+}$	sulphosalicylic acid	0.2	—	453
$Fe(OH_2)_5OH^{2+}$	sulphosalicylic acid	4.1	—	453
$Fe(OH_2)_5OH^{2+}$	HPO_2^-	4.3	—	454
$Fe(OH_2)_6^{3+}$	H_2PO_2	1.1	—	454
$Fe(OH_2)_6^{3+}$	mandelic acid	0.6	—	455
$Fe(OH_2)_5OH^{2+}$	mandelic acid	3.4	6	455
$Fe(OH_2)_6^{3+}$	*p*-NO_2 phenol	1.5	—	456
$Fe(OH_2)_5OH^{2+}$	*p*-NO_2 phenol	3.1	—	456
$Fe(DMSO)_6^{3+}$	SCN^-	2.8	—	397
$Fe(DMSO)_6^{3+}$	sulphosalicylic acid	1.4	—	397

* Temperature = 1.6 °C

Nothing in the recent literature is in conflict with an I_d interpretation as is illustrated by recent discussions of entering ligands ranging from Br^- [447] to $HCrO_4^-$ [448] or phenol[449]. A study of $Fe(DMSO)_6^{3+}$, which is not subject to a 'proton ambiguity', also supports an I_d pathway[397].

A linear free energy relation between rates of aquation and Hammett σ parameters is observed for Fe^{III} complexes of substituted phenols[449]. This is a leaving group variation of the sort expected in *d* reactions. Table 6.14 contains a summary of recent results.

An observed second-order rate constant for a reaction described phenomenologically as between $Fe(OH_2)_6^{3+}$ and X^- may be used to calculate a rate constant for the pathway in equation (6.17), if the acidity constant of HX is known. In Table 6.14 this rate constant is reported where the data clearly warrant its inclusion. Collections of earlier rate constants appear in Basolo and Pearson[2] and Hewkin and Prince[7].

A number of Fe^{III} systems, other than $Fe(OH_2)_6^{3+}$, have been studied recently. The conversion of $Fe(CN)_6^{3-}$ to $Fe(CN)_5OH_2^{2-}$ is subject to a base-catalysed substitution by substituted acetophenones. The rate is increased by electron-withdrawing groups in the acetophenone[458]. The kinetics of the binding of imidazole to ferriprotporphyrin IX suggest an interchange mechanism[459]. Considerable attention has been directed to the kinetics of the formation of bridged binuclear complexes. Bridging ligands include the oxo ligand of the Fe^{III}–EDTA dimer[460, 461], the hydroxo bridge[462, 463], xylenol orange[464] and cyanide[465]. In these reactions, one centre seems to lose water in a rate-determining dissociative step.

6.8.3 Vanadium(III)

The vanadium(III) complexes of the d^2 configuration were considered quite early on as possibly subject to nucleophilic attack and with an associative mechanism of substitution, because of the empty orbital in the t_{2g} sub-shell. Quantitative data having a bearing on this possibility have only recently appeared. Support for the associative pathway is obtained from the quite negative values of $\Delta S^{\ddagger}$ found for reactions of $V(OH_2)_6^{3+}$ with SCN^- [466, 415], and from the comparison of the reactivity of N_3^- toward $V(OH_2)_6^{3+}$ and other metal ions[467]. Large negative values of $\Delta S^{\ddagger}$ were also found in studies of the reactions of $V(DMSO)_6^{3+}$ [468]. However, the span of the rates of reaction of $V(DMSO)_6^{3+}$ with three quite different ligands is similar to the pattern for a dissociative substitution[468], a result which casts some doubt on the interpretation of a few activation parameters.

6.8.4 Manganese(III), Aluminium(III), and Gallium(III)

There have been interesting recent studies on each of the systems discussed in this sub-section. Diebler has suggested that complex formation by Ti^{III} and Mn^{III} is of the normal dissociative type[469]. A Mn^{III} oxo-bridged dimer is found to be formed relatively slowly[470]. Al^{III} complexes show reactivity similar to Fe^{III} complexes[401, 359, 360]. Reactivity of SO_4^{2-} toward Ga^{III} and In^{III} seems too slow for an I_d pathway and an associative character in substitutions at these centres has been suggested[472]. This is consistent with the observation that solvent-exchange at Al^{III} is very similar in DMF and H_2O but that DMF exchanges 10^3-times slower than H_2O at Ga^{III} centres[401]. It is interesting to note that a study of dimerisation kinetics of $Ga(OH_2)_5OH^{2+}$ concludes with the remark, 'unlike $FeOH^{2+}$ . . . dimerisation, persuasive evidence for rate-determining inner coordination sphere water-loss in the dimerisation of $Ga(OH_2)_5OH^{2+}$ is lacking'[473].

6.8.5 Scandium(III) and the lanthanides

Scandium(III) and the lanthanide ions are all extremely labile, a possible result of the ready expansion of the coordination number of these ions beyond 6. In any event, the extreme lability of the systems limits the possible methods of investigation. Geier has successfully studied complexation by murexide using temperature-jump techniques[474] and Marianelli has utilised the favourable magnetic properties of Gd^{III} to obtain a solvent-exchange rate (9×10^8 s^{-1}) from ^{17}O n.m.r. spectroscopic studies[475]. Ultrasonic absorption studies of NO_3^- [476] and SO_4^{2-} [477] complexation have been undertaken, pressure-jump techniques have been applied to oxalate complexation[479], and a temperature-jump study of complexation of lanthanides with the *o*-aminobenzoate ion has appeared[478]. A comparison between murexide and *o*-aminobenzoate complex formation shows some common features which may be interpreted, as suggested by Geier, in terms of the structural properties of the lanthanide ions. If an expansion of the coordination sphere is involved, the reactions must have some associative character. Some authors have interpreted the correspondence among various rates as indicative of a dissociative character, but near the diffusion control limit there is little room for much change in rate with variation in ligand so that this test for a dissociative character becomes virtually meaningless.

6.8.6 Patterns of metal ion reactivity

The fact that complex-formation (and solvent-exchange) rates are similar for a particular octahedral metal ion allows a systematic organisation of the data for that particular ion and enables an attempt to be made at a theoretical interpretation of the behaviour of the ion in terms of its properties. This is a consequence of the overall predominance of dissociative activation, and its use as one of the most powerful generalisations in solution kinetics. Its range as a generalisation is, perhaps, only exceeded by the range of metal ions for which octahedral coordination geometry is common. This last point is important to remember. There are octahedral systems for which definite evidence of specific participation by the attacking ligand exists. These systems are either those of large ionic radius (or alternatively high atomic number and extended valence shell orbitals) or those with few electrons in the d sub-shell. (It is noteworthy that d^0 systems, like Al^{III} and Mg^{II} systems, do not show indications of associative reaction.) The main characteristic patterns of the reactivity of metal ions which follow from an application of the dissociative mechanism may be qualitatively explained in terms of the ionic model notion of charge to radius ratio, after correction for ligand field effects. This, of course, does not necessarily imply an electrostatic picture for these complexes, and predictions based on a molecular orbital analysis of σ-bonding should be qualitatively similar. In fact, where sufficient detail (kinetically and spectroscopically) is available (namely, Co^{III}, Cr^{III}, and Ni^{II}) molecular orbital theory seems the more powerful treatment, although quantitatively not the most simple to interpret. Many points in this review, and even more in the literature, indicate many questions that may be answered by future research. In fairness, it may be stated that many questions about octahedral substitution have already been answered, and that a real opportunity exists to apply these answers to a better understanding of inorganic chemistry in general, and of the processes which occur in electron-transfer reactions and metal photochemistry, in metal biochemistry, in metal geochemical process such as weathering, and of the pathways taken by metals in our environment in particular.

Acknowledgments

We thank Professors J. H. Espenson, R. L. Gay, J. P. Hunt, T. W. Swaddle and R. Wilkins for providing preprints of publications. Barbara Dunlop has earned our gratitude for her assistance in the preparation of the manuscript. It is a pleasure to acknowledge once again the financial support of our work by the National Research Council of Canada.

References

1. Literature for this review was covered up to December 15, 1970 in all journals promptly available to us.
2. Basolo, F. and Pearson, R. H. (1967). *Mechanisms of Inorganic Reactions,* 2nd edn. (New York: John Wiley and Sons)
3. Langford, C. H. and Stengle, T. R. (1968). *Ann. Rev. Phys. Chem.,* **19,** 193
4. Archer, R. D. (1969). *Coord. Chem. Rev.,* **4,** 243
5. Basolo, F. (1967). *Advan. Chem. Ser.,* **62,** 402
6. McAulay, A. and Hill, J. (1969). *Quart. Rev. Chem. Soc.,* **23,** 18
7. Hewkin, D. J. and Prince, R. H. (1970). *Coord. Chem. Rev.,* **5,** 75
8. Amdur, I. and Hammes, G. G. (1966). *Chemical Kinetics,* Chapter 2, (New York: McGraw-Hill)
9. Langford, C. H. (1969). *J. Chem. Educ.* **46,** 557
10. Langford, C. H. and Gray, H. B. (1965). *Ligand Substitution Processes,* Chapter 2, (New York: W. A. Benjamin)
11. Beck, M. T. (1968). *Coord. Chem. Rev.,* **3,** 119
12. Watts, D. W. (1968). *Rec. Chem. Progr.,* **29,** 131
13. Haim, A. and Wilmarth, W. K. (1962). *Inorg. Chem.,* **1,** 573, 583
14. Langford, C. H. and Muir, W. R. (1967). *J. Amer. Chem. Soc.,* **89,** 3141
15. Werner, A. (1912). *Ber.,* **45,** 121
16. Fuoss, R. M. (1958). *J. Amer. Chem. Soc.,* **80,** 5059
17. Eigen, M. (1963). *Pure Appl. Chem.,* **6,** 97
18. Hughes, M. N. and Tobe, M. L. (1965). *J. Chem. Soc.,* 1204
19. Lantzke, I. A. and Watts, D. W. (1966). *Aust. J. Chem.,* **19,** 969
20. Millen, W. A. and Watts, D. W. (1966). *Aust. J. Chem.,* **19,** 43, 51
21. Muir, W. R. and Langford, C. H. (1968). *Inorg. Chem.,* **7,** 1032
22. Duffy, N. V. and Earley, J. E. (1967). *J. Amer. Chem. Soc.,* **89,** 272
23. Kelm, H. and Harris, G. M. (1967). *Inorg. Chem.,* **6,** 706
24. Larson, R. and Johansson, L. (1965). *Proc. Symp. Coord. Chem.,* **43.** (Tihany, Hungary, Akadémai Kiado, Budapest)
25. Stengle, T. R. and Langford, C. H. (1965). *J. Phys. Chem.,* **69,** 3299
26. Masterson, W. L. (1967). *J. Phys. Chem.,* **71,** 2885
27. Masterson, W. L., Munnelly, T. L. and Berka, L. H. (1967). *J. Phys. Chem.,* **71,** 942
28. Tamm, K. (1970). *J. Chem. Phys.,* **52,** 2975
29. Elder, A. and Petrucci, S. (1970). *Inorg. Chem.,* **9,** 19
30. Henney, R. C. (1969). *Inorg. Chem.,* **8,** 389
31. Langford, C. H. (1965). *Inorg. Chem.,* **4,** 265
32. Basolo, F. and Pearson, R. G. (1967). *Mechanisms of Inorganic Reactions,* 2nd ed. (New York: John Wiley and Sons)
33. Jones, W. E., Jordan, R. B. and Swaddle, T. W. (1969). *Inorg. Chem.,* **8,** 2504

34. Swaddle, T. W. and Guastalla, G. (1969). *Inorg. Chem.*, **8,** 1804
35. Swaddle, T. W. and Jones, W. E. (1970). *Can. J. Chem.*, **48,** 1054
36. Jones, W. E. (1969). *Ph. D. Thesis,* University of Calgary, Calgary, Alberta
37. Jones, W. E. and Swaddle, T. W. (1969). *Chem. Commun.*, 998
38. Haim, A. (1970). *Inorg. Chem.*, **9,** 426
39. Kernohan, J. A. and Endicott, J. F. (1970). *Inorg. Chem.*, **9,** 1504
40. Finta, Z., Varhelyi, C. S. and Zsako, J. (1970). *J. Inorg. Nucl. Chem.*, **32,** 3013
41. Ogino, K., Murakami, T. and Saito, K. (1968). *Bull. Chem. Soc., Jap.*, **41,** 1615
42. Basolo, F. and Pearson, R. G. (1967). *Mechanisms of Inorganic Reactions,* 2nd ed. (New York: John Wiley and Sons)
43. Ciampolini, M. and Nordi, N. (1966). *Inorg. Chem.*, **5,** 41
44. Chan, S. C., Chung, C. Y. and Leh, F. (1967). *J. Chem. Soc., A,* 1586
45. Chan, S. C. and Leh, F. (1968). *J. Chem. Soc., A,* 1079
46. Rogers, A. and Staples, P. J. (1965). *J. Chem. Soc.*, 6834
47. Parris, M. and Wallace, W. E. (1969). *Can. J. Chem.*, **47,** 2257
48. Hay, R. W. and Cropp, P. L. (1969). *J. Chem. Soc. A,* 42
49. Chan, S. C. and Lee, C. L. (1969). *J. Chem. Soc. A,* 2649
50. Chan, S. C. and Lau, O. W. (1969). *Aust. J. Chem.*, **22,** 1851
51. Chan, S. C. and Lau, O. W. (1969). *Aust. J. Chem.*, **22,** 1869
52. Langford, C. H. and Gray, H. B. (1965). *Ligand Substitution Processes,* Chapter 3. (New York: W. A. Benjamin)
53. Ingold, C. K., Nyholm, R. S. and Tobe, M. L. (1960). *Nature (London)*, **187,** 477
54. Langford, C. H. (1964). *Inorg. Chem.*, **3,** 228
55. Tobe, M. L. (1965). *Advan. Chem. Ser.*, **49,** 7
56. Pratt, J. M. and Thorp, K. G. (1969). *Advan. Inorg. Radiochem.*, **12,** 375
57. Chan, S. C. and Leung, P. Y. (1969). *J. Chem. Soc., A,* 2650
58. Belluco, U., Cattalini, L. and Turco, A. (1964). *J. Amer. Chem. Soc.*, **86,** 226, 3257
59. Ablov, A. V. and Borykin, B. A. (1965). *Russ. J. Inorg. Chem.*, **10,** 29
60. Syrtsova, G. P., Ablov, A. V. and Korletyanu (1966). *Russ. J. Inorg. Chem.*, **11,** 602
61. Skafranskii, V. N. and Ablov, A. V. (1966). *Zh. Neorg. Khim.*, **11,** 67
62. Cumbliss, A. L. and Wilmarth, W. K. (1970). *J. Amer. Chem. Soc.*, **92,** 2593
63. Randall, W. C. and Alberty, W. A. (1967). *Biochem.*, **6,** 1520
64. Fleischer, E. B., Jacobs, S. and Mestichelli, L. (1968). *J. Amer. Chem. Soc.*, **90,** 2527
65. McAuley, A., Malik, M. N. and Hill, J. (1970). *J. Chem. Soc. A,* 2461
66. Stranks, D. R. and Yandell, J. K. (1970). *Inorg. Chem.*, **9,** 751
67. Langford, C. H. (1964). *Inorg. Chem.*, **3,** 228
68. Parker, A. J. (1967). *Advan. Phys. Org. Chem.*, **5,** 173
69. Tobe, M. L. (1966). *Rec. Chem. Progr.*, **27,** 79
70. Madan, S. K. and Peone, J. (1968). *Inorg. Chem.*, **7,** 824
71. Fitzgerald, W. R., Parker, A. J. and Watts, D. W. (1968). *J. Amer. Chem. Soc.*, **90,** 5474
72. Grunwald, E. and Winstein, S. (1948). *J. Amer. Chem. Soc.*, **20,** 846
73. Kosower, E. (1958). *J. Amer. Chem. Soc.*, **80,** 3253
74. Burgess, J. (1970). *J. Chem. Soc. A,* 2703
75. Jones, G. R. H., Edmondson, R. C. and Taylor, J. H. (1970). *J. Inorg. Nucl. Chem.*, **32,** 1752
76. Ašperger, S. and Pribanic, M. (1968). *J. Chem. Soc. A,* 1503
77. Ašperger, S., Flögel, M. and Murati, J. (1969). *J. Chem. Soc. A,* 569
78. Fitzgerald, W. R. and Watts, D. W. (1968). *Austr. J. Chem.*, **21,** 595
79. Chan, S. C. and Hui, K. Y. (1968). *J. Chem. Soc. A,* 1741
80. Staples, P. J. (1970). *J. Chem. Soc. A,* 811
81. Crossland, B. E. and Staples, P. J. (1970). *J. Chem. Soc. A,* 1305
82. Strel'tsava, E. M., Krestov, G. A. and Markova, N. K. (1970). *Zh. Fiz. Khim.*, **44,** 1357
83. Pearson, R. S., Stellwagen, N. C. and Basolo, F. (1960). *J. Amer. Chem. Soc.*, **82,** 1077
84. Brown, P. M. and Harris, G. M. (1968). *Inorg. Chem.*, **7,** 1872
85. Krishnamurty, K. V. and Harris, G. M. (1961). *Chem. Rev.*, **61,** 213
86. Aggett, J. and Armishaw, R. I. C. (1970). *J. Inorg. Nucl. Chem.*, **32,** 1989
87. Francis, D. J. and Jordan, R. B. (1969). *J. Amer. Chem. Soc.*, **91,** 6626
88. Lincoln, S. F. and Stranks, D. R. (1968). *Aust. J. Chem.*, **21,** 57
89. Buckingham, D. A., Foster, D. M., Marzilli, L. G. and Sargeson, A. M. (1970). *Inorg. Chem.*, **9,** 11

90. Chester, A. W., Heiba, I. E., Dessau, R. M. and Kechl, W. J., Jr. (1969). *Inorg. Nucl. Chem. Lett.*, **5**, 277
91. Basolo, F. and Pearson, R. G. (1970). *Mechanisms of Inorganic Reactions*, 2nd ed. (New York: John Wiley and Sons)
92. Halpern, J., Palmer, R. A. and Blakly, L. M. (1966). *J. Amer. Chem. Soc.*, **88**, 2877
93. Langford, C. H. and Muir, W. R. (1967). *J. Amer. Chem. Soc.*, **89**, 3141
94. Lincoln, S. F. and Stranks, D. R. (1968). *Aust. J. Chem.*, **21**, 1745
95. Burnett, M. G. (1970). *J. Chem. Soc. A*, 2486
96. Burnett, M. G. (1970). *J. Chem. Soc. A*, 2490
97. Poon, C. K. and Tobe, M. L. (1968). *J. Chem. Soc. A*, 1549
98. Stranks, D. R. and Yandell, J. K. (1970). *Inorg. Chem.*, **9**, 751
99. Tsiang, H. G. and Wilmarth, W. K. (1968). *Inorg. Chem.*, **7**, 2535
100. Fleischer, E. B., Jacobs, S. and Mestichelli, L. (1968). *J. Amer. Chem. Soc.*, **90**, 2527
101. Muir, W. R. and Langford, C. H. (1968). *Inorg. Chem.*, **7**, 1032
102. D'Ascenzo, G. and Gore, R. H. (1970). *J. Inorg. Nucl. Chem.*, **32**, 3404
103. Williams, T. J. and Hunt, J. P. (1968). *J. Amer. Chem. Soc.*, **90**, 7210
104. Werner, A. (1912). *Annalen*, **386**, 1
105. Westheimer, F. H. (1968). *Accounts Chem. Res.*, **1**, 70
106. Haake, P. and Pfeiffer, R. M. (1970). *J. Amer. Chem. Soc.*, **92**, 4996
107. Dittmar, E. A. and Archer, R. D. (1968). *J. Amer. Chem. Soc.*, **90**, 1468
108. Farago, M. E., Page, B. and Tobe, M. L. (1969). *Inorg. Chem.*, **8**, 388
109. Tobe, M. L. (1968). *Inorg. Chem.*, **7**, 1260
110. Archer, R. D. (1969). *Coord. Chem. Rev.*, **4**, 249
111. Niththyananthan, R. and Tobe, M. L. (1969). *Inorg. Chem.*, **8**, 1589
112. Ricevuto, V. and Tobe, M. L. (1970). *Inorg. Chem.*, **9**, 1785
113. Alexander, M. D. and Hamilton, H. G. (1969). *Inorg. Chem.*, **8**, 2131
114. Linck, R. G. (1969). *Inorg. Chem.*, **8**, 1016
115. Boucher, L. J. and Paez, N. G. (1970). *Inorg. Chem.*, **9**, 418
116. Cotsoradis, B. P. and Archer, R. D. (1967). *Inorg. Chem.*, **6**, 800
117. York, R. J. and Archer, R. D. (1969). *Coord. Chem. Rev.*, **4**, 234
118. Sargeson, A. M. and Searle, G. H. (1967). *Inorg. Chem.*, **6**, 2172
119. Fitzgerald, W. R. and Watts, D. W. (1968). *J. Amer. Chem. Soc.*, **90**, 1734
120. Poon, C. K. and Tobe, M. L. (1968). *Inorg. Chem.*, **7**, 2398
121. Farago, M. E., Page, B. A. and Mason, C. F. V. (1969). *Inorg. Chem.*, **8**, 2270
122. Girgis, A. Y. and Fay, R. C. (1970). *J. Amer. Chem. Soc.*, **92**, 2061
123. Brady, J. E. (1969). *Inorg. Chem.*, **8**, 1208
124. Chan, S. C. and Chan, F. K. (1970). *Aust. J. Chem.*, **23**, 1175
125. Bailar, J. C., Jr. (1958). *J. Inorg. Nucl. Chem.*, **8**, 165
126. Costa, G., Tanzher, G. and Puxeddu, A. (1969). *Inorg. Chim. Acta*, **3**, 41
127. Syrtsova, G. P. and Nguyen, S. L. (1970). *Zh. Neorg. Khim*, **15**, 470
128. Smith, A. A. and Haines, R. A. (1969). *J. Amer. Chem. Soc.*, **91**, 6280
129. Chan, S. C. and Leh, F. (1966). *J. Chem. Soc. A*, 126
130. Chan, S. C. (1966). *J. Chem. Soc. A*, 1124
131. Buckingham, D. A., Olson, I. I. and Sargeson, A. M. (1968). *Inorg. Chem.* **7**, 174
132. Buckingham, D. A., Marzilli, P. A. and Sargeson, A. M. (1969). *Inorg. Chem.*, **8**, 1595
133. Palmer, J. W. and Basolo, F. (1960). *J. Phys. Chem.*, **64**, 778
134. Buckingham, D. A., Marzilli, L. G. and Sargeson, A. M. (1968). *J. Amer. Chem. Soc.*, **90**, 6028
135. Yamatera, H. and Fujita, M. (1969). *Bull. Chem. Soc., Japan*, **42**, 3043
136. Poon, C. K. and Tobe, M. L. (1968). *Chem. Commun.* 156
137. Aprile, F., Basolo, F., Illuminati, G. and Maspero, F. (1968). *Inorg. Chem.*, **7**, 519
138. Birk, J. P., Chock, P. B. and Halpern, J. (1968). *J. Amer. Chem. Soc.*, **90**, 6959
139. Jones, W. E., Jordan, R. B. and Swaddle, T. W. (1969). *Inorg. Chem.*, **8**, 2504
140. Davis, M. B. and Lalor, G. C. (1969). *J. Inorg. Nucl. Chem.*, **31**, 2189
141. Lalor, G. C. (1969). *J. Inorg. Nucl. Chem.*, **31**, 1206
142. Lincoln, S. F., Jayne, J. and Hunt, J. P. (1969). *Inorg. Chem.*, **8**, 2267
143. Po, L. L. and Jordan, R. B. (1968). *Inorg. Chem.*, **7**, 526
144. Buckingham, D. A., Foxman, B. M. and Sargeson, A. M. (1970). *Inorg. Chem.*, **9**, 1790
145. Chan, S. C. and Lau, O. W. (1969). *Austr. J. Chem.*, **22**, 1851
146. Chan, S. C. and Cheung, T. L. (1970). *Austr. J. Chem.*, **23**, 707

147. Takemoto, J. H. and Jones, M. M. (1970). *J. Inorg. Nucl. Chem.*, **32,** 175
148. Niththyananthan, R. (1970). *Ph.D. Thesis,* University of London, cited in Reference 150
149. Watt, G. N. and Knifton, J. F. (1968). *Inorg. Chem.*, **7,** 1159
150. Tobe, M. L. (1970). *Accounts Chem. Res.*, **3,** 337
151. Chester, A. W. (1969). *Inorg. Chem.*, **8,** 1584
152. Chester, A. W. (1969). *Chem. Commun.*, 865
153. Chester, A. W. (1970). *Inorg. Chem.*, **9,** 1743
154. Chester, A. W. (1970). *Inorg. Chem.*, **9,** 1746
155. Wendt, M. R. and Monk, C. B. (1969). *J. Chem. Soc. A,* 1624
156. Buckingham, D. A., Olsen, I. I. and Sargeson, A. M. (1968). *J. Amer. Chem. Soc.*, **90,** 6539
157. Buckingham, D. A., Olson, I. I. and Sargeson, A. M. (1968). *J. Amer. Chem. Soc.*, **90,** 6654
158. Buckingham, D. A., Creaser, I. I. and Sargeson, A. M. (1970). *Inorg. Chem.*, **9,** 655
159. Buckingham, D. A., Foster, D. M. and Sargeson, A. M. (1969). *J. Amer. Chem. Soc.*, **91,** 4102
160. Nordmeyer, F. R. (1969). *Inorg. Chem.*, **8,** 2780
161. Stevenson, M. B., Mast, R. D. and Sykes, A. G. *J. Chem. Soc. A,* 937
162. Stevenson, M. B., Taylor, R. S. and Sykes, A. G. (1970). *J. Chem. Soc. A,* 1059
163. Harris, G. M. and Gillow, E. W. (1968). *Inorg. Chem.*, **7,** 2652
164. Syrtsova, G. P., Korletyanu, L. N. and Nguyen, S. L. (1970). *Zh. Neorg. Khim.*, **15,** 475
165. Fong, S. W., Stevenson, M. B. and Sykes, A. G. (1970). *J. Chem. Soc. A,* 1064
166. Sykes, A. G. and Taylor, R. S. (1970). *J. Chem. Soc. A,* 1424
167. Mast, R. D. and Sykes, A. G. (1968). *J. Chem. Soc. A,* 1031
168. Panasyuk, V. D. and Falendish, E. R. (1969). *Zh. Neorg. Khim,* **14,** 2768
169. Staples, P. J. (1968). *J. Inorg. Nucl. Chem.*, **30,** 3119
170. Grosse, P. D. and Staples, P. J. (1969). *J. Inorg. Nucl. Chem.*, **31,** 1443
171. Malik, M. N. and McAuley, A. (1969). *J. Chem. Soc. A,* 917
172. Balahura, R. J. and Jordan, R. B. (1970). *Inorg. Chem.*, **9,** 1567
173. Lenz, E. and Murmann, R. K. (1968). *Inorg. Chem.*, **7,** 1880
174. Peloso, A. and Dolcetti, G. (1969). *J. Chem. Soc. A,* 1506
175. Chan, S. C. and Chan, S. F. (1969). *J. Chem. Soc. A,* 202
176. Buckingham, D. A., Forster, D. M., Marzilli, L. G. and Sargeson, A. M. (1970). *Inorg. Chem.*, **9,** 11
177. Bifano, C. and Linck, R. G. (1968). *Inorg. Chem.*, **7,** 908
178. Lalor, G. C. and Rusted, D. S. (1969). *J. Inorg. Nucl. Chem.*, **31,** 3219
179. Tanner, S. P. and Higginson, W. C. E. (1969). *J. Chem. Soc. A,* 1164
180. Adin, A. and Espenson, J. (1971). *Chem. Commun.*, in press
181. Hill, H. A. O., Pratt, J. M., Ridsdale, S., Williams, F. R. and Williams, R. J. P. (1970). *Chem Commun.*, 341
182. Morawetz, H. and Vogel, B. (1969). *J. Amer. Chem. Soc.*, **91,** 563
183. Archer, M. D. and Spiro, M. (1970). *J. Chem. Soc. A,* 68, 73, 78
184. Syrtsova, G. P. and Korletyana, L. N. (1968). *Izv. Sib. Otd. Akad. Nauk SSSR, Ser. Khim. Nauk.*, **3,** 48
185. Dobbins, R. J. and Harris, G. M. (1970). *J. Amer. Chem. Soc.*, **92,** 5104, and other references cited in this paper
186. Das Gupta, T. P. and Harris, G. M. (1971). *J. Amer. Chem. Soc.*, **93,** 91, and other references cited in this paper
187. Sastri, V. S. and Harris, G. M. (1970). *J. Amer. Chem. Soc.*, **92,** 2943
188. Kernohan, J. A. and Endicott, J. F. (1969). *J. Amer. Chem. Soc.*, **91,** 6977
189. Lee Hin-Fat and Higginson, W. C. E. (1970). *J. Chem. Soc. A,* 2836
190. Aggett, J., Oddell, A. L. and Smith, B. E. (1968). *J. Chem. Soc. A,* 1415
191. Yatsimirskii, K. B. (1966). *Theor. Eksp. Khim,* **2,** 451
192. Spees, S. T., Perumareddi, J. R. and Adamson, A. W. (1968). *J. Amer. Chem. Soc.*, **90,** 6626
193. Langford, C. H. (1971). *Can. J. Chem.*, **49,** 1497
194. Levine, M. A., Jones, T. P., Harris, W. E. and Wallace, W. J. (1961). *J. Amer. Chem. Soc.*, **83,** 2453
195. Jones, T. P. and Phillips, J. K. (1968). *J. Chem. Soc. A,* 674
196. Gay, D. L. and Lalor, G. C. (1966). *J. Chem. Soc. A,* 1179

197. Chan, S. C. (1964). *J. Chem. Soc. A,* 2375
198. Sutin, N. (1966). *Ann. Rev. Phys. Chem.,* **17,** 129
199. Gay, D. L. and Nalepa, R. (1970). *Can. J. Chem.,* **48,** 910
200. Parris, M. and Wallace, W. J. (1969). *Can. J. Chem.,* **47,** 2257
201. Swaddle, T. W. and Guastalla, G. (1968). *Inorg. Chem.,* **7,** 1915
202. Espenson, J. H. (1969). *Inorg. Chem.,* **8,** 1554
203. Finholt, J. E. and Deming, J. N. (1967). *Inorg. Chem.,* **6,** 1533
204. (i) Espenson, J. H. and Binau, D. E. (1966). *Inorg. Chem.,* **5,** 1365
(ii) Schroder, K. A. and Espenson, J. H. (1967). *J. Amer. Chem. Soc.,* **89,** 2548
205. Bannerjea, D. and Chaudhari S. Dutta. (1968). *J. Inorg. Nucl. Chem.,* **30,** 871
206. Hartley, F. R. (1968). *Aust. J. Chem.,* **21,** 2723, ibid., **23,** 275
207. Gay, D. L. and Nalepa, R. (1971). *Can. J. Chem.,* in press
208. Swaddle, T. W. Private communication
209. Frankel, L. S., Stengle, T. R. and Langford, C. H. (1965). *Chem. Commun.,* 373
210. Behrendt, S., Langford, C. H. and Frankel, L. S. (1969). *J. Amer. Chem. Soc.,* **91,** 2236
211. Langford, C. H. and White, J. F. (1967). *Can. J. Chem.,* **45,** 3049
212. Sastri, V. S., Henwood, R. W., Behrendt, S. and Langford, C. H. To be published
213. Thomas, G. and Holba, V. (1969). *J. Inorg. Nucl. Chem.,* **31,** 1749
214. Zsako, J., Varhelyi, C., Ganescu, I. and Zoldi, L. (1968). *Monatsh. Chem.,* **99,** 2235
215. Thomas, G. and Treindl, L. (1968). *Z. Phys. Chem.,* **62,** 256
216. Zsako, I., Varhelyi, C., Ganescu, I. and Turos, J. (1969). *Acta. Chim. (Budapest),* **61,** 167; (1969). *C.A.,* **71,** 108610 d
217. Zsako, I. (1970). *Stud. Univ. Babes–Bolyai Ser. Chem.,* **15,** 93
218. Tipping, L. Private communication
219. Matts, T. C. and Moore, P. (1969). *J. Chem. Soc. A,* 219
220. Staples, P. J. (1968). *J. Chem. Soc. A,* 2731
221. Wakefield, D. K. and Schapp, W. B. (1969). *Inorg. Chem.,* **8,** 512
222. Matts, T. C. and Moore, P. (1969). *J. Chem. Soc. A,* 1997
223. Burgess, J., Goodman, B. A. and Raynor, J. B. (1968). *J. Chem. Soc. A,* 501
224. Bustin, D. I., Earley, J. E. and Vlček, A. A. (1969). *Inorg. Chem.,* **8,** 2062
225. Deutsch, E. and Taube, H. (1968). *Inorg. Chem.,* **7,** 1532
226. Matts, T. C. and Moore, P. (1970). *J. Chem. Soc. A,* 2819
227. Ferraris, S. P. and King, E. L. (1970). *J. Amer. Chem. Soc.,* **92,** 1215
228. Coombes, R. G., Johnson, M. D. and Vamplew, D. (1968). *J. Chem. Soc. A,* 2297
229. Doyle, J., Sykes, A. G. and Adin, A. (1968). *J. Chem. Soc. A,* 1314
230. Birk, J. P. and Espenson, J. E. (1968). *Inorg. Chem.,* **7,** 991
231. Orhanovic, M. and Sutin, N. (1968). *J. Amer. Chem. Soc.,* **90,** 538, 4286
232. House, J. E. and Bailar, J. C., Jr. (1969). *Inorg. Chem.,* **8,** 672
233. Birk, J. P. (1970). *Inorg. Chem.,* **9,** 735; (1971). **10,** 66
234. Hoppenjans, D. W., Hunt, J. B. and Gregoire, C. R. (1968). *Inorg. Chem.,* **7,** 2506
235. Williams, T. J. and Garner, C. S. (1970). *Inorg. Chem.,* **9,** 52
236. Childers, R. F., Jr., Vander Zyl, K. G., Jr., House, D. A., Hughes, R. G. and Garner, C. S. (1968). *Inorg. Chem.,* **7,** 749
237. Hughes, R. G. and Garner, C. S. (1968). *Inorg. Chem.,* **7,** 1988
238. Lin, D. K. and Garner, C. S. (1969). *J. Amer. Chem. Soc.,* **91,** 6637
239. Williams, T. J. and Garner, C. S. (1969). *Inorg. Chem.,* **8,** 1639
240. Nazarenko, Yu. P. and Bratushko, Yu. I. (1968). *Izy. Sib. Otd. Akad. Nauk. SSSR Ser. Khim. Nauk.,* **17**; (1969). *C.A.,* **70,** 6930V
241. House, D. A. (1969). *Aust. J. Chem.,* **22,** 647
242. Wakefield, D. K. and Schapp, W. B. (1969). *Inorg. Chem.,* **8,** 811
243. Palmer, D. A. and Watts, D. W. (1968). *Austr. J. Chem.,* **21,** 2895
244. Geher, J. (1970). *Magy Kem. Foly.,* **76,** 150; (1970). *C.A.,* **72,** 125467a
245. Geher, J. and Beck, M. (1970). *Magy. Kem. Foly.,* **76,** 156; (1970). *C.A.,* **72,** 136888a
246. Sykes, A. G. and Thorneley, R. N. F. (1969). *J. Chem. Soc. A,* 655
247. Phatak, G. M., Bhat, T. R. and Shankar, J. (1970). *J. Inorg. Nucl. Chem.,* **32,** 1305
248. Bannerjea, D. and Chaudhuri, P. (1968). *J. Inorg. Nucl. Chem.,* **30,** 3259
249. Bannerjea, D. and Chaudhuri, P. (1970). *Z. Anorg. Allg. Chem.,* **372,** 268
250. Bannerjea, D. and Chaudhuri, P. (1970). *J. Inorg. Nucl. Chem.,* **32,** 2697
251. Krishnamurty, K. V. and Harris, G. M. (1960). *J. Phys. Chem.,* **64,** 346
252. Kelm, H. and Harris, G. M. (1967). *Inorg. Chem.,* **6,** 706

253. Bannerjea, D. and Chatterjee, C. (1968). *J. Inorg. Nucl. Chem.*, **30**, 3354
254. Chang, J. C. (1968). *J. Inorg. Nucl. Chem.*, **30**, 945
255. Furlani, C. and Mantovani, E. (1969). *J. Inorg. Nucl. Chem.*, **31**, 1213
256. Stieger, H., Harris, G. M. and Kelm, H. (1970). *Ber. Bunsenges. Phys. Chem.*, **74**, 262
257. Chang, J. C. (1970). *J. Inorg. Nucl. Chem.*, **32**, 1402
258. Bannerjea, D. and Chaudhuri, S. Dutta (1970). *J. Inorg. Nucl. Chem.*, **32**, 2985
259. Bannerjea, D. and Chaudhuri, S. Dutta (1970). *J. Inorg. Nucl. Chem.*, **32**, 1617
260. Bannerjea, D. and Chatterjee, C. (1969). *J. Inorg. Nucl. Chem.*, **31**, 3845
261. Broomhead, J. A., Kane-Maguire, N. and Lauder, I. (1970). *Inorg. Chem.*, **9**, 1243
262. Kane-Maguire, N., Broomhead, J. A., Lauder, I. and Nimmo, P. (1968). *Chem. Commun.*, 747
263. Aggett, J., Mawtson, I., Odell, A. L. and Smith, B. E. (1968). *J. Chem. Soc. A.*, 1413
264. Kelm, H., Stieger, H. and Harris, G. M. (1969). *Ber. Bunsenges. Phys. Chem.*, **73**, 939
265. Kelm, H., Stieger, H. and Harris, G. M. (1969). *Z. Phys. Chem.*, **67**, 98
266. Brady, J. E. and Thompson, D. E. (1970). *Inorg. Nucl. Chem. Lett.*, **6**, 663
267. Ashley, K. R. and Lane, K. (1970). *Inorg. Chem.*, **9**, 1795
268. Veigel, J. M. (1968). *Inorg. Chem.*, **7**, 69
269. Sastri, V. S. and Langford, C. H. (1970). *J. Phys. Chem.*, **74**, 3945
270. Grant, D. M. and Hamm, R. E. (1958). *J. Amer. Chem. Soc.*, **80**. 4166
271. Connick, R. E. and Thompson, M. G. (1964). 148th A.C.S. meeting, Sept. 1964, papers 23, 24
272. Wolcott, D. and Hunt, J. P. (1968). *Inorg. Chem.*, **7**, 755
273. Hoppenjans, D. W. and Hunt, J. B. (1969). *Inorg. Chem.*, **8**, 505
274. Espenson, J. H. and Bushey, W. R. Private communication
275. Bannerjea, D. and Chatterjee, C. (1968). *Z. Anorg. Allg. Chem.*, 361
276. Davies, R., Evans, G. B. and Jordan, R. B. (1969). *Inorg. Chem.*, **8**, 2025
277. Chan, S. C. and Hui, K. Y. (1968). *J. Chem. Soc. A*, 1741
278. Duffy, N. V. and Earley, J. E. (1967). *J. Amer. Chem. Soc.*, **89**, 272
279. Baltisberger, R. J. and Hanson, J. V. (1970). *Inorg. Chem.*, **9**, 1573
280. Chung, F. M. Private communication
281. Moore, P., Basolo, F. and Pearson, R. G. (1966). *Inorg. Chem.*, **5**, 223
282. Bannerjea, D. and Sarkar, S. (1970). *Z. Anorg. Allg. Chem.*, **373**, 73
283. Lalor, G. C. and Bushnell, G. W. (1968). *J. Chem. Soc. A*, 2250
284. Davies, C. S. and Lalor, G. C. (1968). *J. Chem. Soc. A*, 2328
285. Davies, C. S. and Lalor, G. C. (1970). *J. Chem. Soc. A*, 445
286. Davies, C. S. and Lalor, G. C. (1968). *J. Chem. Soc. A*, 1095
287. Domingos, A. J. P., Domingos, A. M. T. S. and Cabral, J. M. Peixito. (1969). *J. Inorg. Nucl. Chem.*, **31**, 2563
288. Tsayun, G. P. and Yusupova, V. A. (1970). *Zh. Neorg. Khim.*, **15**, 81
289. Kravtsov, V. I., Titova, N. V. and Tsayun, G. P. (1970). *Elektrokhimiya*, **6**, 573
290. Stritar, J. A. and Taube, H. (1969). *Inorg. Chem.*, **8**, 2281
291. Movius, W. and Linck, R. G. (1970). *J. Amer. Chem. Soc.*, **92**, 2677
292. Ford, P. C., Kuempel, J. R. and Taube, H. (1968). *Inorg. Chem.*, **7**, 1976
293. Broomhead, J. A. and Kane-Maguire, L. (1968). *Inorg. Chem.*, **7**, 2519
294. Kane-Maguire, L., Sheridan, P. S., Basolo, F. and Pearson, R. G. (1970). *J. Amer. Chem. Soc.*, **92**, 5865
295. Eliades, T., Harris, R. O. and Reinsalu, P. (1969). *Can. J. Chem.*, **47**, 3823
296. Monacelli, F. (1968). *Inorg. Chim. Acta*, **2**, 263
297. Bott, H. L., Poe, A. J. and Shaw, K. (1968). *Chem. Commun.*, 793
298. Bott, H. L., Poe, A. J. and Shaw, K. (1970). *J. Chem. Soc. A*, 1745
299. Robb, D., Devstyn, M. M. and Kruger, H. (1969). *Inorg. Chim. Acta*, **3**, 383
300. Borghi, E., Monacelli, F. and Prosperi, T. (1970). *Inorg. Nucl. Chem. Lett.*, **6**, 667
301. Kallan, T. W. and Earley, J. E. (1970). *Chem. Commun.*, 851
302. Davies, N. R. and Mullins, T. L. (1968). *Aust. J. Chem.*, **21**, 915
303. Armor, J. N. and Taube, H. (1969). *J. Amer. Chem. Soc.*, **91**, 6874
304. Elson, C. M., Itzkovitch, I. J. and Page, J. A. (1970). *Can. J. Chem.*, **48**, 1639
305. Itzkovitch, I. J. and Page, J. A. (1968). *Can. J. Chem.*, **46**, 2743
306. Shilova, A. K. and Shilov, A. E. (1969). *Kinet. Catal. (USSR)*, **10**, 262; *C.A.*, **71**, 25095p
307. Broomhead, J. A. and Kane-Maguire, L. (1971). *Inorg. Chem.*, **10**, 85
308. Armor, J. N., Scheidegger, H. A. and Taube, H. (1968). *J. Amer. Chem. Soc.*, **90**, 5928

309. Adamson, M. G. (1968). *J. Chem. Soc. A,* 1370
310. Poë, A. J. and Vaughan, D. H. (1968). *Inorg. Chim. Acta,* **2,** 159
311. Poë, A. J. and Vaughan, D. H. (1969). *J. Chem. Soc. A,* 2844
312. Mason, W. R. (1969). *Inorg. Chem.,* **8,** 1756
313. Mason, W. R. (1970). *Inorg. Chem.,* **9,** 1528
314. Johson, R. C. and Berger, E. R. (1968). *Inorg. Chem.,* **7,** 1656
315. Bailey, S. G. and Johnson, R. C. (1969). *Inorg. Chem.,* **8,** 2596
316. Syamal, A. and Johnson, R. C. (1970). *Inorg. Chem.,* **9,** 265
317. Poë, A. J. and Shaw, K. (1970). *J. Chem. Soc. A,* 393
318. Bauer, R. A. and Basolo, F. (1969). *Inorg. Chem.,* **8,** 2237
319. Tirsell, J. B. and Garner, C. S. (1969). *J. Inorg. Nucl. Chem.,* **31,** 2871
320. Ford, P. C. and Sutton, C. (1969). *Inorg. Chem.,* **8,** 1544
321. Illuminati, G. and Maspero, F. (1968). *Ric. Sci.,* **38,** 544; (1969), *C.A.,* **70,** 50964h
322. Armor, J. N. and Taube, H. (1970). *J. Amer. Chem. Soc.,* **92,** 2560
323. Bushnell, G. W. and Lalor, G. C. (1968). *J. Inorg. Nucl. Chem.,* **30,** 219
324. Davies, C. S. and Lalor, G. C. (1968). *J. Chem. Soc. A,* 2328
325. Lalor, G. C. and Carrington, T. (1969). *J. Chem. Soc. A,* 2509
326. Broomhead, J. A. and Kane-Maguire, L. (1969). *Inorg. Chem.,* **8,** 2124
327. Skinner, C. E. and Jones, M. M. (1969). *J. Amer. Chem. Soc.,* **91,** 1984
328. Robb, W. and Harris, G. M. (1965). *J. Amer. Chem. Soc.,* **87,** 4472
329. Connick, R. E. (1965). *Advances in Chemistry of Co-ordination Compounds,* Ed. by Kirschner, S., 15. (New York: MacMillan and Co.)
330. Rund, J. V., Basolo, F. and Pearson, R. G. (1964). *Inorg. Chem.,* **3,** 658
331. Hewkin, D. J. and Prince, R. H. (1970). *Coord. Chem. Rev.,* **5,** 56
332. Wilkins, R. G. (1970). *Accounts Chem. Res.,* **3,** 408
333. Karpel, R. L., Kustin, K. and Pasternak, R. F. (1969). *Biochim. Biophys. Acta* (Amsterdam), 434
334. Hoffman, H. (1969). *Ber. Bunsenges. Phys. Chem.,* **73,** 432
335. Kimura, M. (1969). *Bull. Chem. Soc. Jap.,* **42,** 2844
336. Swift, T. J. and Connick, R. E. (1962). *J. Chem. Phys.,* **37,** 307
337. Brintzinger, H. and Hammes, G. G. (1966). *Inorg. Chem.,* **5,** 1286
338. Davis, A. G. and Smith, W. M. (1961). *Proc. Chem. Soc., London,* 380
339. Eisenstadt, M. (1969). *J. Chem. Phys.,* **51,** 4421
340. Rorabacher, D. B. (1966). *Inorg. Chem.,* **5,** 1891
341. Holyer, R. H., Hubbard, C. D., Kettle, S.F.A. and Wilkins, R. G. (1965). *Inorg. Chem.,* **4,** 929
342. Wilkins, R. G. (1964). *Inorg. Chem.,* **3,** 520
343. Nancollas, G. H. and Sutin, N. (1964). *Inorg. Chem.,* **3,** 360
344. Rorabacher, D. B., Turan, T. S., Defever, J. A. and Nickels, W. G. (1969). *Inorg. Chem.,* **8,** 1498
345. Pearson, R. G. and Anderson, O. P. (1970). *Inorg. Chem.,* **9,** 39
346. Kaden, Th. (1970). *Helv. Chim. Acta,* **53,** 617
347. Margerum, D. W., Rorabacher, D. B. and Clarke, J. F. G. (1963). *Inorg. Chem.,* **2,** 667
348. Margerum, D. W. and Rosen, R. M. (1967). *J. Amer. Chem. Soc.,* **89,** 1088
349. Jones, J. P., Billo, E. J. and Margerum, D. W. (1970). *J. Amer. Chem. Soc.,* **92,** 1875
350. Rablen, D. and Gordon, G. (1969). *Inorg. Chem.,* **8,** 395
351. Kimura, M. (1969). *Bull. Chem. Soc. Jap.,* **42,** 2841
352. Kodama, M. (1970). *Nippon Kagaku Zasshi,* **91,** 134
353. Kodoma, M., Sasaki, C. and Murata, M. (1968). *Bull. Chem. Soc. Jap.,* **41,** 1333
354. Margerum, D. W. and Rosen, H. M. (1968). *Inorg. Chem.,* **9,** 229
355. Kingham, D. J. and Frisbin, A. (1970). *Inorg. Chem.,* **9,** 2034
356. Kolski, G. B. and Margerum, D. W. (1969). *Inorg. Chem.,* **8,** 1125
357. Coombs, L. C. and Margerum, D. W. (1970). *Inorg. Chem.,* **9,** 1711
358. Coombs, L. C., Margerum, D. W. and Nigam, P. C. (1970). *Inorg. Chem.,* **9,** 2087
359. Funahashi, S. and Tanaka, M. (1969). *Inorg. Chem.,* **8,** 2159
360. Funahashi, S. and Tanaka, M. (1970). *Inorg. Chem.,* **9,** 2092
361. Desai, A. G., Dodgen, H. W. and Hunt, J. P. (1970). *J. Amer. Chem. Soc.,* **92,** 798
362. Desai, A. G., Dodgen, H. W. and Hunt, J. P. (1969). *J. Amer. Chem. Soc.,* **91,** 5001
363. Jones, J. P. and Margerum, D. W. (1970). *J. Amer. Chem. Soc.,* **92,** 470

364. Pasternack, R. F., Gibbs, E. and Cassatt, J. C. (1969). *J. Phys. Chem.*, **73**, 3814
365. Pasternack, R. F., Kustin, K., Hughes, L. A. and Gibbs, E. (1969). *J. Amer. Chem. Soc.*, **91**, 4401
366. Lincoln, S. F., Aprile, F., Dodgen, H. W. and Hunt, J. P. (1968). *Inorg. Chem.*, **7**, 929
367. Grant, M., Dodgen, H. W. and Hunt, J. P. (1970). *J. Amer. Chem. Soc.*, **92**, 2321
368. Hoffman, H. and Yeager, E. (1970). *Ber. Bunsenges. Phys. Chem.*, **74**, 641
369. Kluiber, R. W., Kukla, R. and Horrocks, W. D. (1970). *Inorg. Chem.*, **9**, 1319
370. Kluiber, R. W., Thatter, F., Lau, R. A. and Horrocks, W. D. (1970). *Inorg. Chem.*, **9**, 2592
371. Jordan, R. B., Dodgen, H. W. and Hunt, J. P. (1966). *Inorg. Chem.*, **5**, 1906
372. Glaeser, H. H., Lo, G. A., Dodgen, H. W. and Hunt, J. P. (1965). *Inorg. Chem.*, **4**, 206
373. Rice, W. L. and Wayland, B. B. (1968). *Inorg. Chem.*, **7**, 1040
374. Van Geet, A. L. (1968). *Inorg. Chem.*, **7**, 2026
375. Luz, Z. and Meiboom, S. (1964). *J. Chem. Phys.*, **40**, 1058, 1066
376. Ravage, D. K., Stengle, T. R. and Langford, C. H. (1967). *Inorg. Chem.*, **6**, 1252
377. Matwiyoff, N. A. and Hooker, S. V. (1967). *Inorg. Chem.*, **6**, 1127
378. Matwiyoff, N. A. (1966). *Inorg. Chem.*, **5**, 1127
379. Babiec, J. S., Langford, C. H. and Stengle, T. R. (1966). *Inorg. Chem.*, **5**, 1363
380. Thomas, S. and Reynolds, W. L. (1967). *J. Phys. Chem.*, **46**, 4164
381. Frankel, L. S. (1969). *Chem. Commun.*, 1254
382. Frankel, L. S. Private communication quoted in Reference 3
383. Rablen, D. P. Unpublished reference quoted in Reference 386
384. Grant, M., Dodgen, H. W. and Hunt, J. P. (1970). *Chem. Commun.*, 1446
385. Breivogel, F. W. (1969). *J. Phys. Chem.*, **73**, 4203
386. Hunt, J. P. (1971). *Coord. Chem. Rev.*, in press
387. Zeltman, A. H., Matwiyoff, N. A. and Morgan, L. O. (1969). *J. Phys. Chem.*, **73**, 2689
388. Zeltman, A. H. and Morgan, L. O. (1970). *Inorg. Chem.*, **9**, 2522
389. Hoggard, P. D. (1970). *Ph.D. Thesis*, Washington State University, Pullman, Washington (quoted in Reference 386)
390. Luz, Z. (1964). *J. Chem. Phys.*, **41**, 1748, 1756
391. Glaeser, H. H., Dodgen, H. W. and Hunt, J. P. (1965). *Inorg. Chem.*, **4**, 1061
392. Grant, M., Dodgen, H. W. and Hunt, J. P. (1971). *Inorg. Chem.*, **10**, 71
393. Grant, M., Dodgen, H. W. and Hunt, J. P. (1969). *J. Amer. Chem. Soc.*, **91**, 6318
394. Babiec, J. S. (1966). *Ph.D. Thesis*, University of Massachusetts, Amherst, Mass.
395. Levanon, H. and Luz, Z. (1968). *J. Chem. Phys.*, **49**, 2031
396. Purcell, W. L. and Marianelli, R. S. (1970). *Inorg. Chem.*, **9**, 1724
397. Langford, C. H. and Chung, F. M. (1968). *J. Amer. Chem. Soc.*, **90**, 4485
398. Zeltman, A. H. and Morgan, L. O. (1966). *J. Phys. Chem.*, **76**, 2807
399. Matwiyoff, N. A. and Taube, H. (1968). *J. Amer. Chem. Soc.*, **90**, 2796
400. Neely, J. and Connick, R. E. (1970). *J. Amer. Chem. Soc.*, **92**, 3476
401. Movius, W. G. and Matwiyoff, N. A. (1969). *Inorg. Chem.*, **9**, 925
402. Thomas, S. and Reynolds, W. L. (1966). *J. Chem. Phys.*, **44**, 3148
403. Movius, W. G. and Matwiyoff, N. A. (1967). *Inorg. Chem.*, **6**, 847
404. Eigen, M. and Tamm, K. (1962). *Z. Elektrochem.*, **66**, 93, 107
405. Macri, G. and Petrucci, S. (1970). *Inorg. Chem.*, **9**, 1009
406. Bennetto, H. P. and Caldin, E. F. (1969). *Chem. Commun.*, 599
407. Tsiang, H. G. and Langford, C. H. (1970). *Inorg. Chem.*, **9**, 2346
408. Bennetto, H. P., Brilmer, R. and Caldin, E. F. (1968). *Hydrogen Bonded Solvent Systems*, ed. by Covington, A. K., 335, (London: Taylor and Francis)
409. Sanduja, M. L. and Smith, W.MacF. (1969). *Can. J. Chem.*, **47**, 3773
410. Meredith, G. W. and Connick, R. E. (1965). *Abstr.* 149*th Meeting American Chem. Soc.*,
411. Makinen. W. B., Pearlmutter, A. F. and Stuehr, J. E. (1969). *J. Amer. Chem. Soc.*, **91**, 4083
412. Hemmes, P. and Petrucci, S. (1968). *J. Phys. Chem.*, **72**, 3986
413. Kirschenbaum, L. J. and Kustin, K. (1970). *J. Chem. Soc. A*, 684
414. Malin, J. M. and Swinehart, J. H. (1968). *Inorg. Chem.*, **7**, 250
415. Kruse, W. and Thusius, D. (1968). *Inorg. Chem.*, **7**, 464
416. Pearson, R. G. and Gausow, O. A. (1968). *Inorg. Chem.*, **7**, 1373
417. Kowalak, A., Kustin, K. and Pasternak, R. F. (1969). *J. Phys. Chem.*, **73**, 281
418. Simplicio, J. and Wilkins, R. G. (1969). *J. Amer. Chem. Soc.*, **91**, 1325

419. Davis, G., Kustin, K. and Pasternack, R. F. (1969). *Int. J. Chem. Kinet.*, **1,** 45
420. Kustin, K. and Pasternack, R. F. (1968). *J. Amer. Chem. Soc.*, **90,** 2805
421. Kustin, K. and Pasternack, R. F. (1969). *J. Phys. Chem.*, **73,** 1
422. Eigen, M. and Wilkins, R. G. (1965). *Advan. Chem. Ser.*, **49,** 55
423. Rorabacher, D. B. and Moss, D. B. (1970). *Inorg. Chem.*, **9,** 1314
424. Howard, G. P. and Marianelli, R. S. (1970). *Inorg. Chem.*, **9,** 1738
425. Jones, J. P. and Margerum, D. W. (1969). *Inorg. Chem.*, **8,** 1486
426. Miller, F., Simplicio, J. and Wilkins, R. G. (1969). *J. Amer. Chem. Soc.*, **91,** 1962
427. Miller, F. and Wilkins, R. G. (1970). *J. Amer. Chem. Soc.*, **92,** 2687
428. Sokol, C. S. and Brubaker, C. H., Jr. (1968). *J. Inorg. Nucl. Chem.*, **30,** 3267
429. Kustin, K., Taub, I. A. and Weinstock, E. M. (1966). *Inorg. Chem.*, **5,** 1079
430. Pagenhopf, G. K. and Margerum, D. W. (1968). *Inorg. Chem.*, **7,** 2574
431. Burgess, J. (1968). *J. Chem. Soc. A*, 497
432. Burgess, J. (1968). *J. Chem. Soc. A*, 2728
433. Burgess, J. (1969). *J. Chem. Soc. A*, 1899
434. Jones, J. G. and Twigg, M. V. (1969). *Inorg. Chem.*, **8,** 2120
435. Hague, D. N. and Zetter, M. S. (1970). *Trans. Faraday Soc.*, **66,** 1176
436. Kustin, K., Pasternack, R. F. and Weinstock, E. M. (1966). *J. Amer. Chem. Soc.*, **88,** 4610
437. McCain, D. C. and Myers, F. J. (1968). *J. Chem. Phys.*, **72,** 4115
438. Koryta, S. (1960). *Z. Electrochem.*, **64,** 196
439. Kinura, M. (1969). *Nippon Kagaku Zasshi*, **90,** 1246
440. Brisbin, D. A. and Balahura, R. J. (1968). *Can. J. Chem.*, **46,** 3431
441. Williams, R. J. P. (1970). *Quart. Rev. Chem. Soc.*, **24,** 331
442. Lin, C. T. and Bear, J. L. (1969). *J. Inorg. Nucl. Chem.*, **31,** 263
443. Snellgrove, R. and Plane, R. A. (1968). *J. Amer. Chem. Soc.*, **90,** 3185
444. Seewald, D. and Sutin, N. (1963). *Inorg. Chem.*, **2,** 643
445. Pouli, D. and Smith, W. MacF. (1960). *Can. J. Chem.*, **38,** 567
446. Carlyle, D. W. and Espenson, J. H. (1967). *Inorg. Chem.*, **6,** 1370
447. Yasunaga, T. and Harada, S. (1969). *Bull. Chem. Soc. Jap.*, **42,** 2165
448. Espenson, J. H. and Helzer, S. (1969). *Inorg. Chem.*, **8,** 1051
449. Cavasino, F. P. and Dio, E. D. (1970). *J. Chem. Soc. A*, 1151
450. Carlyle, D. W. and Espenson, J. H. (1969). *Inorg. Chem.*, **8,** 575
451. Cavasino, F. P. (1968). *J. Phys. Chem.*, **72,** 1378
452. Baker, F. C. and Smith, W. MacF. (1970). *Can. J. Chem.*, **48,** 3100
453. Saini, G. and Mentasti, E. (1970). *Inorg. Chim. Acta*, **4,** 210
454. Espenson, J. G. and Dushin, D. F. (1969). *Inorg. Chem.*, **8,** 1760
455. Gilmour, A. D. and McAuley, A. (1969). *J. Chem. Soc. A*, 2345
456. Osugi, J., Nakatani, H. and Fuji, T. (1969). *Nippon Kagaku Zasshi*, **90,** 529
457. Duplessis-Legros, J. (1970). *C. R. Acad. Sci. Ser. C*, **270,** 1768
458. Wolfe, S. K. and Swinehart, J. H. (1968). *Inorg. Chem.*, **7,** 1855
459. Hasinoff, B. B., Dunford, H. B. and Horne, D. G. (1969). *Can. J. Chem.*, **47,** 3225
460. Gilmour, A. D. and McAuley, A. (1970). *Inorg. Chim. Acta*, **4,** 158
461. Wilkins, R. G. and Yelin, R. E. (1969). *Inorg. Chem.*, **8,** 1470
462. Wendt, H. (1969). *Inorg. Chem.*, **8,** 1527
463. Sommer, B. A. and Margerum, D. W. (1970). *Inorg. Chem.*, **9,** 2517
464. Budarin, L. I., Rumyantseva, T. A. and Yatsimirskii, K. (1970). *Ur. Khim. Zh.*, **36,** 535
465. Walker, R. G. and Watkins, K. O. (1968). *Inorg. Chem.*, **7,** 885
466. Baker, B. K., Orhanovic, M. and Sutin, N. (1967). *J. Amer. Chem. Soc.*, **89,** 722
467. Espenson, J. H. and Pladziewicz, J. R. (1970). *Inorg. Chem.*, **9,** 1380
468. Langford, C. H. and Chung, F. M. (1970). *Can. J. Chem.*, **48,** 2969
469. Diebler, H. (1969). *Z. Phys. Chem.*, **68,** 64
470. Fenkert, K. and Brubaker, C. H. (1968). *J. Inorg. Nucl. Chem.*, **30,** 3245
471. Thomas, S. and Reynolds, W. L. (1970). *Inorg. Chem.*, **9,** 78
472. Micek, J. and Steahr, J. (1968). *J. Amer. Chem. Soc.*, **90,** 6967
473. Owen, J. D. and Eyring, E. M. (1970). *J. Inorg. Nucl. Chem.*, **32,** 2217
474. Geier, G. (1965). *Ber. Bunsenges. Phys. Chem.*, **69,** 617
475. Mariarrelli, R. (1966). *Ph.D. Thesis*, University of California, Berkeley
476. Garusey, R. and Ebdon, D. W. (1969). *J. Amer. Chem. Soc.*, **91,** 50
477. Purdie, N. and Vincent, A. H. (1967). *Trans. Faraday Soc.*, **63,** 2745

478. Silber, H. B., Farina, R. D. and Swinehart, J. H. (1969). *Inorg. Chem.*, **8**, 819
479. Graffeo, A. J. and Bear, J. L. (1968). *J. Inorg. Nucl. Chem.*, **30**, 1577

7
Mechanism of Square-Planar Substitution

L. CATTALINI
University of Messina, Italy

7.1 INTRODUCTION

This article deals with the information which has been collected in the past few years on substitution reactions at planar tetra-coordinate complexes. The general account of what is known in the field can be found in a number of other publications[1–6] and therefore will be not discussed here in detail. A great deal of the data reported refer to platinum(II) complexes which undergo nucleophilic displacement of the coordinated ligands at rates which usually allow the use of conventional (mainly spectrophotometric) techniques. However, the amount of data relating to complexes of other d^8 transition metal ions, especially palladium(II) and gold(III) is increasing and much new information is becoming available.

7.2 THE GENERAL KINETIC EXPRESSION

The displacement of a group X from a planar tetra-coordinate complex of a d^8 metal ion by an incoming group Y can be usually represented by the equation,

$$\text{T—M(L}_1\text{)(L}_2\text{)—X} + \text{Y} \longrightarrow \text{T—M(L}_1\text{)(L}_2\text{)—Y} + \text{X}$$

where the ligands T and L_1, L_2 are *trans* and *cis* to the leaving group. As a rule, all the substrates undergo substitution with complete retention of the geometric configuration.

The general kinetic expression for these reactions is the two-term rate law:

$$\text{rate} = -\text{d[substrate]}/\text{d}t = (k_1 + k_2[\text{Y}])[\text{substrate}] \qquad (7.1)$$

The kinetics are usually carried out in the presence of an excess Y ([Y]$\gg$[substrate]) and pseudo first-order rate constants are obtained, which obey the expression

$$k_{\text{obsd}} = k_1 + k_2[\text{Y}]$$

The plot of k_{obsd} *v.* [Y] allows the identification of k_1 and k_2 as intercept and slope of a straight line.

The second-order contribution to the rate law, expressed by the k_2 term is simply related to the bimolecular attack of Y at the substrate. It has been proved also that the k_1 term is related to a bimolecular process, i.e. to the displacement of X by a solvent molecule (S), followed by the fast entry of Y displacing the solvent. The general reaction scheme is therefore:

$$\text{—M—X} + \text{S} \underset{k_-}{\overset{k_1}{\rightleftharpoons}} \text{—M—S} + \text{X}$$

$$\text{—M—X} \xrightarrow[k_2]{(+\text{Y}, -\text{X})} \text{—M—Y} \xleftarrow[\text{fast}]{(+\text{Y}, -\text{S})} \text{—M—S}$$

The presence of a k_{-1} contribution has been observed only in a few cases[7] and is usually negligible. There are examples of substrates which undergo substitution with $k_1 \gg k_2$ and others where $k_2 \gg k_1$; however, both the k_1 and k_2 terms are often detectable.

A number of indications strongly suggest that the best description of the transition state for these processes is in terms of a (distorted) trigonal-bipyramidal structure:

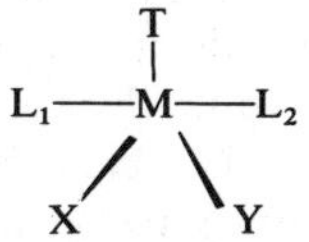

$$\begin{array}{c} \mathrm{T} \\ | \\ \mathrm{L_1}\text{—}\mathrm{M}\text{—}\mathrm{L_2} \\ \diagup \quad \diagdown \\ \mathrm{X} \qquad \mathrm{Y} \end{array}$$

where the entering and the leaving groups and the *trans* group occupy the equatorial positions, while the ligands L_1 and L_2, originally *cis* to X in the substrate, lie in the axial positions.

These substitutions are therefore associative and can be classified according to the suggestion of Langford and Gray, as processes with A or I_a mechanisms. This point is discussed in some detail in Section 7.10. However, we must point out here that the nature of the entering group is always important in determining the kinetic behaviour.

The fairly simple rate law and general mechanism allow one to compare data obtained from various systems, when a systematic change is made in the parameters controlling the kinetic behaviour, such as the nature of Y, X, L_1, L_2, T, M and solvent. In the last few years it has become quite clear that any discussion of the behaviour in terms of the rate of the reactions alone, even if it is a very simple approach, can lead to some difficulty in rationalising the data. Quite often the parameter that is most used is the ability of the substrate to discriminate between the various entering nucleophiles. A relatively large amount of experiments is required in order to investigate a number of reactions of the same substrate with various reagents. However, because of the importance of Y in determining the kinetic behaviour, the measure of the discriminating ability of the substrate can often be related to the electrophilicity of the metal which is the reaction centre, thus allowing a discussion of the distribution of charge on the complex.

7.3 NUCLEOPHILICITY SEQUENCES

The approach of the incoming nucleophile to the metal in the substrate can be described as a Lewis acid–base interaction, between Y, an electron donor and the metal, an electron acceptor. However, the 'proton basicity' of the nucleophile is not necessarily primarily involved, since it generally refers to the thermodynamic equilibrium

$$YH^+ \rightleftarrows Y + H^+$$

and the proton donor properties of Y are usually measured by the appropriate equilibrium constants (or pK_a values), in water. There are at least two reasons for this lack of correspondence, namely, (a) the electronic structure of the metal differs from that of H^+ and (b) the rate of the substitution reaction refers to a kinetic property and is related to the free energy change on going from the ground state (substrate $+Y$) to the transition state (substrate...Y), and not to a thermodynamic property such as the standard free energy change on going from reagents to products, as measured by the equilibrium constants.

Apart from steric hindrance, which will be discussed separately (Section 7.8), various factors influencing the reactivity of Y towards planar tetracoordinate complexes have been identified and examined in some detail.

The first one which we are considering here is the basicity of the nucleophile, Y, as measured by its pK_a value. For a group of nucleophiles, such as Cl^-, Br^-, I^-, OH^-, NO_2^-, N_3^-, SCN^- etc., there is no relationship between their pK_a values and their reactivity towards d^8 planar complexes. Moreover, the reactivity sequence is often the reverse of the basicity sequence. For instance, OH^- and CH_3O^-, which are strong bases, do not react with a second-order contribution with platinum(II) complexes, and the reactivity of halide ions is always $I^- > Br^- > Cl^-$.

7.3.1 Reactivity of amines

If one considers a class of reagents of the same type, such as heterocyclic amines having no (or the same amount of) steric hindrance and different basicity, free energy relationships of the general type

$$\log k_2 = A(pK_a) + \text{constant} \qquad (7.2)$$

have been observed in a number of cases and in various solvents. The parameter A in the above expression is a measure of the ability of the substrate to discriminate between the various amines, and is therefore a measure of the sensitivity of the system with respect to proton basicity.

The use of amines (and of strong bases in general) as nucleophiles in substitution processes presents some difficulty when the substrate can behave as an acid. It is known that coordination enhances the acidity of some ligands, as in the case of the complex $[Au(dien)Cl]^{2+}$ (dien = diethylenetriamine), which easily forms the conjugate base $[Au(dien{-}H)Cl]^{+}$ [8]. When this is possible particular care is obviously required in any interpretation.

As far as platinum(II) complexes are concerned, the values of A seem to be fairly small as, for example, with the systems

$$[Pt(bipy)Cl_2] + am \rightarrow [Pt(bipy)Cl(am)]^+ + Cl^- \qquad (7.3)$$

and

$$trans\text{-}[Pt(py)_2Cl_2] + am \rightarrow trans\text{-}[Pt(py)_2Cl(am)]^+ + Cl^-$$

in methanol, where the A values are 0.06 [9] and 0.05 [10] respectively. This is consistent with a low influence of the proton-basicity of the incoming nucleophile in determining the reactivity at platinum(II) complexes. However, on going from platinum(II) to complexes of other d^8 metal ions, A values have been measured which lie in the range between 0 and 0.9. The data available in the case of gold(III) derivatives are summarised in Table 7.1.

The interest of the results in Table 7.1 arises from the large variation of the discrimination parameter A with the changing nature of the substrate. Apart from the consideration that even $[AuCl_4]^-$, in spite of the negative charge, gives a value of A which is greater than those obtained with neutral platinum(II) complexes, thereby suggesting a greater role of basicity in the reactions at gold(III) derivatives, there is a dramatic change of A in the series of cationic complexes. The increase of A from 0.22 to 0.46 and 0.89, on going from the cationic complex containing 1,10-phenanthroline to those containing 2,2′-bipyridyl and 5-nitro-1,10-phenanthroline, seems to

Table 7.1 Kinetic parameters for the replacement of chloride in gold(III) complexes by amines (data obtained in acetone)

Substrate	*A*	Δ*	*Reference*
$[AuCl_4]^-$	0.15†	0.95	11
$[Au(phen)Cl_2]^+$	0.22	1.2	12
$[Au(bipy)Cl_2]^+$	0.46	1.2	12
$[Au(5\text{-}NO_2\text{-}phen)Cl_2]^+$	0.89	1.2	12
$[Au(tmen)Cl_2]^+$	0.14	—	10

Abbreviations: phen = 1,10-phenanthroline; bipy = 2,2′-bipyridyl; tmen = *N,N,N′,N′*-tetra-methyl-ethylenediamine
*The Δ values reported are a measure of steric retardation effects, discussed in Section 7.8
†The same values of *A* and Δ have been observed in a series of hydroxylic solvents[13]

indicate quite clearly that there is a delocalisation of negative charge from the gold atom to the chelate ligand when coordination leads to the formation of a 5-membered pseudo-aromatic chelate ring containing the metal. The cationic complex containing the aliphatic chelating diamine (tmen) which has no pseudo-aromatic character gives a relatively low value of A (0.14).

Data of this type have also been collected for a number of reactions at palladium(II) complexes and are reported in Table 7.2.

It is of interest to note that, on going from the anionic substrate $[PdCl_4]^{2-}$ to the cationic $[Pd(dien)X]^+$, the discriminating ability decreases from

Table 7.2 Kinetic parameters for the entry of amines in some palladium(II) complexes (data at 25 °C)

Substrate	*Solvent*	*Leaving group*	*A*	Δ*	*Reference*
$[PdCl_4]^{2-}$	H_2O	Cl^-	0.24	1.5	14
§$[Pd(dien)X]^+$	H_2O	X^-†	<0.01	1.6	15
$[Pd(S\text{—}S)Cl_2]$	$(CH_3OCH_2)_2$	S‡	0.22	2.4	16
trans-$[Pd(Pr^i_2S)_2Cl_2]$	$(CH_3OCH_2)_2$	i-Pr_2S	0.13	2.3	17
trans-$[Pd(4Cl\text{-}py)_2Cl_2]$	$(CH_3OCH_2)_2$	4Cl-py	0.075	—	18
trans-$[Pd(4Cl\text{-}py)_2Br_2]$	$(CH_3OCH_2)_2$	4Cl-py	0.24	—	18
trans-$[Pd(4Cl\text{-}py)_2I_2]$	$(CH_3OCH_2)_2$	4Cl-py	0.34	—	18

Abbreviations: dien = diethylenetriamine; S–S = 1,2-di(phenylthio)ethane
*The Δ values reported are a measure of steric retardation effects discussed in Section 7.8
†X = NO_2, N_3, SCN, I
‡The data refer to the opening of the chelate ring
§The data relative to these processes have been obtained in acid buffers in order to avoid the possible deprotonation of the ligand

$A = 0.24$ to $A < 0.01$, thus suggesting that the total charge of the complex is not directly related to the electrophilicity of the reaction centre.

The data relating to the complex $[Pd(S—S)Cl_2]$ reported in Table 7.2, refer to the processes

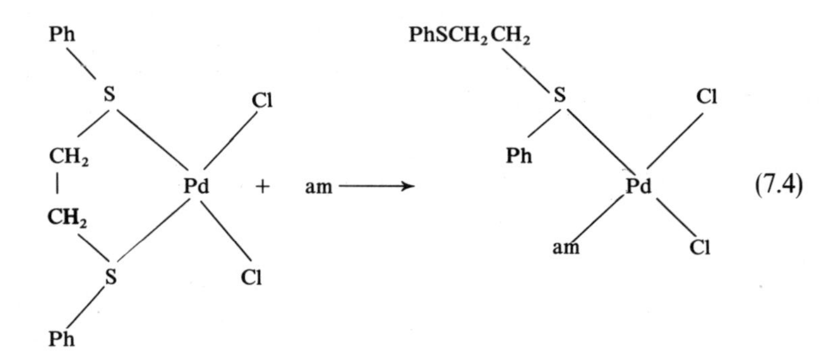

i.e. to the opening of the chelate ring. They can be compared with the corresponding data, obtained under the same experimental conditions, with the platinum(II) complex $[Pt(S—S)Cl_2]$ [19] ($A = 0.14$ and $\Delta = 1.6$) and will be used in discussing the role of the metal (Section 7.10).

The data relative to the substrate *trans*-$[Pt(Pr^i_2S)_2Cl_2]$ are also available[17] ($A = 0.13$ and $\Delta = 2.4$), and will be discussed in relation to steric hindrance problems, in Section 7.8.

Finally, the values of A obtained from complexes of the type *trans*-[Pd $(4\text{-Cl-py})_2X_2$] will allow a discussion of the *cis* effect in palladium(II) complexes (Section 7.6).

7.3.2 Reactivity of thioethers

A second homogeneous group of nucleophiles is formed by the thioethers, R—S—R′. In this case the pK_a values are not available, and a relative measure of their donor tendency can be found in the $-\Sigma\sigma^*$, that is the sum of the Taft σ^* constants relative to the radicals R and R′ attached to the sulphur atom. The discrimination ability of a given substrate between various entering thioethers can be measured from the coefficient A' in the expression

$$\log k_2 = A'(-\Sigma\sigma^*) + \text{constant} \tag{7.5}$$

A remarkable point is that a linear relationship has been proved between the pK_a and $-\Sigma\sigma^*$ values for amines[20]: $pK_a = -3.2(\Sigma\sigma^*) + \text{constant}$.

The existence of linear relationships of the types depicted in equations (7.2) and (7.5) should also prove that the relative σ donor tendency of amines and thioethers is independent of the nature of the solvent. However, since the actual σ donor tendency is obviously related to the nature of the reaction medium, comparison between the behaviour of different substrates is only possible when the reactions are carried out in the same solvent.

The entry of thioethers has been studied in a number of cases and the values of A' which have been measured are summarised in Table 7.3.

Table 7.3 Discrimination parameters of planar complexes for the entry of thioethers (data at 25°C)

Substrate	*Leaving group*	*Solvent*	A'	*Reference*
$[Pt(bipy)Cl_2]$	Cl^-	CH_3OH	1.2_4	21
$[Pt(bipy)(N_3)Cl]$	Cl^-	CH_3OH	0.47	21
$[Pt(bipy)NO_2)Cl]$	Cl^-	CH_3OH	2.7	21
$[Pt(bipy)(NO_2)_2]$	NO_2^-	CH_3OH	2.6	21
$[Pt(bipy)(NO_2)(N_3)]$	N_3^-	CH_3OH	2.6	21
trans-$[Pd(4CN\text{-}py)_2Cl_2]$	4CN-py	$(CH_3OCH_2)_2$	0.9_0	22
trans-$[Pd(3CN\text{-}py)_2Cl_2]$	3CN-py	$(CH_3OCH_2)_2$	0.9_1	22
trans-$[Pd(py)_2Cl_2]$	py	$(CH_3OCH_2)_2$	0.9_2	22
trans-$[Pd(4CH_3\text{-}py)_2Cl_2]$	$4CH_3$-py	$(CH_3OCH_2)_2$	0.9_0	22
trans-$[Pd(4Cl\text{-}py)_2Cl_2]$	4Cl-py	$(CH_3OCH_2)_2$	0.29	18
trans-$[Pd(4Cl\text{-}py)_2Br_2]$	4Cl-py	$(CH_3OCH_2)_2$	0.18	18
trans-$[Pd(4Cl\text{-}py)_2I_2]$	4Cl-py	$(CH_3OCH_2)_2$	0.02	18
trans-$[Pd(py)_2Br_2]$	py	$(CH_3OCH_2)_2$	0.28	18
trans-$[Pd(py)_2I_2]$	py	$(CH_3OCH_2)_2$	<0.02	18

A comparison of the discriminating ability of the complex $[Pt(bipy)Cl_2]$ between entering amines and entering thioethers shows that the kinetics are much more influenced by the nature of the substituents on the sulphur than they are by those on the nitrogen atom [21].

7.3.3 Reactivity of various groups

Apart from the homogeneous classes of ligands such as amines and thioethers, linear free energy relationships have been obtained with a wide range of different nucleophiles, by using as a reference the reactivity at the standard substrate *trans*-$[Pt(py)_2Cl_2]$ [23]. In these cases the relationship does not involve free energy variations related to equilibria but to free energies of activation. The n^0_{Pt} (or n_{Pt}) scale refers to the processes:

$$trans\text{-}[Pt(py)_2Cl_2] + Y^- \rightarrow trans\text{-}[Pt(py)_2Cl(Y)] + Cl^- \quad (7.6)$$

carried out in methanol at 30 °C. The index n^0_{Pt}, which is a characteristic of each nucleophile, was defined[24] by the expression:

$$n^0_{Pt} = \log(k_Y/k_S)$$

where k_Y is the second-order rate constant for the reaction of Y at the standard complex, and k_S the k_1 term, divided by the concentration of the solvent itself in order to get the index dimensionless. In the same way n_{Pt} is simply defined by the expression

$$n_{Pt} = \log(k_2/k_1)$$

and the two scales only differ because of a shift of constant value.

The n^0_{Pt} values for the common reagents are reported in Table 7.4. A large set of values (44 entering groups) is available in the literature[25].

There is no simple correlation between the n^0_{Pt} values and other properties of the reagents (such as basicity, redox potentials, etc.). They are, however, useful, since on plotting the values of log k_2 for a nucleophile against its n^0_{Pt} value for substitutions at any platinum(II) complex, it appears that the majority of the data obeys a linear relationship. It is therefore possible to obtain and compare discrimination parameters, which are always relative to the discrimination of the standard complex, according to the expression

$$\log k_2 = s(n^0_{Pt}) + \text{constant} \tag{7.7}$$

The parameter s is usually reported as the nucleophilic discrimination factor and a number of comparisons will be made later in this article.

Table 7.4 Values of the n^0_{Pt} index for some common nucleophiles

Reagent (Y)	n^0_{Pt}	*Reference*
Cl^-	3.04	23,24
NO_2^-	3.22	23,24
N_3^-	3.58	23,24
Br^-	3.96	23,24
I^-	5.42	23,24
SCN^-	5.65	23,24
$SeCN^-$	7.10	23,24
Thiourea	7.17	23,24
CN^-	7.14	25
$(C_6H_5)_3P$	8.93	25

The n^0_{Pt} sequence has been interpreted[26] in the sense that, throughout the series, the micropolarisability of the ligand seems to be the most important factor in determining the reactivity. Micropolarisability is considered to be the polarisability of the entering group in the anisotropic, and relatively strong, electric field which Y encounters in the transition state. In this sense it can be related to the concept of 'softness'[27]. It is not surprising that this property is very important since platinum(II) is a typical 'soft' or 'class B' metal ion.

Several values of the 'nucleophilic discrimination factor' s have been collected and will be discussed in the following sections. An important point to be considered is the intersection of the plots of log k_2 against n^0_{Pt} for different complexes. These show quite clearly that any general rationalisation in terms of reaction rates alone is impossible. For instance, on comparing the displacement of chloride from *trans*-[Pt(py)$_2$Cl$_2$] and *trans*-[Pt(PEt$_3$)$_2$Cl$_2$] by Br^- in methanol at 30 °C, the greater reactivity of the pyridine complex seems to suggest that the *cis* effect of pyridine on the reactivity is larger than the *cis* effect of the phosphine. However, on using I^- instead of Br^- as the entering nucleophile the reverse is true, and the effect appears to be dependent on the nature of the entering group. This arises from the wide difference between the nucleophilic discrimination factors for the two complexes.

A remarkable point emerges on comparing the results discussed up to now. Since micropolarisability is the major cause of the reactivity of Y, the existence of linear free-energy relationships within the homogeneous classes of nucleophiles (amines and thioethers) indicates that the micropolarisability is constant within a homogeneous class. For instance, amines as different from each other as pyridine and ammonia fit the same linear relationship when reacting with the substrate $[Pt(bipy)Cl_2]$[9]. Thus, micropolarisability is related to the nature of the donor atom (e.g. nitrogen in amines and sulphur in thioethers) and not to the molecule Y, as a whole.

As reported above, in the substitution reactions at platinum(II) complexes the majority of nucleophiles obey the n^0_{Pt} relationship, according to equation (7.7). However, the values for some of them do not lie on the straight lines obtained in the log k_2 *v.* n^0_{Pt} plots. A systematic study[26] of the deviations from the expected values showed that, whereas CH_3OH, pyridine, Cl^-, Br^-, N_3^-, SCN^-, SO_3^{2-} and $S_2O_3^{2-}$ obey equation (7.7), NO_2^-, $SeCN^-$ and thiourea are more reactive than expected when reacting with anionic species such as $[PtCl_4]^{2-}$, and less reactive than expected when reacting with cationic species such as $[Pt(dien)Cl]^+$. In addition, deviations have been also observed in neutral complexes containing π-bonded ligands, e.g. *trans*-$[Pt(PEt_3)_2Cl_2]$. This peculiar behaviour can be explained if one take into account the possibility that these 'biphilic' reagents can delocalise, by means of π-interactions, negative charge from the filled d orbitals of the metal. This leads to an extra stabilisation of the transition state, and the effect increases with the increasing availability of negative charge.

It is not surprising that $SeCN^-$ behaves as a biphilic reagent, whereas N_3^- and SCN^-, which have the same type of electronic configuration, closely obey equation (7.7). This can only mean that, even if the orbital symmetry is suitable to permit π-interactions, the disparity in the energy of the orbitals involved is such that significant overlap does not occur. It has been shown that N_3^- behaves as a biphilic reagent towards gold(III) complexes[28].

Apart from the reagents which have been studied in connection with the n^0_{Pt} relationship, other entering groups have also been investigated more recently. For instance, in a study of the formation of the platinum–olefine bond catalysed by $SnCl_3^-$, the reactivity sequence

$$\text{allyl-NH}_3^+ > \text{butyl-NH}_3^+ > \text{allyl-SO}_3^- > \text{allyl alcohol} > \text{pentenyl-NH}_3^+$$

has been reported[112] for the uncatalysed attack of the olefine at $[PtCl_4]^{2-}$.

On going from platinum(II) to palladium(II) and nickel(II) there is a large increase of reactivity[29], roughly in the order $1{:}10^{4-5}{:}\ 10^{6-7}$, and gold(III) complexes, as a rule, are more reactive than their platinum(II) analogues.

It is impossible to apply the n^0_{Pt} sequence to gold(III) complexes[30], and it is also impossible to obtain a new sequence (n_{Au}) since the relative reactivity of various nucleophiles towards gold(III) derivatives is dependent on the nature and charge of the substrate. The difference in the behaviour of the two metals is discussed in Section 7.10.3. There are not, at the moment, enough data to decide whether it is possible to obtain a self-consistent internal sequence of reactivity towards palladium(II) derivatives or other d^8 complexes.

Apart from the studies with amines and thioethers, the nucleophilicity of different entering groups towards palladium(II) has been measured in a few cases. The relative order of reactivity seems to be similar to that obtained with platinum(II) complexes. Thus, the rate of opening of the acetylacetonate ring in $[Pd(acac)_2]$ by reagents Y decreases in the order[31]

$$SCN^- > I^- > Br^- > Cl^- > H_2O,$$

and the sequence

$$I^- > Br^- > Cl^-$$

has been also reported[32] for the displacement of the diarsine $Ph_2AsCH_2CH_2AsPh_2$ (=EDA) from the complex $[Pd(EDA)_2]^{2+}$. These data suggest that micropolarisability is also very important in determining the reactivity towards palladium(II) substrates.

7.4 THE ROLE OF THE LEAVING GROUP

The influence of the nature of the leaving group on the rate of these substitutions cannot, in general, be expected to be as large as the effect of the nature of the entering nucleophile. In fact, since these reactions occur with an associative mechanism, one could expect that on going from the ground to the transition state the bond M—X between the metal and the displaceable ligand can either be partially stretched or, in the limit, be completely unaffected. Although this problem will be discussed more deeply in Section 7.10, it is clear that, in the limiting case, the variation of the M—X bond energy will only be related to the different geometry of the two states and, possibly, to the variations of charge. Therefore, in some cases, the nature of X could be relatively unimportant in determining the reactivity.

In the reactions

$$[Pt(dien)X]^+ + py \rightarrow [Pt(dien)py]^{2+} + X^-$$

(dien = diethylenetriamine), carried out[33] in water at 25 °C, the rate changes with the nature of X in the order:

$$\begin{aligned} &NO_3 > H_2O > Cl(1.0) > Br(0.67) > I(0.3) > \\ &> N_3(2.5 \times 10^{-2}) > SCN(9.2 \times 10^{-3}) > NO_2(1.4 \times 10^{-3}) > \\ &> CN(5 \times 10^{-4}) \end{aligned}$$

where the numbers in parentheses are an index of relative lability, calculated from the ratio $k_2(X):k_2(Cl)$. In this case the variation of reactivity over the whole series of substrates is large; however the range of X covered is also very extensive, ranging from the very labile ligands NO_3 and H_2O to the groups NO_2 and CN for which a π contribution to the M—X bond seems very reasonable.

Kinetic investigations have been carried out with platinum(II) substrates, in which the nature of both the leaving and entering groups are changed systematically. In the group of reactions[34]

$$[Pt(dien)X]^+ + Y^- \rightarrow [Pt(dien)Y]^+ + X^-$$

carried out in water at 30 °C, the lability sequence of X is

$$I(2.22) > Br(2.12) > Cl(1.0) > N_3(0.03)$$

irrespective of the nature of the entering group Y^-. In the group of reactions

$$[Pt(bipy)(NO_2)X]+Y^- \rightarrow [Pt(bipy)(NO_2)Y]+X^- \quad (7.8)$$

in methanol at 25 °C,

$$I(6.4) > Br(1.7) > Cl(1.0) > NO_2(5\times 10^{-2}) > N_3(7\times 10^{-3})$$

and for

$$trans\text{-}[Pt(pip)_2(NO_2)X]+Y^- \rightarrow trans\text{-}[Pt(pip)_2(NO_2)Y]+X^-$$

carried out in methanol at 30 °C, the lability sequence is

$$Br(1.7) > Cl(1.0) > NO_2(0.1) > N_3(0.071)$$

also independent on the nature of Y^- [35, 36, 37].

The same is true for the processes

$$[Pt(bipy)(NO_2)X]+RSR' \rightarrow [Pt(bipy)(NO_2)(RSR')]^+ + X^-$$

(methanol, 25 °C), where the observed sequence of lability[21] is

$$Cl(1.0) > NO_2(0.286) > N_3(0.143)$$

The differences in the sequences of lability reported above indicate that the mechanistic role of the nature of X is largely dependent on the nature of the substrate, and always (in platinum(II) chemistry) independent of the nature of the incoming nucleophile Y. In other words, on plotting the values of log k_2 *v.* n^0_{Pt} for the nucleophile Y, with each group of substrates differing only in the nature of the leaving group X, parallel straight lines are obtained, one for each substrate (see Section 7.10).

As far as reactions of the type depicted in equation (7.8) are concerned, the data relative to the entry of thiourea are available. As has already been discussed in Section 7.3, thiourea is a biphilic reagent, showing deviations from the n^0_{Pt} relationship. It is therefore possible, in these systems, to examine if and how the anomalous behaviour of thiourea is related to the nature of the leaving group. The deviation of thiourea from linearity in the n^0_{Pt} diagram can be quantitatively expressed as the difference between the experimental value of log k_2 and the value calculated from the n^0_{Pt} index and the parameters of the substrate. The rate of the processes depicted in equation (7.8) decrease on changing the nature of X in the order given by the sequence reported above ($I > Br > Cl > NO_2 > N_3$), and the corresponding deviations for thiourea are -0.34, -0.43, -0.15, $+0.80$, $+0.24$ respectively. Since thiourea is expected to increase its relative reactivity on increasing the negative charge at the metal, the data seem to indicate that the negative charge is more available for π-interaction in the transition state when less labile leaving groups are present in the system. The relatively high deviation ($+0.80$) observed when the leaving group is NO_2 can be explained by assuming that the NO_2 group is also biphilic. In this case the presence of two groups of this type in the transition state allow a better distribution of charge in a large π-system, thereby leading to the relative stabilisation of the transition state[35]. This was also observed in the entry of NO_2^- into *trans*-$[Pt(pip)_2(NO_2)Cl]$ [38], where two NO_2 groups are present in the equatorial plane of the trigonal-bipyramidal transition state.

More recently, a study has been carried out of the displacement of various thioethers coordinated to platinum(II)[39] by following the reactions

$$[Pt(bipy)Cl(RSR')]^+ + Y^- \rightarrow [Pt(bipy)Cl(Y)] + RSR'$$

($Y^- = Cl^-, Br^-, I^-$, in methanol at 25 °C). The order of relative lability is

$$C_6H_5SCH_3 > C_6H_5CH_2SCH_3 > C_2H_5SCH_3 > C_2H_5SC_2H_5$$

corresponding to the decreasing σ donor ability of the sulphur donors, as measured by the $-\Sigma\sigma^*$ values.

On going from platinum(II) to palladium(II) complexes, data are available to compare the behaviour of the substrates $[Pd(dien)X]^+$ and $[Pt(dien)X]^+$ in their reactions with pyridine. The comparison shows that in the case of the palladium(II) substrate[15], apart from the large increase in reactivity, the lability sequence is qualitatively the same ($Cl,Br > I > N_3 > SCN > NO_2$).

The processes

$$trans\text{-}[Pd(am)_2X_2] + RSR' \rightarrow trans\text{-}[Pd(am)(RSR')X_2] + am$$

which have been studied[22] in 1,2-dimethoxyethane at 25 °C can also be considered in this section, even though the amine ligand represents both the leaving group and the *trans* partner. Increasing the basicity of the amine leads to a parallel decrease in reactivity and this can be interpreted as a consequence of the decreasing electrophilicity of the metal (or as a consequence of the increasing Pd—N bond strength).

The stability of the M—X bond does not have to be opposite to the lability of X. For instance, in a recent work[113] concerning the relative stabilities and rates of exchange of chelating dienes(1–9) coordinated to rhodium(I) in the binuclear complexes $[Rh_2(diene)_2Cl_2]$, the most stable complexes are the kinetically most labile. This was accounted for in terms of a strong π-bonding character of the most labile dienes, favouring penta-coordination.

The role of the leaving group has also been studied in gold(III) complexes, mainly in processes involving the displacement of heterocyclic amines. The processes

$$[AuCl_3(am)] + Y^- \rightarrow [AuCl_3Y]^- + am \tag{7.9}$$

($Y^- = Cl^-, Br^-, N_3^-, NO_2^-$) have been investigated in methanol at 25 °C[28,40]. The values of $\log(k_2/k_1)$, which are a measure of the relative free energy of activation, and the corresponding pK_a values of the leaving amine are reported in Table 7.5. The data show a clear difference between the entry of NO_2^- and N_3^- on the one hand and that of Cl^- and Br^- on the other. In the second case, the values of $\log(k_2/k_1)$ decrease linearly with increasing pK_a of the amine, according to the expressions

$$\log(k_2/k_1) = -C(pK_a) + \text{constant}$$

where $C = 0.61$ and 0.35 for $Y^- = Cl^-$ and Br^- respectively. This behaviour can be interpreted, as reported above for palladium(II) complexes, as a consequence of the changing electrophilicity of the metal and the difficulty of bond-breaking. This is not inconsistent with the smaller influence of the nature of the amine when $Y^- = Br^-$, as compared to the entry of Cl^-, since it seems logical to believe that the formation of the new bond (Au—Br) is relatively more important.

In the case of the entry of N_3^- and NO_2^-, the reactivity at first increases with the basicity of the amine, reaches a maximum and then decreases. This

Table 7.5 Values of $\log(k_2/k_1)$ for the reactions $[AuCl_3(am)]+Y \rightarrow [AuCl_3Y] +$ am in methanol at 25°C

Leaving amine	pK_a	$Y^- = Cl^-$	$Y^- = Br^-$	$Y^- = NO_2^-$	$Y^- = N_3^-$
Quinoline	4.95	2.91	4.47	3.69	4.99
Isoquinoline	5.14	2.62	—	4.07	5.09
Pyridine	5.17	2.74	4.30	3.96	5.16
3-methylpyridine	5.68	2.44	4.01	4.77	5.04
4-methylpyridine	6.02	2.20	3.97	4.85	—
3,5-dimethylpyridine	6.34	2.16	—	4.69	4.86
2,6-dimethylpyridine	6.75	1.80	3.75	4.62	4.60
2,4-dimethylpyridine	6.99	1.61	3.63	4.29	4.06

behaviour has been interpreted in terms of π-interactions between the metal and the entering group. The increasing reactivity can be ascribed to a progressively greater availability of negative charge at the metal, but further increase of basicity leads to a decrease in reactivity as the bond-breaking aspects of these processes become dominant.

Exactly the same type of behaviour has been also observed[40] in the displacement by Cl^-, NO_2^- and N_3^- of the amine ligand in complexes of the type *trans*-$[Au(CN)_2Cl(am)]$.

A study of the role of the leaving group has been carried out[70] with nickel(II) planar substrates, by following the reactions

$$trans\text{-}[Ni(PR_3)_2(NCS)_2] + 2\ \text{bipy} \rightarrow$$
$$\rightarrow trans\text{-}[Ni(\text{bipy})_2(NCS)_2] + 2\ PR_3$$

which produce the hexa-coordinate nickel(II) species. The rate-determining step in these processes seems to be the bimolecular replacement of the first phosphine ligand by bipy. A linear decrease of reactivity ($\log k_2$) accompanies the increasing basicity of the phosphine ligands, as measured by the $-\Sigma\sigma^*$ values of the radicals attached to the phosphorus atom. This can be ascribed to difficulty in the breaking of the Ni—P bond, and to the different localisation of negative charge at the nickel atom, and can be explained assuming a synergic $\sigma+\pi$ Ni—P bond. Data dealing with the displacement of phosphine ligands are not available for other planar complexes.

7.5 ACTIVATION PARAMETERS

The number of kinetic studies carried out over a range of temperatures in order to discuss the enthalpy and entropy changes which accompany the formation of the transition state in these processes is not very large, but there are enough data to point out some characteristic features.

A general observation (coming mainly from data relating to platinum(II) complexes) is that the rate is mainly controlled by the enthalpy changes. This is important if one wishes to speak of the reactivity in terms of electronic effects and, as a rule, a high reactivity is paralleled by a low enthalpy of activation.

A second point, which is also general, is that the entropy of activation is largely negative (apart from some very special cases which will be discussed later), for both the reaction paths controlled by the k_1 and k_2 terms (see Section 7.2).

In the systems $[Pt(dien)X]^+$, where the halide ion is displaced by various reagents[34] including the solvent water, the $\Delta S^‡$ values lie in the range -17 to -31 cal deg^{-1} mol^{-1}, and similar values have also been found in dimethylsulphoxide[41]. Systems of the type *trans*-$[PtL_2Cl_2]$ (L = PEt_3 and piperidine) give entropy values, for the displacement of one Cl^- by various entering groups including the solvent methanol, ranging from -11 to -33 cal deg^{-1} mol$^-$, and the same is true for the mixed complex *trans*-$[Pt(PEt_3)(pip)Cl_2]$ [42–44, 69]. The hydrolysis of $[PtCl_4]^{2-}$ [45], $[Pt(NH_3)Cl_3]^-$ [46], *cis*-$[Pt(NH_3)_2Cl_2]$ [47] and $[Pt(NH_3)_3Cl]^+$ [47] have entropies of activation of magnitude -8, -15, -14 and -18 cal deg^{-1} mol^{-1} respectively.

An explanation for these large negative entropy values is not easy; they are far more negative than expected simply for the formation of the transition state from two separate particles, and do not depend upon the charge (cationic, neutral or anionic) of the substrate or of the entering group (cationic, neutral or anionic). An example can be found in processes of the type

$$[PtCl_4]^{2-} + ol \rightarrow [Pt(ol)Cl_3]^- + Cl^-$$

where the change from ol = allyl-ammonium cation to ol = allyl alcohol and ol = allyl-sulphonate anion give rise to entropy values of $\Delta S^‡ = -22.5$, -22.6 and -24.3 cal deg^{-1} mol^{-1} respectively[114]. Moreover, the $\Delta S^‡$ values do not depend significantly upon the nature of the solvent. It seems reasonable to associate these entropy values with a net increase in the tightness of bonding in the formation of the transition state, since thermodynamic entropy changes in equilibria involving an increase in the coordination number are also large and negative[48].

Similar observations can be made about the entropies of activation of some reactions of nickel(II) [70], palladium(II) [49] and gold(III) [30, 77, 78] substrates, although the changes for the gold(III) complexes are usually not so large as those found for the complexes of the other metals.

Finally, one significant feature is that the entropies of activation which accompany the entry of the biphilic reagents are always relatively more negative than those observed for the entry of nucleophiles. This fact can be related to a relatively more compact transition state and has been observed for the reactions of NO_2^- and thiourea with platinum(II) [34] and of NO_2^- and N_3^- with gold(III) substrates[30].

7.6 *trans* AND *cis* EFFECTS

The role of the ligands which are *trans* (T) and *cis* (L_1,L_2) to the leaving group X in the planar substrate has been studied and discussed in several papers. A characteristic (and perhaps surprising) feature of these studies is that whereas a relatively large amount of rate data about the *cis* effect is available, the same is not true for the *trans* effect. The effect of T on the reactivity is usually quite large in comparison to the effect of L_1 and L_2, and studies involving systematic changes in the nature of T in a series of substrates usually require the use of a range of kinetic techniques including those designed for fast reactions. For this reason the attention of various authors has been drawn to the *trans* effect, and a number of theories on this subject have been developed[1, 2, 49–56].

The actual difference between the *trans* and *cis* partners may be related to the non-equivalent positions that they occupy in the transition state as well as in the ground state, the ligand T lying in the equatorial plane of the trigonal-bipyramid whereas L_1 and L_2 occupy the axial positions.

7.6.1 The *trans*-directing influence

Preparative chemistry offered the first approach to the study of *trans* ligands and for this reason the term '*trans* effect' has been used to describe reaction features which are really due to a '*trans*-directing influence'. A number of experimental observations have been made which indicate that the following sequence of *trans*-directing influences apply:

$$\text{CO, CN, } C_2H_4 > PR_3\text{, H} > CH_3\text{, thiourea, } > NO_2\text{, I, SCN} > \text{Br} > \text{Cl} > \text{py, } NH_3 > H_2O\text{, } CH_3OH \qquad (7.10)$$

The concept of *trans*-directing influence has been used to distinguish between *cis*- and *trans*-platinum(II) complexes (Kurnakov's test[62]) and, more recently, the Russian school has reported data relating to the *trans*-directing influence of dimethylsulphoxide[63].

The above sequence, which is based on non-kinetic data, may be related to the strength of the *trans* bond, and for this reason the *trans* influence has been investigated by infrared, n.m.r. and x-rays methods, and also studied from a theoretical point of view[61, 115].

However, several examples exist which show that sequence (7.10) above cannot always be used to predict which of the four ligands around the metal will be the leaving group. Thus, for example, in the complex $[Pt(C_2H_4)Cl_3]^-$, ethylene can be replaced by other olefins[59] in spite of its position relative to that of Cl in sequence (7.10). In the reactions of Y^- with *trans*-$[Pt(PEt_3)(pip)Cl_2]$ in methanol[57], chloride is displaced instead of piperidine, even though PR_3 is one of the strongest *trans*-effect ligands in sequence (7.10). In *trans*-$[Pt(RSR')_2Cl_2]$ the leaving group can be either RSR′ or Cl_2 depending on the nature of the nucleophile and solvent used[17, 58].

7.6.2 The *trans*-labilising effect

For kinetic studies any definition of the *trans*-labilising effect should take into account both the ground and transition states involved in the reaction. The majority of kinetic data refer to the definition proposed by Basolo and Pearson[1] of a *trans*-labilising effect, namely that it is '... the effect of a coordinated group upon the rate of the substitution reactions of ligands opposite to it in a metal complex'. Qualitatively, the *trans*-labilising effect occurs in the following sequence:

$$C_2H_4, \sim NO, \sim CO, \sim CN > R_3P, \sim H, \sim \text{thiourea} > CH_3 > C_6H_5 > SCN > NO_2 > I > Br > Cl > NH_3 > OH > H_2O$$

A theoretical molecular orbital study has been carried out[61] to explain the sequence for σ-bonded ligands.

The labilising effect of the *trans* partner, T, is usually large in comparison to that of the *cis* groups, L_1, L_2, but not always so, as has been proved in the reactions of chloro-ammine-platinum(II) complexes[68].

Data concerning the *trans*-labilising effect have also been obtained for unusual ligands and by various methods. The kinetics of the processes[29]

$$trans\text{-}[Pt(PEt_3)_2(T)Cl] + py \rightarrow trans\text{-}[Pt(PEt_3)_2(T)(py)]^+ + Cl^-$$

give the following *trans*-labilising effect sequence for the groups T,

$$H > CH_3 > C_6H_5 > p\text{-}ClC_6H_4 > p\text{-}CH_3OC_6H_4 > p\text{-}C_6H_5C_6H_4 > Cl$$

with a change of reactivity of four orders of magnitude along the sequence. Studies of the isotopic exchange processes of $[^{14}C]NHEt_2$ with *trans*-$[Pt(T)(NHEt_2)Cl_2]$ [64] showed the *trans*-labilising sequence,

$$C_2H_4 > Pr^iCH{=}CHMe > Me_2C(OH)C{\equiv}C(OH)Me_2 > Et_3Sb > Ph_3Sb > Me_3P > Et_3P > Pr^n_3P > Ph_3P > Et_3As > Ph_3As > Pr^n_2S \quad (7.11)$$

in various solvents. Other authors have claimed, from n.m.r. studies on *trans*-1,3-dichloro-2-olefin-4-(4-*Z*-pyridine)platinum(II), that ethylene is a better *trans*-labilising olefine than either *cis*- or *trans*-2-butene[65].

There is a similarity between the *trans*-labilising effect depicted in sequence (7.11) and the order of nucleophilicity towards platinum(II) substrates. This is not surprising, since both the entering group Y and the *trans* partner T occupy equivalent positions in the transition state; however, one must also consider that, as far as the ligand T is concerned, the effect on the rate is related to the differing ability of the ligand to stabilise the ground state in relation to the transition state, whereas for Y such a comparison must be made between the transition state and the free reagent in the reaction mixture. The reasons for sequence (7.11) are certainly related to various factors, such as micropolarisability and σ and π bonding. The relative σ and π contributions have been estimated in the ground state by a ^{19}F n.m.r. study[66] on complexes of the type *trans*-$[Pt(PEt_3)_2(T)L]$, where $L = p\text{-}FC_6H_4$ or $m\text{-}FC_6H_4$. A comparison[67] of the roles of T in the ground and transition state has also been made by means of the isotope effect in the reactions of *trans*-$[Pt(PEt_3)_2HCl]$ and *trans*-$[Pt(PEt_3)_2DCl]$.

Data relating to palladium(II) and gold(III) complexes seem to indicate the same type of *trans*-labilising influence as is found in platinum(II), but only a small amount of data on this subject is available.

Attempts to rationalise the available kinetic data concerning aquation and chloride substitution reactions of $[Pd(NH_3)_4]^{2+}$ and $[Pd(NH_3)_3Cl]^+$ in terms of *cis* and *trans* effects of the ligands have been also carried out[116, 49], although the results were not conclusive. However, the data concerning the kinetics of the successive ammonation of $[PdCl_4]^{2-}$ indicate[117] larger *trans* and *cis* effects for Pd relative to Pt. In gold(III) chemistry the *trans*-labilising effect Br > Cl has been recently observed[118] from a study of the displacement of the two bromide ligands in *trans*-$[Au(CN)_2Br_2]$ by chloride ion.

7.6.3 *trans* effect in terms of discrimination

The concept of *trans* effect in terms of labilising power is usually accepted and generally useful. There are, however, a few data showing that care must be used in discussing the experimental data. Thus, Belluco *et al.*[69] showed that in the processes

$$trans\text{-}[Pt(PEt_3)_2(T)Cl] + Y^- \rightarrow trans\text{-}[Pt(PEt_3)_2(T)Y] + Cl^-$$

the rate is largely dependent on the nature of T; however, while the usual sequence of *trans*-labilising effect $CH_3 > C_6H_5 > Cl$ was observed for the entry of the poor nucleophiles NO_2^-, N_3^- and Br^-, the entry of I^- and thiourea led to the sequence $CH_3 > Cl > C_6H_5$. Apparently the order of *trans*-labilising effect is dependent upon the nature of the entering group. What really happens is that when the nature of T is changed this affects the ability of the substrate to discriminate between the various reagents (the *s* parameters in equation (7.7)) and an inversion of reactivity is observed, depending upon the nature of the entering group. Because of this it was suggested that it might be possible to define the *trans* effect as the influence of the *trans* ligand on the ability of the substrate to discriminate between entering nucleophiles.

7.6.4 The *cis*-labilising effect

On moving from *trans* to *cis* effects, it is found that the latter have only been examined and discussed as '*cis*-labilising effects', and in terms of the effect of the nature of the *cis* partner upon the discriminating ability of the substrate. The rate is usually not as much affected by L_1 and L_2 as it is by the nature of T, but there are data showing that the *cis*-labilising sequence can be the same, at least qualitatively, as the *trans* effect. For instance, in the reactions of *cis*-$[Pt(PEt_3)_2(L)Cl]$ with pyridine, the rate changes with the nature of L according to the sequence $CH_3 > C_6H_5 > Cl$, as reported for the *trans* effect[29]. On the other hand, in the case of the acid hydrolysis of various chloroamine complexes of platinum(II) it has been observed that 'the *cis*-neighbour has a somewhat greater influence on the reactivity than the *trans*-neighbour[68]. In a study on the isotopic exchange of X^- with the complexes *trans*-$[Pt(PEt_3)(am)X_2)](X = Cl,Br)$, no systematic difference was observed between the effect of a phosphine or an amine in the *cis* position[71].

7.6.5 *cis*-effect in terms of discrimination

It is now quite clear[75] that the role of the *cis* partners in the processes discussed in Section 7.6.4 cannot be rationalised in terms of reaction rate alone, since inversions of reactivity have often been observed on changing the nature of the entering nucleophile. The comparison of discrimination parameters, such as '*s*' in equation (7.7), *A* in equation (7.2) and *A'* in equation (7.5), seems to be the best way of discussing the *cis* effect of the various ligands.

In reactions of the type[23]

$$trans\text{-}[PtL_2Cl_2]+Y^- \rightarrow trans\text{-}[PtL_2(Cl)Y]+Cl^-$$

carried out in methanol at 30 °C, the nucleophilic discrimination parameter *s* changes with the nature of L according to the sequence:

$$PEt_3(1.62) > AsEt_3(1.3) > SeEt_2(1.1) > \text{pyridine}(1.00) > \text{piperidine}(0.95) \quad (7.12)$$

and in the reactions[72]

$$[Pt(bipy)(L)Cl]+Y^- \rightarrow [Pt(bipy)(L)Y]+Cl^-$$

in methanol at 25 °C the corresponding sequence of *s*, depending on the nature of L, is

$$NCS(1.3) > N_3(0.95) > NO_2(0.87) > Cl(0.75) \quad (7.13)$$

Both of these two sequences can be interpreted by saying that the presence of one (or two) *cis*-neighbours, which are able to delocalise negative charge away from the metal, either because of π-interactions as in the case of PEt_3 in sequence (7.12), or because they are high in the scale of micropolarisability as in the case of NCS in sequence (7.13), enhances the electrophilicity of the reaction centre leading to a relatively easier attack by the nucleophile. The actual rates depend on both bond-formation and bond-rupture, and therefore inversions of reactivity sequences can be observed in some cases.

The general idea that a relatively large value of the *s* parameter is related to a relatively large delocalisation of negative charge away from the metal in the substrate is confirmed by the data relating to the entry of the biphilic thiourea into the substrate [Pt(bipy)(L)Cl]. A comparison between the *s* values and the deviations of thiourea from linearity (see Section 7.4) in the log k_2 *v.* n^0_{Pt} plot is shown in Table 7.6, and indicates that nucleophilic attack, as measured by *s*, is easier when there is relatively less charge at the metal, as indicated by the relatively larger negative deviation of thiourea[72].

Table 7.6 Nucleophilic discrimination parameters, *S*, and deviations for the reactivity of thiourea in the complexes [Pt(bipy)(L)Cl] (methanol, 25 °C)

cis Partner, L	*Discrimination, s*	*Deviation* Δ
Cl	0.75	−0.10
NO_2	0.87	−0.15
N_3	0.95	−0.3
NCS	1.3	−1.2

Further support for this concept has been obtained from studies[73] of the reactions

$$trans\text{-}[Pt(am)_2Cl_2]+Y^- \rightarrow trans\text{-}[Pt(am)_2Cl(Y)]+Cl^-$$

where the values of *s* increase as the basicity of the amine decreases.

As far as palladium(II) complexes are concerned, data relating to the reaction[18]

$$trans\text{-}[Pd(4Cl\text{-}py)_2X_2] + am \rightarrow trans\text{-}[Pd(4Cl\text{-}py)(am)X_2] + 4Cl\text{-}py$$

carried out in 1,2-dimethoxyethane at 25 °C, show that the *cis* effect of X on the discrimination is I > Br > Cl, i.e., the same as the order of micropolarisability. The same is true ($NCS > I > Br > N_3$) for the *cis* effect of X in the reactions

$$cis\text{-}[Pd(am)_2X_2] + am \rightarrow [Pd(am)_3X]X$$

which is believed to be the rate-determining step in the *cis-trans* isomerisation of these species[74] (see Section 7.11).

An interesting case has been observed[18] in the reactions

$$trans\text{-}[Pd(am)_2X_2] + RSR' \rightarrow trans\text{-}[Pd(am)(RSR')X_2] + am \quad \text{(1,2-dimethoxyethane, 25 °C)},$$

where the order of the *cis* effect of X on the discrimination is Cl > Br > I. This anomaly has been attributed to a stereo-electronic interaction between the lone pair of electrons of the sulphur atom which is not used in bonding and the *cis* partners X. Another anomaly of this type ($N_3 > Cl$) has been observed for the entry of thioethers in the complexes [Pt(bipy)(L)Cl].

However, in the opening of the chelate ring in complexes of the type $[Pd(PhSCH_2CH_2SPh)X_2]$ by amines under the same experimental conditions

$$[Pd(S\frown S)X_2] + am \longrightarrow [Pd(am)(S\frown S)X_2] \qquad (7.14)$$

where the ligand X plays the role of both *cis* and *trans* partner[76], the sequence of the dependence of discrimination on the nature of X is the 'normal' one, i.e.,

$$SCN(0.42) > I(0.36) > N_3(0.3) > Br(0.28) > Cl(0.23)$$

(in this case the number in parenthesis is the A parameter of equation (7.2), i.e., the slopes of the lines in the plot of $\log k_2$ *v.* pK_a). This is not inconsistent with the interpretation of the 'anomalous' behaviour reported above, since, in these reactions, any stereo-electronic interaction must be present in both the ground and transition states, so that the effect cancels out in the kinetics (see Section 7.8 for an analogous case for sterically-hindered gold(III) complexes).

On going to gold(III) substrates, the only data available about the *cis* effects come from a comparison of the two reactions[28, 40]:

$$[AuCl_3(am)] + Y^- \rightarrow [AuCl_3Y]^- + am$$

$$trans\text{-}[Au(CN)_2Cl(am)] + Y^- \rightarrow trans\text{-}[Au(CN)_2Cl(Y)]^- + am$$

In the latter reactions the discrimination between entering reagents is decreased to some extent, as compared to the former, and this can be interpreted as a decrease of electrophilicity at the metal on going from the chloro to the cyano complexes.

7.7 SOLVENT EFFECTS

The nature of the solvent is probably of importance for various reasons in these processes. First, it determines the actual substitution which occurs in the reaction mixture. Thus, complexes of the type *trans*-[Pd(am)$_2$Cl$_2$] undergo substitution of the amine ligand when treated with neutral nucleophiles such as thioethers in 1,2-dimethoxyethane, whereas in methanol they may be used as starting materials for the preparation of *trans*-[Pd(am)$_2$X$_2$] by the replacement of Cl with other halide ions[18].

7.7.1 Nucleophilicity of solvents

Quite often the solvent behaves as a nucleophile towards planar complexes, leading to substitution via the k_1 controlled path of the reaction (see Section 7.2). The values of k_1 do not depend upon the nature of the nucleophile Y, but change with the nature of the substrate and solvent, as expected in accordance with the general reaction scheme reported in Section 7.2.

Comparisons of the nucleophilic reactivity of various solvents are difficult, because, on the one hand, the solvent is also the reaction medium, and therefore solvation effects could also be of importance, and, on the other hand, the k_2 term is usually large enough to introduce an appreciable error in the determination of k_1. Furthermore, since the solvents are usually chosen to facilitate the study of the reactivity of the various reagents Y, they are, as a rule, poor nucleophiles. A possible exception can be found in the solvents $(CH_3)_2SO$ and CH_3NO_2 which, probably because of π-bonding, are relatively good reagents towards platinum(II) complexes[79, 80].

The k_1 values have been compared, in a number of cases when the nature of the solvent has been systematically changed. For example, when the chloride of *trans*-[Pt(py)$_2$Cl$_2$] is the leaving group, the sequence of k_1 values is:

$$(CH_3)_2SO > CH_3NO_2 > C_2H_5OH > n\text{-}C_3H_7OH$$

and, with *trans*-[Pt(pip)$_2$Cl$_2$] [81] is:

$$(CH_3)_2SO > HCON(CH_3)_2 > CH_3CN > CH_3NO_2 > (CH_3)_2CO > CH_3OH$$

whereas in the case of *trans*-[Pt(pip)$_2$(NO$_2$)Cl] the sequence is:

$$HCON(CH_3)_2 > (CH_3)_2SO > CH_3OH$$

The k_1 values for dimethylsulphoxide are roughly one order of magnitude greater than the k_1 values for alcohols.

7.7.2 Solvation effects

Perhaps the most remarkable point concerning the role of the solvent is the absence of a large effect on the rate of substitution and on the discriminating ability of the substrate in platinum(II) complexes, as found by Belluco *et* al. in various studies[36, 81–83]. For instance[81], the values of the nucleophilic discrimination parameter *s* (see equation (7.7)) changes with the nature of the solvent in reactions involving the complex *trans*-$[Pt(pip)_2Cl_2]$ in the order:

$$(CH_3)_2SO(0.61) < CH_3NO_2(0.64) < CH_3CN(0.74) < HCON(CH_3)_2(0.74) < (CH_3)_2CO(0.78) < CH_3OH(0.91)$$

whereas the k_2 rate constants for displacement of Cl^- by Br^- change in the order:

$$(CH_3)_2SO(5.2 \times 10^{-3}) < CH_3OH(6.16 \times 10^{-3}) < HCON(CH_3)_2 (6.7 \times 10^{-3}) < CH_3CN(1.28 \times 10^{-2}) < CH_3NO_2(1.5 \times 10^{-2}) < (CH_3)_2CO(8 \times 10^{-2})$$

(rate constants expressed in $M^{-1}\ s^{-1}$).

As far as the solvent effect is concerned, a distinction must be made between general and specific solvation. The second could be a characteristic of planar complexes, since they are coordinatively unsaturated and can possess two solvent molecules weakly coordinated at the axial positions. An adduct of this type, namely the complex $[Pt(NH_3)_4(CH_3CN)_2]Cl_2$, has been isolated[84], and circular dichroism studies also indicate[85] this form of specific solvation. (More recently[119] the association of the planar substrate with one solvent molecule to give a distorted 5-coordinate structure has also been suggested.) It seems likely that the two axial solvent molecules are lost during the formation of the transition state, and if this is energetically important it should be experimentally detectable from kinetic measurements.

A study in this direction[13] has been carried out in the reaction between amines and $[AuCl_4]^-$, and in the reverse processes, namely the displacement of the coordinated amine from $[AuCl_3(am)]$ by chloride, in various hydroxylic solvents. The effect of the solvent on the rate of the forward reaction is opposite to that observed in the reverse process, and may be easily explained by general solvation phenomena, whereas a relatively strong axial solvation would require the same trend in both series of substitutions.

7.8 STERIC EFFECTS

Steric hindrance can be introduced in the reacting system by using either bulky ligands coordinated to the metal or hindered nucleophiles. Data are available in both cases, and the consequences of the hindrance on the kinetic behaviour can be of two different types. In the first, the rate of substitution is decreased, as normally expected for steric retardation in bimolecular processes. In the second case, the hindrance at the substrate is such ('pseudo-octahedral complexes') as to prevent the usual bimolecular attack of the incoming group Y, and in this case either the reaction is greatly slowed down or does not occur, or else an alternative mechanism develops[89–92].

7.8.1 Steric retardation

As a consequence of the non-equivalent positions occupied by the *trans* and *cis* partners in the trigonal-bipyramidal transition state, the effect of sterically hindered T or L_1, L_2 ligands is different. This was proved[29] in the displacement, by pyridine, of the chloride ion in *trans*-[Pt(PEt_3)$_2$(T)Cl] and *cis*-[Pt(PEt_3)$_2$(L)Cl], where T and L are phenyl, *o*-tolyl and mesityl groups. The kinetic data indicate that the retardation due to the *cis*-hindered ligand is much greater than that due to the *trans*-hindered ligand and, in fact, on going from the *cis*-phenyl to the *cis*-mesityl derivative the rate decreases by four orders of magnitude, whereas the corresponding retardation is only a factor of 35 when these groups are *trans* to the leaving chloride.

The two methyls in the phenyl ring of the mesityl group bonded to the metal interfere in the ground state with the two ligands that are *cis* to it. On going to the transition state the interference remains the same as in the ground state when the mesityl group occupies an equatorial position in the trigonal-bipyramid, but increases when the originally *cis*-mesityl group occupies an axial position, since, in this second case, interference develops between the two *ortho*-methyls of the mesityl group and the three ligands in the equatorial plane. According to this description a small, or negligible, steric retardation can be expected in the case of the *trans*-hindered mesityl, as has been observed, and, in fact, the small effect reported above might even be due to the electronic difference between phenyl and mesityl arising from the inductive effects of the methyl groups.

A recent detailed re-examination of the reaction between *trans*-[Pt(PEt_3)$_2$(mesityl)Cl] and pyridine[86] seems to indicate that the mechanism of substitution of the chloride is somewhat more complicated than originally suggested, involving the accumulation of the intermediate *trans*-[Pt(PEt_3)$_2$(mesityl)(CH_3OH)]$^+$ (the solvent was methanol). However this fact is not inconsistent with the conclusions about steric hindrance reported above.

Steric hindrance has also been used recently to gain evidence of a trigonal-bipyramidal transition state in amine-substitution reactions of complexes of the type *trans*-[Pt(R^1_3P)(NHR^2_2)X_2] [120].

The steric hindrance of the incoming nucleophile has been studied in some depth in the reactions of heterocyclic amines towards planar substrates (see Section 7.3). Substituted pyridines of different basicity were used, and could be divided into three groups, containing derivatives of pyridine, 2-methylpyridine and 2,6-dimethylpyridine respectively. In the plots of log k_2 *v.* pK_a the amines fit three parallel straight lines, separated from each other by a constant value (Δ). The decrease of reactivity (measured in terms of log k_2) due to the presence of two methyl groups *ortho* to the nitrogen in the pyridine ring is twice as much (2Δ) as that due to only one *ortho* methyl group (Δ). The values of Δ are therefore a measure of the steric retardation, and the data relative to palladium(II) and gold(III) have already been summarised in Tables 7.1 and 7.2. The available values of Δ for platinum(II) complexes relate to the processes[9]

$$[\mathrm{Pt(bipy)Cl_2}] + \mathrm{am} \rightarrow [\mathrm{Pt(bipy)Cl(am)}]^+ + \mathrm{Cl^-}$$

(methanol, 25 °C), $\Delta = 0.96$; the opening of the chelate ring[16]

$$[\mathrm{Pt(S{-}S)Cl_2}] + \mathrm{am} \rightarrow [\mathrm{Pt(S)(am)Cl_2}]$$

(with the S of the product linked by a vertical bond to a second S written below)

(1,2-dimethoxyethane at 25 °C), $\Delta = 1.6$; and the substitutions[17]

$$trans\text{-}[Pt(i\text{-}Pr_2S)_2Cl_2] + am \rightarrow trans\text{-}[Pt(i\text{-}Pr_2S)(am)Cl_2] + i\text{-}Pr_2S$$

(1,2-dimethoxyethane, 25 °C), $\Delta = 2.48$.

Examination of the data indicates that, as expected, the same amount of steric retardation has been found for the series of complexes of the type $[Au(N—N)Cl_2]^+$, where N—N is 2,2′-bipyridyl, 1,10-phenanthroline and $5\text{-}NO_2$-1,10-phenanthroline; furthermore, the same steric hindrance has been found for the displacements at $[AuCl_4]^-$ in different solvents[11,13].

In addition, it seems that the increase in the effect of steric retardation on changing the nature of the metal, parallels the increase of reactivity. The greater reactivity of gold(III) as compared to platinum(II) complexes, in the reactions of the substrates $[Pt(bipy)Cl_2]$ and $[Au(bipy)Cl_2]^+$, corresponds to Δ values which are 0.96 and 1.2 respectively[12]. For the rate of the opening of the chelate ring in complexes of the type $[M(PhSCH_2CH_2SPh)Cl_2]$, the greater reactivity of palladium(II) in comparison to platinum(II) is reflected in the increase of Δ from 1.6 to 2.4. The fact that the complexes of the type *trans*-$[M(Pr^i_2S)_2Cl_2]$ provide virtually the same values of Δ and A for M = Pd and Pt is not inconsistent with this observation, since in this peculiar case the effect due to the nature of the metal is almost cancelled, at least as far as bond formation is concerned, by the large steric hindrance within the substrate, due to the presence of the bulky Pr^i_2 ligands[17].

A comment is required about the absence of any detectable steric effect in the reaction expressed in equation (7.9), involving the displacement of amines from $[AuCl_3(am)]$ complexes[28,40]. In these reactions there is no indication of any steric retardation even when the amine belongs to the 2,6-dimethyl-pyridine group. This fact can be easily understood if one bears in mind that, in this case, the amine is the leaving group, and the steric disturbance is therefore present in both the ground and transition states so that it is un-detectable kinetically (a similar case is quoted in the discussion about the reaction expressed in equation (7.14) in Section (7.6).

A relatively small retardation has been also observed with pyridines having substituents in the 3- and 5- positions in reactions of the type[74]

$$cis\text{-}[Pd(am)_2X_2] + am \rightarrow [Pd(am)_3X]X$$

and the same observation was reported for the reactions of amines towards $[Pt(dien)Br]^+$ in water[88]. This is unusual, and probably relates to the very bulky transition states.

An attempt has been made recently[121,122] to evaluate the importance of steric blocking by gradually changing the size of bulky substituents of the ligands in the complex during the examination of substrates of the type $[Pt(A—A)(SCN)_2]$ (A—A = bidentate amine) in their thiocyanate-exchange reactions.

7.8.2 Pseudo-octahedral complexes

The study of highly-hindered substrates has been carried out mainly by Basolo and co-workers[89–92], with 'pseudo-octahedral complexes' of the type

$[M(Et_4dien)X]^+$, where Et_4dien is the polydentate ligand $Et_2NCH_2CH_2NHCH_2CH_2NEt_2$, or with compounds of the same type containing $Et_2NCH_2CH_2N(Me)CH_2CH_2NEt_2$ or $Me_2NCH_2CH_2N(Me)CH_2CH_2NMe_2$. In the ground state the terminal ethyl groups of the ligand lie in such a position as to block the positions above and below the plane of the complex (hence the term 'pseudo-octahedral'), so that the usual attack of Y cannot occur. The rate of displacement of X (Cl or Br) by various reagents is reduced by many powers of 10, and it was found that when M = Pd this displacement is independent of the concentration of the nucleophile[90]. Hydroxide, which is usually unreactive, displaces X according to a two-term rate law (similar to that expressed in equation (7.1)) in the Et_4dien complex. The second-order contribution for the reaction of OH^- is probably related to the deprotonation of the coordinated ligand, and disappears on going to complexes containing $MeEt_4dien$ or Me_5dien.

Perhaps the most interesting results obtained are those related to the bimolecular displacement of water from the complex $[Pd(Et_4dien)(H_2O)]^{2+}$ by various nucleophiles[90]. Water is very labile in these planar substrates, but the order of reactivity is:

$$S_2O_3^{2-} > HSO_3^- > SCN^- > NO_2^- > Cl^- > Br^- > I^- > CH_3COO^- > \text{thiourea}$$

i.e., it differs completely from the usual sequence of reactivity. It could be that in these bimolecular processes, the mechanism of substitution does not involve an attack at the empty p_z orbital of the metal (with the consequent possible increase of coordination number during the course of the reaction), but is due to attack on an orbital already used for bonding, so that the process becomes a concerted one. An alternative possibility has been suggested by Burmeister and Lim[123]. In the study of the reaction

$$[Pd(Et_4dien)XCN]^+ + Br^- \rightarrow [Pd(Et_4dien)Br]^+ + XCN^-$$

in dimethylformamide, they report that for X = S an S_N1 or solvent-assisted ligand interchange is operative, whereas for X = Se a path involving the opening of the chelate ring is also involved. This seems to be an example of mechanistic control through the nature of the leaving group.

In the gold(III) substrate $[Au(Et_4dien)Cl]^{2+}$ deprotonation of the coordinated ligand occurs easily to form the conjugate base $[Au(Et_4dien{-}H)Cl]^+$, and the attack of Y^- at this substrate leads to the displacement of the coordinated chloride without an appreciable second-order contribution to the rate law. When the entering group is N_3^-, the behaviour is unusual, since the entry of the first N_3^- is accompanied by the opening of one of the chelate rings[92].

Very interesting results have been recently obtained[93,94] from systematic studies with complexes of the type *trans*-$[Pt(PEt_3)_2(R)Cl]$ (R = phenyl, *o*-tolyl, mesityl) where the entry of the nucleophiles N_3^-, Br^-, NO_2^-, I^-, Et_2S, $C_6H_5S^-$, $S_2O_3^{2-}$, in methanol, occurs at a rate that is independent of the concentration of the entering reagent, whereas strong biphilic reagents (CN^-, $SeCN^-$, thiourea) displace the chloride with a second-order contribution to the rate law. The value of k_1 is the same in all cases; on changing the solvent from methanol to dimethylsulphoxide the value of k_1 is decreased to some extent, and a second-order contribution is also detectable for the entry of SCN^- and I^-.

It seems that in the reactions of these substrates the nucleophilicity of the solvent and the reactivity of the biphilic reagents appears to increase with respect to the nucleophilicity of the usual reagents, but even so the rates are very much slower than those for displacements at comparable complexes which are not sterically hindered.

7.9 POLYDENTATE LIGANDS

The use of polydentate ligands as nucleophiles towards planar complexes has been studied mainly with chelating nitrogen and sulphur donors (N—N and S—S) of normal ring size. As a rule, the rate is determined by the first attack, the closure of the chelating ring being fast since it does not require a further collision between the reagents. Results relating to the formation of the chelate ring, starting from the complex $[Pt(NH_3)Cl_2(H_2NCH_2CH_2NH_2)]$, indicate that the closure of the ring is a fast process, especially with ethylenediamine[96]. This has also been proved recently in a study of the ring-closing process starting from the complex *trans*-$[Pt(H_2NCH_2CH_2NH_3^+)_2Cl_2]$[124].

The same is true for the reaction of ethylenediamine with gold(III) complexes. The reaction

$$[AuCl_3(OH)]^- + enH^+ \rightarrow [AuCl_3(en)] + H_2O$$

has been reported[125] to be followed by the fast process

$$[AuCl_3(en)] \rightarrow [AuCl_2(en)]^+ + Cl^-$$

and also the slow reaction resulting in the formation of $[AuCl(en)(enH)]^{3+}$ from $[Au(en)Cl_2]^+$ and enH^+ is followed by the fast formation of $[Au(en)_2]^{3+}$. The closing of the chelate ring is also fast in reactions of $[AuCl_4]^-$ with dien in aqueous acid medium[126].

Exceptions are found when the polydentate ligand has some of its coordination sites in an unreactive form, as in the case of the mixed ligand, $H_2NCH_2CH_2OH$, where the formation of $[Pt(H_2NCH_2CH_2O)_2]$ from *trans*-$[Pt(H_2NCH_2CH_2OH)_2Cl_2]$, by way of HCl-elimination, is controlled by the OH^- concentration in the reaction mixture[95], and in the displacement of Br^- from $[PtBr_4]^{2-}$ with dien over a range of pH, where a partial protonation of the entering reagent can prevent ring closure[97].

Some interesting results emerge from a very recent study[98] which compares the rate constants and activation parameters for reactions involving the displacement of one chloride from $[Pt(bipy)Cl_2]$ by monodentate and bidentate amines. It was generally believed that the second donor atom does not participate in the first stage of the entry of a bidentate reagent, but it appears that, whereas for monodentate amines the $\Delta H^\ddagger$ and $\Delta S^\ddagger$ values are 15–17 kcal mol^{-1} and $-12, -17$ cal deg^{-1} mol^{-1} respectively, in the case of the bidentate amines the corresponding values are 11–12 kcal mol^{-1} and $-27, -28$ cal deg^{-1} mol^{-1}. This is thought to be due to an interaction of the 'free' amine group through its lone pair of electrons with the metal, and/or by means of hydrogen bonding with one of the two chloride ligands.

A general observation is that chelation in the ground state has a greater effect on the stability of the complex than other factors, such as softness or micropolarisability. Complexes containing chelate oxygen donors, e.g., $[Pt(C_2O_4)_2]^{2-}$ are well known, whereas monodentate oxygen ligands are usually very labile. The stability of chelated structures of the type

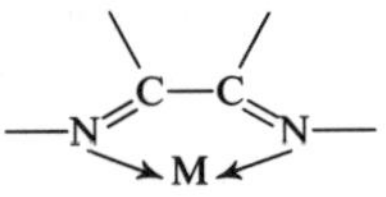

$$\begin{array}{c} \diagdown \quad \diagup \\ C{-}C \\ {-}N^{\nearrow} \quad {}^{\searrow}N{-} \\ \searrow M \swarrow \end{array}$$

is well known, and it has been suggested by Haake and Cronin[102] that this is due to the formation of a 5-membered pseudo-aromatic ring containing the metal. The very low lability of such a group has been used in various kinetic studies on the mechanistic role of the leaving group and of the *cis* partner in planar substitutions[9, 12, 21, 36, 37, 39]. A kinetic test of the aromaticity of this chelate structure can be found in the work[12] on the complexes of the type $[Au(N{-}N)Cl_2]^+$ (see Section 7.3).

The displacement of polydentate ligands has also been studied. In the case of polydentate amines the acidity of the reaction mixture can play a most important part; in fact, the displacement of polydentate amines from palladium(II) complexes by chloride only takes place in acid solution, and at a rate that depends on the hydrogen ion concentration[99, 100]. It seems that the acidity influences the mechanism by preventing the closing of the ring during the substitutions, since a rate dependence on $[H^+]$ is uncommon for the displacement of monodentate amines. The rate of exchange of $C_2O_4^{2-}$ with $[Pt(C_2O_4)_2]^{2-}$ has been shown to be independent of the acid concentration[101].

The mechanism of the displacement, by amines, of the chelating sulphur donor $PhSCH_2CH_2SPh$, (S—S), from the complex $[Pd(S{-}S)Cl_2]$ in 1,2-dimethoxyethane at 25 °C has been studied in some detail[16]. The entry of two amine molecules leads to the formation of *cis*-$[Pd(am)_2Cl_2]$, which then isomerises to the *trans* form. The usual two-term rate law of Equation (7.1) was observed for the entry of non-hindered pyridines, but the entry of the hindered pyridines is accompanied by a more complicated rate law which takes into account reversible ring closure and solvolyses.

7.10 THE MECHANISM AND THE ROLE OF THE METAL

It was pointed out, in Section 7.2, that the general mechanism of substitution of a ligand X coordinated to a d^8 transition metal ion in a planar complex involves the bimolecular attack of the reagent Y at the substrate. These d^8 complexes are unsaturated from the point of view of coordination, and so it is not surprising that the nucleophilic attack is accompanied by a transient increase in the coordination number of the metal during the course of the substitution, involving a partial change in the hybridisation of the metal from dsp^2 to dsp^3. In other words, these reactions are typical associative processes, according to the definition of Langford and Gray[2]. This fact must be reflected in the reaction profile; the free energy of the ground state (substrate + Y) changes to give first a maximum, corresponding to the transition state for the formation of the new bond M—Y, which is followed by a relative minimum (the 5-coordinate intermediate) and then by a second maximum corresponding to the breaking of the M—X bond. The higher maximum is the rate-determining transition state.

The two problems which come out from such a situation are (a) the need to establish the relative position of the minimum, i.e., the relative stability

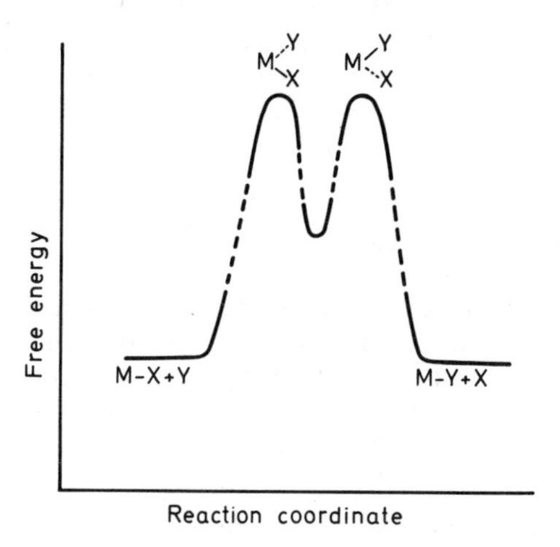

Figure 7.1

of the intermediate and its mechanistic role and (b) the necessity of deciding which of the two maxima is the higher and therefore represents the rate-determining transition state.

7.10.1 Existence and stability of the intermediate

The existence of the intermediate can be easily demonstrated, since there are very many examples of stable 5-coordinate complexes of d^8 transition metal ions. Studies of the kinetics of the displacement of SbR_3 from complexes of the type [Rh(diolefin)(SbR_3)X] by amines[103] indicate that the rate of the process,

$$[\text{Rh(diolefin)(SbR}_3\text{)X}] + \text{am} \rightarrow [\text{Rh(diolefin)(am)X}] + \text{SbR}_3$$

is independent of the concentration of the entering amine, but dependent upon its nature. The bimolecular attack of the amine leads to the rapid formation of a 5-coordinated complex which then slowly eliminates SbR_3. Such a situation can be described in terms of a single reaction profile in which the free energy level of the minimum, corresponding to the intermediate, is lower than the level of the ground state, but higher than that of the products. The first maximum is very much lower than the second.

The accumulation of a stable 5-coordinate intermediate has been also claimed[127] in an explanation of the results obtained from the reaction of bipyridyl with potassium trichloro(2,5-dimethyl-3-hexene-2,5-diol)platinate (II), which leads to the displacement of the acetylenic glycol. In this case the authors suggest a mechanistic path involving the accumulation of the species [Pt(ac)(bipy)Cl_2], which slowly rearranges to the cationic complex $[Pt(ac)(bipy)Cl]^+$. It is remarkable that although accumulation of the intermediate seems unlikely in the reaction of bipyridyl with $[Pt(C_2H_4)Cl_3]^-$, kinetic evidence has been presented[128] for the accumulation of the unstable cationic complex $[Pt(C_2H_4)(bipy)Cl]^+$.

In general, it is difficult to obtain sufficient accumulation of the 5-coordinate intermediate to enable it to be observed kinetically, even when substrates containing a relatively inert group such as in $[Pt(dien)NO_2]^+$ are used, as well as reagents such as I^- and thiourea which might be expected to associate well with the substrate[129].

7.10.2 The transition states

If the difference between the energies of the intermediate and the transition state is large enough, the bond-making and bond-breaking aspects of the substitution could be considered as completely separate from each other, even if the intermediate were not detectable. This limiting case appears to exist in the substitution reactions of platinum(II) complexes; for instance, in the processes of the type

$$[Pt(bipy)(NO_2)X] + Y \rightarrow [Pt(bipy)(NO_2)Y] + X$$

(methanol, 25 °C)[35], where the discriminating ability of the various substrates is completely independent of the nature of the leaving group X. This means that the $\Delta G^{\ddagger}$ value for any reaction of this group can be obtained by adding to a constant value contributions due to the nature of the entering group and of the leaving group. These two contributions can be varied independently of each other, and therefore bond-making and bond-breaking are completely separated. This treatment has been successful with many other platinum(II) complexes and also with palladium(II) derivatives, but it is not generally true for gold(III) complexes (see Section 7.4).

Bearing in mind that kinetic measurements deal only with free energy changes from the ground state to the transition state, it is clear that, irrespective of whether the first or the second maximum in the reaction profile is the rate-determining transition state, the formation of the new bond, M—Y, is always important in these processes. On the other hand, the breaking of the M—X bond would be extremely important if the second maximum was rate-determining, whereas if the first (bond-making) maximum represents the transition state the mechanistic role of X is that of a non-participating ligand (see the beginning of Section 7.4).

Deciding which of the two maxima is the rate-determining transition state can be quite difficult, and until now this problem has been solved only for a few cases.

In the reactions[11, 28, 40]

$$[AuCl_4]^- + am \rightarrow [AuCl_3(am)] + Cl^-$$

the retardation due to the presence of sterically-hindered amines has been only observed in the forward reactions (see Section 7.8). This seems to suggest that the Au—N bond is almost completely formed in the transition state, so that the second maximum or bond-breaking must be the transition state for the forward reaction.

Since the system must traverse the same reaction profile in both the directions, the reverse reaction has the bond-making maximum as the rate-determining transition state. This point of view has been applied in the study of the reversible reactions[39]:

$$[\mathrm{Pt(bipy)Cl(RSR')}]^{+} + \mathrm{Y}^{-} \rightarrow [\mathrm{Pt(bipy)Cl(Y)}] + \mathrm{RSR'}$$

(methanol, 25 °C). In the forward process, the reactivity is dependent on both the nature of the entering group Y^- and the leaving thioether, and whereas the effect due to the nature of the leaving Y is relatively small in the reverse reaction, the dependence on the nature of the entering thioether is significant. These results indicate that, in the forward reaction, the second maximum in the reaction profile is of higher energy than the first.

7.10.3 The role of the metal

An understanding of the role of the metal in determining the kinetic behaviour requires the comparison of a large amount of data. Such a comparison has been made with special regard to the differences between platinum(II) and gold(III) derivatives[30], but the conclusions, which deal largely with the relative energy of the intermediate in comparison to that of the transition state, have also been applied to complexes of other d^8 transition metal ions.

In the case of a bimolecular displacement occurring at a reaction centre where no empty orbitals are available, the entering and leaving groups must of necessity share the same orbital for bonding in the transition state, as for example in nucleophilic attack at an aliphatic carbon atom. In this case the formation of the new bond and the breaking of the old one are simultaneous, and the reaction profile has only one maximum. The processes can be described as 'concerted' or synchronous. On the other hand, if an accessible empty orbital is suitable for the entry of the nucleophile, bond formation and bond breaking can be completely separated in the limit, and the process can be described as being non-synchronous.

On going from platinum(II) to gold(III) complexes, the increase of reactivity is accompanied by an increase of all the factors which favour the formation of the new bond, i.e., micropolarisability, basicity and π-bonding. In addition, the effect of the nature of the leaving group is also, as a rule, important in determining the kinetic behaviour. Moreover, the effects due to the nature of the entering and leaving groups are not independent of one another, as can be seen by the fact that the nature of the leaving group influences the ability of the substrate to discriminate between different nucleophiles[30].

All these observations can be explained if it is assumed that substitution in gold(III) complexes occurs through a mechanism which is relatively more synchronous than that of substitution in platinum(II) complexes. This has been related to the greater effective nuclear charge[104, 105] and oxidation state of gold(III), for which the use of a dsp^3 hybrid for covalent bonding is more difficult than for platinum(II).

Further confirmation of this point of view can be found in the kinetic behaviour of the derivatives of other d^8 metal ions, the first one being the accumulation of the 5-coordinate intermediate in the case of the rhodium(I) complex reported above.

Data available for palladium(II) complexes are also consistent with this idea. The role of steric hindrance (see Section 7.8), which is greater in palladium(II) than in platinum(II) complexes, and also the relatively greater dependence of the reactivity on the nature of X in $[Pd(dien)X]^+$ than in $[Pt(dien)X]^+$ (see Section 7.4), indicates that the differences between palladium(II) and platinum(II) are opposite to those between platinum(II) and gold(III).

It seems possible to conclude, therefore, that the 'synchronicity' of the substitution processes change with the nature of the metal, according to the sequence

$$Au^{III} > Pt^{II} > Pd^{II} > Rh^{I} \tag{7.15}$$

Data concerning the effect of the nature of the ligands on the intimate mechanism, and the position in Sequence (7.15) of other d^8 transition metal ions, are not yet available. However, a recent study of Pearson and Sweigart[130] on the substitution reactions of square-planar nickel(II) dithiolate complexes indicates that the formation of a 5-coordinate intermediate is rapid in comparison to subsequent steps.

7.11 ISOMERISATION

The *cis–trans* isomerisation of planar 4-coordinate complexes can be either a thermal or a photochemical process. Apart from photochemical studies on this subject[106–108] which are not within the scope of this review, it is known that thermal isomerisation is catalysed by certain free ligands in solution. For example, complexes of the type $[PtL_2Cl_2]$ (L = tertiary phosphine, arsine or stibine) rapidly reach isomeric equilibrium in the presence of a trace of L[109, 110]. Since substitutions at these planar complexes occur with complete retention of the geometric configuration, isomerisation through substitution requires at least two subsequent reactions[110].

The isomerisation of *cis*-$[Pd(am)_2X_2]$, prepared by the reaction of am on $[Pd(PhSCH_2CH_2SPh)X_2]$, is catalysed by the excess amine present in the reaction mixture[74].

The rate of isomerisation obeys the rate law

$$\text{rate} = k[cis\text{-isomer}][\text{amine}]$$

and the following scheme has been suggested for isomerisation

$$cis\text{-}[Pd(am)_2X_2] + am \rightarrow [Pd(am)_3X]X$$

$$[Pd(am)_3X]X \xrightarrow{\text{fast}} trans\text{-}[Pd(am)_2X_2]$$

The fast entry of the anionic group X (X = N_3, Br, I, CNS) is probably related to the fact that the reactions were carried out in the solvent 1,2-dimethoxyethane.

Such a double-displacement mechanism for isomerisation does not always operate. In the *cis–trans* isomerisation of *cis*-[Pt(Bu^n_3P)$_2$$Cl_2$] catalysed by Bu^n_3P is cyclohexane, Haake and Pfeiffer[131] suggested a mechanism involving pseudo-rotation in a 5-coordinate intermediate. However, more recently, the same authors report[132] that in these systems isomerisation must proceed much faster than exchange of the phosphine ligands. The results are interpreted by considering an association between the substrate *cis*-[PtL_2X_2] (L = phosphine) and the catalyst L_1; during this association a fluxional change interconverts the position of L and X and, at the same time, L_1 must retain its identity.

A very interesting study has been made recently[111] on the uncatalysed *cis–trans* isomerisation of *cis*-[Pt(PEt_3)$_2$(*o*-tolyl)Cl] in methanol and ethanol. Whereas the substitution of the coordinated chloride occurs with the usual two-term rate law, (Equation (7.1)), the isomerisation takes place at a rate that is dependent only on the concentration of the starting isomer, and with a first-order rate constant that is nearly two orders of magnitude smaller than the k_1 for the substitution reactions of the *cis* complex. The isomerisation is sensitive to mass-law retardation by chloride ions. The entropy changes which accompany the formation of the transition state during isomerisation is largely positive ($+21$ cal deg^{-1} mol^{-1} in methanol).

The authors suggest a dissociative mechanism where the rate-determining step is the breaking of the Pt—Cl bond to form a 3-coordinate intermediate. This is consistent with the experimental data, and could be the first clear-cut example of a dissociative path in planar 4-coordinate complexes. Even so, it only provides a very minor path for substitution in this complex.

References

1. Basolo, F. and Pearson, R. G. (1969). *Mechanism of Inorganic Reaction,* 2nd edn. Chap. 5. (New York: Wiley)
2. Langford, C. H. and Gray, H. B. *Ligand Substitution Processes.* (1965), Chap. 2. (New York: W. A. Benjamin)
3. Edwards, J. O. (1965). *Inorganic Reaction Mechanism,* (New York: W. A. Benjamin)
4. Basolo, F. (1965). *Advan. Chem. Ser.,* **49,** 81
5. Martin, D. S. (1967). *Inorg. Chim. Acta Rev.,* **1,** 87
6. Cattalini, L. (1970). *Inorganic Reaction Mechanism,* (J. O. Edwards, editor). (New York: Wiley)
7. Belluco, U., Cattalini, L. and Turco, A. (1964). *J. Amer. Chem. Soc.,* **86,** 226
8. Baddley, W. H. and Basolo, F. (1964). *Inorg. Chem.,* **3,** 1087
9. Cattalini, L., Orio, A. and Doni, A. (1966). *Inorg. Chem.,* **5.** 1517
10. Cattalini, L., unpublished results.
11. Cattalini, L., Nicolini, M. and Orio, A. (1966). *Inorg. Chem.,* **5,** 1674
12. Cattalini, L., Doni, A. and Orio, A. (1967). *Inorg. Chem.,* **6,** 280
13. Cattalini, L., Ricevuto, V., Orio, A. and Tobe, M. L. (1968). *Inorg. Chem.,* **7,** 51
14. Cattalini, L., Orio, A. and Martelli, M. (1967). *Chim. Ind. (Milan),* **49,** 625
15. Orio, A., Ricevuto, V. and Cattalini, L. (1967). *Chim. Ind. (Milan),* **49,** 1337
16. Cattalini, L., Martelli, M. and Marangoni, G. (1968). *Inorg. Chim. Acta,* **2,** 405
17. Marangoni, G., Martelli, M. and Cattalini, L. (1968). *Gazz. Chim. Ital.,* **98,** 1038
18. Cattalini, L., Marangoni, G. and Martelli, M. (1970). *3rd Coordination Chemistry Conference* (Debrecen, Hungary)
19. M. Martelli, Marangoni, G. and Cattalini, L. (1968). *Gazz. Chim. Ital.,* **98,** 1031
20. Hall, H. K. (1957). *J. Amer. Chem., Soc.,* **79,** 5422
21. Cattalini, L., Martelli, M. and Kirschner, G. (1968). *Inorg. Chem.,* **7,** 1488
22. Cattalini, L., Marangoni, G. and Martelli, M. (1968). *Inorg. Chem.,* **7,** 1495
23. Belluco, U., Cattalini, L., Basolo, F., Pearson, R. G. and Turco, A. (1965). *J. Amer. Chem. Soc.,* **87,** 241

24. Belluco, U. (1966). *Coord. Chem. Rev.,* **1,** 111
25. Pearson, R. G., Sobel, H. and Songstad, J. (1968). *J. Amer. Chem. Soc.,* **90,** 319
26. Cattalini, L., Orio, A. and Nicolini, M. (1966). *J. Amer. Chem. Soc.,* **88,** 5734
27. Pearson, R. G. (1963). *J. Amer. Chem. Soc.,* **85,** 3533
28. Cattalini, L. and Tobe, M. L. (1966). *Inorg. Chem.,* **5,** 1145
29. Basolo, F., Chatt, J., Gray, H. B., Pearson, R. G. and Shaw, B. L. (1961). *J. Chem. Soc.,* 2207
30. Cattalini, L., Orio, A. and Tobe, M. L. (1967). *J. Amer. Chem. Soc.,* **89,** 3130
31. Pearson, R. G. and Johnson, D. A. (1964). *J. Amer. Chem. Soc.,* **86,** 3983
32. Watt, G. W. and Layton, R. (1962). *Inorg. Chem.,* **1,** 496
33. Basolo, F., Gray, H. B. and Pearson, R. G. (1960). *J. Amer. Chem. Soc.,* **82,** 4200
34. Belluco, U., Ettorre, R., Basolo, F., Pearson, R. G. and Turco, A. (1966). *Inorg. Chem.,* **5,** 591
35. Cattalini, L. and Martelli, M. (1967). *Gazz. Chim. Ital.,* **97,** 488
36. Belluco, U., Graziani, M., Nicolini, M. and Rigo, P. (1967). *Inorg. Chem.,* **6,** 721
37. Martelli, M., Orio, A. and Graziani, M. (1965). *Ric. Sci., II-A,* **35,** 361
38. Belluco, U., Cattalini, L. and Turco, A. (1964). *J. Amer. Chem. Soc.,* **86,** 3257
39. Cattalini, L., Marangoni, G., Degetto, S. and Brunelli, M. (1971). *Inorg. Chem.,* in press
40. Cattalini, L., Orio, A. and Tobe, M. L. (1966). *Inorg. Chem.,* **6,** 75
41. Ettorre, R., Graziani, M. and Rigo, P. (1967). *Gazz. Chim. Ital.,* **97,** 58
42. Belluco, U., Cattalini, L., Martelli, M. and Ettorre, R. (1964). *Gazz. Chim. Ital.,* **94,** 733
43. Cattalini, L., Belluco, U., Martelli, M. and Ettorre, R. (1965). *Gazz. Chim. Ital.,* **95,** 567
44. Belluco, U., Cattalini, L. and Orio, A. (1963). *Gazz. Chim. Ital.,* **93,** 1422
45. Grantham, L. F., Elleman, T. S. and Martin, D. S. (1955). *J. Amer. Chem. Soc.,* **77,** 2965; Elleman, T. S., Reishus, J. W. and Martin, D. S. (1958). *J. Amer. Chem. Soc.,* **80,** 536
46. Reishus, J. W. and Martin, D. S. (1961). *J. Amer. Chem. Soc.,* **83,** 2457
47. Aprile, F. and Martin, D. S. (1962). *Inorg. Chem.,* **1,** 551
48. Sacconi, L. and Lombardo, G. (1960). *J. Amer. Chem. Soc.,* **82,** 6267 and references therein; Boschi, T., Rigo, P., Pecile, C. and Turco, A. (1967). *Gazz. Chim. Ital.,* **97,** 1392
49. Reinhardt, R. A. and Sparkes, R. K. (1967). *Inorg. Chem.,* **6,** 2190
50. Chatt, J., Duncanson, L. A. and Venanzi, L. M. (1955). *J. Chem. Soc.,* 4456
51. Orgel, L. E. (1956). *J. Inorg. Nucl. Chem.,* **1,** 137
52. Grinberg, A. A. (1935). *Acta Physiochim., (USSR),* **3,** 573
53. Cardewell, H. M. E. (1955). *Chem. Ind., (London),* 422
54. Bersuker, I. B. (1964). *Russ. J. Inorg. Chem., (English),* **9,** 18
55. Chernyaev, I. I. (1957). *Zh. Neorg. Khim.,* **2,** 475
56. Oleari, L., Di Sipio, L. and De Michelis, G. (1965). *Ric. Sci., II-A,* **35,** 413
57. Martelli, M. and Orio, A. (1963). *Ric. Sci., II-A,* **35,** 1089
58. Petren, J. (1898). *Akad. Afh. Lund.,* 52
59. Chernyaev, I. I. and Gel'man, A. D. (1935). *Izv. Plat.,* **14,** 77
60. Cramer, R. (1965). *Inorg. Chem.,* **4,** 445
61. Zumdahl, S. S. and Drago, R. S. (1968). *J. Amer. Chem. Soc.,* **90,** 6669 and references therein
62. Kurnakov, N. S. (1894). *J. Prakt. Chem.,* **50,** 481
63. Kukuskin, Yu. N., Vyaz'menskii, Yu. E. and Zorina, L. I. (1968). *Russ. J. Inorg. Chem.,* **13,** 11.
64. Cheesman, T. P., Odell, A. L. and Raethel, H. A. (1968). *Chem. Commun.,* 1496
65. Kaplan, P. D., Schimdt, P., Brause, A. and Orchin, M. (1969). *J. Amer. Chem. Soc.,* **91,** 85
66. Parshall, G. W. (1964). *J. Amer. Chem. Soc.,* **86,** 5367
67. Falk, C. D. and Halpern, J. (1965). *J. Amer. Chem. Soc.,* **87,** 3003
68. Tucker, M. A., Colvin, C. B. and Martin, D. S. (1964). *Inorg. Chem.,* **3,** 1373
69. Belluco, U., Graziani, M. and Rigo, P. (1966). *Inorg. Chem.,* **5,** 1123
70. Cattalini, L., Martelli, M. and Rigo, P. (1967). *Inorg. Chim. Acta,* **1,** 149
71. Carturan, G. and Martin, D. S. (1970). *Inorg. Chem.,* **9,** 258
72. Cattalini, L. and Martelli, M. (1967). *Inorg. Chim. Acta,* **1,** 189
73. Cattalini, L., Marangoni, G. and Cassol, A. (1969). *Inorg. Chim. Acta,* **3,** 74
74. Cattalini, L. and Martelli, M. (1969). *J. Amer. Chem. Soc.,* **91,** 312
75. Cattalini, L., Croatto, U. and Marangoni, G. (1969). *XII International Coordination Chemistry Conference,* Sydney

76. Cattalini, L., Marangoni, G., Coe, J. S., Vidali, M. and Martelli, M. (1971). *J. Chem. Soc. A,* in press
77. Baddley, W. H. and Basolo, F. (1964). *Inorg. Chem.,* **3,** 1087
78. Robb, W. (1967). *Inorg. Chem.,* **6,** 382
79. Cotton, F. A. and Francis R. (1960). *J. Amer. Chem. Soc.,* **82,** 2986; Meek, D. W., Straub, D. K. and Drago, R. S. (1960). *J. Amer. Chem. Soc.,* **82,** 6013
80. Pearson, R. G., Gray, H. B. and Basolo, F. (1960). *J. Amer. Chem. Soc.,* **82,** 787
81. Belluco, U., Orio, A. and Martelli, M. (1966). *Inorg. Chem.,* **5,** 1370
82. Belluco, U., Martelli, M. and Orio, A. (1966). *Inorg. Chem.,* **5,** 582
83. Belluco, U., Rigo, P., Graziani, M. and Ettorre, R. (1966). *Inorg. Chem.,* **5,** 1125
84. Harris, C. M. and Stephenson, N. C. (1957). *Chem. Ind. (London),* **14,** 426
85. Bosnich, B. (1966). *J. Amer. Chem. Soc.,* **88,** 2606
86. Faraone, G., Ricevuto, V., Romeo, R. and Trozzi, M. (1968). *Atti Soc. Peloritana Science MM. FF. NN.,* **14,** 341
87. Hudson, R. F. and Klopman, G. (1962). *J. Chem. Soc.,* 1062, and references therein.
88. Chan, S. C. and Wong, F. T. (1968). *Aust. J. Chem.,* **21,** 2873
89. Baddley, W. H. and Basolo, F. (1966). *J. Amer. Chem. Soc.,* **88,** 2944
90. Goddard, J. B. and Basolo, F. (1968). *Inorg. Chem.,* **7,** 936
91. Baddley, W. H. and Basolo, F. (1964). *J. Amer. Chem. Soc.,* **86,** 2075
92. Weick, C. F. and Basolo, F. (1966). *Inorg. Chem.,* **5,** 576
93. Faraone, G., Ricevuto, V., Romeo, R. and Trozzi, M. (1969). *Inorg. Chem.,* **8,** 2207
94. Faraone, G., Ricevuto, V., Romeo, R. and Trozzi, M. (1970). *Inorg. Chem.,* **9,** 1525
95. Basolo, F. and Stephen, K. H. (1966). *Inorg. Nucl. Chem. Lett.,* **2,** 23
96. Monsted, O. and Bjerrum, J. (1968). *XI International Conference on Coordination Chemistry,* 103
97. Teggins, J. E. and Woods, T. S. (1968). *Inorg. Chem.,* **7,** 1424
98. Baracco, L., Cattalini, L., Coe, J. S. and Rotondo, E. (1971). *J. Chem. Soc. A,* 1800
99. Poe, A. J. and Vaughan, D. H. (1967). *Inorg. Chim. Acta,* **1,** 255
100. Coe, J. S., Lyons, J. R. and Hussain, M. D. (1970). *J. Chem. Soc. A,* 90
101. Teggins, J. E. and Milburn, R. M. (1964). *Inorg. Chem.,* **3,** 364
102. Haake, P. and Cronin, P. A. (1963). *Inorg. Chem.,* **2,** 879
103. Cattalini, L., Ugo, R. and Orio, A. (1968). *J. Amer. Chem. Soc.,* **90,** 4800
104. Nyholm, R. S. (1961). *Proc. Chem. Soc. London,* 273
105. Nyholm, R. S. and Tobe, M. L. (1965). *Experimentia, Supp.,* **9,** 112
106. Haake, P. (1962). *J. Amer. Chem. Soc.,* **84,** 3764
107. Balzani, V., Carassiti, V., Moggi, L. and Scandola, F. (1965). *Inorg. Chem.,* **4,** 1243
108. Scandola, F., Traverso, O., Balzani, V. and Zucchini, G. L. (1967). *Inorg. Chim. Acta,* **1,** 76
109. Quagliano, J. V. and Shubert, L. (1952). *Chem. Rev.,* **50,** 201
110. Drew, H. D. K. and Wyatt, G. H. (1934). *J. Chem. Soc.,* 56
111. Faroone, G., Ricevuto, V., Romeo, R. and Trozzi, M. *J. Chem. Soc. A,* in the press
112. Pietropaolo, R., Dolcetti, G., Giustiniani, N. and Belluco, U. (1970). *Inorg. Chem.,* **9,** 549
113. Volger, H. C., Gaasbeek, M. M. P., Hogeveen, H. and Vrieze, K. (1969). *Inorg. Chim. Acta,* **3,** 145
114. Milburn, R. M. and Venanzi, L. M. (1968). *Inorg. Chim. Acta,* **2,** 97
115. Booth, G. (1964). *Advan. Inorg. Radiochem.,* **6,** 1
116. Coe, J. S., Hussain, M. D. and Malik, A. A. (1968). *Inorg. Chim. Acta,* **2,** 65
117. Reinhardt, R. A. and Monk, W. N. (1970). *Inorg. Chem.,* **9,** 2026
118. Mason, W. R. (1970). *Inorg. Chem.,* **9,** 2688
119. Haake, P. and Pfeiffer, R. M. (1970). *J. Amer. Chem. Soc.,* **92,** 5243
120. Odell, A. L. and Raethel, H. A. (1969). *Chem. Commun.,* 87
121. Robb, A. W. and Mureinik, R. J. (1969). *Inorg. Chim. Acta,* **3,** 575
122. Mureinik, R. J. and Robb, W. (1969). *Inorg. Chim. Acta,* **3,** 580
123. Burmeister, J. L. and Lim, J. C. (1969). *Chem. Commun.,* 1154
124. Carter, M. J. and Beattie, J. K. (1970). *Inorg. Chem.,* **9,** 1233
125. Louw, W. J. and Robb, W. (1969). *Inorg. Chim. Acta,* **3,** 29
126. Louw, W. J. and Robb, W. (1969). *Inorg. Chim. Acta,* **3,** 303
127. Hubert, J. and Theophanides (1969). *Inorg. Chim. Acta,* **3,** 391
128. Uguagliati, P., Belluco, U., Croatto, U. and Pietropaolo, R. (1967). *J. Amer. Chem. Soc.,* **89,** 1336
129. Haake, P., Chan, S. C. and Jonas, V. (1970). *Inorg. Chem.,* **9,** 1925

130. Pearson, R. G. and Sweigart, D. W. (1970). *Inorg. Chem.*, **9**, 1167
131. Haake, P. and Pfeiffer, R. M. (1969). *Chem. Commun.*, 1330
132. Haake, P. and Pfeiffer, R. M. (1970). *J. Amer. Chem. Soc.*, **92**, 4296

8
Rates and Mechanisms of Oxidation-Reduction Reactions of Metal Ion Complexes

R. G. LINCK
University of California, San Diego

8.1 INTRODUCTION

The subject of this review is the kinetics and mechanism of oxidation–reduction reactions of metal ions in aqueous solution. During the past several years a number of significant results have been reported; this circumstance justifies the appearance of this article so soon after previous reviews[1,2]. The subject has been restricted to the reaction between two metal ions in solution in order that the focus can be fine enough to allow a correlation of the available information. The contents of this review are divided into two major sections. Firstly, the conceptual developments of the mechanisms of electron-transfer processes are discussed. The aim of this material is to establish a framework on which results may be placed in order to direct attention to several of the major problems in redox reactions. Several of these problems date to Taube's review[3], and remain unresolved. The second major section examines much of the data that has been reported within the past several years. It is intended that this coverage will serve to relieve some of the literature retrieval pressure on others. In order to gain optimum advantage of this survey, all data reported are accompanied by the conditions of temperature and ionic strength, as well as the activation parameters if these are available. An attempt has been made, within space limitations, to cover the literature thoroughly, but there are surely lapses. The last papers seen arrived in California in early 1971. Earlier reviews and a data compilation published in 1969[4] should be consulted for older data.

The conventions used herein are as follows. M^{n+} will imply a metal ion with its normal complement of water molecules of coordination, $Cr(H_2O)_6^{3+} \equiv Cr^{3+}$. Similarly, $CrCl^{2+}$ should be read as $Cr(H_2O)_5Cl^{2+}$. Cr^{III} should be taken as implying all forms of trivalent Cr in solution. The ligand abbreviations used are bip, 2,2′-bipyridine; phen, 1,10-phenanthroline; tmp, 3,5,6,8-tetramethylphenanthroline; terpy, 2,2′,2″-terpyridine; py, pyridine; EDTA, ethylene-diaminetetra-acetic acid, tetra-anion; PDTA, propylene-diamine-tetra-acetic acid, tetra-anion; CyDTA, cyclohexane-diaminetetra-acetic acid, tetra-anion; NTA, nitrilotriacetic acid, tri-anion; en, ethylenediamine; ox, oxalate ion; FuOH, fumaric acid, mono-anion.

8.2 CONCEPTUAL DEVELOPMENTS

The attempt to collect scattered data and compile it to yield generalised models is the ultimate aim of mechanistic studies. For those who study electron-transfer reactions there has been the added impetus of *a priori* theoretical work[5]. Nevertheless, a great deal of effort has been expended on a basic question of mechanism, does a given reaction utilise the inner-sphere or outer-sphere path[6]? with, to some extent, a neglect of other factors. In this review an attempt has been made to look beyond this question. In doing so, the danger of discussing and comparing inner-sphere rate constants with outer-sphere rate constants without *explicit* concern for mechanism arises, if, for no other reason, than that the mechanisms of most reactions are not known with certainty. Two points are pertinent: (1) this danger must be faced if other features are to be developed at all; (2) many of the features of outer-sphere reactions have been shown to hold for inner-sphere reactions as well.

8.2.1 General structural aspects

The concept of inner-sphere and outer-sphere transition states for electron-transfer reactions has been an important one. Since Taube and Myers[7] established proof in one system of a transition state with bonds linking the two metal centres, a great deal of effort has been expended attempting to classify other systems. This clearly stated question, 'What are the gross geometrical relationships between nuclei in the transition state?' has not always been satisfactorily answered. (The adjective 'gross' is meant to imply that, at this juncture, the detailed positions of the atoms undergoing bond length and angle changes are *not* in question.) In this section some of the possibilities and experimental features of the general structural aspects are examined.

8.2.1.1 Monobridged systems

The usual representation of a simple inner-sphere transition state is that of the first known example,

$$[(NH_3)_5Co—Cl—Cr(H_2O)_5^{4+}]^{\ddagger}$$

A simple extension of this picture to polgatomic ligands leads to an ambiguity of binding sites. The concept of 'remote' attack, taken here to mean binding of the second metal ion centre to an atom not directly linked to the first metal ion, was developed early, although its history in the case of conjugated organic ligands has been marred[8]. It has been definitely established, however, that for iso-nicotinamide as a ligand on the $Co(NH_3)_3^{3+}$ residue, reduction by Cr^{2+} leads to a transition state of structure[9]

$$\left[(NH_3)_5Co{-}N\langle C_5H_4\rangle{-}\underset{\displaystyle NH_2}{\underset{|}{C}}{-}O{-}Cr(H_2O)_5^{5+}\right]^{\ddagger}$$

One of the conjugated organic ligands originally postulated as demonstrating remote attack was the methylfumarate complex,

$$\left[(NH_3)_5Co{-}O{-}\overset{O}{\overset{\|}{C}}{-}CH{=}CH{-}\overset{O}{\overset{\|}{C}}{-}OCH_3\right]^{2+}$$

Hurst and Taube[10] have now established that the dominant path for reduction is not accompanied by ester hydrolysis. Nevertheless, the data obtained indicates attack at the remote carboxyl group is likely. In support of this contention is evidence indicating that at low $[Cr^{2+}]$ more free monomethylfumaric acid is formed as a product at the expense of the dominant product

$$\left[Cr{-}O{-}\overset{O}{\overset{\|}{C}}{-}CH{=}CH{-}\overset{O}{\overset{\|}{C}}{-}OCH_3\right]^{2+}$$

The mono-ester presumably arises from competition between reactions (8.1) and (8.2)

$$\left[Cr{-}O{=}C(OCH_3){-}CH{=}CH{-}C({=}O)OH\right]^{3+} \longrightarrow Cr^{3+} + HO\overset{O}{\overset{\|}{C}}CH{=}CH\overset{O}{\overset{\|}{C}}{-}OCH_3 \quad (8.1)$$

$$Cr^{2+} + \left[Cr{-}O{=}C(OCH_3){-}CH{=}CH{-}C({=}O)OH\right]^{3+} \longrightarrow \underset{\mathbf{A}}{\left[CrO\overset{O}{\overset{\|}{C}}CH{=}CH\overset{O}{\overset{\|}{C}}OCH_3\right]^{2+}} + Cr^{2+} + H^+ \quad (8.2)$$

The complex **A** is formed by remote attack on the Cr^{III}–carbonyl complex. In addition, reduction of various penta-amine fumarate derivatives have a term in the rate law of the form

$$k_2[H^+][Co^{III}][Cr^{II}]$$

where the first-order $[H^+]$ dependence is interpreted as arising because of protonation of the adjacent carboxyl oxygen, thus improving conjugation

$$\left[(NH_3)_5Co{-}O{=}C(OH){-}CH{=}CH{-}C(OH){=}O{-}Cr\right]^{5+\ddagger}$$

B

Support for this concept is found in the relatively rapid reduction of

$$\left[(NH_3)_5CoOC(NH_2){-}CH{=}CH{-}C({=}O)OCH_3\right]^{3+}$$

where the amide function serves the same role as the H^+ in species **B**.

Recently, Stritar and Taube[11], working on the reduction of Ru^{III}–amine complexes of carboxylic acids, have taken advantage of the relatively long lifetime of the binuclear product of reduction by Cr^{II} to argue that remote attack occurs on the carbonyl oxygen in these systems

$$\left[(NH_3)_5Ru{-}O{-}C(CH_3){=}O{-}Cr(H_2O)_5\right]^{4+\ddagger}$$

and by relative-rate ratios, in the Cr^{2+}-reductions of Co^{III} complexes as well. The formation of linkage-isomers by oxidation–reduction reactions also implies remote attack. Table 8.1 lists some of these reactions. It is to be noted that the reaction

$$5H^+ + Co(NH_3)_5N_3^{2+} + Cr^{2+} = Cr(H_2O)_5N_3^{2+} + Co^{2+} + 5NH_4^+ \quad (8.3)$$

has not been established as being of remote attack. Recent work on the site of protonation of this complex[19], $(NH_3)_5CoN(H)NN^{3+}$, (although see Reference 20 for another example of this type) and its importance in relative-rate comparisons make such information important.

Balahura and Jordan[16,21] have argued that it is necessary, at least for N, that the atom serving as the binding site contain a pair of electrons for this binding, and have available another pair, either lone or of π-symmetry, to allow remote attack. This feature is illustrated by data[16] on the reduction of $(NH_3)_5Co{-}NH_2CHO^{3+}$, and the linkage-isomer $(NH_3)_5CoOCHNH_2^{3+}$.

The former is reduced by Cr^{2+} by a path involving the loss of a proton to $(NH_3)_5CoNHCHO^{2+}$, with $k = 1.74\ M^{-1}\ s^{-1}$, and the product in the Cr^{III} O–bonded formamide complex; the latter is reduced with $k = 8.5 \times 10^{-3}\ M^{-1}\ s^{-1}$ and Cr^{3+} is the product. Presumably ionisation of $(NH_3)_5Co\ OCHNH_2^{3+}$ does not occur at low pH because the Co^{III} centre is more remote. The implication is that the transition state for $(NH_3)_5CoNHCHO^{2+}$ involves remote attack and that the loss of a proton is what enables the

Table 8.1 Reactions proceeding by remote attack as shown by product analysis

Reaction	k^* $M^{-1}s^{-1}$	I, M	*Product*	*Reference*
$Co(CN)_5^{3-} + Co(NH_3)_5CN^{2+}$	2.9×10^2†	0.2	$Co(CN)_5NC^{3-}$	12
$Co(CN)_5^{3-} + Co(NH_3)_5NO_2^{2+}$	3.4×10^4†	0.2	$Co(CN)_5ONO^{3-}$	12
$Cr^{2+} + CrSCN^{2+}$	4.2×10^1	1.0	$CrNCS^{2+}$	13
Cr^{2+} + *trans*-$Co(en)_2H_2ONCS^{2+}$ §	1.4×10^3	1.0	$CrSCN^{2+}$	14
$Cr^{2+} + Co(NH_3)_5CN^{2+}$ §	3.6×10^1 ‖	0.15	$CrNC^{2+}$	15
$Cr^{2+} + Co(NH_3)_5L^{3+}$ ¶	3.3×10^{-2} ¶	1.0	CrL^{4+} §§	9
$Cr^{2+} + Co(NH_3)_5NHC(=O)H^{2+}$	1.7**	0.92	$CrOC(=O)NH_2^{2+}$	16
$Cr^{2+} + Ru(NH_3)_5L^{3+}$††	3.9×10^5††	0.1	$Ru(NH_3)_5LCr^{5+}$	17
$Fe^{2+} + Co(NH_3)_5SCN^{2+}$	1.2×10^{-1}	1.0	$FeNCS^{2+}$	18

*At 25 °C
†This is only one term of the rate law.
§Other complexes of this type also yield remote attack products.
‖$\Delta H^* = 6.9$ kcal mol^{-1}; $\Delta S^* = -28$ cal mol^{-1} deg^{-1}
¶L is nicotinamide, pyridine bound; $\Delta H^* = 10.2$ kcal mol^{-1}; $\Delta S^* = -31$ cal mol^{-1} deg^{-1}
**$\Delta H^* = 12.0$ kcal mol^{-1}; $\Delta S^* = -17$ cal mol^{-1} deg^{-1}
††L is iso-nicotinamide, pyridine bound; $\Delta H^* = 0.0$ kcal mol^{-1}; $\Delta S^* = -46$ cal mol^{-1} deg^{-1}
§§L is nicotinamide, carbonyl bound.

Table 8.2 Reduction of Co^{III} complexes containing R—NH_2 ligands by Cr^{II} $T = 25\ °C$, $I = 1.0$ M

Complex	k, $M^{-1}s^{-1}$	$\Delta H^‡$, kcal mol^{-1}	$\Delta S^‡$, cal mol^{-1} deg^{-1}	*Reference*
$Co(NH_3)_5OC(NH_2)_2^{3+}$	1.9×10^{-2}	10.6	−31	21
$Co(NH_3)_5NCNC(NH_2)_2^{3+}$	2.9×10^{-2}	8.3	−37	21
$Co(NH_3)_5OC(=O)-NH_2^{2+}$	2.4	11.4	−19	21
$Co(NH_3)_5NCNH^{2+}$	3.3×10^3	4.3	−28	21

reaction to proceed via this path. Further support for this argument comes from other systems examined by these workers – see Table 8.2. It is to be noted that neither the urea, *N*-cyanoguanidine, or carbamate complex take advantage of the potential remote site of an $-\ddot{N}H_2$ group: the former two presumably choose an outer-sphere path (by product analysis) whereas the latter chooses attack at the carbonyl. In the case of the cyanamide complex, the remote nitrogen has both a lone pair and is in position to conjugate; the rate constant indicates inner-sphere attack although direct evidence is unfortunately lacking because only Cr^{3+} is found as the Cr^{III} product.

8.2.1.2 Chelated systems

Another geometry that perforce belongs to the category of remote attack, although here the intervening ligand may be innocent in terms of actual electron transfer, is that of a transition state in which the ligand bound to one metal ion centre forms a chelate with the second metal ion. The reduction of $Co(NH_3)_5OCOCH_2COOH^{2+}$ by Cr^{2+} proceeds by the rate law[22]

$$\frac{d[Co^{III}]}{dt} = (k_1[H^+]+k_2+k_3[H^+]^{-1})[Co^{III}][Cr^{II}] \qquad (8.4)$$

Both the k_1 and k_3 paths involve chelated malonatochromium(III) as a product[23] and imply a chelated transition state (since ring closure of monodentate malonatochromium(III) is relatively slow[23]), for instance for the k_3 path,

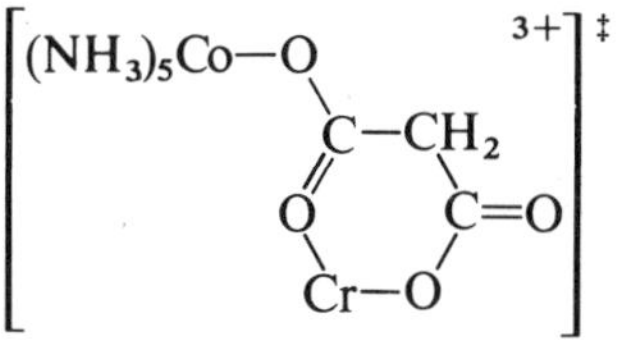

$$\left[\begin{array}{l}(NH_3)_5Co-O \\ \qquad\quad \backslash \\ \qquad\quad C-CH_2 \\ \quad\;\; // \qquad \backslash \\ \quad\; O \qquad\;\; C=O \\ \quad\;\; \backslash \qquad / \\ \qquad Cr-O\end{array}\right]^{3+\,\ddagger}$$

The influence of chelation on rate is discussed in Section 8.2.5.1.

8.2.1.3 Double-bridged systems

The search for 'double-bridged' transition-state geometry has also been pursued for some time. Snellgrove and King[24] first observed such a process in the exchange of ^{51}Cr isotopes between Cr^{2+} and *cis*-$Cr(N_3)_2^+$. Haim[25] has postulated this geometry as an explanation for the lack of H^+ catalysis of the reduction of *cis*-$Co(en)_2(N_3)_2^+$ by Fe^{2+}; reduction of the *trans*-isomer is catalysed by H^+. Further work established that *cis*-$Cr(N_3)_2^+$ is a partial product in the Cr^{2+}-reduction of *cis*-$Co(en)_2(N_3)_2^+$ and *cis*-$Co(NH_3)_4(N_3)_2^+$ [26]. Recently Ward and Haim[27] have shown the Cr^{2+}-reduction of *cis*-$Co(en)_2(OCOH)_2^+$ proceeds partially by the double-bridged mechanism, and Huchital[28] has given evidence for double-bridged paths in Cr^{2+} reactions with Cr^{III}-oxalate complexes. Double-bridged geometry has been used to explain several sets of data in which inverse hydrogen ion dependence on the rate is found, although another ligand is transferred. A summary of the reactions in which double-bridging has been postulated is given in Table 8.3. An alternative explanation of the results with paths inverse in $[H^+]$ is that an OH^- *cis* to the bridging ligand acts as an H^+ acceptor for a hydrogen on

Table 8.3 Reactions in which double-bridged transition states have been postulated or established $T = 25\,°C$, $I = 1.0$ M

Reaction A. $[H^+]^0$ paths	k, $M^{-1}\,s^{-1}$	$\Delta H^\ddagger$, kcal mol^{-1}	$\Delta S^\ddagger$, cal mol^{-1} deg^{-1}	*Reference*
Cr^{2+} + *cis*-$Cr(N_3)_2^+$	5.9×10^{1}*			24
Fe^{2+} + *cis*-$Co(en)_2(N_3)_2^+$†	1.85×10^{-1}			25
Cr^{2+} + *cis*-$Co(en)_2(N_3)_2^+$	$> 10^3$			26
Cr^{2+} + *cis*-$Co(NH_3)_4(N_3)_2^+$	$> 10^3$			26
Cr^{2+} + *cis*-$Co(en)_2(O\overset{O}{C}H)_2^+$	3.3×10^2	3.7	−35	27
Cr^{2+} + $Cr(ox)_3^{3-}$	1.3×10^{-1}§	8.2	−35	28
Cr^{2+} + *trans*-$Cr(ox)_2(H_2O)_2^-$	1.1×10^{-1}‖	11.9	−23	28
Cr^{2+} + *cis*-$Cr(ox)_2(H_2O)_2^-$	1.7×10^{-1}¶			28

B. $[H^+]^{-1}$ paths	k, s^{-1}	$\Delta H^\ddagger$, kcal mol^{-1}	$\Delta S^\ddagger$, cal mol^{-1} deg^{-1}	*Reference*
Cr^{2+} + *cis*-$Co(NH_3)_4(OH_2)(CH_3COO)^{2+}$	2.8			29
Cr^{2+} + $CrSCN^{2+}$	2.0			14
Cr^{2+} + $CrH_2PO_2^{2+}$	6.1×10^{-4}	19.7	−7.2	30
Cr^{2+} + $CrCN^{2+}$	4.2×10^{-3}	17	−8	31
Cr^{2+} + $CrOCOCH_3^{2+}$	5.8×10^{-4}	19	−10	20
Cr^{2+} + *cis*-$Co(NH_3)_4H_2OFuOH^{2+}$	2.5	4.5	−42	32
Cr^{2+} + *cis*-$Co(en)_2H_2OFuOH^{2+}$	7.2×10^{-1}**	6.2	−38	32
Cr^{2+} + $CrFuOH^{2+}$	2.5	12.6	−15	32

*At 0 °C, I = 0.5
†See, however, footnote 13 of Reference 26.
§I = 1.2 M
‖For isomerisation to the *cis*-isomer
¶For Cr exchange, I = 1.1 M
**I = 1.5 M

a *cis* water of the reducing agent. For example, in the Cr exchange between Cr^{2+} and $Cr(CN)^{2+}$ by the inverse H^+ path[31],

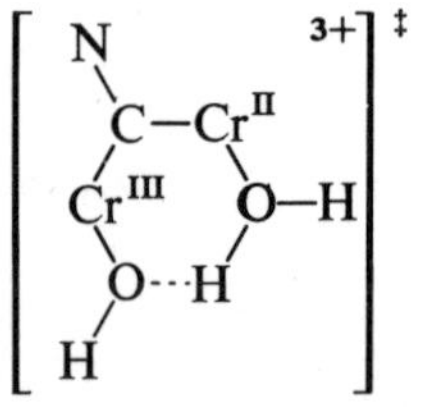

```
[ N              3+ ] ‡
   \
    C—Cr^II
   /     \
 Cr^III   O—H
   \     /
    O···H
   /
  H                   ]
```

This role takes cognisance of the large stabilising effect an OH^- would have on the trivalent chromium product[20, 33]. The direct evidence of product identification has been used to claim multiple-bridged geometry in some other reactions: the Cr^{2+}-reduction of $Co(ox)_3^{3-}$ [34] and $Co(CH_3COCHCOCH_3)_3$ [35].

8.2.1.4 *Abnormal transition-state structures*

It is appropriate at this juncture to discuss 'abnormal' transition-state structures. In this class are placed those reactions that proceed by a transition state of expanded coordination number, those in which interaction of ligands without metal–ligand bond disruption lead to the transition state. This latter class was postulated many years ago[36] when it was suggested that the reaction of Fe^{2+} with $FeOH^{2+}$ proceeded by hydrogen atom transfer; this has not yet been established. The former has been discussed several times in the literature; the most definitive case against such a mechanism has been discussed by Connick and Taube[37]. It must be noted that although the 'working' conception of transition-state structures rules out this model, it does so without firm evidence. Such evidence, either in support of or against this structure, is difficult to obtain. One experimental result has been explained on the basis of an expanded coordination sphere. Deutsch and Taube[20] have shown that a term in the V^{2+}-catalysed loss of acetate from $Cr(CH_3COO)^{2+}$ has the form $k_1[V^{2+}][Cr(CH_3COO)^{2+}][H^+]^{-1}$ with $k_1 = 6.7 \times 10^{-3}\ s^{-1}$ at 25 °C and I = 1.0 M. Since loss of proton from either ion results in a second order k of $> 100\ M^{-1}\ s^{-1}$, a value too large to allow substitution on either metal centre, this term must reflect either an outer-sphere path with OH^- stabilising the incipient V^{III} ion, or, as Deutsch and Taube suggest, a transition state in which a hydroxide attached to Cr^{III} penetrates the trigonal face of the V^{II} octahedron.

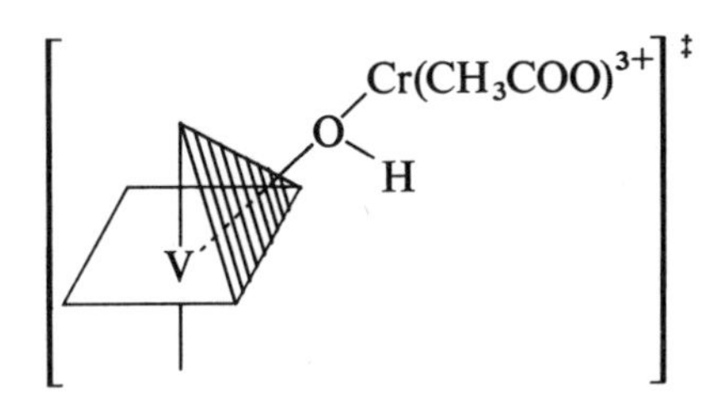

It is interesting to note that the aquation of $Ru(NH_3)_6^{2+}$ is catalysed by H^+ [38]. Of the commonly employed reducing agents the Ru^{II} amine species exhibits by its chemistry[39] the most pronounced ability to donate t_{2g} electrons. Presumably this explains the H^+-catalysis of substitution. It might also explain the reported H^+-catalysis of the reduction of $Co(NH_3)_5F^{2+}$ by $Ru(NH_3)_6^{2+}$ [40]. One can envisage an expanded coordination-shell transition state for the Ru^{II} centre being favoured by protonation of the coordinated fluoride. Such a H^+-catalysis is absent in the reduction of $Co(NH_3)_5F^{2+}$ by V^{2+} [41], a reaction thought by some to be outer-sphere. Further experiments are needed to explore these questions.

8.2.1.5 Outer-sphere transition-state structures

The structures of outer-sphere transition states have only begun to be investigated. When the reactants are both strictly octahedral in symmetry, the question of structure may be a subject of little importance (although it seems reasonable that the nature and energy of orbital overlap will differ if the two metal ion centres approach each other by superposition of their C_4 axis, by C_4—C_3 or C_3—C_3 superposition—see, for instance, Reference

42), but for complexes of lower symmetry differences in structure could easily affect relative-rate arguments. Some scattered information is relevant to this question.

In 1966, Bruning and Weissman[43] observed that the rate of an outer-sphere reaction (between an aromatic molecule and its radical anion) was dependent on the optical isomer used in the reaction. Although attempts to seek optical isomeric effects in some other outer-sphere reactions were not successful[44] (see Table 8.11), Sutter and Hunt[45] have reported asymmetric induction in the oxidation of $Cr(phen)_3^{2+}$ with D-$Co(phen)_3^{3+}$. About 90% of the product is L-$Cr(phen)_3^{3+}$. This result implies an alignment in the transition state that is sensitive to the lack of spherical symmetry in the reagents.

Some indirect arguments about the geometrical arrangements in outer-sphere transition states have been made. Doyle and Sykes have noted that the ratio of the anion-catalysed rate constant to the uncatalysed rate constant, k_2/k_1, in the reduction of $((NH_3)_5Co)_2NH_2^{5+}$ by V^{2+} [46],

$$-\frac{d[V^{2+}]}{dt} = \{k_1 + k_2[X^-]\}\{[Co^{III}][V^{II}]\} \tag{8.5}$$

are nearly the same as the ratios reported earlier by Dodel and Taube[47] for $Co(NH_3)_6^{3+}$ ($F^- > SO_4^{2-} > Cl^- > I^-$). They interpret these data to imply that the anion must be associated with the vanadium centre since preferential affinity of the 5+ Co^{III} ion over the 3+ one would be predicted, and that the anion is remote from the Co^{III} ion: the anion thus serves a role more important than just charge neutralisation (as Dodel and Taube have suggested[47]). It has been argued that outer-sphere reactions of complexes of the type CoL_5Cl^{n+}, that show similar relative rates when L is changed as do inner-sphere reactions, most probably have a geometrical structure independent of the nature of L [48].

8.2.2 Paths for formation and destruction of inner-sphere transition states

There have been various paths found for the relative energetics of achieving and destroying the activated complex of inner-sphere electron-transfer reactions. Such paths are most conveniently discussed in terms of precursor and postcursor complexes[49]. These species shall be treated as metastable entities, although in only a few cases has their existence as true intermediates been established. Those systems in which the highest point on the free energy–reaction coordinate diagram corresponds to the electron-transfer act itself will be considered first; this will be followed by systems in which precursor formation dominates the process, and finally reactions in which postcursor destruction is of importance will be discussed. These three cases are schematically illustrated in Figure 8.1. It should be noted that the composition of all three transition states is the same. The further complications introduced when rate data require multiple, consecutive transition states of the same or varied composition are considered at the close of this section. At the present time only selected systems can be classified in this manner:

the discussion is representative, not inclusive. Further, in this section the aim is not principally concerned with energetic considerations. This feature is discussed below (Section 8.2.5.1).

8.2.2.1 *Scheme I*

Reactions in which precursor complexes are readily formed, and their rates of transformation to postcursor complexes are relatively slow are usually believed to be the most common reaction pattern. An example is the inner-sphere reduction[50] of *trans*-$Co(en)_2H_2OCl^{2+}$ by Fe^{2+}. This 'typical' reaction

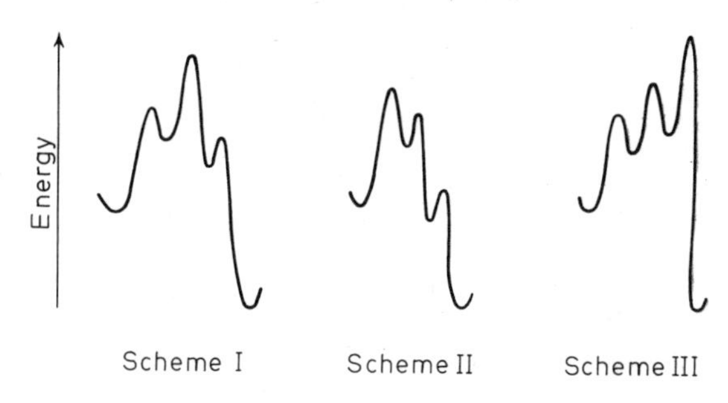

Figure 8.1 Potential energy as a function of reaction coordinate.

proceeds by substitution on the labile Fe^{2+} ion, electron transfer, and subsequent Co—Cl bond rupture to yield $FeCl^{2+}$ and Co^{2+} as initial products. A simple modification of this pathway occurs when the bridging ligand is brought to the transition state by the reductant[51].

$$HCrO_4^- + 3Fe(CN)_6^{4-} + 7H^+ = 2Fe(CN)_6^{3-} + CrFe(CN)_6 + 4H_2O \quad (8.6)$$

Another feature of Scheme I reactions on which data have been reported involves a sufficiently stable binuclear complex to observe kinetically. This reaction between Fe^{2+} and the nitrilotriacetate complex of penta-ammine cobalt(III) involves saturation of the pseudo-first-order rate constant at high $[Fe^{II}]$ [52]. The data fit the mechanism

$$Co(NH_3)_5NTA + Fe^{2+} \overset{K}{=} [Co(NH_3)_5NTAFe]^{2+} \quad (8.7)$$

$$[Co(NH_3)_5NTAFe]^{2+} \xrightarrow{k} FeNTA + Co^{2+} + \ldots \quad (8.8)$$

with $K = 1.1 \times 10^6\ M^{-1}$ and $k = 9.4 \times 10^{-2}\ s^{-1}$ at 25 °C and I = 1.0 M.

Similar reaction types within Scheme I apply to the means by which the postcursor complex breaks down. The usual rupture – that which results in ligand transfer – is not always followed[52a]

$$Co(EDTA)^{2-} + Fe(CN)_6^{3-} \rightarrow [(EDTA)CoNCFe(CN)_5]^{5-} \rightarrow Fe(CN)_6^{4-} + Co(EDTA)^- \quad (8.9)$$

Another more recent example involves the net reaction[53]

$$OH^- + Co(CN)_5^{3-} + IrCl_6^{2-} \rightarrow Co(CN)_5OH^{3-} + IrCl_6^{3-} \quad (8.10)$$

A similar bond-breaking has been suggested[54] for some reductions by Eu^{2+} because of the high lability of Eu^{3+} [55], but the evidence for an inner-sphere path in those reductions is only indirect.

It is also possible for a postcursor complex to decompose by bond-rupture at both metal ion sites before observation of either the binuclear complex or its fragments. This is, of course, the observed result when the oxidant–reductant pair does not have the peculiar labile-inert characteristics of, for instance, the Cr^{2+}-reductions of Co^{III} complexes, yet proceed through the inner-sphere mechanism. A probable* example is

$$5H^+ + Fe^{2+} + Co(NH_3)_5Cl^{2+} = Fe^{3+} + Co^{2+} + Cl^- + 5NH_4^+ \quad (8.11)$$

But even Cr^{2+}-oxidations by some Co^{III} complexes involve this path[56]

$$6H^+ + (NH_3)_5Co{-}O\overset{O}{\overset{\|}{C}}{-}C_6H_4{-}C{\overset{O}{\underset{H}{\lessgtr}}}^{2+}$$

$$+ Cr^{2+} = Co^{2+} + Cr^{3+} + HL + 5NH_4^+$$

where HL is *p*-formylbenzoic acid. This case is an example of a more subtle ramification of aquation lability. As the leaving-group changes the rate of metal–ligand bond fracture can change over a wide range. The rates of aquation of $Co(NH_3)_5O\overset{O}{\overset{\|}{C}}CH_3^{2+}$ and $Co(NH_3)_5O{=}C(CH_3)_2^{3+}$ are $1.2 \times 10^{-7}\ s^{-1}$ and $2 \times 10^{-2}\ s^{-1}$, respectively[57,58]. This comparison is pertinent to the question of the position of initial bond rupture in a postcursor complex such as

$$\left[(NH_3)_5Co^{II}{-}O{-}\overset{O}{\overset{\|}{C}}{-}C_6H_4{-}C{\overset{O-Cr^{III}}{\underset{H}{\lessgtr}}}\right]^{4+}$$

If the polarising effect of the Co^{II} centre and the lability of the Cr^{III}–carbonyl bond are sufficiently high, the initial bond rupture could occur at Cr—O. Subsequently Co—O bond rupture would occur. Although in this case the mode of decomposition has no special ramifications, in cases where both Cr^{3+} and CrL^{n+} (where L is the bridging ligand) are found, the dual interpretation of parallel inner-sphere and outer-sphere paths versus competitive Co—L and Co^{II}-assisted Cr—L bond rupture are possible, as is Cr^{2+}-catalysed aquation. Parallel inner-sphere and outer-sphere paths have been claimed in the V^{2+}-reduction of VO^{2+} [59] and the Cr^{2+}-reduction of $IrCl_6^{2-}$ [60] by direct observation of the postcursor complex and by indirect arguments in the Fe^{2+}-reduction of PuO_2^{2+} [61]. The observation of both Cr^{3+} and CrL^{n+} as products has been used to suggest parallel electron-transfer paths in several other reactions; and although the rate data support such claims, the ambiguity referred to above has not been eliminated. Table 8.4 lists the rate data for the available systems.

*Probable, because without observation of the postcursor complex or its fragments arguments about mechanism are indirect.

Table 8.4 Rate data for reactions in which outer-sphere and inner-sphere paths are believed to be in competition

Reaction	*k (inner)* $M^{-1}s^{-1}$	*k (outer)* $M^{-1}s^{-1}$	T	*I*, M	*Reference*
$V^{2+} + VO^{2+}$*	6.7×10^{-2}	3.9×10^{-2}	0	1.0	62, 63
$Fe^{2+} + PuO_2^{2+}$†	$\cong 5.0 \times 10^3$	$\cong 1.0 \times 10^3$	25	2.0	61
$Cr^{2+} + (NH_3)_5CoL^{3+}$§,‖	3.3×10^{-2}	1.4×10^{-2}	25	1.0	9
$Cr^{2+} + (NH_3)_5CoL^{3+}$¶	1.4×10^{-2}	9.6×10^{-3}	25	1.3	64
$Cr^{2+} + (NH_3)_5RuL^{3+}$§,**	1.3×10^4	4.8×10^3	4	1.0	17

*Overall activation parameters are $\Delta H^{\ddagger} = 12.3$ kcal mol^{-1} and $\Delta S^{\ddagger} = -16.5$ cal mol^{-1} deg^{-1}.
†$\Delta H^{\ddagger}$ (inner) = 8.6 kcal mol^{-1} and $\Delta H^{\ddagger}$ (outer) = 4.4 kcal mol^{-1}; $\Delta S^{\ddagger}$ (inner) = -13 cal mol^{-1} deg^{-1} and $\Delta S^{\ddagger}$ (outer) = -30 cal mol^{-1} deg^{-1}.
§L is nicotinamide bonded through the pyridine nitrogen.
‖$\Delta H^{\ddagger}$ (inner) = 10.2 kcal mol^{-1} and $\Delta H^{\ddagger}$ (outer) = 9.2 kcal mol^{-1}; $\Delta S^{\ddagger}$ (inner) = -31 cal mol^{-1} deg^{-1} and $\Delta S^{\ddagger}$ (outer) = -36 cal mol^{-1} deg^{-1}.
¶L is 4-pyridone, oxygen bonded.
**Overall activation parameters are 1.1 kcal mol^{-1} and -35 cal mol^{-1} deg^{-1}.

8.2.2.2 Scheme II

The prototype inner-sphere reaction, the Cr^{2+}-reduction of $Co(NH_3)_5Cl^{2+}$, cannot be used as an example of a Scheme I pathway because doubt exists concerning the pathway of the Cr^{2+} reaction. It has been observed, repeatedly, that the fastest inner-sphere Cr^{2+}-reductions occur with rate constants about 10^6–10^7 M^{-1} s^{-1}. Table 8.5 illustrates these data. Halpern[69], and

Table 8.5 Rates of systems possibly reacting via Scheme II; Cr^{2+}-reductions, $T = 25$ °C

Oxidant	*k*, $M^{-1}s^{-1}$	$\Delta H^{\ddagger}$, kcal mol^{-1}	$\Delta S^{\ddagger}$, cal mol^{-1} deg^{-1}	*I*, M	*Reference*
$Co(NH_3)_5F^{2+}$	2.5×10^5	—	—	0.1	65
$Co(NH_3)_5Cl^{2+}$	6×10^5	—	—	0.1	65
$Co(NH_3)_5Br^{2+}$	1.4×10^6	—	—	0.1	65
$Co(NH_3)_5I^{2+}$	3.4×10^6	—	—	0.1	65
cis-$Co(en)_2NH_3Cl^{2+}$	2.3×10^5	-2 ± 3	-41 ± 4	0.1	66
cis-$Co(en)_2H_2OCl^{2+}$	4.2×10^5	1 ± 4	-29 ± 11	0.1	66
cis-$Co(en)_2pyCl^{2+}$	9.4×10^5	8 ± 7	-3 ± 24	0.1	66
cis-$Co(en)_2(NH_2CH_2C_6H_5)Cl^{2+}$	4.5×10^5	-17 ± 5	-90 ± 15	0.1	66
cis-$Co(en)_2(NH_2\text{–}CH_2COOH)Cl^{2+}$	2.6×10^6	—	—	0.1	66
cis-$Co(en)_2Cl_2^+$	7.7×10^5	-6 ± 4	-51 ± 10	0.1	66
cis-$Co(en)_2FCl^+$	9×10^5	-10 ± 3	-65 ± 13	0.1	66
trans-$Co(en)_2Cl_2^+$	6.6×10^6	-8 ± 6	-53 ± 30	0.1	66
trans-$Co(en)_2F_2^+$	4.3×10^5	0 ± 3	-33 ± 11	0.1	66
cis-$Co(en)_2H_2OF^{2+}$	1.4×10^5	0 ± 6	-34 ± 20	0.1	66
$FeOH^{2+}$	3.3×10^6	4.6	-13	1.0	54
FeF^{2+}	7.6×10^5	—	—	1.0	54
$FeNCO^{2+}$	7.2×10^5	2.9	-22	1.0	54
FeN_3^{2+}†	2.9×10^7	—	—	1.0	67
$FeNCS^{2+}$	2.8×10^7	—	—	1.0	67
$FeCl^{2+}$	1.0×10^7	—	—	1.0	68

*T = 18.6 °C
†T = 1.6 °C
§Carlyle and Espenson[54] give lower limits of $> 2 \times 10^7$ at 1.6 °C and I = 1.0 for these reactions.

Sutin and co-workers[70] have suggested that the reason for the similarity in rate constant is that substitution on the Cr^{2+} centre limits the rate. The rate of electron transfer occurs rapidly as soon as the precursor complex is formed – Scheme II in Figure 8.1. The most well-established model for this type of behaviour are some reductions by V^{2+}; it has recently been suggested by Davies and Watkins[71] that the Co^{3+}-oxidation of several substances also proceeds by Scheme II. These latter sets of data are collected in Tables 8.6 and 8.7. The case for Scheme II in the several reductions by V^{2+} listed

Table 8.6 Rates of some possible Scheme II reductions by V^{2+} T=25 °C, I=1.0 M

Complex	k, $M^{-1}\,s^{-1}$	$\Delta H^{\ddagger}$, kcal mol^{-1}	$\Delta S^{\ddagger}$, cal mol^{-1} deg^{-1}	Reference
$Co(NH_3)_5C_2O_4H^{2+}$	12.5	12.2	−13	72
$Co(NH_3)_5C_2O_4^{+}$	45	—	—	72
$Co(NH_3)_5O\overset{O}{C}\overset{O}{C}H^{2+}$	8.2	11.0	−17	72
$Co(NH_3)_5O\overset{O}{C}\overset{O}{C}CH_3^{2+}$	10.2	11.6	−15	72
$Co(NH_3)_5O\overset{O}{C}\overset{O}{C}C(CH_3)_3^{2+}$	2.1	11.9	−17	72
$Co(NH_3)_5O\overset{O}{C}\overset{O}{C}NH_2^{2+}$	20.7	11.6	−14	72
$Co(NH_3)_5N_3^{2+}$	13	11.7	−14	73
$Co(en)_2(N_3)_2^{+}$	32.9	—	—	74
cis-$Co(en)_2NH_3N_3^{2+}$	10.3	12.6	−12	75
cis-$Co(NH_3)_4H_2ON_3^{2+}$	16.6	12.1	−12	75
trans-$Co(en)_2(N_3)_2^{+}$	26.6	12.2	−11	75
trans-$Co(en)_2H_2ON_3^{2+}$	18.1	11.0	−16	75
$Co(NH_3)_5SCN^{2+}$	30	—	—	18
$Co(NH_3)_4C_2O_4^{+}$	45.4	12.3	−9.8	76
Cu^{2+}	26.6	11.4	−13.8	77
$Cr(H_2O)_5SCN^{2+}$	8.0	13.0	−11	78
$Co(CN)_5N_3^{3-}$	1.1×10^2	—	—	79

Table 8.7 Rate of Scheme II oxidations by Co^{III}

Reductant	k, s^{-1}*	$\Delta H^{\ddagger}$, kcal mol^{-1}	$\Delta S^{\ddagger}$, cal mol^{-1} deg^{-1}	I, M	*Reference*
HN_3	3.3×10^1	24.8	32	2.0	80
ClO_2	1.1×10^2	22.2	24	2.1	81
H_2O_2	4.6×10^1	23.5	25	3.0	71
HNO_2	3.4×10^1	21.7	19	3.0	71
Br^-	3.0×10^1	26.1	37	3.0	71
Br^-	2.8×10^1†	18.5†	10†	0.5	82
NCS^-	1.6×10^1§	25.6	37	3.0	71
H_2ox	1.1×10^1	24.0	28	3.0	83
Hox^-	8.2×10^1	25.9	39	3.0	83

*The rate law is $\frac{-d[Co^{III}]}{dt} = k[Co^{III}][Reductant][H^+]^{-1}$

†Computed from original data by the non-linear least-squares programme Celia[84].

§This value may be too large by a factor of two because there is an ambiguity between Equation 6 and Figure 1 in Reference 71. See also Reference 83.

in Table 8.6 is compelling. The outstanding feature is the agreement between the rate of reduction, the rate of substitution of H_2O in the first coordination sphere of V^{2+} (100 s^{-1}, $\Delta H^{\ddagger} = 16.4$ kcal mol^{-1}, $\Delta S^{\ddagger} = 5.5$ cal mol^{-1} deg^{-1})[85], and the rate of substitution of NCS^- into the coordination sphere of V^{2+} ($k = 28$ M^{-1} s^{-1}, $\Delta H^{\ddagger} = 13.5$ kcal mol^{-1}, $\Delta S^{\ddagger} = -7$ cal mol^{-1} deg^{-1})[86]; the activation parameters are also in agreement. Similar considerations hold for the oxidations by Co^{3+}. Rate constants agree with those determined in substitution reactions: Cl^-, 40 s^{-1}, I = 3.0 M[87]; malic acid, 6.9 s^{-1} at 7 °C and I = 0.25 M[88]. Unfortunately, activation parameters are not available. (Other data for some of the reactions of Co^{3+} in Table 8.7 have been reported[89,90]. In one case this conflicting data has been carefully considered and experimentally checked[91]. It would appear the data of Wells and co-workers are not consistent with that obtained elsewhere.)

The case for the Cr^{2+}-reductions is less compelling. Firstly, the rate of water exchange (10^{10} s^{-1} [92]) greatly exceeds the limit found for oxidation–reduction, even with a correction for charge repulsion (compare the two numbers in the V^{2+} case); secondly, the activation parameters, where observed, fluctuate rather markedly from complex to complex. (It should be noted that because of the rapidity of the reaction these experiments are difficult and large errors are associated with the reported[66] activation parameters.) These fluctuations have been rationalised in terms of a delicate balance between rate-determining precursor-formation and rate-determining electron transfer[66]. Taube and co-workers have pointed out that the rate limit of 10^6–10^7 for Cr^{2+}-reductions could be determined by the large distortion necessary at the Cr^{II} centre to satisfy the Franck–Condon principle[32]. At least one system that apparently violates the 10^7 M^{-1} s^{-1} limit rule for reductions with Cr^{2+} has been reported[93]. The rate law for reduction of salicylato-penta-amminecobalt(III) is

$$-\frac{d[Co^{III}]}{dt} = \left(k_1 + \frac{k_2}{[H^+]}\right)[Cr^{II}][Co^{III}] \tag{8.13}$$

with $k_1 = 0.11$ M^{-1} s^{-1} and $k_2 = 0.03$ s^{-1} at 25 °C, I = 1.33 M. If one ascribes this second term to reaction of Cr^{2+} with

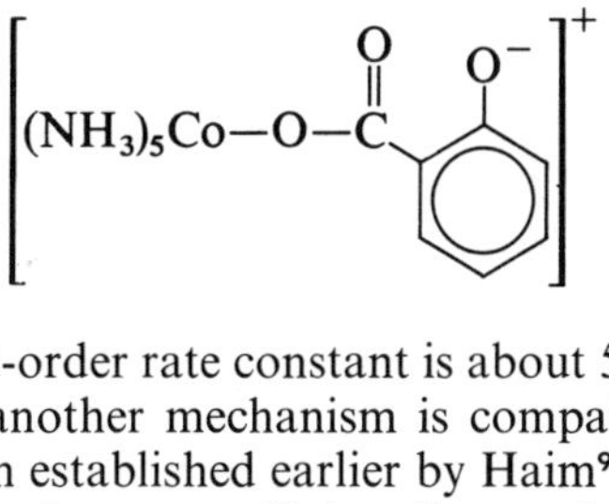

$$\left[(NH_3)_5Co{-}O{-}C(=O){-}C_6H_4{-}O^-\right]^+$$

then the specific second-order rate constant is about 5×10^8 M^{-1} s^{-1}. Liang and Gould point out another mechanism is compatible with this form of the rate law as had been established earlier by Haim[94] for a somewhat more complex system. These features will be discussed in Section 8.2.2.4. At present then, the upper limit of about 10^7 M^{-1} s^{-1} for inner-sphere reductions by Cr^{2+} has not been unequivocally violated, nor has the correct explanation for the limit been established.

The utility of the identification of Scheme II for V^{2+}-reductions has been great. The establishment of a rate limit beyond which reductions by V^{2+}

Table 8.8 Rate data for outer-sphere oxidations of V^{2+} $T = 25\ °C$

Oxidant	k, $M^{-1}\,s^{-1}$	$\Delta H^{\ddagger}$, kcal mol^{-1}	$\Delta S^{\ddagger}$, cal mol^{-1} deg^{-1}	I	*Reference*
Fe^{3+}	1.8×10^4	—	—	1.0	95
$FeCl^{2+}$	4.6×10^5	—	—	1.0	95
$FeNCS^{2+}$	6.6×10^5	—	—	1.0	95
FeN_3^{2+}	5.2×10^5	—	—	1.0	95
Br_2	3.0×10^4	3.5	−26	1.0	96
I_2	7.5×10^3	5.9	−21	1.0	96
I_3^-	9.7×10^2	9.1	−14	1.0	96
$[Ru(H_2O)_5Cl]^{2+}$	1.9×10^3	—	—	1.0	97
$[trans\text{-}Ru(NH_3)_4Cl_2]^+$	8.3×10^2	—	—	0.1	98
$[trans\text{-}Ru(NH_3)_4H_2OCl]^{2+}$	1.4×10^3	—	—	0.1	98
$[cis\text{-}Ru(NH_3)_4Cl_2]^+$	9.8×10^3	—	—	0.1	98
$[Ru(NH_3)_5OC(O)CH_3]^{2+}$	1.3×10^3	—	—	0.5	11
$[Ru(NH_3)_5OC(O)H]^{2+}$	4.1×10^3	—	—	0.5	11
$[Ru(NH_3)_5OC(O)CF_3]^{2+}$	2.9×10^3	—	—	0.5	11
$[Ru(NH_3)_5Cl]^{2+}$	3.0×10^3	3.0	−30	0.11	11
$[Ru(NH_3)_5Br]^{2+}$	5.1×10^3	2.8	−34	0.11	11
$[Ru(NH_3)_5NC_5H_5]^{3+}$	1.2×10^5	—	—	2.0	17
$[Ru(NH_3)_5NC_5H_4\text{-}3\text{-}C(O)OCH_3]^{3+}$	3.6×10^5	—	—	1.0	17
$[Ru(NH_3)_5NC_5H_4\text{-}3\text{-}C(O)NH_2]^{3+}$	9.5×10^5	—	—	1.0	17
$[Ru(NH_3)_5NC_5H_4\text{-}4\text{-}C(O)OCH_3]^{3+}$	6×10^5	—	—	1.0	17
$[Ru(NH_3)_5NC_5H_4\text{-}4\text{-}C(O)NH_2]^{3+}$	1.25×10^6	$\cong 0$	—	1.0	17
$IrCl_6^{2-}$	$>4 \times 10^6$	—	—	—	99
VO_2^+	2.6×10^3	1.9	−37	1.0	100
$VO_2^+ + H^+$	2.2×10^3*	1.8	−37	1.0	100

*This third-order term has units of $M^{-2}\,s^{-1}$

must proceed by the outer-sphere mechanism has been a very useful tool in assigning V^{2+}-reductions as outer-sphere or inner-sphere[49]. A summary of outer-sphere reductions by V^{2+} assigned by this criterion is given in Table 8.8.

8.2.2.3 Scheme III

Scheme III is a mechanism in which reactants and the postcursor complex are in equilibrium, with the overall reaction rate dependent on the rate of bond-rupture in the postcursor complex. To find systems in which Scheme III is operative requires consideration of those pairs of metal ions which are generated in the appropriate valence-states by oxidation–reduction reactions. Adamson and Gonick[52] first observed that the reaction of $Co(EDTA)^{2-}$ with $Fe(CN)_6^{3-}$ involved an intermediate that was in equilibrium with reactants. This reaction has been re-investigated recently[101]. The values found at 25 °C and I = 0.6 M are

$$Co(EDTA)^{2-} + Fe(CN)_6^{3-} \underset{86\ s^{-1}}{\overset{1.2\times10^5\ M^{-1}\ s^{-1}}{\rightleftharpoons}} [EDTA\ CoNCFe(CN)_5]^{5-} \tag{8.14}$$

The 5– ion can dissociate to final products; it has also been found that the dimer is oxidisable to $[(EDTA)CoNCFe(CN)_5]^{4-}$ by $Fe(CN)_6^{3-}$. Movius and Linck reported that the Cr^{2+}-reduction of *cis*-$Ru(NH_3)_4Cl_2^+$ [102], as well as that of *cis*-$Ru(NH_3)_4H_2OCl^{2+}$ and $Ru(NH_3)_5Cl^{2+}$ [98] proceed by the mechanism of Scheme III. Their data fit the rate law

$$-\frac{dRu^{III}}{dt} = \frac{k_1K}{1+K[Cr^{2+}]}\cdot[Cr^{2+}][Ru^{III}] \tag{8.15}$$

which is consistent with the mechanism

$$Ru^{III} + Cr^{2+} \overset{k}{=} (RuCr)^{n+} \tag{8.16}$$

$$(RuCr)^{n+} \xrightarrow{k_1} Ru^{II} + CrCl^{2+} \tag{8.17}$$

Similar results were reported by Seewald, Sutin, and Watkins[97] for the Cr^{2+}-reduction of $Ru(H_2O)_5Cl^{2+}$; because of the magnitude of K, however, these authors took advantage of an elegant trapping experiment to establish the reversibility of the first step.

$$Ru(H_2O)_5Cl^{2+} + Cr^{2+} = [Ru(H_2O)_5ClCr]^{4+} \tag{8.18}$$

$$V^{2+} + Ru(H_2O)_5Cl^{2+} \rightarrow V^{3+} + Ru^{3+} + Cl^- \tag{8.19}$$

A summary of the data for these Ru^{III} reductions is in Table 8.9. Firstly, it is clear that the observations as reported herein do not require Scheme III. They require only a species in equilibrium with reactants that decomposes to final products. The assignment of this intermediate as a Ru^{II}–Cr^{III} species rather than Ru^{III}–Cr^{II} is made as an assumption. In the reported experiments both spectral and substitution kinetic arguments make the former assignment quite plausible. Secondly, the trend in K and k_1 values in Table 8.9 are

Table 8.9 Values for the equilibrium constant of equation (8.16), K and rate constant for equation (8.17), k_1 $T = 25$ °C

Ru^{III} *complex*	K, M^{-1}	k_1, s^{-1}	I, M	*Reference*
cis-$Ru(NH_3)_4Cl_2^+$	465	1.5×10^2	0.1	98
cis-$Ru(NH_3)_4H_2OCl^{2+}$	360	1.2×10^2	0.1	98
$Ru(NH_3)_5Cl^{2+}$	70	4.6×10^2	0.1	98
$Ru(H_2O)_5Cl^{2+}$	(7700)*	$1.2 + \frac{0.36†}{H^+}$	1.0	97

*Calculated indirectly: see text and Reference 97.
†M s^{-1}

reasonable for this model. Since the free energy change for the reduction of $Ru(H_2O)_6^{3+}$ [103] is greater than that for $Ru(NH_3)_5OH_2^{3+}$ [104], the assumption that both species show a parallel affinity for Cl^- leads to a larger K for the penta-aquo ion than for the pentammine, although the observed increase is ten-times as large as that accounted for by this thermodynamic factor. The difference may lie in ionic strength effects. Also, the rate of Ru^{II}—ClCr bond fission, k_1, in these binuclear complexes appears to parallel the rate of Ru^{III}—Cl^- bond rupture in the corresponding complexes[105]. A comparison with the 'd-isoelectronic' Co^{III} system is satisfactory for the amine complexes[98], but is questionable when the penta-aquo ion is included, because substitution reactions on $Co(H_2O)_6^{3+}$ are apparently more rapid than on $Co(NH_3)_5Cl^{2+}$, although the inverse H^+ path dominates[87].

8.2.2.4 *Complex pathways*

In several cases reported to date the observed rate law for one-to-one oxidation–reduction reactions involves complications that render assignment to the three schemes in Figure 8.1 impossible. Examples are to be found in the Fe^{2+}-reduction of PuO_2^{2+} [61], the V^{3+}-reduction of UO_2^{2+} [106], the Cr^{2+}-reduction of V^{3+} [107], and the Cr^{2+}-reduction of *trans*-$CoL(H_2O)_2^{3+}$ [108,109] where L is 5,7,7,12,14,14-hexamethyl-1,4,8,11-tetra-azacyclotetradeca-4,11-diene or 5,7,7,12,14,14-hexamethyl-1,4,7,11-tetra-azacyclotetradecane. The Cr^{2+}-reduction of V^{3+} serves as an adequate example. Espenson[107] first reported the rate law

$$-\frac{d[V^{3+}]}{dt} = \frac{k[V^{3+}][Cr^{2+}]}{k' + [H^+]} \tag{8.20}$$

which although questioned later[110], has now been established as the correct experimental description of the kinetics of the reaction[111,112]. Haim[94] has shown that this rate law is consistent with either of two mechanisms; in one, the transition state of composition $[VCr^{5+}]^‡$ is first formed, followed by one of composition $[VOHCr^{4+}]^‡$

$$V^{3+} + Cr^{2+} \underset{k_2}{\overset{k_1}{\rightleftharpoons}} VOHCr^{4+} + H^+ \tag{8.21}$$

$$VOHCr^{4+} \xrightarrow{k_3} V^{2+} + CrOH^{2+} \xrightarrow[H^+]{\text{fast}} V^{2+} + Cr^{3+} \tag{8.22}$$

In the other, the state of composition $[VOHCr^{4+}]^{\ddagger}$ is formed first, followed by that of composition $[VCr^{5+}]^{\ddagger}$

$$V^{3+} \overset{k}{=} VOH^{2+} + H^{+}$$

$$VOH^{2+} + Cr^{2+} \underset{k_5}{\overset{k_4}{\rightleftharpoons}} VOHCr^{4+} \tag{8.24}$$

$$VOHCr^{4+} + H^{+} \xrightarrow{k_6} V^{2+} + Cr^{3+} \tag{8.25}$$

These two mechanisms cannot be distinguished by the kinetic experiment. As Haim[94] and Newton and Baker[113] point out, the inability to determine the order of occurrence of transition states is general in multi-term rate laws. Clearly the occurrence of complicated rate laws and multiple transition states leads to a situation wherein assignment of the mechanistic scheme (Figure 8.1) becomes quite difficult.

In some cases indirect arguments have been made. Liteplo and Endicott[108] have shown that the Cr^{2+}-reduction of the $CoL(H_2O)_2^{3+}$ complex proceeds by the rate law

$$-\frac{dCo^{III}}{dt} = k_{obs}[Cr^{II}][Co^{III}] \tag{8.26}$$

where

$$k_{obs} = \frac{a + b[H^{+}]^{-1}}{[H^{+}] + c} \tag{8.27}$$

They interpret these H^{+}-dependencies as acid-base reactions of a precursor complex (although it appears that their mechanism contains more information than the rate law warrants). This complex therefore contains Cr^{II} and Co^{III}. The argument follows from consideration of the net ΔG^{0} of reaction and the high lability of the Co^{II} complex of the macrocyclic ligand. Both features argue against the ability of the postcursor Co^{II}–Cr^{III} binuclear complex to influence the rate law.

Another interesting case of some importance is that reported by Liang and Gould[93] as indicated in equation 8.13. The argument presented by these authors is best cast as the ambiguity associated with the order of transition states of reactants forming the precursors to the oxidation–reduction transition state. Since this case is most important in reactions of aquo complexes, it is discussed in these terms. Three mechanisms can be imagined for the general case of reaction of $A^{III}OH_2$ with B^{II}.

I. Pre-dissociation of the proton

$$A^{III}OH_2 \overset{K_H}{=} A^{III}OH^{-} + H^{+} \tag{8.28}$$

$$A^{III}OH^{-} + B^{II} \overset{K_P}{=} A^{III}\overset{H}{O}B^{II-} \tag{8.29}$$

$$A^{III}\overset{H}{O}B^{II-} \xrightarrow{k} A^{II}\overset{H}{O}B^{III-} \rightarrow \text{Products} \tag{8.30}$$

II. Dissociation of the proton from a binuclear intermediate

$$A^{III}OH_2 + B^{II} \overset{K'_P}{=} A^{III}\overset{H}{\underset{H}{O}}B^{II} \tag{8.31}$$

$$A^{III}\overset{H}{\underset{H}{O}}B^{II} \overset{K'_H}{=} A^{III}\overset{H}{O}B^{II-} + H^+ \tag{8.32}$$

$$A^{III}\overset{H}{O}B^{II-} \xrightarrow{k} A^{II}\overset{H}{O}B^{III-} \rightarrow \text{Products} \tag{8.33}$$

III. Concerted mechanism

$$A^{III}OH_2 + B^{II} \overset{K}{=} A^{III}\overset{H}{O}B^{II-} + H^+ \tag{8.34}$$

$$A^{III}\overset{H}{O}B^{II-} \xrightarrow{k} A^{II}\overset{H}{O}B^{III-} \rightarrow \text{Products} \tag{8.35}$$

While it is true that these mechanisms are not distinguishable under the usual range of $[H^+]$ investigated, they do illustrate an important feature. All three mechanisms react from the same precursor complex, $A^{III}\overset{H}{O}B^{II-}$; the free energy of activation from this species to the oxidation–reduction transition state is the parameter on which interest centres. But because the stability of precursor complexes relative to separated reactants is not generally known, most rate comparisons of observed second-order rate constants involve both precursor stability and k. For systems that do not involve the proton ambiguity illustrated above, this comparison is, while not rigorous, of some value. But a comparison of the $B^{II} + A^{III}OH_2$ reaction by the inverse $[H^+]$ path with, for instance, the $B^{II} + A^{III}Cl$ reaction is of great uncertainty. If pre-dissociation of the proton does not take place, but mechanism II or III is applicable, then dividing k_{obs} by K_H leads to a fictitious value for K_pk to be compared with k_{obs} for the $A^{III}Cl$ reaction. In terms of the net activation process, mechanism I assumes a substantial part of the observed $\Delta G^‡$ for the reaction

$$A^{III}OH_2 + B^{II} = [A\overset{H}{O}B]^‡ + H^+ \tag{8.36}$$

arises because ΔG for the process

$$A^{III}OH_2 = A^{III}OH^- + H^+ \tag{8.37}$$

is very unfavourable; thus the remaining free energy is small and OH^- is an efficient bridge. Mechanisms II and III do not feature this 'initial cost' of energy and the role of the free energy needed for the conversion of $A^{III}\overset{H}{O}B^{II-}$ into the transition state for electron-transfer, $[\ A\overset{H}{O}B\text{-}\ ]^‡$, is unknown. Comparisons involving $[H^+]^{-1}$ terms in rate laws are thus quite ambiguous.

8.2.3 Detailed considerations of transition-state structures

Although little work has been reported concerning the detailed nature of the bond length and angle changes in achieving the transition state, that which is known is of significance. This knowledge comes from isotopic fractionation experiments.

The stretching of the bridging ligand in inner-sphere reactions is expected to be large, especially when the oxidant is adding an electron to an anti-bonding σ-orbital as in reduction of Co^{III} complexes. Since stretching is easier with the lighter isotope, an isotope effect greater than one is expected. This suggestion has been substantiated as the data in Table 8.10 for the Cr^{2+}-reduction of $Co(NH_3)_5OH^{2+}$ indicate. It is also of note that the

Table 8.10 Isotopic fractionation effects

Reaction	*Isotopic rate ratio*	*Reference*
	$k_{(^{14}N)}/k_{(^{15}N)}$	
$Cr^{2+} + Co(NH_3)_5Cl^{2+}$	1.001	114
$Cr^{2+} + Co(NH_3)_5OH^{2+}$	1.002	114
Cr^{2+} + *cis*-$Co(en)_2NH_3Cl^{2+}$	1.001	114
Cr^{2+} + *trans*-$Co(en)_2NH_3Cl^{2+}$	1.001	114
Cr^{2+} + *cis*-$Co(en)_2NH_3OH^{2+}$	1.004	114
Cr^{2+} + *trans*-$Co(en)_2NH_3OH^{2+}$	1.004	114
	$k_{(^{14}N)}/k_{(^{15}N)}$	
$Cr^{2+} + Co(NH_3)_5OH^{2+}$	1.056	115
$Ru(NH_3)_6^{2+} + Co(NH_3)_5OH^{2+}$	1.017	115
$Ru(NH_3)_6^{2+} + Co(NH_3)_5OH_2^{3+}$	1.021	115
$V^{2+} + Co(NH_3)_5OH_2^{3+}$	1.020	115
$Eu^{2+} + Co(NH_3)_5OH_2^{3+}$	1.019	115
Cr^{2+} + *trans*-$Cr(NH_3)_4H_2OCl^{2+}$	1.017	116
Cr^{2+} + *cis*-$Cr(NH_3)_4H_2OCl^{2+}$	1.007	116

reduction of both $Co(NH_3)_5OH_2^{3+}$ and its conjugate base by $Ru(NH_3)_6^{2+}$ show significantly smaller values of $k_{(^{16}O)}/k_{(^{18}O)}$. These data have been used to argue that the V^{2+} and Eu^{2+} reactions with $Co(NH_3)_5OH_2^{3+}$ are outer-sphere[115]. The effect on the non-bridging ligands in achieving the transition state is less marked. It is to be noted that whereas the difference in N-isotopic fraction between *cis*- and *trans*-$Co(en)_2(NH_3)Cl^{2+}$ or $Co(en)_2(NH_3)OH^{2+}$ is negligible, the reduction of *cis*- and *trans*-$Cr(NH_3)_4H_2OCl^{2+}$ shows a larger ^{16}O preference for the *trans* position than for the *cis*. All of these data are reconciled if a difference in the structure of the Co^{III} centre[114] and Cr^{III} centre[116] in the transition state is allowed. It is also interesting that the H_2O in *trans*-$Cr(NH_3)_4H_2OCl^{2+}$ and that in $Co(NH_3)_5OH_2^{3+}$ in the $Ru(NH_3)_6^{2+}$ reduction show similar $k_{(^{16}O)}/k_{(^{18}O)}$ ratios. Experiments designed to measure N-isotopic fractionation (sterospecifically) on these complexes would be very enlightening.

8.2.4 Energetics of transition-state formation

In this section the energy required for the net activation process

$$A + B = [AB]^{\ddagger} \tag{8.38}$$

is considered. As was remarked above, for inner-sphere reactions it has become usual to divide this energy into two components: the energy required to form the precursor complex and that required to form the transition state for the actual electron-transfer act from this precursor complex. This latter energy, which includes bond length and angle changes in both the first coordination shell, and in the surrounding solvent, is designated rearrangement energy. The rearrangement energy can presumably be calculated by either the pre-dissociation model of Schmidt[117], by the Marcus Theory including the possibility of strong overlap[118, 119], or by Hush's intervalence transfer treatment[120]. These calculations are in principle similar to those used to compute $\Delta G^{\ddagger}$ for outer-sphere reactions. The reader is referred to these articles, the review by Marcus[121], and the book by Reynolds and Lumry[5] for the details of these treatments. Here attention is focused on a more qualitative understanding.

8.2.4.1 Outer-sphere transition states

The applications of the results of the Marcus theory[121] have been discussed in earlier reviews[1, 2, 122] and references therein. The 'cross-reaction' equation is well known:

$$k_{12} = (k_{11}k_{22}K_{12}f_{12})^{\frac{1}{2}} \tag{8.39}$$

where k_{12} is the rate of reaction

$$A^{3+} + B^{2+} \xrightarrow{k_{12}} A^{2+} + B^{3+} \tag{8.40}$$

K_{12} is the equilibrium constant for this reaction, k_{11} and k_{22} are the exchange rate constants for A and B respectively,

$$A^{3+} + {}^{*}A^{2+} \xrightarrow{k_{11}} A^{2+} + {}^{*}A^{3+} \tag{8.41}$$

and

$$\log f = \frac{(\log K_{12})^2}{4 \log (k_{11}k_{22}/Z^2)} \tag{8.42}$$

where Z is the number of collisons per unit time between particles in solution. In the past several years a number of tests of this equation have been made. Table 8.11 lists several of these recently studied as well as some older data where applicable. Another type of comparison allowable by equation (8.39) is obtained by dividing the cross-reaction constant for a common reagent with two other reagents, k_{12}/k_{13}, and comparing this relative-rate value with that obtained by a similar procedure for reagent '4', k_{42}/k_{43}. If the f terms are small, the two ratios should be the same. Table 8.12 offers the data to test this proposition with the commonest reducing agents.

The data in Table 8.11 illustrate, in general, reasonable agreement. The outstanding exception is the $Co(terpy)_2^{2+}$-reduction of Co^{3+}; the latter being a reagent whose rate has been unsuccessfully predicted with equation

Table 8.11 Comparison between observed and calculated values for outer-sphere reactions*

Reaction	k_{obs} $M^{-1} s^{-1}$	$\Delta H^‡$	$\Delta S^‡$	I, M	k_{calc} $M^{-1} s^{-1}$	*Reference*
L-Co[(−)PDTA]$^{2-}$ + Fe(bip)$_3^{3+}$	8.1×10^4	5.7	−17	0.5	$\geq 10^5$	44
L-Fe[(−)PDTA]$^{2-}$ + Co(EDTA)$^-$†	1.3×10^1	7.0	−30	0.5	1.3×10^1	44
L-Fe[(−)PDTA]$^{2-}$ + Co(ox)$_3^{3-}$†	2.2×10^2	3.0	−37	0.5	1.0×10^3	44
Cr(EDTA)$^{2-}$ + Fe(EDTA)$^-$	$\geq 10^6$			?	10^9	123
Cr(EDTA)$^{2-}$ + Co(EDTA)$^-$	$\cong \times 10^5$			?	4×10^7	123
Fe(EDTA)$^{2-}$ + Mn(CyDTA)$^-$	$\cong 4 \times 10^5$			0.25	6×10^6	123
Co(EDTA)$^{2-}$ + Mn(CyDTA)$^-$	9×10^{-1}	4.8	−42	?	2.1	123
Fe(PDTA)$^{2-}$ + Co(CyDTA)$^-$	1.2×10^1			?	1.8×10^1	123
Co(terpy)$_2^{2+}$ + Co(bip)$_3^{3+}$	6.4×10	8.4	−19	0.05	3.2×10	124
Co(terpy)$_2^{2+}$ + Co(phen)$_3^{3+}$	2.8×10^2	6.9	−21	0.05	1.1×10^2	124
Co(terpy)$_2^{2+}$ + Co(bip)$^{3+}$	6.8×10^2	10.8	−6	0.05	6.4×10^4	124
Co(terpy)$_2^{2+}$ + Co(phen)$^{3+}$	1.4×10^3	10.9	−4	0.05	6.4×10^4	124
Co(terpy)$_2^{2+}$ + Co^{3+}	7.4×10^4	3.4	−23	1.0	2×10^{10}	124
Fe^{2+} + IrCl$_6^{2-}$ §	3.0×10^6	2.3	−28	1.0	1.8×10^4	125

*The original reference should be consulted for the values of k_{11} and k_{22} used.
†Reactivity is about equal for the D and L forms of the Co^{III} complex.
§This reaction could be inner-sphere.

(8.39) before[132]. Other studies reported on outer-sphere reactions that are unable to use equation (8.39) because of the lack of available data, but which are of interest, have been reported[133, 134].

The agreement in Table 8.12 is good. With the exception of the value for reduction of Ru(NH$_3$)$_6^{3+}$, the Cr^{2+}-reductions are about 2% of the rate of those using V^{2+} as the reductant. The Eu^{2+} numbers are difficult to compare because of the scarcity of values. Comparison between Ru(NH$_3$)$_6^{2+}$ and V^{2+} as reducing agents is also good: the former reagent reacts about 30-times faster. It is to be noted that the value for Co(NH$_3$)$_5$Cl^{2+} fits the model. It has been suggested by Guenther and Linck[131] that this oxidant utilises the outer-sphere path, although Haim[135] and Orhanovic, Po and Sutin[78] have argued that the mechanism is inner-sphere. If the ratios reported in Table 8.12 are accepted, the interesting calculation that the rate of outer-sphere reduction of Co(NH$_3$)$_5$Cl^{2+} by Cr^{2+} is about 0.1 $M^{-1} s^{-1}$ can be made. The inner-sphere path for Cr^{2+}-reductions is quite strongly favoured.

8.2.4.2 *Inner-sphere transition states*

The division of the free energy of activation into two components, the energy to form the precursor complex and rearrangement free energy, has been carried out in the interpretation of most inner-sphere reactions. Such a division must be approached with some caution as consideration of the reaction of Co(NH$_3$)$_5$H$_2$O^{3+} with Cr^{2+} establishes. This reaction proceeds by transfer of oxygen from Co^{III} to the Cr^{III} [136] product and the composition of one of the transition states is [Co(NH$_3$)$_5$OH$_2$Cr^{5+}]‡ [137] (the acid-dependent path is of no concern here). This species presumably arises from

Table 8.12 Relative rate comparisons for outer-sphere reactions $T = 25\,°C$, $I = 1.0$ M

Oxidant	Cr^{2+}*			*Reference*	V^{2+}*			*Reference*	Eu^{2+}*			*Reference*	$Ru(NH_3)_6^{2+}$*			*Reference*
$Co(NH_3)_6^{3+}$	8.9×10^{-5}† (0.024)			126	3.7×10^{-3} (1.0)	9.1	−40	47	2×10^{-2}** or 1.7×10^{-3}§ (0.5–5.4)	8.8	−42	73, 128	1.1×10^{-2}§ (3.0)			129
$Co(en)_3^{3+}$	$\cong 2 \times 10^{-5}$† (0.10)			73	2×10^{-4} (1.0)			73	5×10^{-3} (25)			73				
$Co(NH_3)_5py^{3+}$	4.3×10^{-3} (0.018)	9.8	−36	127	2.4×10^{-1} (1.0)	10.2	−27	130								
$[Co(NH_3)_5]_2NH_2^{5+}$	3.1×10^{-3}† (0.021)			128	1.5×10^{-1}† (1.0)	8.9	−33	46	5.8×10^{-2}§ (0.39)	9.6	−32	128				
$Ru(NH_3)_6^{3+}$	2×10^{2}§ (2.5)			129	8×10^{1}¶ (1.0)			129					4.4×10^{3}†† (50)			104
$Ru(NH_3)_5py^{3+}$	3.4×10^{3} (0.028)	3.0	−32	17	1.2×10^{5} (1.0)			17								
Fe^{3+}	‖				1.8×10^{4} (1.0)			95	‖				4.2×10^{5}** (23)	3.2	−22	104
$FeOH^{2+}$	‖				$< 4 \times 10^{5}$ (1.0)			95	‖				5×10^{4}§§ (≥ 0.1)			104
$FeCl^{2+}$	‖				4.6×10^{5} (1.0)			95	‖				$\geq 2 \times 10^{7}$§§ (≥ 43)			104
$Co(NH_3)_5Cl^{2+}$	‖				7.6‖ (1.0)	7.4	−30	75, 131	‖				2.6×10^{2}¶ (34)	6.9**	−27**	129

*The values listed at each element of the table are the rate constants in units of M^{-1}, s^{-1}, ΔH^* in kcal mol^{-1} and ΔS^* in cal mol^{-1} deg^{-1}; the number in parentheses is the relative rate constant for that *row* of the table.

†I = 0.4 M
§I = 0.2 M
‖Reaction is or may be inner-sphere
¶I = 0.18 M
**I = 0.10 M
††I = 0.16 M
§§T = 10 °C, I = 0.10 M

a precursor complex which contains an oxygen atom bonded to two protons, a Co^{III} centre and a Cr^{II} centre – a molecule that must be an extremely strong acid! Yet before a proton can be lost from this precursor complex, it passes over the transition state to oxidation–reduction products. Since the rate of proton-transfer is anticipated to be very rapid in this molecule, the result implies a very low rearrangement energy. The geometrical requirements of the precursor complex must be very much like those required for transfer of an electron. Distinction between the precursor complex and the transition state becomes fuzzy. Nevertheless, in several systems this breakdown appears useful.

(a) *Precursor stability* – The classic example of an influence of precursor stability on $\Delta G^{\ddagger}$ is the Cr-exchange of CrN_3^{2+} and $CrNCS^{2+}$ with $^{51}Cr^{2+}$[138]. The slowness of the latter reaction was ascribed to unfavourable binding of Cr^{2+} to the sulphur, the 'soft' end of thiocyanate. This feature has been often used as a criterion for the inner-sphere reaction mechanism[139]. Another reaction in which precursor-stability has been used to explain rate behaviour involves the Fe^{2+}-reduction of $Co(NH_3)_4ox^+$ and $Co(NH_3)_5ox^+$. Hwang and Haim[76] have compared their measured value for the former complex ($k = 4.6 \times 10^{-4}$ M^{-1} s^{-1}, $\Delta H^{\ddagger} = 18.5$ kcal mol^{-1}, $\Delta S^{\ddagger} = -12$ cal mol^{-1} deg^{-1} at 25 °C and I = 1.0 M) with the value of the latter complex calculated from Espenson's measured rate[139] of the complex $Co(NH_3)_5oxH^{2+}$ that is dependent on the inverse of $[H^+]$, and the value for the acid dissociation constant for coordinated binoxalate[140]: for $Co(NH_3)_5ox^+$ they find $k = 4.3 \times 10^{-1}$ M^{-1} s^{-1}. The increased rate in the latter case is believed to originate in a more stable precursor complex resulting from greater basicity[76]. Another phenomenon that is well rationalised by the concept of precursor stability involves chelated transition states. Some studies of reactions that proceed through chelated transition states have been reported[72, 76, 93, 141]. Earlier studies are recorded in Sykes's review[2]. The reaction of the maleato-[141] and salicylato-penta-amminecobalt(III)[93] complexes illustrate that chelate formation need not provide a path greatly lower in energy: in both cases, there is a competitive monodentate path. The maleato complex is especially interesting in that the monodentate Cr^{III} product is more stable than chelated product by about 1.6 kcal mol^{-1} at 1 M H^+. Thus, the reaction preferentially chooses a reaction path that does *not* correspond to the path with largest ΔG^0 (see below). Comparisons of the effect of chelation on rate constant are more difficult in the other systems studied because of the absence of suitable model systems. This is especially true when consideration is given to the effect of ligand structure on the rearrangement free energy.

(b) *Rearrangement energy* – After consideration of, and qualitative correction for, precursor-effects, a substantial variation in rate constant still exists in some systems. This is most clearly recognised when the oxidant or reductant is changed. For instance, it is very likely that both $Co(NH_3)_5N_3^{2+}$ and $Cr(NH_3)_5N_3^{2+}$ have equal affinities for Cr^{2+}, their precursor-complexes are probably of similar stability, yet the former is reduced about 10^7-times as rapidly[73, 142]. Effects of this type are most readily discussed in terms of two models – the chemical mechanism and the resonance mechanism.

It has been suggested that a number of reactions, especially those in which unsaturated organic compounds serve as the bridging ligand, proceed by reduction of the ligand, followed by reduction of the oxidant by this coordinated ligand – the chemical mechanism[143]

$$A^{III}\text{—}X\text{—}B^{II} \rightleftharpoons A^{III}\text{—}X^{-}\text{—}B^{III} \quad (8.43)$$
$$A^{III}\text{—}X^{-}\text{—}B^{III} \rightarrow A^{II}\text{—}X\text{—}B^{III} \rightarrow \text{Products.} \quad (8.44)$$

Taube and Gould have reviewed this subject[8]. The first systematic investigation seeking evidence for such a process was reported by Gould and Taube[56], who observed that some of the organic bridging ligands used in their study were reduced by Cr^{2+}. A similar result was reported later by Gould[144] who noted that substituted nitrobenzoic acid penta-ammine cobalt complexes reacted with Cr^{2+} in two ways – by reduction of Co^{III} and reduction of the NO_2 group. With *m*-nitrobenzoic acid as the ligand, the yield of Co^{II} was very small, thus implying conjugation is needed between the NO_2 site that is reduced and the Co^{III} centre in order to reduce the latter. Nordmeyer and Taube[9] first cited strong evidence in favour of the radical-intermediate mechanisms. The iso-nicotinamide ligand on the $Co^{III}(NH_3)_5$ and $Cr^{III}(H_2O)_5$ residues leads to oxidation of Cr^{2+} at rates differing by about ten (Co^{III} faster). In contrast, $Co(NH_3)_5X^{2+}$ is reduced much faster than $Cr(H_2O)_5X^{2+}$ by Cr^{2+} where X^- is F^- (Table 8.16). If equation (8.43) determines the rate of reduction by the radical ion path, then that rate should be only slightly perturbed by the oxidant centre. The iso-nicotinamide complexes fit this model. Further insight into this model comes from reductions of the corresponding $Ru(NH_3)_5X$ complexes[17]. Table 8.1 shows the value for X = iso-nicotinamide. The Ru^{III}-reduction is much more rapid. Gaunder and Taube[17] explain this result on the basis of variation of the electronic structure of the oxidant. The hole on Co^{III} exists in an orbital of σ-symmetry whereas that on Ru^{III} is of π-symmetry. Since the electron of the radical ion presumably exists in an orbital of π-symmetry, interaction with the Ru^{III} centre is strong and the lifetime of the radical ion intermediate becomes too short to have chemical meaning: the mechanism changes to resonance transfer. Other data have been accumulated that yield rate constants for reduction of $Co(NH_3)_5X^{n+}$ and $Cr(H_2O)_5X^{n+}$. These are compared in Table 8.13. The results there show that at best only aromatic ligands and those containing conjugated double-bond systems (fumarate and maleate) seem to utilise the chemical mechanism as determined by the $Co(NH_3)_5X^{2+}/Cr(H_2O)_5X^{2+}$ rate constant ratio. Other ligands that are reducible seem not to utilise the chemical mechanism, at least in both Cr^{III} and $Co(NH_3)_5X^{2+}$ complexes. It is possible that Cr^{III} as an oxidising agent reacts via the chemical mechanism in several cases; however, to establish this a criterion is needed that does not rely on a comparison with $Co(NH_3)_5X^{2+}$ complexes. Recently Davies and Jordan[148] have investigated the oxalatotetra-amminechromium(III) and maleatopenta-amminechromium(III) reactions with Cr^{2+}. In both cases the rates are considerably less than the corresponding Co^{III}–amine or Cr^{III}–aquo complexes. The results in this case may be interpreted as showing the importance of the reverse of Equation (8.43).

If it is assumed that most electron-transfer processes do take place by resonance transfer, then the available data on rearrangement energies can be looked upon from a theoretical, if qualitative, approach. This approach

Table 8.13 Rate data to test utilisation of the chemical mechanism by Cr^{2+} for electron exchange $T = 25\,°C$, $I = 1.0$ M

Complex	M = $Co(NH_3)_5^{3+}$				M = $Cr(H_2O)_5^{3+}$			
	k, $M^{-1}\,s^{-1}$	$\Delta H^\ddagger$, kcal mol^{-1}	$\Delta S^\ddagger$, cal mol^{-1} deg^{-1}	*Reference*	k, $M^{-1}\,s^{-1}$	$\Delta H^\ddagger$, kcal mol^{-1}	$\Delta S^\ddagger$, cal mol^{-1} deg^{-1}	*Reference*
M fumarate monoanion^{2+}	1.32	6.7	−36	32	3.0	6.3	−35	32
M fumarate monoanion^{2+} *	6.1×10^1	5.1	−33	32				
Mox^+ †	2×10^5			76	1.3×10^{-1}§	10.8	−26	145
M maleate monoanion^{2+} ‖	1.6×10^2	2.9¶	−38	141	4.0	9	−25	141
$M(ox)_3^{3-}$ **	4×10^6††			34	1.3×10^{-1}	8.2	−35	28
MN_3^{2+}	3×10^5			73	5.4§§	9.6§§	−23§§	146
MCH_3COO^{2+}	3.5×10^{-1}	8.2	−33	147	6×10^{-5}‖‖			20
MCN^{2+}	3.6×10^1¶¶	6.9¶¶	−28¶¶	15	7.7×10^{-2}***	9.3	−32	31

*M for Co in this case is *cis*-$Co(NH_3)_4H_2O$.
†M in this case is the tetra-aminecobalt(III) and tetra-aquochromium(III).
§I = 2.0 M
‖This refers to the path that leads to chelated Cr^{III} product.
¶Activation parameters contain a contribution from the path leading to monodentate product.
**M is Co^{III} and Cr^{III}.
††T = 20 °C; determined by competition experiments.
§§I = 0.5 M
‖‖By the acid-independent path.
¶¶I = 0.15 M
***Assuming this path for Cr-exchange proceeds via attack at N, as the $Co(NH_3)_5CN^{2+}$ reaction does.

uses as its theoretical model the treatment developed by Marcus for outer-sphere reactions (Section 8.2.4.1). The most significant feature of this approach is the division of the rearrangement energy into an intrinsic factor and a thermodynamic factor[49]. The intrinsic factor is that associated with the exchange reactions of each of the reacting partners – the energy required to cause the appropriate coordination sphere changes in the absence of any net ΔG^0 for the reaction – whereas the thermodynamic factor measures the effect of the driving force of the reaction on the energy of the transition state. When inner-sphere reactions are considered, the intrinsic factor must account for the effect of the bridging ligand on the self-exchange rate and the thermodynamic driving force. This approach has been used numerically by Haim and Sutin[1,14], by Newton and Baker[149], and by Patel and Endicott[40]. Implicit in this treatment is the assumption that the two reacting centres do not 'communicate' well with each other: activation on each centre requires an energy that is independent of the reacting partner. This assumption allows a Marcus-type expression for the reactivity of the cross-reaction,

$$\log k_{12} = \tfrac{1}{2}\log k_{11} + \tfrac{1}{2}\log k_{22} + \tfrac{1}{2}\log K_{12} \qquad (8.45)$$

where the f terms of equation (8.39) are assumed equal to one and the self-exchange terms are taken to include the bridging ligand. Hence, in order to calculate the rate of the Fe^{2+}-reduction of $Co(NH_3)_5F^{2+}$, Patel and Endicott[40] use the $Ru(NH_3)_6^{2+}$-reduction of $Co(NH_3)_5F^{2+}$ to calculate $\tfrac{1}{2}\log k_{11} + \tfrac{1}{2}\log K_{12}$ where k_{11} is the self-exchange rate for

$$^*Co(NH_3)_5H_2O^{2+} + Co(NH_3)_5F^{2+} = {}^*Co(NH_3)_5F^{2+} + Co(NH_3)_5H_2O^{2+}$$

Combination of this value with the value of $\tfrac{1}{2}\log k_{22}$ for the reaction

$$^*Fe^{2+} + FeF^{2+} \overset{k_{22}}{\rightleftharpoons} {}^*FeF^{2+} + Fe^{2+} \qquad (8.46)$$

and correction for the difference in thermodynamic driving force for oxidation of $Ru(NH_3)_6^{2+}$ to $Ru(NH_3)_6^{3+}$ compared to that for Fe^{2+} to FeF^{2+} allows calculation of k_{12} for $Fe^{2+} + Co(NH_3)_5F^{2+}$. They obtain a value of $1.3 \times 10^{-4}\ M^{-1}\ s^{-1}$ [40], compared to an experimental value of $6.6 \times 10^{-3}\ M^{-1}\ s^{-1}$ [139]. This error of 2.3 kcal mol^{-1} in $\Delta G^{\ddagger}$ is large, but at least some of the difference might be ascribable to difference in ionic strength of the measured data put into the calculation. Reference to the original papers[14, 40, 149] is instructive in terms of the anticipated differences between calculated and observed rate constants.

If attention is turned from a quantitative to qualitative approach, several other systems can be compared. The much greater ability of F^- compared to Cl^- to act as a catalyst in a third-order rate term,

$$k[\text{oxidant}][\text{reductant}][X^-]$$

has been considered by several workers. In the V^{2+}-reduction of $Co(NH_3)_6^{3+}$ [47], $Co(NH_3)_5OH_2^{3+}$ [47], and $[Co(NH_3)_5]_2NH_2^{5+}$ [46], Fe^{2+}-reduction of $Co(NH_3)_5F^{2+}$ [150], and Cr^{2+}-catalysed substitution of X^- for I^- in CrI^{2+} [151], F^- is much more efficient than Cl^-. Since the role of F^- and Cl^- in the transition states for these reactions is not as a bridging ligand, the intrinsic rate of electron transfer may not be highly dependent on the nature of F^- and Cl^-; but the thermodynamic driving force of the reactions is much more negative for F^- than Cl^- if the anion is located in a non-bridging ligand position on the reductant. Hence the more rapid reaction for F^- might be viewed as a thermodynamic effect. Fay and Sutin[18] have summarised the data on the reactivity of $Co(NH_3)_5NCS^{2+}$, $Co(NH_3)_5SCN^{2+}$, and $Co(NH_3)_5N_3^{2+}$ toward Fe^{2+}, Cr^{2+}, and V^{2+}, and $CrNCS^{2+}$, $CrSCN^{2+}$, and CrN_3^{2+} toward Cr^{2+} and V^{2+}. In all cases where rate constants are known the sulphur-bonded thiocyanate oxidant reacts faster than the azide complex, which in turn is faster than the nitrogen-bonded thiocyanate. These data are explained on the basis of increased free energy of reaction along the series thiocyanate, azide, iso-thiocyanate. For a semi-quantitative fit, however, a correction must be applied for added precursor-instability in the case of $MNCS^{n+}$ as the oxidant. These authors believe the intrinsic factor, the rearrangement energy, is equal for N_3^- or NCS^- as a bridge. Another example of the influence of thermodynamic aspects has been presented by Davies, Sutin and Watkins[33]. They rationalise the observed data on the rate of reduction of H_2O_2 by several reducing agents on the basis of the free energy change associated with precursor to postcursor complex (as distinct from final products). Those data are given in Table 8.14.

Table 8.14 Reduction of H_2O_2 by various metal ions $T = 25\,°C$, $I = 1.0$ M

Reducing agent	k, $M^{-1}\,s^{-1}$	$\Delta H^‡$, kcal mol^{-1}	$\Delta S^‡$, cal mol^{-1} deg^{-1}	*Reference*
$Co(CN)_5^{3-}$	7.4×10^2*	4.2	−31	152
$Fe(H_2O)_6^{2+}$	5.8×10^1	8.8†	−21†	153
$Cu(bip)_2^+$	8.5×10^2§			156
$Cr(CN)_6^{4-}$	3.3×10^2			33
$Cr(CN)_5OH_2^{3-}$	3.6×10^3			33

*I = 0.5 M
†From References 154 and 155; I not controlled.
§I = ?

The postcursor complex is assumed to be $[Cr(H_2O)_5OH^{2+} + OH]$ and $[Fe(H_2O)_5OH^{2+} + OH]$ for Cr^{2+} and Fe^{2+} as the reducing agents. This production of a hydroxy-complex of the 3-positive ion makes the oxidation potential for $Cr^{II} \rightarrow Cr^{III} + e^-$ more negative, increases the driving force, and hence allows Cr^{2+} to be as effective as $Cr(CN)_6^{4-}$ in reducing H_2O_2 [33], although the latter is a more powerful reductant if the total free energy change is considered.

Perhaps the most serious question concerning the approach discussed in the last two paragraphs is the assumption of a constant intrinsic factor as variations in the molecule are made. This assumption is especially critical for Cr^{II} oxidations. In these reactions, the bridging ligand's motion is presumably large (Section 8.2.3) and it seems likely that the energy required for that motion will be dependent on the immediate environment of the metal centre. Unfortunately, the separation of intrinsic factors from thermodynamic factors and the understanding of the variables important in determining intrinsic factors have not advanced much. A firmer understanding of electron-transfer processes will be gained as these features are probed more fully. There is another approach to the determination of rearrangement energies; this method does not *a priori* divide the rearrangement into intrinsic and thermodynamic factors, but rather attempts to understand the influence of electronic–structural variations on rearrangement energies. Such an approach retains a thermodynamic factor, since electronic factors cause the variation in net driving force, but does not explicitly divide the thermodynamic factor out. In the next section this electronic approach is examined.

8.2.5 The effect of electronic configuration on rate

One of the most striking features of transition-metal chemistry is the excellent correlations of physical properties with electronic structure. Examples range from structure – the Cr^{II}, Cu^{II}, and Rh^{II} acetates are of unusual, yet similar, structure[157], and each is characterised by a single electron or hole in the e_g set – to rates of substitution[105,158]. It is not unexpected, then, that attempts to find reactivity patterns characteristic of electronic structures have been initiated.

8.2.5.1 Co^{III} reductions

One of the first attempts to explain electron-transfer rate behaviour on the basis of electronic configuration involved the $Co(NH_3)_6^{2+}-Co(NH_3)_6^{3+}$-exchange reaction. It was suggested[3] that the large electronic configuration change necessary to convert Co^{II} $(t_{2g}^5e_g^2)$ to Co^{III} (t_{2g}^6) was responsible for the slowness of this reaction. Recent data has suggested that this is not the complete picture. In Table 8.15 are presented some data pertinent to the question. The rate of electron transfer is seen to depend very greatly on the nature of the ligands, but the $Co(terpy)_2^{2+}$-reduction of $Co(py)_4Cl_2^+$, the case in which the Co^{II} species is in a $(t_{2g}^6e_g^1)$ configuration, is not very different from other aromatic ligand complexes. These data do not, however, completely establish the absence of a strong rate-dependence on electronic

Table 8.15 Rate of reaction of some Co^{II} complexes with Co^{III} complexes

*Reductant**	*Oxidant†*	*k*§ $M^{-1}s^{-1}$	$\Delta H^{\ddagger}$ kcal mol^{-1}	$\Delta S^{\ddagger}$ cal deg^{-1} mol^{-1}	*I* M	*Reference*
$Co(bip)_3^{2+}$	*trans*-$Co(py)_4Cl_2^+$	1.1×10^4	5.2	−20	1.6×10^{-4}	124
$Co(phen)_3^{2+}$	*trans*-$Co(py)_4Cl_2^+$	9.1×10^3	8.4	−12	1.6×10^{-4}	124
$Co(terpy)_2^{2+}$	*trans*-$Co(py)_4Cl_2^+$	3.0×10^5	5.2	−14	1.6×10^{-4}	124
Co^{2+}	$CoCl^{2+}$	1.0			3.0	87
Co^{2+}	Co^{3+}	≅4§	12	−17	1.0	159
$Co(NH_3)_6^{2+}$	$Co(NH_3)_6^{3+}$	$<10^{-9}$‖	—	—	?	160
$Co(phen)_3^{2+}$	$Co(phen)_3^{3+}$	≅40¶	5.1	−34	0.1	161
$Co(bip)_3^{2+}$	$Co(bip)_3^{3+}$	≅16¶	7.7	−27	0.1	161

*Electronic configuration assuming octahedral geometry is $(t_{2g}^5e_g^2)$ in all cases but $Co(terpy)_2^{2+}$ in which it is $(t_{2g}^6e_g^1)$.
†All Co^{III} systems are t_{2g}^6.
§At 25 °C
‖At 65 °C
¶Calculated from activation parameters.

configuration. The Co^{II}—Co^{III}-exchange reactions involving complexes with unsaturated ligands may be special in that the molecular orbitals describing the t_{2g} electrons have considerable ligand character. This has at least two possible ramifications. (1) Metal ion electrons can easily be moved to the surface of the complex; (2) the bond length changes of the two oxidation states required to reach the activated complex are lessened because of stronger π-bonding in the lower oxidation state and stronger σ-bonding in the higher oxidation state. If these factors change the mechanism (in an admittedly unspecified fashion), then the comparison between $Co(terpy)_2^{2+}$ and $Co(bip)_3^{2+}$ reacting with *trans*-$Co(py)_4Cl_2^+$ may be unrelated to the slowness of the $Co(NH_3)_6^{2+}$–$Co(NH_3)_6^{3+}$-exchange. The hypothesis that the Co^{2+}–Co^{3+} reaction is faster than the $Co(NH_3)_6^{2+}$–$Co(NH_3)_6^{3+}$ reaction because of the availability of a low-lying high-spin state of Co^{III} [162] remains untested, although this hypothesis appears unnecessary to explain substitution data[87]. Waltz and Pearson[163] have published the results of a pulse-radiolysis study in which they interpret the data in terms of an excited electronic state of $Co(bip)_3^{2+}$.

8.2.5.2 *Electronic configuration effects on variation of metal centre*

More recently, other arguments have been advanced in an attempt to establish a correlation of observed rate constants with electronic structure. In what follows these attempts and some others are examined. The data chosen hopefully reflect rearrangement energies, not precursor effects. Further, for many of the reactions chosen the question of whether the reagents choose an inner-sphere or outer-sphere path is unanswered. At the present time, these difficulties are unavoidable; therefore scepticism of the correlations is necessary and desirable.

The effect of electronic structure-variation must take into account nuclear structural-variation. Oxidations of Cr^{2+} are accompanied by tetragonal to octahedral structural changes; this is true of other reagents in which an e_g electron is lost. On this basis Stritar and Taube[11] have discussed the reactivity of Ru^{III} and Co^{III} amines toward Cr^{2+}. Although their original arguments involve comparisons utilising the Cr^{2+}-reduction of $Ru(NH_3)_5Cl^{2+}$, a reaction since shown to proceed by an unusual mechanism[98] (Section 8.2.2.3), the point can be established with other complexes. In Table 8.12 are given the data for the reduction of $Ru(NH_3)_5py^{3+}$ and $Co(NH_3)_5py^{3+}$ by Cr^{2+}. Comparison of these data with those for reduction of the penta-ammine acetates of Ru^{III} ($k = 2.6 \times 10^4$ M^{-1} s^{-1}, $\Delta H^{\ddagger} = 1$ kcal mol^{-1}, $\Delta S^{\ddagger} = -34$ cal mol^{-1} deg^{-1} at 25 °C, I = 0.1 M)[11] and Co^{III} (Table 8.14) show that the Co^{III} system gains about a factor of ten more in rate on going from an outer-sphere mechanism to an inner-sphere mechanism. This result has been explained on the complementary motion needed for the coordination sphere rearrangements when a reductant is to lose an e_g electron and an oxidant is to gain an e_g electron. Comparison of the outer-sphere reactions in Table 8.12 illustrate a similar point. Reductions by V^{2+} are faster than those by Cr^{2+} because the electronic structure of Cr^{2+}

requires more substantial bond length changes upon oxidation than does V^{2+} [11]. (These same factors also make Cr^{II} a more powerful reductant in spite of the ionisation potential difference; apparently, the rearrangement energy is more influenced by the bond deformation aspect than the thermodynamic.) An examination of electronic structure effects on the intrinsic factor has been published by Endicott[42].

In Table 8.16 are gathered data on the reduction of several complexes with the organising feature the characteristic that there is a common ligand on each oxidant capable of serving as a bridging ligand. These common ligands are chosen mostly from the practical point of view: data are available. The first three reductants, A—C, presumably donate an e_g electron. (It is generally believed that Cu^+ reacts by an inner-sphere mechanism[168].) It can be seen that for oxidants that accept an e_g electron, the reactivity order for X is $Cl^- > N_3^- \gg CH_3COO^- \cong H_2O$. Data for reaction of these reductants with oxidants that accept a t_{2g} electron are less available. The Cr^{2+}-reduction of $Fe(H_2O)_5X^{n+}$ indicates $N_3^- > Cl^- \gg H_2O$, but the first two ligands may react near the limit for Cr^{2+} (Section 8.2.2.2). The data with $Ru(NH_3)_5X^{n+}$ verify that $Cl^- \gg H_2O$, and also show $CH_3COO^- > H_2O$. The data for Eu^{2+} fit the above inequalities relatively well. Reductants E → H lose a t_{2g} electron, but the interpretation of these is clouded by the inner-sphere/outer-sphere comparison. Fe^{2+} is thought to react by an inner-sphere path and shows, for 'e_g oxidant', $N_3^- > Cl^-$; for oxidants that accept a t_{2g} electron, $N_3^- \gg Cl^- > H_2O$. The reduction by V^{2+} of $Co(NH_3)_5N_3^{2+}$ is thought to be inner-sphere[49, 75], and that of $Co(NH_3)_5Cl^{2+}$ outer-sphere[131] (see also Table 8.12). Thus, for a comparison of inner-sphere paths, V^{2+} exhibits $N_3^- > Cl^-$ for 'e_g oxidants'. The outer-sphere reactions of reductants E → H show $Cl^- > N_3^- \cong H_2O > CH_3COO^-$ for e_g oxidants. Reductions by V^{2+} fit this if the $Co(NH_3)_5N_3^{2+}$ reaction goes at least 90% of an inner-sphere path and the reduction of $Co(NH_3)_5O\overset{O}{C}CH_3^{2+}$ is inner-sphere, a conclusion consistent with the activation data[147] and with Deutsch and Taube's interpretation of Cr^{2+}-, V^{3+}-, and CH_3COOH-inhibition of the V^{2+}-catalysed aquation of $CrO\overset{O}{C}CH_3^{2+}$ [20]. For outer-sphere reductions of oxidants that accept a t_{2g} electron, the data are limited to V^{2+}-reductions and the $Ru(NH_3)_6^{2+}$-reductions of Fe^{3+} and $FeCl^{2+}$. These data indicate $Cl^- \cong N_3^- \cong CH_3COO^- > H_2O$.

The trends in inner-sphere reactions would seem explicable on the basis of the matching of electronic configurations. When the oxidant has its lowest empty orbital of e_g symmetry and the reductant has a donor electron in an e_g orbital, overlap of these wave functions to cause strong interaction in the transition state, and hence to lower its energy, is easier with Cl^- as a ligand than with N_3^- or CH_3COO^-. where this interaction is through three atoms. The ineffective nature of H_2O as a bridging ligand is probably caused by poor precursor-stability, for the argument in Section 8.2.4.2 indicates OH_2 may be very effective at the actual electron transfer act once the precursor-complex is formed. Whenever a mixed electronic configuration (t_{2g} oxidant, e_g reductant, or vice versa) is encountered, N_3^- becomes more effective than Cl^- and acetate more effective than H_2O (in observed rate), a result whose

Table 8.16 Reduction of various M—X complexes $T = 25\,°C$, $I = 1.0$ M

Oxidant, A	A—Cl k*	A—Cl $\Delta H^\ddagger$†	A—Cl $\Delta S^\ddagger$§	A—Cl *Reference*	A—N_3 k*	A—N_3 $\Delta H^\ddagger$†	A—N_3 $\Delta S^\ddagger$§	A—N_3 *Reference*	A—$OC(O)CH_3$ k*	A—$OC(O)CH_3$ $\Delta H^\ddagger$†	A—$OC(O)CH_3$ $\Delta S^\ddagger$§	A—$OC(O)CH_3$ *Reference*	A—OH_2 k*	A—OH_2 $\Delta H^\ddagger$†	A—OH_2 $\Delta S^\ddagger$§	A—OH_2 *Reference*
						A. Cr^{2+} *as Reductant*										
$Co(NH_3)_5^{3+}$	2.6×10^6	—	—	65	3.0×10^5	—	—	73	3.5×10^{-1}	8.2	−33	147	5×10^{-1}‖	2.9	−52	137
$Cr(NH_3)_5^{3+}$	5.1×10^{-2}	11.1	−23	164	2.1×10^{-2}	—	—	142	—	—	—	—	—	—	—	—
$Cr(H_2O)_5^{3+}$	9¶	8.0**	−27**	138	1.3††	9.6	−23	146	$<6\times10^{-5}$	—	—	20	2×10^{-5}	—	—	20, 166
$Fe(H_2O)_5^{3+}$	1×10^7	—	—	68	2.9×10^7	—	—	67	—	—	—	—	2.3×10^3	5.2	−28	54
$Ru(NH_3)_5^{3+}$	$>3\times10^5$	—	—	98	—	—	—	—	2.6×10^4	1	−34	11	5×10^2§§	—	—	11
						B $Co(CN)_5^{3-}$ *as Reductant*										
$Co(NH_3)_5^{3+}$	$\cong5\times10^7$	—	—	167	1.6×10^6	—	—	167	$<1\times10^2$‖‖	—	—	167	—	—	—	—
						C. Cu^+ *as Reductant*										
$Co(NH_3)_5^{3+}$	4.9×10^4¶¶	5.3	−19	168	1.5×10^3¶¶	5.4	−26	168	—	—	—	—	1×10^{-3}¶¶	—	—	168
$Fe(H_2O)_5^{3+}$	—	—	—	—	$>1\times10^8$***	—	—	169	—	—	—	—	$<3\times10^{-1}$	—	—	169
						D. Eu^{2+} *as Reductant*										
$Co(NH_3)_5^{3+}$	3.9×10^2	5.0	−30	73	1.9×10^2	5.5	−30	73	1.8×10^{-1}	—	—	170	7.4×10^{-2}	9.3	−32	128
$Cr(H_2O)_5^{3+}$	2.2×10^{-3}	17.1	−14†††	171	—	—	—	—	—	—	—	—	$\cong1.7\times10^{-5}$	—	—	171
$Ru(NH_3)_5^{3+}$	2.4×10^4	—	—	11	—	—	—	—	1.7×10^5	—	—	11	—	—	—	—
$Fe(H_2O)_5^{3+}$	2.0×10^6***	—	—	172	1.2×10^7***	—	—	54	—	—	—	—	3.4×10^3***	3.5	−29	172
						E. Fe^{2+} *as Reductant*										
$Co(NH_3)_5^{3+}$	1.35×10^{-3}	12.5	−30	139	8.7×10^{-3}	—	—	173	—	—	—	—	—	—	—	—
$Fe(H_2O)_5^{3+}$	5.4§§§	11.0	−15	174	1.9×10^3¶¶¶	13.3****	7****	176	—	—	—	—	8.7×10^{-1}¶¶¶	9.3	−25	177
						F. $Cr(bip)_3^{2+}$ *as Reductant*										
$Co(NH_3)_5^{3+}$	8×10^5	—	—	73	4.1×10^4	—	—	73	1.2×10^3	—	—	73	5×10^4	—	—	73
						G. V^{2+} *as Reductant*										
$Co(NH_3)_5^{3+}$	7.6	7.4	−30	75, 131	1.3×10^1	11.7	−14	73	1.2	11.6	−19	147	5.3×10^{-1}	8.2	−32	47
$Cr(H_2O)_5^{3+}$	3.8×10^{-2}††††		—	111	—	—	—	—	1.6×10^{-1}	—	—	20	see Section 8.2.2.4			
$Fe(H_2O)_5^{3+}$	see Table 8.8															
$Ru(NH_3)_5^{3+}$	see Table 8.8															
						H. $Ru(NH_3)_6^{2+}$ *as Reductant*										
$Co(NH_3)_5^{3+}$	2.6×10^2¶¶	6.9§§§§	−27§§§§	66, 129 129	1.2¶¶			129	—	—	—	—	3¶¶	—	—	129
$Fe(H_2O)_5^{3+}$	$>2\times10^7$	—	—	104	—	—	—	—	—	—	—	—	3.1×10^5	3.2	−22	104

*Units are $M^{-1}\,s^{-1}$
†Units are kcal mol^{-1}
§Units are cal $mol^{-1}\,deg^{-1}$
‖At 20 °C
¶At 0 °C
**I = 0.2 M
††At 0 °C, I = 0.5 M
§§There is a path inverse in $[H^+]$, $k = 2.2\times10^2\,s^{-1}$.
‖‖The outer-sphere path has $k = 1.1\times10^4\,M^{-2}\,s^{-1}$.
¶¶I = 0.2 M
***T = 1.6 °C
†††The published value has an apparent incorrect sign
§§§For a detailed analysis of this reaction, see Reference 175.
¶¶¶At 0 °C, I = 0.55 M.
****Valid only below 15 °C; see Reference 176.
††††I = 2.5 M; there is also a term in $[H^+]^{-1}$; $k = 2.5\times10^{-3}\,s^{-1}$.
§§§§I = 0.1 M

direction is to be anticipated on the basis of the t_{2g}-oxidant–t_{2g}-reductant systems, wherein N_3^- becomes highly effective. In such a system the optimal overlap is achieved through the π-system of azide.

8.2.5.3 Non-bridging ligand effects

The change in rate with a change in the ligand environment of a pair of metal ions has long been studied, but systematic variation of the non-bridging ligands is a recent development. (A review has been prepared on this subject, see Reference 178, but was not available at this writing.) Such studies are very useful in understanding the factors contributing to rearrangement energies because many of the other important factors – precursor stability, bridging ligand role in facilitating electron transfer – at least approximately cancel out[179].

The non-bridging ligands in an inner-sphere reaction are easily defined: the ligands that do not bind to both metal centres. However, it has also been found to be useful to consider 'non-bridging' ligand effects on outer-sphere reactions. Since the role of non-bridging ligands appears to be electronic in nature, it follows that their effect should be similar in inner-sphere and outer-sphere reactions. The qualification that should be supplied for this outer-sphere consideration is that the geometry of the outer-sphere transition state is constant. In practice the non-bridging ligands in outer-sphere reactions or reactions of unknown mechanism are defined as those that are not varied.

The first extensive study of non-bridging ligand effects was made by Benson and Haim[180]. They found that the rate of oxidation of Fe^{2+} by some Co^{III} complexes containing chloride as the common ligand showed tremendous sensitivity to a change in non-bridging ligand. The data from this study and others involving the Fe^{II}–Co^{III} pair are shown in Table 8.17. The rationalisation of these data by Benson and Haim was that the ligand-field strength of the non-bridging ligands and the ease of stretching them, as measured by substitution reactions, were important parameters in determining the rate variations. This argument was based on some remarks of Orgel[181]. In reductions of Cr^{III} complexes, Orgel argued, the *trans* ligand will stretch so as to lower the energy of the d_{z^2} orbital, the orbital that must receive the electron. Benson and Haim's[180] argument extended this concept: Co^{III}, which requires an e_g electron should behave similarly. Further, the larger the ligand-field strength, the greater the ligands must be stretched to lower sufficiently the energy of the orbital.

Because ligand-field strength measures the energy difference between the t_{2g} and e_g sets of orbitals, and the former are affected by π-bonding effects, Bifano and Linck[182] reasoned that a more appropriate measure of the energy needed would be achieved by consideration of the σ-bonding strength of the ligands. This approach is best seen as a modification of Hush's comparison of intervalence-transfer absorption[120] and thermal electron transfer. The former energy is measured by the vertical (optical) transition

$$\underset{(A)}{[L_5Co^{III}ClFe^{II}]} \rightarrow \underset{(B)}{[L_5Co^{II}ClFe^{III}]}; \qquad (8.47)$$

Table 8.17 The effect of non-bridging ligands on the rate of reduction of Co^{III} complexes by Fe^{II}; = 25 °C, $[ClO_4^-]$ = 1.0 M

Co^{III} *Complex*	$X = Cl^-$ k, $M^{-1}\,s^{-1}$	*Reference*	$X = Br^-$ k, $M^{-1}\,s^{-1}$	*Reference*	$X = N_3^-$ k, $M^{-1}\,s^{-1}$	*Reference*
cis-$Co(en)_2NH_3X^{2+}$	1.8×10^{-5}	180	6.1×10^{-6}¶	84	2.3×10^{-4}¶¶	41
trans-$Co(en)_2NH_3X^{2+}$	6.6×10^{-5}	180				
cis-$Co(en)_2H_2OX^{2+}$	4.6×10^{-4}	180	2.8×10^{-4}**	84	6.9×10^{-3}	41
cis-$Co(en)_2ClX^{+}$	1.6×10^{-3}	180				
$Co(NH_3)_5X^{2+}$	1.4×10^{-3}*	139	8.9×10^{-4}††	84	8.7×10^{-3}***	25
cis-$Co(NH_3)_4H_2OX^{2+}$	3.5×10^{-2}†	48			3.6×10^{-1}†††	173
trans-$Co(en)_2ClX^{+}$	3.2×10^{-2}	180				
trans-$Co(en)_2BrX^{+}$	3.6×10^{-2}§	180	1.8×10^{-2}§§	84		
trans-$Co(en)_2H_2OX^{2+}$	2.4×10^{-1}	180	9.4×10^{-2}‖‖	84	8.0×10^{-1}	41
trans-$Co(NH_3)_4ClX^{+}$	2.2‖	48				
trans-$Co(NH_3)_4H_2OX^{2+}$	$\cong 1.0 \times 10^{1}$	48			2.4×10^{1}†††	173

*$\Delta H^\ddagger$ = 12.5 kcal mol^{-1}; $\Delta S^\ddagger$ = −30 cal mol^{-1} deg^{-1}
†$\Delta H^\ddagger$ = 14.1 kcal mol^{-1}; $\Delta S^\ddagger$ = −18 cal mol^{-1} deg^{-1}
ΓIt is not clear that Cl^- is the bridging ligand. See Reference 84.
‖$\Delta H^\ddagger$ = 8.5 kcal mol^{-1}; $\Delta S^\ddagger$ = −29 cal mol^{-1} deg^{-1}
¶$\Delta H^\ddagger$ = 17.6 kcal mol^{-1}; $\Delta S^\ddagger$ = −23 cal mol^{-1} deg^{-1}
**$\Delta H^\ddagger$ = 15.4 kcal mol^{-1}; $\Delta S^\ddagger$ = −23 cal mol^{-1} deg^{-1}
††$\Delta H^\ddagger$ = 15.5 kcal mol^{-1}; $\Delta S^\ddagger$ = −20 cal mol^{-1} deg^{-1}. Other values have been reported – see References 139 and 150.
§§$\Delta H^\ddagger$ = 14.8 kcal mol^{-1}; $\Delta S^\ddagger$ = −17 cal mol^{-1} deg^{-1}
‖‖$\Delta H^\ddagger$ = 12.6 kcal mol^{-1}; $\Delta S^\ddagger$ = −21 cal mol^{-1} deg^{-1}
¶¶T = 33.3 °C
***I = 0.89 M
†††$[ClO_4^-]$ = 0.5 M

the difference in energy of this transition as L is changed is caused by the influence of L on the lowest-lying empty orbital, the σ-antibonding orbitals of the Co^{III} centre. (For changing L does not affect the energy of (A) to first-order.) Now if, as Hush argues, the thermal electron-transfer energy and this vertical transfer energy are related, then it is clear that reactivity will parallel the energy of the σ-antibonding orbital relative to some arbitrary, but constant, zero of energy. This parallel will not, in general, follow the ligand field strength of L.

Bifano and Linck[182] tested this argument by comparing the rate of reduction of *cis*-$Co(en)_2NH_3Cl^{2+}$ and *cis*-$Co(en)_2pyCl^{2+}$ with Fe^{2+}. The difference in non-bridging ligand, NH_3 compared to py, keeps the ligand-field strength constant, but since $NH_3 >$ py in σ-bonding strength, at least as measured by acidity, the σ-bonding theory predicts the pyridino complex will have the greater reactivity. The value for $Co(en)_2NH_3Cl^{2+}$, Table 8.17, is to be compared with 7.9×10^{-4} M^{-1} s^{-1} for the pyridino complex[182]. The result is in complete accord with the σ-bonding theory.

The other data in Table 8.17 illustrate the most striking feature of the non-bridging ligands: the response to a change in non-bridging ligand is more or less independent of the bridging ligand. It has been shown that the response of L_5CoCl^{n+} and L_5CoBr^{n+} is such that a given change in non-bridging ligand makes an identical change in $\Delta G^\ddagger$ in both systems[84]. These data also show that the non-bridging ligand effect is dominated by a change in $\Delta H^\ddagger$, at least in the Fe^{2+}-reduction of CoL_5Br^{n+}. The dominance of $\Delta H^\ddagger$ in determining non-bridging ligand effects has been shown in other systems as well[183, 184]. The reduction of $L_5CoN_3^{n+}$ complexes by Fe^{2+} are less sensitive to a change in non-bridging ligands than is L_5CoCl^{n+}. Thus, whereas *cis*-$Co(en)_2H_2ON_3^{2+}$ is reduced 15 times faster than the chloride analogue, *trans*-$Co(NH_3)_4H_2ON_3^{2+}$ is reduced only 2.5-times as rapidly as is the chloride analogue.

Several other studies of non-bridging ligand effects on reduction rate have been made recently. These include the V^{2+} [131] and $Ru(NH_3)_6^{2+}$ [40] reductions of CoL_5Cl^{n+} complexes. Data for the V^{2+}-reduction of RuL_5Cl^{n+} [98] complexes are to be found in Table 8.8. Earlier studies, including the Cr^{2+}-reduction of CoL_5NCS^{n+} [14], CrL_5OH^{n+} [185], and CrL_5Cl^{n+} complexes[116, 183], have been summarised earlier with some other data[6]. With the exception of the V^{2+}-reduction of Ru^{III} complexes, these data all show the same dependence on non-bridging ligands: a *trans*-H_2O is a better non-bridging ligand group than is a *trans*-Cl^-, which is better than a *trans*-NH_3, etc. Ethylenediamine complexes react somewhat more slowly than do ammonia complexes. Although it has been suggested that this rate difference may be due to conformational effects of the chelated system[116, 184], it is equally consistent with the σ-bonding model in that ethylenediamine is a slightly stronger bonding ligand than is NH_3. It is especially pertinent to note that these relative rates are found even when the reaction is outer-sphere ($Ru(NH_3)_6^{2+}$ as reductant). The extreme importance of the symmetry of the orbital to be populated in the reduction is illustrated by the V^{2+}-reduction of Ru^{III} complexes[98]. In this case, that orbital is of π-type and its energy is dependent on interaction with the π-orbitals of the ligands. Therefore the relative effectiveness of the non-bridging ligands is different in Ru^{III} reductions compared with Co^{III} or Cr^{III} reductions. Ammonia as a non-bridging ligand, without π-type electrons, is expected to be more reactive than Cl^- or H_2O in Ru^{III} systems. Reference to Table 8.8 confirms this hypothesis.

Recently Williams and Garner[184] have taken advantage of a great deal of careful synthetic work by Garner and co-workers[186] to study the reduction of a series of related Cr^{III} complexes. Table 8.18 presents these data as well as an estimate of the Cr^{2+}-reduction of $CrCl^{2+}$ at 25 °C and I = 1.5 M. Williams and Garner have interpreted these data in terms of a bond-stretching model. They assume the ligand-field of the Cr^{III} centre must become equal to that of a $Cr(H_2O)_5Cl^{2+}$ complex (that is, the ligand environment of the Cr^{II} centre). By allowing only the *trans* ligand to stretch and hence to influence the ligand field on the oxidant, they are able to compute the relative energy needed to reach the transition state. Their calculated values are compared to the literature values in Table 8.18.

Another means of understanding these data utilises the σ-bonding model. Comparison of the rate of *trans*-$Cr(NH_3)_4H_2OCl^{2+}$ and $Cr(H_2O)_5Cl^{2+}$ indicates the replacement of four *cis*-NH_3's by four *cis*-H_2O's leads to a gain in reactivity of a factor of 25, or a free energy factor of

$$\log \frac{33}{1.2} = 1.40 \tag{8.48}$$

If one assumes each *cis*-H_2O contributes equally to this value, then the free energy factor for replacement of one *cis*-NH_3 by a *cis*-H_2O is 0.35. The value of replacement of a *trans*-NH_3 by a *trans*-H_2O can be computed from the relative rate constants for $Cr(NH_3)_5Cl^{2+}$ and *trans*-$Cr(NH_3)_4H_2OCl^{2+}$,

$$\log \frac{1.3}{8.8 \times 10^{-2}} = 1.17 \tag{8.49}$$

These two values can be used to determine the perturbation on the lowest-lying empty σ-orbital by any arbitrary replacement of NH_3's by H_2O. This procedure has been used to predict the rate constants for the reactions in Table 8.18 under the column labelled $k_{(calc)}$. It is important to note that replacement of a *cis* ligand has less effect than a *trans* ligand because of the directional properties of the antibonding σ-orbital. Nevertheless, the non-bridging ligand effect is not a pure '*trans*' effect; replacement of *cis*- ligands

Table 8.18 Rate of Cr^{2+} reaction with chloro–Cr^{III} complexes*

Complex†	k $M^{-1}\,s^{-1}$	$\Delta H^{\ddagger}$, kcal mol^{-1}	$\Delta S^{\ddagger}$, cal mol^{-1} deg^{-1}	$\Delta H^{\ddagger}$, (calc.)	$\Delta H^{\ddagger}$ §, (calc.)	k, (calc.)
$Cr(NH_3)_5Cl^{2+}$	8.8×10^{-2}	11.1	−26	11.0	11.4	—
cis-$Cr(NH_3)_4H_2OCl^{2+}$	1.3×10^{-1}	10.4	−28	10.6	10.3	2.0×10^{-1}
trans-$Cr(NH_3)_4H_2OCl^{2+}$	1.3	9.2	−28	9.3	9.5	—
$2,6(H_2O)_2Cr(NH_3)_3Cl^{2+}$	2.2	9.3	−26	8.9	8.7	2.9
$2,4,6(H_2O)_3Cr(NH_3)_2Cl^{2+}$	6.9	8.5	−27	8.5	8.4	6.5
$2,3,4,6(H_2O)_4CrNH_3Cl^{2+}$	1.9×10^{1}	7.8	−27	7.9	8.1	1.5×10^{1}
$Cr(H_2O)_5Cl^{2+}$	3.3×10^{1}†					

*From Reference 184.
†Estimated from data in Reference 165.
§Two models were used in Reference 184.

is effective in changing the rate. The values calculated by the σ-bonding, $k_{(calc)}$, agree with the observed values as well as the calculated $\Delta H^{\ddagger}$'s do with the observed $\Delta H^{\ddagger}$'s (note that an error of 0.3 kcal mol^{-1} in $\Delta H^{\ddagger}$ produces a twofold change in k).

Several studies of non-bridging ligand effects on the reductant have been reported, but few systematic studies have been undertaken. This type of experiment, while needed for a thorough understanding of the activation on the reductant, is difficult because of the usual lability of the reductant. Cannon and Earley[187] interpret the rate law for the Cr^{2+}-catalysis of substitution of methyl iminodiacetate (L) into Cr^{3+},

$$\frac{d[CrL^+]}{dt} = 2.2 \times 10^{-4}[Cr^{2+}][Cr^{3+}][HL^-][H^+]^{-2} \tag{8.50}$$

as arising from the non-bridging ligand effect of L on Cr^{II}. This accelerates the reaction; CrL is a better reductant for $CrOH^{2+}$ than is Cr^{2+}. Pennington and Haim have reported on the Cr^{2+}-reduction of $Co(NH_3)_5Cl^{2+}$ in Cl^- media. This reaction shows a third-order rate law term, $k[Cr^{2+}][Co(NH_3)_5Cl^{2+}][Cl^-]$ [188]. This term leads to equal quantities of *cis*-$CrCl_2^+$ and *trans*-$CrCl_2^+$ as products, thus showing, after statistical correction, a slight preference for a *trans*-Cl^- over a *cis*-Cl^- on the Cr^{II} reductant in the transition state. Reactions of Pt^{II} complexes will be useful in further studies on the non-bridging ligand effect on the reductant (Section 8.3.7). Such studies are needed.

8.3 RATE PATTERNS

In this section the rates of reaction of several other systems are examined in terms, where possible, of the concepts developed above. In oxidation–reduction reactions, either reductants or oxidants can be chosen as the focus; the former are rather arbitrarily picked here, although some oxidants will be discussed separately.

8.3.1 Reductions by vanadium reagents

8.3.1.1 Vanadium(II)

Data are to be found in References 78, 79, 147, 189–194. A substantial fraction of the oxidation–reduction chemistry of V^{2+} is dominated by substitution into the first coordination sphere. The rate data for reaction of V^{2+} with $Co(CN)_5X^{3-}$ complexes have been interpreted[79] as due to rate-limiting substitution into the first coordination sphere of V^{2+}

$$V^{2+} + Co(CN)_5X^{3-} \overset{K}{\rightleftharpoons} V^{2+}, Co(CN)_5X^{3-} \tag{8.51}$$

$$V^{2+}, Co(CN)_5X^{3-} \xrightarrow{k} [Co(CN)_5XV^-]^\ddagger \rightarrow \text{Products} \tag{8.52}$$

The formation of the ion-pair is saturated because of the opposing charges on the two reagents. At least in the case of $X^- = N_3^-$ and $X^- = NCS^-$, the reaction has been shown to utilise an inner-sphere path by direct observation of VN_3^{2+} and $VNCS^{2+}$. Not surprisingly, $Co(CN)_5^{3-}$ is the initial Co^{II} product[193]. The reaction of V^{2+} with UO_2^{2+} [194] shows how a rate constant can indicate a mechanism ($k = 40\ M^{-1}\ s^{-1}$ at I = 1.0 M is indicative of an inner-sphere mechanism) whereas the activation data ($\Delta H^\ddagger = 7.1$ kcal mol^{-1}, $\Delta S^\ddagger = -26$ cal mol^{-1} deg^{-1} at I = 2.0 M) do not support the conclusion. Based on a comparison of the activation data in Table 8.6, this reaction would seem to utilise the outer-sphere path. In the case of the reaction with Hg^{2+}, Green and Sykes[190] have found two paths, each of the same composition, $[VHg^{4+}]^\ddagger$. They claim one path, that characterised by a higher $\Delta H^\ddagger$ (15.2 kcal mol^{-1}) and more positive $\Delta S^\ddagger$ (-7 cal mol^{-1} deg^{-1}), leads to Hg^I and V^{3+}. The other path leads directly to V^{IV} and Hg^0.

$$V^{2+} + Hg^{2+} \rightarrow VO^{2+} + Hg^0 \tag{8.53}$$

[NOTE ADDED IN PROOF: The data on which these conclusions were based have been re-evaluated. The authors[259] conclude that there are paths of composition $[VHg^{4+}]^\ddagger$ and $[VHgOH^{3+}]^\ddagger$. The first leads to V^{III} and Hg^I whereas the second produces V^{IV} and Hg^0 directly. The revised rate parameters are respectively $k = 1.0\ M^{-1}\ s^{-1}$, $\Delta H^\ddagger = 15.8$, $\Delta S^\ddagger = -6$; $k = 8.7\ s^{-1}$, $\Delta H^\ddagger = 14.8$, $\Delta S^\ddagger = -5$].

Several reactions of V^{2+} are sufficiently rapid to be unambiguously classified as utilising the outer-sphere path. These have been listed in Table 8.8. It is to be noted that the reductions of $Ru(NH_3)_5X^{2+}$, where X^- is acetate, formate, or trifluoroacetate, by either V^{2+} or Eu^{2+}, lead to intermediates independent of the reducing agent, presumably species such as $Ru(NH_3)_5O\overset{O}{\overset{\|}{C}}CH_3^+$.

Bennett and Taube[133] have examined the reactions of V^{II} complexes of polypyridines. The reactions are peculiar in that V^{IV} is the first observed vanadium product. The apparent reducing power of $V(bip)_3^{2+}$ is not great (Bennett and Taube estimate the oxidising potential is $-0.8 < E_0 < -0.3$); the driving force for reduction of oxidants such as $Ru(NH_3)_6^{3+}$ is to be found in the secondary reaction of V^{III} to V^{IV}. Nevertheless, the reactivity is moderate: $k = 1.2 \times 10^{-1}$ at 25 °C, I = 0.1 M in trifluoroacetate. Perchlorate ion causes a rate acceleration.

8.3.1.2 Vanadium(III) and vanadium(IV)

The reaction of V^{3+} with Cu^{2+} has been studied as the Cu^{2+}-catalysis of the V^{3+}-reduction of $Co(NH_3)_5Cl^{2+}$ and other Co(III) complexes[195]. Parker and Espenson find the rate law

$$\frac{-dV^{3+}}{dt} = \left(1.8 \times 10^{-2} + \frac{3.4 \times 10^{-1}}{[H^+]}\right)[V^{3+}][Cu^{2+}] \qquad (8.54)$$

in 3.0 M $LiClO_4$ at 25 °C. For the acid-independent and acid-dependent terms, $\Delta H^{\ddagger} = 21.3$ and 19.4 kcal mol^{-1}, and $\Delta S^{\ddagger} = 5$ and 4 cal mol^{-1} deg^{-1} respectively. The V^{3+}-reduction of $IrCl_6^{2-}$ proceeds by an inverse acid term,

$$\frac{-dV^{3+}}{dt} = 5.2 \times 10^1[V^{3+}][IrCl_6^{2-}][H^+]^{-1} \qquad (8.55)$$

with $\Delta H^{\ddagger} = 12.4$ kcal mol^{-1} and $\Delta S^{\ddagger} = -9$ cal mol^{-1} deg^{-1} [125]. In contrast to the V^{2+}-reduction of $Co(CN)_5X^{3-}$ discussed above, there is no evidence in this case of ion-pair formed in observable concentrations.

Rosseinsky and Nicol have published the results of several reductions by VO^{2+} [196–199]. They have made the point that the rates of these reactions parallel those of the same oxidants with Fe^{2+} [200, 201]. The rates with VO^{2+} are somewhat less than those of Fe^{2+}, as expected on the basis of reducing power. A summary appears in Reference 198.

8.3.2 Reductions by chromium(II)

In addition to the extensive data in tables in Section 8.2, other data on Cr^{2+}-reductions have appeared[11, 72, 123, 202–212]. Almost all reactions of Cr^{2+} utilise an inner-sphere path if one is available – the outer-sphere activated complex with Cr^{2+} is generally of much higher energy.

Lane and Bennett's[207] report of the Cr^{2+}-reduction of glycollato-bis-ethylenediaminecobalt(III) and mercaptoacetato-bis-ethylenediaminecobalt (III) represent the first report of a comparison between a sulphur-bridging ligand and oxygen analogue. The sulphur-bridged system reacts much more rapidly; the same result of greater reactivity with a second-row element compared to the analogous first-row element is to be seen in the Cr^{2+}-reduction of $Co(NH_3)_5X^{2+}$ [65] and $Cr(NH_3)_5X^{2+}$ [164], although the discrimination is not nearly so large in these latter cases. The Cr^{2+}-catalysed aquations of Cr^{III} complexes have been reviewed recently[6]. To that compilation should be added the data on the reaction

$$CrN_3^{2+} + H^+ = Cr^{3+} + HN_3$$

which obeys the rate law[213]

$$\frac{-d \ln[CrN_3^{2+}]}{dt} = \left[5.2 \times 10^{-4}\ M^{-1}\ s^{-1} + \frac{3.4 \times 10^{-5}\ s^{-1}}{[H^+]}\right][Cr^{2+}]$$

at $I = 2.0$ M, $T = 40$ °C.

Two careful studies of activation parameters have recently been reported. Vriesenga[205], working with Cobble, has investigated the Cr^{2+}-reduction of $Co(NH_3)_5O\overset{O}{C}CH_3^{2+}$ over a temperature range of 0–120 °C. No curvature of the activation energy plot is seen: $\Delta C_p^{\ddagger} = 0 \pm 5$ cal mol^{-1} deg^{-1}. James and King[165] have examined the activation parameters for chromium-exchange between Cr^{2+} and $CrCl^{2+}$ via the chloride-bridged transition state as a function of solvent composition (dioxane-water) and ionic strength. The effect of electrolytes on the rate constant is large (0.5 M^{-1} s^{-1} at I = 0 versus 45 M^{-1} s^{-1} at I = 5 M $HClO_4$ at 0 °C).

Another report of oxidation–reduction reactions of $Cr(bip)_3^{2+}$ has appeared[209]. The $Cr(bip)_3^{2+}$ reductions of several Pt^{IV} complexes proceed by second-order rate laws. As in reduction of Co^{III} complexes, the Pt^{IV} complexes with halogen ligands are reduced much faster than the hexamine complex. Reduction with other Cr^{II} complexes have also been reported. Cannon[214] has studied the reduction of $Cr(NH_3)_5Cl^{2+}$ by Cr^{II} in acetic acid–acetate buffer. The rate law is

$$\frac{-d[Cr(NH_3)_5Cl^{2+}]}{dt} = k[Cr(NH_3)_5Cl^{2+}][Cr^{II}]^{\frac{1}{2}} \qquad (8.56)$$

where k is 9.7×10^{-3} $M^{-\frac{1}{2}}$ s^{-1} at 25 °C and I = 1.0 M. The activation energy is 14 kcal mol^{-1}. This rate law is interpreted as reflecting reaction between a Cr^{II} monomer, $Cr(O\overset{O}{C}CH_3)_3^-$, and the Cr^{III} complex with a second-order rate constant of 1.2 M^{-1} s^{-1}. This value is consistent with a reasonable non-bridging ligand effect on the reducing agent. Comparison of the above rate constant with that for the Cr^{2+} [164] and $Cr(EDTA)^{2-}$ [204] with $Cr(NH_3)_5Cl^{2+}$ illustrates this point. Other studies using $Cr(EDTA)^{2-}$ as a Cr^{II} reducing agent[203,204] show that the reactivity of Cr^{2+} and $Cr(EDTA)^{2-}$ exhibit a parallel differentiation of the nature of the bridging ligand.

The studies of reduction of non-metal-containing substances by Cr^{II} continues to be actively pursued. It has been estimated that the rate of reaction of Cr^{II}–ethylenediamine complexes in DMF–H_2O mixtures with alkyl radical proceeds with a rate constant of about 4×10^7 M^{-1} s^{-1} [215]. A fascinating case of activation of a ligand for reduction by Cr^{2+} by coordination to another metal ion is illustrated in the reaction[216],

$$2H^+ + Ru(NH_3)_5N_2O^{2+} + 2Cr^{2+} = Ru(NH_3)_5N_2^{2+} + 2Cr^{3+} \quad (8.57)$$

in which $k > 10^2$ for a second-order reaction.

8.3.3 Reduction by iron(II)

Birk[217] has studied the reduction of VO_2^+ with $Fe(CN)_6^{4-}$. The reaction is quite rapid ($k = 1.2 \times 10^6$ M^{-1} s^{-1} at 8 °C and I = 0.5) and forms a binuclear intermediate. The $Fe(CN)_6^{4-}$–$Fe(CN)_6^{3-}$ exchange has been examined[218] under a wide variety of conditions of accompanying cation. At zero ionic strength, $k = 5.9$ M^{-1} s^{-1} (0.1 °C), $\Delta H^\ddagger = 8.5$ kcal mol^{-1} and $\Delta S^\ddagger = -24$ cal mol^{-1} deg^{-1}. It has been claimed that a diffusion measurement has yielded the rate constant for the ferrocene–ferrocinium exchange at room temperature[219].

8.3.4 Reduction by cobalt(II)

Cobalt(II) is not a reasonable reducing agent unless coordinated with ligands which will stabilise the three-valent state. Several reactions of Co^{II} have been used to test the Marcus relation (Table 8.12). Other data is to be found in those references and References 220 and 221. Farina and Wilkins[124] have reported that the rates of reduction with $Co(terpy)_2^{2+}$ of $Co(phen)_3^{3+}$ and $Co(tmp)_3^{3+}$ are nearly equal (4.8×10^1 and 6.8×10^1 M^{-1} s^{-1} at 25 °C and $I = 0.02$ M in 95% ethanol); thus steric bulk does not significantly affect this outer-sphere reaction. Another interacting feature is the lack of drastic charge-effect on $Co(EDTA)^{2-}$ reaction with $Fe(phen)_3^{3+}$ and $Fe(SO_3phen)_3$[220]. It should be noted that the reduction of $Co(py)_4Cl_2^+$ by either $Co(EDTA)^{2-}$ or $Co(EDTAH)H_2O^-$ proceeds by an inner-sphere path[222]. A recent publication[223] has suggested that a Co^{II} redox path accounts for the previous reports that $Co(bip)_2Cl_2^+$ aquates very rapidly. The exchange of labelled Co between $Co(NH_3)_5^{2+}$ and $Co(NH_3)_5OH^{2+}$ proceeds without exchange of oxygen, and hence is an inner-sphere reaction[221]. Only 2.5 to 4.0 nitrogens exchange per exchange of Co!

8.3.5 Reductions by copper (I)

Espenson and co-workers[168, 169, 224] have argued that this reagent chooses an inner-sphere mechanism, where possible, because of the similarity of the rate patterns with Cr^{2+} reactions. Cu^+ has a potentially interesting feature in its redox chemistry: to achieve activation it must presumably elongate a pair of *trans* ligands. If one assumes that the bond to the bridging ligand is one of these, the stretching needed in the oxidant–bridging ligand bond should be somewhat less than in Cr^{2+}-reductions, where this bond presumably stretches greatly. There is little indication in the data of ramifications of such an effect. The extraordinary difference in reactivity between $Co(NH_3)_5F^{2+}$ and $Co(NH_3)_5Cl^{2+}$ is paralleled in reductions by $Co(CN)_5^{3-}$ [167] and may be rationalised by 'hard-soft' considerations of precursor-stabilities or on a basis similar to that used above for oxygen versus sulphur bridges in Cr^{2+}-reductions (Section 8.3.2).

8.3.6 Reductions by ruthenium(II)

This reagent has not been as widely investigated as it should be: it is one of only a few reagents that 'must' utilise the outer-sphere path. It is to be noted that the variation in rate (with a change in X) for

$$Co(NH_3)_5O\overset{O}{\overset{\|}{C}}\overset{O}{\overset{\|}{C}}X^{n+}$$

as oxidant represents about 1 kcal mol^{-1} in free energy of activation, but $\Delta H^\ddagger$ varies over 10 kcal mol^{-1} [72]. Until variations of this sort are understood, our knowledge of this reagent will remain incomplete. To further complicate this picture, data reported on the ionic strength dependence of the activation parameters for the $Ru(NH_3)_6^{2+}$-reduction of $Co(NH_3)_5Br^{2+}$ indicate extreme sensitivity: at $I = 1.2 \times 10^{-3}$ M, $\Delta H^\ddagger = 8.6$ kcal mol^{-1} and $\Delta S^\ddagger = -20$ cal mol^{-1} deg^{-1}, whereas at $I = 0.4$ M, $\Delta H^\ddagger = 13.5$ kcal mol^{-1} and $\Delta S^\ddagger = 2$ cal mol^{-1} deg^{-1} [66].

The reactions of $Ru(NH_3)_5OH_2^{2+}$ have been examined in some cases. Stritar and Taube[11] have estimated the reactivity of this reagent with $Ru(NH_3)_5I^{2+}$ and $Ru(NH_3)_5Br^{2+}$ as about 10^4 and 10^3 M^{-1} s^{-1} respectively. Pulse-radiolysis studies[225] have allowed a determination of the reaction

$$Ru(NH_3)_5OH_2^{2+} + Ru(NH_3)_5Cl^{2+} \rightarrow Ru(NH_3)_5OH_2^{3+} + Ru(NH_3)_5Cl^{+} \quad (8.58)$$

as 1.0×10^3 at 20 °C, I = 3×10^{-4} M. The reactions of $Ru(NH_3)_5OH_2^{2+}$ with Fe^{3+} and $FeOH^{2+}$ are unusual in that the latter is faster (8.0×10^4 M^{-1} s^{-1} and 4.8×10^5 M^{-1} s^{-1} at 10 °C, $I = 0.1$) [104] in contrast to other outer-sphere reactions. A proton transfer has been suggested as a possible explanation.

A feature of Ru^{II} chemistry is the ease with which Ru^{II} complexes reduce ClO_4^-. The rate at which Ru^{2+} reduces ClO_4^- to ClO_3^- follows the rate law[226]

$$\frac{d[Ru^{3+}]}{dt} = 2k[Ru^{2+}][ClO_4^-] \quad (8.59)$$

where $k = 3.2 \times 10^{-3}$ M^{-1} s^{-1}, $\Delta H^\ddagger = 19$ kcal mol^{-1}, $\Delta S^\ddagger = -5$ cal mol^{-1} deg^{-1} at 25 °C and $I = 0.3$ M. This rate law and constant are to be compared with the values for the rate of the Ru^{2+}-catalysed anation of Ru^{3+},

$$\frac{d[RuX^{2+}]}{dt} = k[Ru^{2+}][X^-] \quad (8.60)$$

where $k = 8.5\times10^{-3}\ M^{-1}\ s^{-1}$, $\Delta H^{\ddagger} = 2.02$ kcal mol^{-1}, $\Delta S^{\ddagger} = 0$ kcal $mol^{-1}\ deg^{-1}$ under the same conditions. This comparison strongly indicates that the mechanism of reduction involves substitution of ClO_4^- into the coordination sphere, after which reduction occurs[226].

8.3.7 Reductions by platinum(II)

Most Pt^{II} oxidations involve catalysed substitution of Pt^{IV} complexes and have been reviewed[6]. More recently, several other studies have appeared. Mason[227] has examined the reaction

$$XPt(CN)_4Y^{2-} + Pt(CN)_4^{2-} + Z^- \rightarrow Pt(CN)_4^{2-} + YPt(CN)_4Z^{2-} + X^- \tag{8.61}$$

for X, Y, Z = Cl^-, Br^-. He finds that Br^- is a better bridging ligand (Y) than is Cl^-; the change in rate resides almost entirely in the enthalpy of activation, $\Delta S^{\ddagger} \cong -46$ cal $mol^{-1}\ deg^{-1}$. A variation of *cis* non-bridging ligands has been reported[228]. On the oxidant NO_2^- ($k = 9.3\times10^2\ M^{-2}\ s^{-1}$, $\Delta H^{\ddagger} = 5.0$ kcal mol^{-1} and $\Delta S^{\ddagger} = -28$ cal $mol^{-1}\ deg^{-1}$ at 25 °C, $I = 0.2$ M) enhances the rate somewhat more than docs Br^- or NH_3 ($k = 2.0\times10^2$, $\Delta H^{\ddagger} = 6.2$ kcal mol^{-1}, $\Delta S^{\ddagger} = -27$ cal $mol^{-1}\ deg^{-1}$; $k = 1.7\times10^2$, $\Delta H^{\ddagger} = 8.0$ kcal mol^{-1}, $\Delta S^{\ddagger} = -21$ cal $mol^{-1}\ deg^{-1}$, respectively) for the reaction

$$ClPt(dien)XCl^+ + Pt(dien)Br^+ + Br^- = Pt(dien)BrCl^+ + Pt(dien)X + Cl^- \tag{8.62}$$

(where dien is diethylenetriamine). A reverse efficiency is found on the reductant. The reaction of $Pt(diars)_2^{2+}$ (diars is *o*-phenylene-bis-dimethylarsine) with *trans*-$Pt((C_2H_5)_3P)_2Cl_4$ in methanol proceeds by the normal third-order rate law, but saturation of the coordination of the free anion is found[229]

$$\frac{-d[Pt^{IV}]}{dt} = \frac{kK[Pt^{II}][Y^-][Pt^{IV}]}{1+K[Y^-]} \tag{8.63}$$

The efficiency order $I^- > Br^- > Cl^-$ applies for these three anions as Y^-. The rate difference again arises from enthalpy effects. Other studies reported have been concerned with Pt^{II} oxidation by halogens[230, 231], and Pt^{IV}-reductions by halides[232].

Halpern and Pribanic[233] have found kinetic evidence for Pt^{III} in the Pt^{II} reduction of $IrCl_6^{2-}$. For either $PtCl_4^{2-}$ or $Pt(en)_2^{2+}$ the rate law is

$$\frac{-d[IrCl_6^{2-}]}{dt} = \frac{2k_1k_2[Pt^{II}][IrCl_6^{2-}]^2[Cl^-]}{k_1[IrCl_6^{3-}]+k_2[IrCl_6^{3-}][Cl^-]} \tag{8.64}$$

k is $6.2\times10^{-1}\ M^{-1}\ s^{-1}$ for $PtCl_4^{2-}$ ($I = 1.0$ M, 25 °C) and 1.40×10^5 for $Pt(en)_2^{2+}$. These data are consistent with a strong non-bridging ligand effect.

8.3.8 Reductions by lanthanide and actinide elements

This subject has been thoroughly and excellently reviewed several years ago[113]. Recent reports on actinides have featured an extension of the M^{3+}

$+M'O_2^{2+}$ reactions[234–236]. At 25 °C and $I = 1.0$ M, the rate of reduction of UO_2^{2+} by Np^{3+} and NpO_2 by Pu^{3+} are very similar. The former reaction also features a path first-order in $[H^+]$. The U^{3+}-reductions of UO_2^{2+} [236], $Co(NH_3)_5X^{2+}$ complexes[237, 238], and CrX^{2+} complexes[237] have been reported. The reduction of Fe^{III} by Np^{3+} features an efficient path containing Cl^- coordinated to Fe^{III}: $FeCl^{2+}$ reacts 100-times as fast as does Fe^{3+} [239]. Sullivan and Thompson[240] have reported the unusual observations that Co^{III} oxidises Np^V by the rate law

$$\frac{d[Np^{VI}]}{dt} = k[Co^{III}][Np^V] \tag{8.65}$$

with $k = 3.5 \times 10^2$ M^{-1} s^{-1}, $\Delta H^\ddagger = 2.3$ kcal mol^{-1} and $\Delta S^\ddagger = -5$ cal mol^{-1} deg^{-1} at 25 °C and $I = 2.1$; but when the Np^V is coordinated to Cr^{III}, the rate law becomes

$$\frac{d[Np^{VI}]}{dt} = k[Co^{III}][CrNpO_2^{4+}][H^+]^{-1} \tag{8.66}$$

with $k = 2.3$ s^{-1}, $\Delta H^\ddagger = 17.6$ kcal mol^{-1} and $\Delta S^\ddagger = 2$ cal mol^{-1} deg^{-1}. What causes this change in preferred path is puzzling, although an activated complex of +7 charge would form if the proton were not lost from the transition state of reaction (8.66) [140].

Reductions by Eu^{2+} have been studied for Ru^{III} [11], Fe^{III} [54, 172], VO^{2+} [241], V^{3+} [171], and CrX^{2+} [242] as oxidant. The inverse $[H^+]$ path is efficient for both Fe^{3+} and V^{3+}. A preliminary report of the Yb^{2+} reduction of Co^{III} and Cr^{III} complexes has appeared[243].

8.3.9 Oxidations by chromium(VI)

Since quite early in the study of oxidation–reduction reactions in aqueous media the chemistry of Cr^{VI} has been investigated. This multi-electronic reagent offers a number of fascinating possibilities for reaction. It appears that most Cr^{VI} oxidations by one-equivalent reductants proceed through Cr^V and Cr^{IV}, and either reduction of Cr^{VI} or Cr^V may be rate-limiting. A great deal of careful experimental work has been done by Espenson and co-workers and he has summarised that work and other relevant studies[244].

8.3.10 Miscellaneous reactions

The outer-sphere reactions of Mn–isonitrile complexes have been studied by n.m.r. techniques[245]. The effect of changing from NCC_2H_5 to $NCC(CH_3)_3$ on the reaction

$$*Mn(NCR)_6^+ + Mn(NCR)_6^{2+} = *Mn(NCR)_6^{2+} + Mn(NCR)_6^+$$

is a decrease in k by a factor of 16 ($k = 6.4 \times 10^5$ M^{-1} s^{-1}, $\Delta H^\ddagger = 1.7$ kcal mol^{-1}, $\Delta S^\ddagger = -26$ cal mol^{-1} deg^{-1} and $k = 4 \times 10^4$ M^{-1} s^{-1}, $\Delta H^\ddagger = 4.6$ kcal mol^{-1} and $\Delta S^\ddagger = -21$ cal mol^{-1} deg^{-1} for R = C_2H_5 and $C(CH_3)_3$ respectively at 7 °C in CH_3CN). This factor is most probably steric in nature.

The exchange process with phenylisocyanide as a ligand is remarkable – see Reference 245.

Several other interesting, yet unusual, metal ion systems have been studied. The reduction of H^+ by In^I [246], and the oxidation of I^-, Br^- and Cl^- by Bi^V [247] have been reported. The latter reaction is first-order in Bi^V and independent of halide; it has been suggested that OH is formed. The rate data for the Sn^{2+}-reduction of V^V in ClO_4^- media exhibits several unusual features[248].

Several new studies of Tl^+ reactions have been published. There is a claim that the old discrepancy of the effect of $[H^+]$ on the Tl^+–Tl^{3+} exchange is removed if Li^+ is used instead of Na^+ to control the ionic strength[249]. The exchange reaction has been studied as a function of pressure[250] ($\Delta V^{\ddagger} = -13$ cm^3) and in the presence of carboxylic acids[251]. Thompson and Sullivan find the Tl^+ reduction of Np^{VII} is first-order in H^+ [252].

The reduction of several dimeric complexes have been examined by Sykes and co-workers[253–255] and by Taube and co-workers[256, 257]. A report on the reduction of the chloroerythro ion has appeared[258].

ACKNOWLEDGMENTS

I wish to thank Drs. C. L. Perrin and D. L. Toppen for helpful discussions.

References

1. Sutin, N. (1966). *Ann. Rev. Phys. Chem.*, **17,** 119
2. Sykes, A. G. (1967). *Adv. Inorg. Chem. Radiochem.*, **10,** 153
3. Taube, H. (1959). *Adv. Inorg. Chem. Radiochem.*, **1,** 1
4. Matusek, M., Jaros, F. and Tockstein, A. (1969). *Chem. Listy.*, **63** (a) 188, (b) 317, (c) 435
5. Reynolds, W. L. and Lumry, R. W. (1966). *Mechanisms of Electron Transfer* (New York, The Ronald Press)
6. Linck, R. G. (1971). *Homogeneous Catalysis*, Ed. by Schrauzer, G. N. (New York: Marcel Dekker, Inc.)
7. Taube, H. and Myers, H. (1954). *J. Amer. Chem. Soc.*, **76,** 2103
8. Taube, H. and Gould, E. S. (1969). *Accounts Chem. Res.*, **2,** 321
9. Nordmeyer, F. and Taube, H. (1968). *J. Amer. Chem. Soc.*, **90,** 1163
10. Hurst, J. K. and Taube, H. (1968). *J. Amer. Chem. Soc.*, **90,** 1178
11. Stritar, J. A. and Taube, H. (1969). *Inorg. Chem.*, **8,** 2281
12. Halpern, J. and Nakamura, S. (1965). *J. Amer. Chem. Soc.*, **87,** 3002
13. Haim, A. and Sutin, N. (1965). *J. Amer. Chem. Soc.*, **87,** 4210
14. Haim, A. and Sutin, N. (1966). *J. Amer. Chem. Soc.*, **88,** 434
15. Birk, J. P. and Espenson, J. H. (1968). *J. Amer. Chem. Soc.*, **90,** 1153
16. Balahura, R. J. and Jordan, R. B. (1970). *J. Amer. Chem. Soc.*, **92,** 1533
17. Gaunder, R. G. and Taube, H. (1970). *Inorg. Chem.*, **9,** 2627
18. Fay, D. P. and Sutin, N. (1970). *Inorg. Chem.*, **9,** 1291
19. Monacelli, F., Mattogno, G., Gattegno, D. and Maltese, M. (1970). *Inorg. Chem.*, **9,** 686
20. Deutsch, E. and Taube, H. (1968). *Inorg. Chem.*, **7,** 1532
21. Balahura, R. J. and Jordan, R. B. (1971). *J. Amer. Chem. Soc.*, **93,** 625
22. Svatos, G. and Taube, H. (1961). *J. Amer. Chem. Soc.*, **83,** 4172
23. Huchital, D. H. and Taube, H. (1965). *Inorg. Chem.*, **4,** 1660
24. Snellgrove, R. and King, E. L. (1962). *J. Amer. Chem. Soc.*, **84,** 4609
25. Haim, A. (1963). *J. Amer. Chem. Soc.*, **85,** 1016
26. Haim, A. (1966). *J. Amer. Chem. Soc.*, **88,** 2324

27. Ward, J. R. and Haim, A. (1970). *J. Amer. Chem. Soc.*, **92,** 475
28. Huchital, D. H. (1970). *Inorg. Chem.*, **9,** 486
29. Kopple, K. D. and Miller, R. R. (1962). *Proc. Chem. Soc.*, 306
30. Schroeder, K. A. and Espenson, J. H. (1967). *J. Amer. Chem. Soc.*, **89,** 2548
31. Birk, J. P. and Espenson, J. H. (1968). *J. Amer. Chem. Soc.*, **90,** 2266
32. Diaz, H. and Taube, H. (1970). *Inorg. Chem.*, **9,** 1304
33. Davies, G., Sutin, N. and Watkins, K. O. (1970). *J. Amer. Chem. Soc.*, **92,** 1892
34. Wood, P. B. and Higginson, W. C. E. (1966). *J. Chem. Soc., A,* 1645
35. Linck, R. G. and Sullivan, J. C. (1967). *Inorg. Chem.*, **6,** 171
36. Dodson, R. W. and Davison, N. (1952). *J. Phys. Chem.*, **56,** 866
37. Taube, H. (1962). *Proceedings of the Robert A. Welch Foundation Conferences on Chemical Research, Houston, Texas, VI,* 7
38. Ford, P. C., Kuempel, J. R. and Taube, H. (1968). *Inorg. Chem.*, **7,** 1976
39. Ford, P. C. (1970). *Coord. Chem. Rev.*, **5,** 75
40. Patel, R. C. and Endicott, J. F. (1968). *J. Amer. Chem. Soc.*, **90,** 6364
41. Linck, R. G., unpublished observations.
42. Endicott, J. F. (1969). *J. Phys. Chem.*, **73,** 2594
43. Bruning, W. and Weissman, S. I. (1966). *J. Amer. Chem. Soc.*, **88,** 373
44. Grossman, B. and Wilkins, R. G. (1967). *J. Amer. Chem. Soc.*, **89,** 4230
45. Sutter, J. H. and Hunt, J. B. (1969). *J. Amer. Chem. Soc.*, **91,** 3107
46. Doyle, J. and Sykes, A. G. (1967). *J. Chem. Soc. A,* 795
47. Dodel, P. H. and Taube, H. (1965). *Z. Phys. Chemie,* **44,** 92
48. Linck, R. G. (1968). *Inorg. Chem.*, **7,** 2394
49. Sutin, N. (1968). *Accounts Chem. Res.*, **1,** 225
50. Haim, A. and Sutin, N. (1966). *J. Amer. Chem. Soc.*, **88,** 5343
51. Birk, J. P. (1969). *J. Amer. Chem. Soc.*, **91,** 3189
52a. Adamson, A. W. and Gonick, E. (1963). *Inorg. Chem.*, **2,** 129
53. Grossman, B. and Haim, A. (1970). *J. Amer. Chem. Soc.*, **92,** 4835
53. Grossman, B. and Haim, A. (1971). *J. Amer. Chem. Soc.*, **92,** 4835
54. Carlyle, D. W. and Espenson, J. H. (1969). *J. Amer. Chem. Soc.*, **91,** 599
55. Eigen, M. (1963). *Rev. Pure Appl. Chem.*, **6,** 105
56. Gould, E. S. and Taube, H. (1964). *J. Amer. Chem. Soc.*, **86,** 1318
57. Monacelli, F., Basolo, F. and Pearson, R. G. (1962). *J. Inorg. Nucl. Chem.*, **24,** 1241
58. Hurst, J. K. and Taube, H. (1968). *J. Amer. Chem. Soc.*, **90,** 1174
59. Newton, T. W. and Baker, F. B. (1964). *Inorg. Chem.*, **3,** 569
60. Sykes, A. G. and Thorneley, R. N. V. (1970). *J. Chem. Soc., A,* 232
61. Newton, T. W. and Baker, F. B. (1963). *J. Phys. Chem.*, **67,** 1425
62. Newton, T. W. and Baker, F. B. (1964). *J. Phys. Chem.*, **68,** 228
63. Newton, T. W. and Baker, F. B. (1964). *Inorg. Chem.*, **3,** 569
64. Gould, E. S. (1968). *J. Amer. Chem. Soc.*, **90,** 1740
65. Candlin, J. P. and Halpern, J. (1965). *Inorg. Chem.*, **4,** 766
66. Patel, R. C., Ball, R. E., Endicott, J. F. and Hughes, R. G. (1970). *Inorg. Chem.*, **9,** 23
67. Barile, R. and Sutin, N. (quoted in Reference 70)
68. Dulz, G. and Sutin, N. (1964). *J. Amer. Chem. Soc.*, **86,** 829
69. Halpern, J., quoted in Green, M., Schug, K. and Taube, H. (1965). *Inorg. Chem.*, **4,** 1184
70. Orhanovic, M. and Sutin, J. (1968). *J. Amer. Chem. Soc.*, **90,** 4286
71. Davies, G. and Watkins, K. O. (1970). *J. Phys. Chem.*, **74,** 3388
72. Price, H. J. and Taube, H. (1968). *Inorg. Chem.*, **7,** 1
73. Candlin, J. P., Halpern, J. and Trimm, D. L. (1964). *J. Amer. Chem. Soc.*, **86,** 1019
74. Espenson, J. H. (1967). *J. Amer. Chem. Soc.*, **89,** 1276
75. Hicks, K., Toppen, D. L. and Linck, R. G. (to be submitted)
76. Hwang, C. and Haim, A. (1970). *Inorg. Chem.*, **9,** 500
77. Parker, O. J. and Espenson, J. H. (1969). *Inorg. Chem.*, **8,** 185
78. Orhanovic, M., Po, H. N. and Sutin, N. (1968). *J. Amer. Chem. Soc.*, **90,** 7224
79. Davies, K. M. and Espenson, J. H. (1969). *J. Amer. Chem. Soc.*, **91,** 3093
80. Murmann, R. K., Sullivan, J. C. and Thompson, R. C. (1968). *Inorg. Chem.*, **7,** 1876
81. Thompson, R. C. (1968). *J. Phys. Chem.*, **72,** 2642
82. Malik, M. N., Hill, J. and McAuley, A. (1970). *J. Chem. Soc., A,* 643
83. Davies, G. and Watkins, K. O. (1970). *Inorg. Chem.*, **9,** 2735
84. Linck, R. G. (1970). *Inorg. Chem.*, **9,** 2529

85. Olson, M. V., Kanazawa, Y. and Taube, H. (1969). *J. Chem. Phys.,* **51,** 289
86. Malin, J. M. and Swinehart, J. H. (1968). *Inorg. Chem.,* **7,** 250 and 2678
87. Conocchioli, T. J., Nancollas, G. H. and Sutin, N. (1966). *Inorg. Chem.,* **5,** 1
88. Hill, J. and McAuley, A. (1968). *J. Chem. Soc., A,* 1169
89. Wells, C. F. and Mays, D. (1969). *J. Chem. Soc., A,* 2175
90. Wells, C. F. and Kuritsyn, L. V. (1969). *J. Chem. Soc., A,* 2930
91. Thompson, R. C. and Sullivan, J. C. (1970). *Inorg. Chem.,* **9,** 1590
92. Merideth, C. W. and Connick, R. E. (1965). *Abstracts of 149th Amer. Chem. Soc. Meeting, Detroit, Mich.,* Paper 106M
93. Liang, A. and Gould, E. S. (1970). *J. Amer. Chem. Soc.,* **92,** 6791
94. Haim, A. (1966). *Inorg. Chem.* **5,** 2081
95. Baker, B. R., Orhanovic, M. and Sutin, N. (1967). *J. Amer. Chem. Soc.,* **89,** 722
96. Malin, J. M. and Swinehart, J. H. (1969). *Inorg. Chem.,* **8,** 1407
97. Seewald, D., Sutin, N. and Watkins, K. O. (1969). *J. Amer. Chem. Soc.,* **91,** 7307
98. Movius, W. G. and Linck, R. G. (1970). *J. Amer. Chem. Soc.,* **92,** 2677
99. Thorneley, R. N. F. and Sykes, A. G. (1970). *J. Chem. Soc., A,* 1036
100. Espenson, J. H. and Krug, L. A. (1969). *Inorg. Chem.,* **8,** 2633
101. Huchital, D. H. and Wilkins, R. G. (1967). *Inorg. Chem.,* **6,** 1022
102. Movius, W. G. and Linck, R. G. (1969). *J. Amer. Chem. Soc.,* **91,** 5394
103. Buckley, R. R. and Mercer, E. E. (1966). *J. Phys. Chem.,* **70,** 3103
104. Meyer, T. J. and Taube, H. (1968). *Inorg. Chem.,* **7,** 2369
105. Basolo, F. and Pearson, R. G. (1967). *Mechanisms of Inorganic Reactions,* 2nd ed. (New York: John Wiley and Sons, Inc.)
106. Newton, T. W. and Baker, F. B. (1966). *J. Phys. Chem.,* **70,** 1943
107. Espenson, J. H. (1965). *Inorg. Chem.,* **4,** 1025
108. Liteplo, M. P. and Endicott, J. F. (1969). *J. Amer. Chem. Soc.,* **91,** 3982
109. Liteplo, M. T. and Endicott, J. F. (1970). Private communication.
110. Sykes, A. G. (1965). *Chem. Commun.,* 442
111. Espenson, J. H. and Parker, D. J. (1968). *J. Amer. Chem. Soc.,* **90,** 3689
112. Adin, A. and Sykes, A. G. (1968). *J. Chem. Soc., A,* 351
113. Newton, T. W. and Baker, F. B. (1967). *Adv. Chem.,* **71,** 268
114. Green, M., Schug, K. and Taube, H. (1965). *Inorg. Chem.,* **4,** 1184
115. Diebler, H., Dodel, P. H. and Taube, H. (1966). *Inorg. Chem.,* **5,** 1688
116. DeChant, M. J. and Hunt, J. B. (1968). *J. Amer. Chem. Soc.,* **90,** 3695
117. Schmidt, P. O. (1970). *Australian J. Chem.,* **23,** 1287
118. Marcus, R. A. (1968). *J. Phys. Chem.,* **72,** 891
119. Cohen, A. O. and Marcus, R. A. (1968). *J. Phys. Chem.,* **72,** 4249
120. Hush, N. S. (1967). *Progress in Inorganic Chemistry,* Vol. 8, 391, Ed. by Cotton, F. A. (New York: Interscience Pub.)
121. Marcus, R. A. (1964). *Ann. Rev. Phys. Chem.,* **15,** 155
122. Sutin, N. (1965). *Exchange Reactions,* International Atomic Energy Agency, Vienna, 7
123. Wilkins, R. G. and Yelin, R. E. (1968). *Inorg. Chem.,* **7,** 2667
124. Farina, R. and Wilkins, R. G. (1968). *Inorg. Chem.,* **7,** 514
125. Thorneley, R. N. F. and Sykes, A. G. (1970). *J. Chem. Soc., A,* 1036
126. Zwickel, A. and Taube, H. (1961). *J. Amer. Chem. Soc.,* **83,** 793
127. Toppen, D. L. and Linck, R. G. Unpublished observations
128. Doyle, J. and Sykes, A. G. (1968). *J. Chem. Soc., A,* 2836
129. Endicott, J. F. and Taube, H. (1964). *J. Amer. Chem. Soc.,* **86,** 1686
130. Nordmeyer, F. Private communication
131. Guenther, P. R. and Linck, R. G. (1969). *J. Amer. Chem. Soc.,* **91,** 3769
132. Campion, R. J., Purdie, N. and Sutin, N. (1964). *Inorg. Chem.,* **3,** 1091
133. Bennett, L. E. and Taube, H. (1968). *Inorg. Chem.,* **7,** 254
134. Stasiw, R. and Wilkins, R. G. (1969). *Inorg. Chem.,* **8,** 156
135. Haim, A. (1968). *Inorg. Chem.,* **7,** 1475
136. Kruse, W. and Taube, H. (1960). *J. Amer. Chem. Soc.,* **82,** 526
137. Zwickel, A. and Taube, H. (1959). *J. Amer. Chem. Soc.,* **81,** 1288
138. Ball, D. L. and King, E. L. (1958). *J. Amer. Chem. Soc.,* **80,** 1091
139. Espenson, J. H. (1965). *Inorg. Chem.,* **4,** 121
140. Andrade, C. and Taube, H. (1966). *Inorg. Chem.,* **5,** 1087
141. Olson, M. V. and Taube, H. (1970). *Inorg. Chem.,* **9,** 2072

142. Gearon, I. F., Fisher, R. and Hunt, J. B. Private communication
143. George, P. and Griffith, J. (1959). *Enzymes,* **1,** 347
144. Gould, E. S. (1966). *J. Amer. Chem. Soc.,* **88,** 2983
145. Spinner, T. and Harris, G. M. (1968). *156th Amer. Chem. Soc. Meeting,* Paper INOR-36
146. Snellgrove, R. and King, E. L. (1964). *Inorg. Chem.,* **3,** 288
147. Barrett, M. (1969). *Ph.D. Thesis,* Stanford University, *Diss. Abs.,* **29,** 2333
148. Davies, R. and Jordan, R. B. Private communication
149. Newton, T. W. and Daugherty, N. A. (1967). *J. Phys. Chem.,* **71,** 3768
150. Diebler, H. and Taube, H. (1965). *Inorg. Chem.,* **4,** 1029
151. Pennington, D. E. and Haim, A. (1967). *Inorg. Chem.,* **6,** 2138
152. Chock, P. B., Dewar, R. B. K., Halpern, J. and Wong, L.-Y. (1969). *J. Amer. Chem. Soc.,* **91,** 82
153. Po, H. N. and Sutin, N. (1968). *Inorg. Chem.,* **7,** 621
154. Barb, W. G., Baxendale, J. H., George, P. and Hargrave, K. R. (1951). *Trans. Faraday Soc.,* **47,** 462
155. Hardwick, T. J. (1957). *Can. J. Chem.,* **35,** 428
156. Pecht, I. and Anbar, M. (1968). *J. Chem. Soc., A,* 1902
157. Cotton, F. A., DeBoer, B. G., LaPrade, M. D., Pipal, J. R. and Ucko, D. A. (1970). *J. Amer. Chem. Soc.,* **92,** 2926 and references therein
158. Taube, H. (1952). *Chem. Rev.,* **50,** 69
159. Habib, H. S. and Hunt, J. P. (1966). *J. Amer. Chem. Soc.,* **88,** 1668
160. Stranks, D. R. (1960). *Disc. Faraday Soc.,* **29,** 73
161. Neumann, H. M. Quoted in reference 124
162. Friedman, H. L., Taube, H. and Hunt, J. P. (1950). *J. Chem. Phys.,* **18,** 759
163. Waltz, W. L. and Pearson, R. G. (1969). *J. Phys. Chem.,* **73,** 1941
164. Ogard, A. E. and Taube, H. (1958). *J. Amer. Chem. Soc.,* **80,** 1084
165. James, R. V. and King, E. L. (1970). *Inorg. Chem.,* **9,** 1301
166. Anderson, A. and Bonner, N. A. (1954). *J. Amer. Chem. Soc.,* **76,** 3826
167. Candlin, J. P., Halpern, J. and Nakamura, S. (1963). *J. Amer. Chem. Soc.,* **85,** 2517
168. Parker, O. J. and Espenson, J. H. (1969). *J. Amer. Chem. Soc.,* **91,** 1968
169. Parker, O. J. and Espenson, J. H. (1969). *Inorg. Chem.,* **8,** 1523
170. Fraser, R. T. M. (1961). *Advances in the Chemistry of Coordination Compounds,* Ed. by Kirschner, S. 287 (New York: Macmillan)
171. Adin, A. and Sykes, A. G. (1966). *J. Chem. Soc., A,* 1230
172. Carlyle, D. W. and Espenson, J. H. (1968). *J. Amer. Chem. Soc.,* **90,** 2272
173. Haim, A. (1964). *J. Amer. Chem. Soc.,* **86,** 2352
174. Sutin, N., Rowley, J. K. and Dodson, R. W. (1961). *J. Phys. Chem.,* **65,** 1248
174a. Moorhead, E. G. and Sutin, N. (1967). *Inorg. Chem.,* **6,** 428
175. Campion, R. J., Conocchioli, T. J. and Sutin, J. (1964). *J. Amer. Chem. Soc.,* **86,** 4591
176. Bunn, D., Dainton, F. S. and Duckworth, S. (1961). *Trans. Faraday Soc.,* **57,** 1131
177. Silverman, J. and Dodson, R. W. (1952). *J. Phys. Chem.,* **56,** 846
178. Earley, J. E. (1970). *Progress in Inorg. Chem.,* **13,** 243
179. Falk, L. C. and Linck, R. G. (1971). *Inorg. Chem.,* **10,** 215
180. Benson, P. and Haim, A. (1965). *J. Amer. Chem. Soc.,* **87,** 3826
181. Orgel, L. E. (1956). *Report of the Tenth Solvay Conference, Brussels,* 289
182. Bifano, C. and Linck, R. G. (1967). *J. Amer. Chem. Soc.,* **89,** 3945
183. Pennington, D. E. and Haim, A. (1966). *Inorg. Chem.,* **5,** 1887
184. Williams, T. J. and Garner, C. S. (1970). *Inorg. Chem.,* **9,** 2058
185. Cannon, R. D. and Earley, J. E. (1965). *J. Amer. Chem. Soc.,* **87,** 5264; (1966) **88,** 1872
186. Garner, C. S. and House, D. A. (1970). *Transition Metal Chem.,* **6,** 200
187. Cannon, R. D. and Earley, J. E. (1968). *J. Chem. Soc., A,* 1102
188. Pennington, D. E. and Haim, A. (1968). *Inorg. Chem.,* **7,** 1659
189. Espenson, J. H. and Berge, L. A. Unpublished observation quoted in reference 15
190. Green, M. and Sykes, A. G. (1970). *Chem. Commun.,* 241
191. Baker, F. B., Brewer, W. D. and Newton, T. W. (1966). *Inorg. Chem.,* **5,** 1294
192. Burkhart, M. J. and Newton, T. W. (1969). *J. Phys. Chem.,* **73,** 1741.
193. Davies, K. M. and Espenson, J. H. (1969). *Chem. Commun.,* 111
194. Newton, T. W. and Baker, F. B. (1965). *J. Phys. Chem.,* **69,** 176
195. Parker, O. J. and Espenson, J. H. (1969). *J. Amer. Chem. Soc.,* **91,** 1313
196. Rosseinsky, D. R. and Zlotnick, J. (1970). *J. Chem. Soc., A,* 1200

197. Rosseinsky, D. R. and Nicol, M. J. (1966). *Electrochim. Acta,* **11,** 1069
198. Rosseinsky, D. R. and Nicol, M. J. (1968). *J. Chem. Soc., A,* 1022
199. Rosseinsky, D. R. and Nicol, M. J. (1970). *J. Chem. Soc., A,* 1196
200. Rosseinsky, D. R. and Nicol, M. J. (1968). *Trans. Faraday Soc.,* **24,** 2410
201. Rosseinsky, D. R. and Nicol, M. J. (1969). *J. Chem. Soc., A,* 2887
202. Espenson, J. H. and Slocum, S. G. (1967). *Inorg. Chem.,* **6,** 906
203. Orino, H. and Tanaka, N. (1968). *Bull. Chem. Soc. Japan,* **41,** 1622
204. Ogino, H. and Tanaka, N. (1968). *Bull. Chem. Soc. Japan,* **41,** 2411
205. Vriesenga, J. R. (1968). *Diss. Abstr. B,* **29,** 973
206. Balahura, R. J. and Jordan, R. B. (1971). *Inorg. Chem.,* **10,** 198
207. Lane, R. H. and Bennett, L. G. (1970). *J. Amer. Chem. Soc.,* **92,** 1089
208. Shaw, K. and Espenson, J. H. (1968). *Inorg. Chem.,* **7,** 1619
209. Beattie, J. K. and Basolo, F. (1967). *Inorg. Chem.,* **6,** 2069
210. Doyle, J. and Sykes, A. G. (1968). *J. Chem. Soc., A,* 215
211. Thompson, R. C. and Sullivan, J. C. (1967). *J. Amer. Chem. Soc.,* **89,** 1096
212. Thompson, R. C. and Sullivan, J. C. (1967). *J. Amer. Chem. Soc.,* **89,** 1098
213. Doyle, J., Sykes, A. G. and Adin, A. (1968). *J. Chem. Soc., A,* 1314
214. Cannon, R. C. (1968). *J. Chem. Soc., A,* 1098
215. Kochi, J. K. and Powers, J. W. (1970). *J. Amer. Chem. Soc.,* **92,** 137
216. Armor, J. N. and Taube, H. (1969). *J. Amer. Chem. Soc.,* **91,** 6874
217. Birk, J. P. (1970). *Inorg. Chem.,* **9,** 125
218. Campion, R. J., Deck, C. F., King, Jr., P. and Wahl, A. C. (1967). *Inorg. Chem.,* **6,** 672
219. Ruff, I. and Korösi-Ódor, I. (1970). *Inorg. Chem.,* **9,** 186
220. Wilkins, R. G. and Yelin, R. E. (1970). *J. Amer. Chem. Soc.,* **92,** 1191
221. Williams, T. J. and Hunt, J. P. (1968). *J. Amer. Chem. Soc.,* **90,** 7213
222. Wilkins, R. G. and Yelin, R. (1967). *J. Amer. Chem. Soc.,* **89,** 5496
223. Josephsen, J. and Schaffer, C. E. (1970). *Chem. Commun.,* 61
224. Shaw, K. and Espenson, J. H. (1968). *J. Amer. Chem. Soc.,* **90,** 6622
225. Baxendale, J. H., Rodgers, M. A. J. and Ward, M. D. (1970). *J. Chem. Soc., A,* 1246
226. Kallen, T. W. and Earley, J. E. (1970). *Chem. Commun.,* 851
227. Mason, W. R. (1970). *Inorg. Chem.,* **9,** 1528
228. Bailey, S. G. and Johnson, R. C. (1969). *Inorg. Chem.,* **8,** 2596
229. Peloso, A. and Ettorre, R. (1968). *J. Chem. Soc., A,* 2253
230. Skinner, C. E. and Jones, M. M. (1969). *J. Amer. Chem. Soc.,* **91,** 4405
231. Poe, A. J. and Vaughan, D. H. (1969). *J. Chem. Soc., A,* 2844
232. Poe, A. J. and Vaughan, D. H. (1970). *J. Amer. Chem. Soc.,* **92,** 7537
233. Halpern, J. and Pribanic, M. (1968). *J. Amer. Chem. Soc.,* **90,** 5942
234. Fulton, R. B. and Newton, T. W. (1970). *J. Phys. Chem.,* **74,** 1661
235. Newton, T. W. (1970). *J. Phys. Chem.,* **74,** 1655
236. Newton, T. W. and Fulton, R. B. (1970). *J. Phys. Chem.,* **74,** 2797
237. Espenson, J. H. and Wang, R. T. (1970). *Chem. Commun.,* 207
238. Wang, R. T. and Espenson, J. H. (1971). *J. Amer. Chem. Soc.,* **93,** 280
239. Newton, T. W., McCrary, G. E. and Clark, W. G. (1968). *J. Phys. Chem.,* **72,** 4333
240. Sullivan, J. C. and Thompson, R. C. (1967). *Inorg. Chem.,* **6,** 1795
241. Espenson, J. H. and Christensen, R. J. (1969). *J. Amer. Chem. Soc.,* **91,** 7311
242. Adin, A. and Sykes, A. G. (1968). *J. Chem. Soc., A,* 354
243. Christensen, R. J. and Espenson, J. H. (1970). *Chem. Commun.,* 756
244. Espenson, J. H. (1970). *Accounts Chem. Res.,* **3,** 347
245. Matteson, D. S. and Bailey, R. A. (1969). *J. Amer. Chem. Soc.,* **91,** 1975
246. Taylor, R. S. and Sykes, A. G. (1969). *J. Chem. Soc., A,* 2419
247. Ford-Smith, M. H. and Habeeb, J. J. (1969). *Chem. Commun.,* 1445
248. Schiefelbein, B. and Daugherty, N. A. (1970). *Inorg. Chem.,* **9,** 1716
249. Gelsema, W. J., DeLigny, C. L., Blijleven, H. A. and Hendriks, E. J. (1969). *Rec. Trav. Chem.,* **88,** 110
250. Adamson, M. G. and Stranks, D. R. (1968). *Chem. Commun.,* 648
251. McGregor, R. G. and Wiles, D. R. (1970). *J. Chem. Soc., A,* 323
252. Thompson, R. C. and Sullivan, J. C. (1970). *J. Amer. Chem. Soc.,* **92,** 3028
253. Davies, R. and Sykes, A. G. (1968). *J. Chem. Soc., A,* 2831
254. Taylor, R. S., Thorneley, R. N. F. and Sykes, A. G. (1970). *J. Chem. Soc., A,* 856
255. Taylor, R. S. and Sykes, A. G. (1970). *J. Chem. Soc., A,* 1991

256. Hoffman, A. B. and Taube, H. (1968). *Inorg. Chem.,* **7,** 903
257. Hoffman, A. B. and Taube, H. (1968). *Inorg. Chem.,* **7,** 1971
258. Hoppenjans, D. W., Hunt, J. B. and Penzhorn, L. (1968). *Inorg. Chem.,* **7,** 1467
259. Sykes, A. G. (1971). Private communication

9
Nucleophilic Displacement at Some Main Group Elements

R. H. PRINCE
University of Cambridge

9.1 GENERAL

The last two decades have seen interest in the mechanisms of reactions in solution spread from carbon as reaction centre to include nearly all the elements, both transitional and non-transitional. Perhaps the most remarkable finding for widely diverse systems is that fundamentally only a few basic mechanisms are recognisable by the experimental criteria normally applied to assign mechanisms.

We note at the outset that for many of the p-block elements the usual triad of mechanistic techniques applied to a well-characterised solute reacting by a well-characterised reaction, namely, kinetics, stereochemical course, and isotopic labelling have not been applied together. This is often inevitable, e.g. the stereochemical course of the reactions of high-labile systems is difficult to follow (although n.m.r. spectroscopy is often able to provide useful information in this area, see below, Section 9.5.1). Often mechanisms have been assigned from minimal experimental evidence, or by analogy. In this Review we shall usually be concerned with reactions in which results of more than one experimental mechanistic tool converge to the same conclusion: this, together with space restrictions, means that this Review is selective rather than exhaustive, and much of it will be concerned with silicon and phosphorus chemistry.

Although studies on free-radical and gas-phase reactions of many of the systems described are becoming more numerous (see for instance Reference 1), we shall be concerned here mainly with nucleophilic displacement, i.e. ligand interchange, at Main Group elements in solution systems.

There are broadly two types of ligand interchange process: associative, in which the entering ligand may be bound to the reacting centre before the departing ligand has completely left the coordination shell of the centre, and dissociative, in which the departing ligand leaves the coordination shell of the reacting centre before the entering group becomes bonded to it[2–4]. These are extremes and intermediate behaviour is possible. It is important to ask what factors determine whether a given system, MX, will select an associative or dissociative pathway when it reacts with a nucleophile, Y. To say, in terms of transition state theory, that the system will react by whichever pathway has least activation free energy conceals many factors, any of which could be decisive in tipping the free-energy scales in favour of one mechanism or the other. The complexity of the problem is illustrated in Figure 9.1 where factors contributing to the observed activation free energy $\Delta G^{\ddagger}_{\text{exp.}}$ are summarised diagrammatically in cyclic form for an extreme dissociative (D) and an extreme associative (A) process.

Some of the quantities (e.g. solvation energies of reactants) can be measured experimentally, others cannot. Clearly, the preference for a dissociative over an associative process is determined by the difference,

$$(\Delta G^{\ddagger}_{\text{exp.}})_{\text{D}} - (\Delta G^{\ddagger}_{\text{exp.}})_{\text{A}} = -(\Delta G^{0}_{\text{solv.}})_{\text{Y}} + (\Delta G^{\ddagger}_{\text{D}} - \Delta G^{\ddagger}_{\text{A}}) + (\Delta G^{\ddagger}_{\text{solv.}})_{\text{D}} - (\Delta G^{\ddagger}_{\text{solv.}})_{\text{A}}$$

This preference is independent of the solvation free energy of the initial state, but depends on the activation free energies for the two processes in isolation ($\Delta G^{\ddagger}_{\text{D}}$ and $\Delta G^{\ddagger}_{\text{A}}$), on the solvation free energies of the two transition states $(\Delta G^{\ddagger}_{solv.})_D$, $(\Delta G^{\ddagger}_{solv.})_A$, and on the solvation free energy of the attacking reagent Y. (Note that arguments, frequently used in a mechanistic context concerning electron release or withdrawal or steric effects, confine attention solely to $\Delta G^{\ddagger}_D$ and $\Delta G^{\ddagger}_A$: the other terms could be very important in determining the shift of rate with change of substituent groups, for example.) From this aspect there appears no reason why, given suitably balancing free-energy terms, both mechanisms should not operate simultaneously, and a situation could be envisaged in which the mechanism is determined just by the solvation free energy of Y. It is worth bearing in mind that a difference in activation free energy of ~ 1.4 kcal mol^{-1} changes a rate constant by an

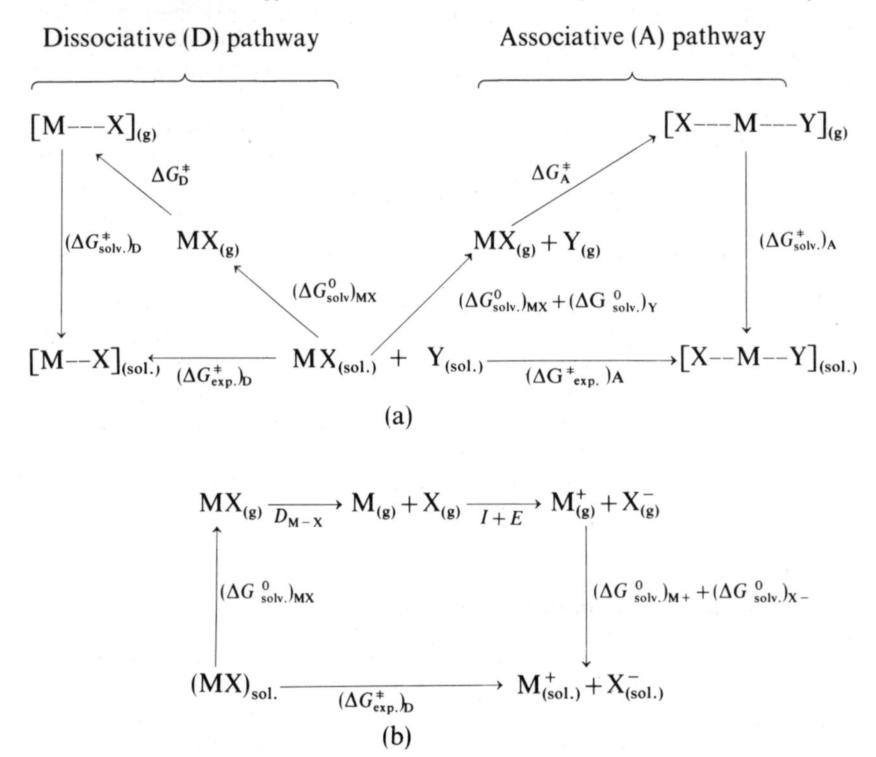

Figure 9.1 A dissection (a) of the observed activity free energy $(\Delta G^{\ddagger}_{exp.})_{A,D}$ for associative and dissociative mechanisms for the process $MX + Y \rightarrow MY + X$ (see text). Most terms are self-explanatory: sol. refers to solution (standard state); $(\Delta G^{\ddagger}_{solv.})_Z$ is the solvation free energy of Z. For a completely dissociative process, giving a transition state approximating to free M^+ and X^- ions, $(\Delta G^{\ddagger}_{exp.})_D$ may be split up as in (b); $D_{M—X}$ is the dissociation energy of the M—X bond; I = ionisation potential of M; E = electron affinity of X

order of magnitude at 25 °C, and *heats* of solution—let alone solvation energies—can differ in structurally closely-related systems by amounts of this order. We shall see also that the steric course of a reaction can be strongly affected by the solvent[5]. This emphasises the difficulty and danger of making detailed mechanistic arguments in terms of steric and polar effects from data on change of reactivity with structure when the rate data show relatively small differences of an order of magnitude or so.

We shall be concerned in this Review not only with rate data showing large differences but also with stereochemical and isotopic-labelling experiments used in combination to draw mechanistic conclusions. We note in passing that the effect of size, and the nature of Y and X on the quantities $\Delta G^{\ddagger}_{D}$ and $\Delta G^{\ddagger}_{A}$ has been discussed many times before (see for instance Reference 2) and will not be repeated here in detail. We merely note that in a fully-dissociative process with given Y, X and solvent, as one descends a Group $\Delta G^{\ddagger}_{D}$ should decrease, since $D_{M—X}$ and I should both decrease (Figure 9.1); for the associative process similarly $AG^{\ddagger}_{A}$ for given Y should decrease, M—X usually becoming weaker and M larger and usually more polarisable as the Group is descended. In general terms then, we expect reactivity in nucleophilic displacements to increase as a Group is descended, an expectation borne out in practice though, as we shall see, there are interesting exceptions.

So far we have not considered the electronic structure of M: when M is changed from a first row to a second row element we may anticipate that not only size and polarisability but also d orbital availability will become an additional factor when transition state and initial state can be differentially affected by participation of d orbitals in σ- and π-bonds. We shall examine this question below.

Lastly, we have assumed above that the solvent is constant: if this is allowed to vary then obviously for a reaction occurring by a given mechanism the solvation free energy of the initial state becomes an important additional term, and techniques exist (see for example References 6–13) for determining whether solvent change affects rate by changing initial-state solvation or transition-state solvation, but there have been few applications outside carbon chemistry, although solvent effects, e.g. in silicon and germanium chemistry, can be very spectacular, for instance in chloride-exchange reactions[14].

9.2 METHODS OF INVESTIGATION

Many systems can be studied kinetically by classical techniques, especially those of the middle Groups. However, many reactions of interest are rapid, and flow, relaxation, and n.m.r. techniques[15–17] are useful, particularly in the study of ligand interchange in systems of the extreme Groups of the Periodic Table (e.g. water-exchange reactions of Group IA metal ions, and hydration of the halogens), though the techniques are not confined to these Groups, and many interesting reactions of elements in the middle Groups (e.g. of Si and P and later elements of these Groups) are rapid.

N.M.R. spectroscopy has yielded much useful information concerning not only rates of ligand interchange but also information about the intimate details of stereochemical change, including optical stability (e.g. of organotin derivatives[18–20]) and details of inversion[21] and pseudorotation in Group V compounds (see for instance References 22–31). From the viewpoint of availability of techniques, it is now probably true to say that the first (and often the second) of the triad of mechanistic techniques, kinetics, stereochemical course, and isotopic labelling, can be applied to any system in the Periodic Table.

9.3 SUBSTITUTION AT ELEMENTS IN GENERAL

Compounds of elements containing atoms or groups which readily exist as anions are readily attacked by nucleophiles; complex formation usually involves an attack by ligand (a nucleophile), and nucleophilic displacement is perhaps the commonest type of reaction. We shall first briefly consider studies on elements in general.

9.3.1 Groups IA, IIA, IIIA

Studies in detail giving comparative data are confined mainly to water-exchange and complex-formation reactions with a restricted range of ligands.

The metal ions fall into three fairly well-defined categories[32].

(a) Very rapid ligand-dependent reactions are observed for complex formation by alkali metals (Li^+ to Cs^+) with multidentate ligands such as EDTA, NTA and ATP and also by the alkaline earths Ca^{2+}, Sr^{2+}, Ba^{2+} with $k > 10^7\ s^{-1}$ with all ligands and some variation from ligand to ligand. The rate is correlated with decreasing hydration energy of the metal ion.

(b) Magnesium(II) ion (and also transition metal(II) cations and lanthanide(III) ions) – $k < 10^7 s^{-1}$, and independent of the nature of L.

(c) A few metal ions display lower rates of ligand entry and water exchange which vary with the basicity of the entering group, e.g. Al^{3+}, Be^{2+} (and also Fe^{3+}).

It has been pointed out that water loss at the alkali metal cations ((a) above) is sufficiently facile that ligand attack is the rate-determining step, but it is doubtful that conventional considerations of nucleophilicity of attacking atoms are relevant to a discussion of rates of ligand attack[3]. Eigen suggests[32] that the rate-determining step is the replacement of several water molecules by the multidentate ligand. With reactions of category (b) metals, the entering group is not important and these may well be of the dissociative type. These findings may also be looked at from the viewpoint of the familiar general scheme for a dissociative process:

$$MX \underset{k_{-1}}{\overset{k_1}{\rightleftharpoons}} M + X \qquad X = H_2O$$

$$M + L \xrightarrow{k_2} ML \qquad L \overset{k_2}{=} \text{ligand}$$

for which the usual steady-state treatment for M gives

$$\text{Rate} = k_1 k_2[MX][L]/(k_{-1}[X] + k_2[L])$$

When [X] is large, as in water, or is buffered, and the water molecules have a high affinity for M then $k_{-1}[X] \gg k_2[L]$ and ligand-dependent rates are observed (as in category (a)); on the other hand, when M has a greater preference for L compared with X then $k_2[L] > k_{-1}[X]$ and the rate tends to $k_1[MX]$, the classical dissociative case. For metals of category (c) the ligand may act as a proton acceptor: the rate of hydrolysis is much greater than the rate of water substitution. Studies of rate as a function of pH show that hydrolysed species are generally labile, so reaction may proceed by H^+-transfer to the entering group in the outer sphere, followed by water loss, but this is a difficult mechanism to characterise[33].

More recent work on substitution at beryllium has shown that the formation of $Be(OH_2)_3F^+$ (from $Be(OH_2)_4^{2+}$ plus HF and F^-) is an S_N1 process involving dissociation of a water molecule from the ion pair (S_N1 IP)[34] like the formation of $BeSO_4$ from the parent ions. Also the mechanism of the fluoride exchange with BeF_4^{2-} involves preliminary dissociative equilibria of the form $BeF_4^{2-} \rightleftharpoons BeF_3^- + F^-$, the BeF_3^- then takes part in reversible reactions with other species including the starting material, BeF_4^{2-} [35]. It is interesting that BF_4^- reacts by a similar mechanism[36].

The relative importance of bond-making and -breaking has also been discussed in connection with water exchange between $Al(OH_2)_6^{3+}$ and $Ga(OH_2)_6^{3+}$ and solvent water. The experimental technique used was ^{17}O n.m.r. spectroscopy using a paramagnetic Co^{II} salt to separate bulk and coordinated water resonances. The widely-different activation entropies suggest that the mechanism is close to a limiting S_N1 process for aluminium(III), but closer to an S_N2 for gallium(III)[37].

This behaviour of aluminium is also consistant with the isomerisation of $Al(acac)_2(DMF)_2^+$ which apparently involves a rate-controlling loss of a DMF molecule[38].

Recent work highlights the importance of ensuring that the species under investigation in kinetic studies are fully characterised: there has been considerable discussion concerning the nature of aluminium ions in aqueous solution[41–44], dinuclear and higher cationic species being readily formed,

e.g. $$2[(H_2O)_5AlOH]^{2+} \rightleftharpoons [(H_2O)_4AlOH]_2^{4+}$$

Aluminium is rather prone to behaviour of this kind. It also occurs under the widely-different conditions of oxine (HQ) chelation in ethanol, e.g. the chelates AlQ_2^+ and AlQ^{2+} appear to exist in two forms which differ in the composition of the coordination shell of the aluminium, and here also polynuclear species are indicated[46].

It has been found that aqueous solutions of R_2GaF (R = Me or Et)[39] and Me_2GaX ($X = NO_3$, OH, or ClO_4)[40] are relatively stable with respect to the hydrolysis of the Ga—C bonds. The aqueous species are believed to be $Me_2Ga(OH_2)_2^+$, $[Me_2GaOH]_4$ and $Me_2Ga(OH)_2^-$. Spectroscopic evidence[40] suggests a bent Me—Ga—Me unit, the C—Ga—C angle of which decreases in the above order. Rapid exchange of water molecules between aquated Me_2Ga^+ ions and bulk solvent is evident from n.m.r. spectroscopic studies.

The re-distribution behaviour and optical inversion of some aluminium β-diketonates have been examined by ^1H n.m.r. techniques[45].

9.3.2 Groups IV and V

Silicon and phosphorus continue to dominate studies in this area, and the substitution reactions of these elements, particularly those of silicon, are dealt with below.

9.3.3 Groups VI and VII

Some studies of substitution at oxygen have been reported, e.g. attack of triethyl phosphite on the oxygen in compound (1),

$(EtO)_3P:$ → $(EtO)_3P^+$–O–C_6H_4–Me + CBr_3^- → $(EtO)_2P(=O)$–O–C_6H_4–Me + $EtCBr_3$

(1) [Me, CBr_3]

is believed to proceed by an S_N2' mechanism[47].

Reactions of peroxides can also involve nucleophilic attack on oxygen (e.g. with $MePhPr^nP$ as nucleophile)[48]. Substitution reactions at oxygen and sulphur have been surveyed from the viewpoint of the theory of hard and soft acids and bases[49–53].

Substitution at sulphur has been recently reviewed[54] and we shall not consider it here: perhaps the most interesting feature of mechanisms of substitution at sulphur is the fact that there is as yet no unambiguous case of an S_N1 mechanism for sulphur. We shall find that this is true of silicon as well.

Miscellaneous studies of halogen substitution have been reported[48, 55–57]. The classic studies in this area are probably those on halogen hydrolysis studied by rapid-reaction techniques[58–60]. The overall reaction is $X_2 + OH_2 \rightleftharpoons X^- + H^+ + XOH$, but the kinetic scheme suggested from measurements as a function of concentration and pH is complex:

$$X_2 + H_2O \rightleftharpoons H_2OX^+ + X^- \rightleftharpoons XOH + H^+ + X^-$$

$$X_2 + H_2O \rightleftharpoons X_2OH^- + H^+ \rightleftharpoons XOH + H^+ + X^-$$

$$X_2 + H_2O \rightleftharpoons X_2 + OH^- + H^+$$

Some of these steps are very fast, e.g. k_2 for $I_2 + OH^- \rightarrow I_2OH^-$ is greater than $5 \times 10^9\ M^{-1}\ s^{-1}$.

We shall now consider some recent developments in the study of nucleophilic substitution at silicon and phosphorus.

9.4 NUCLEOPHILIC SUBSTITUTION AT SILICON, GERMANIUM AND TIN

9.4.1 Kinetic features

There are few recent reviews dealing specifically with nucleophilic substitution at silicon apart from the stereochemical aspects, though a review dealing with displacement at silicon in terms of linear free-energy relationships has appeared[64], and mechanisms of displacement at silicon from kinetic and stereochemical points of view have been surveyed[65].

We shall be concerned mainly with reactions of compounds of the type R_3MX, where R is an alkyl or aryl group, M is Si, Ge or Sn, and X is a functional group. These compounds form a series which is uniquely useful for studying effects of changing electronic structure on reactivity; in this sequence the stereochemistry is constant, reaction involves the breakage of only one bond, complications due to instability or side or consecutive reactions are minimal, and completely analogous reactions can be found for different M. The use of different R groups permits optically-active compounds to be made, and the steric course of their reactions to be examined.

9.4.2 Relevant physical properties

These have been the subject of a comprehensive review[61].

9.4.2.1 The size factor

The radii of C, Si, Ge, Sn and Pb, when these elements are in a tetrahedral, molecular environment are 0.77, 1.17, 1.22, 1.40 and 1.52 Å respectively. The large difference between carbon and silicon and the small one between silicon and germanium are noteworthy. Increasing size should decrease steric shielding of the central atom and also lead to a decrease in bond energy: much evidence exists for steric effects, and this has been recently summarised and reviewed[62]. Physical evidence is consistent with the differences in radii already referred to.

9.4.2.2 Bond energies and polarisabilities

Because of increasing diffuseness of orbitals with increasing atomic number, bond strengths normally decrease down Groups of the Periodic Table (see, for example, Reference 63), but bonds of second-row elements with the highly electronegative elements are often stronger than those of the first row elements. Bond dissociation energies of CX increase in the order for X: $I<Br<N<Si<Cl<C<O<H<F$; for the SiX bond the sequence is $Si<I<Br<C<H<N<O<F$. Although in some respects these orders reflect decreasing reactivities of the various bonds, we expect many anomalies and find them: the Si—C bond is much more resistant to cleavage than expected. Reference to Figure 9.1 underlines that the bond energy term is only one among many, and that only for highly endothermic reactions is it possible that some relation between bond energy and reactivity might be expected, although this will depend precisely on the mechanism. Bond polarisabilities, as determined from refractive index data, show the expected increase from carbon to silicon[65]: polarisability could well be an important factor influencing the $\Delta G^{\ddagger}_{A}$ or $\Delta G^{\ddagger}_{D}$ term, and this has been examined implicitly from the viewpoint of soft and hard acids and bases[49–53] in other contexts.

9.4.2.3 Use of d orbitals

The role of d orbitals in the ground states of silicon and some of the heavier members of Group IV has been a fertile field of discussion for many years, and the data, particularly for ground states, have been evaluated critically[61].

All the Group IVb elements except carbon have vacant d orbitals of the same principle quantum number as the valence s and p orbitals, and the presence of energetically-accessible d orbitals has long been used to account for many of the differences between elements of the Group. These modes of d-orbital participation will be briefly considered, namely, in σ-bonding, d_π-acceptor and d_π-donor bonding, since these also can influence $AG_A^\ddagger$ or $\Delta G_D^\ddagger$.

(a) *σ-bonding* – σ-Bonding involving d orbitals is often assumed to imply an increase in coordination number above the common maximum of four when only s and p orbitals are used, but this assumption is not necessary[61, 210–215]. Higher coordination numbers become progressively commoner down the Group, and are usually found only with highly electronegative substituents for silicon, e.g. SiF_6^{2-}, $SiF_4{\cdot}NMe_3$, whereas a coordination number of five is the commonest one for tin, even with substituents of such weak donor power as perchlorate.

Craig has examined theoretically the effect of ligands on the energy of d orbitals using an electrostatic approximation and SCF wave functions[66, 67]. He concluded that (i) the presence of electronegative groups can cause d orbital contraction to dimensions where effective bonding becomes possible for second-row elements, (ii) the effectiveness of ligands in this respect is in the order $F>Cl>C>H$ and the difference between fluorine and carbon is not as great as the electronegativity difference might suggest, and (iii) the first ligand produces most of the contraction. Furthermore, for heavier elements, better overlap is likely for Cl and Br than F with metal s and p orbitals. (Photo-electron spectroscopic data for CF_4 and SiF_4 indicate substantial d-orbital involvement[214].) Particularly interesting is the finding that carbon can cause appreciable d-orbital contraction. The existence of compounds such as PPh_5 and $P_3(CH_3)_6N_3$ may well be connected with this fact. Of the triorgano derivatives of Group IVb, 5- and 6-coordinate complexes are known with suitable donors for tin and lead. For example, 1:1 adducts are formed between trialkyl tin and trialkyl lead chloride and tetramethylene sulphoxide, *N–N* dimethylformamide and *N,N*-dimethylacetamide[68]. The ion $Me_3SnBr_2^-$ is known[69]; $Ph_3PbCl_2^-$ has been postulated on the basis of conductance studies, but such complexes are much less stable than those of mono- or di-organo derivatives[70].

Several other studies have been reported, but the conclusion is that only for tin and lead does there seem to be any possibility of isolating 5- or 6-coordinate intermediates in a nucleophilic substitution reaction of triorgano-substituted compounds. We must note, however, that the existence of such an intermediate does not *necessarily* mean that it lies on the reaction coordinate[71]. Nevertheless, σ-participation of d orbitals can, in principle, have a profound effect on reactivity. A number of types of reaction profile can be envisaged for the reaction $R_3MX+Y \rightarrow R_3MY+X$ (Figure 9.2). Path a, with a single maximum, represents reaction via a transition state similar to the S_N2 transition state for nucleophilic substitution at carbon; path b represents d-orbital stabilisation of the transition state, in which bonds to the incoming and leaving groups have some d-orbital character; paths c and d, with two maxima, represent a more extreme case of b in which d-orbital participation in the transition state has proceeded sufficiently to permit the existence of definite intermediates, C. For c the rate-determining step is formation of the intermediate, for d its decomposition is rate-determining. All these mechanisms will show second-order kinetics in classical kinetic experiments when the stationary state has been reached, but if the

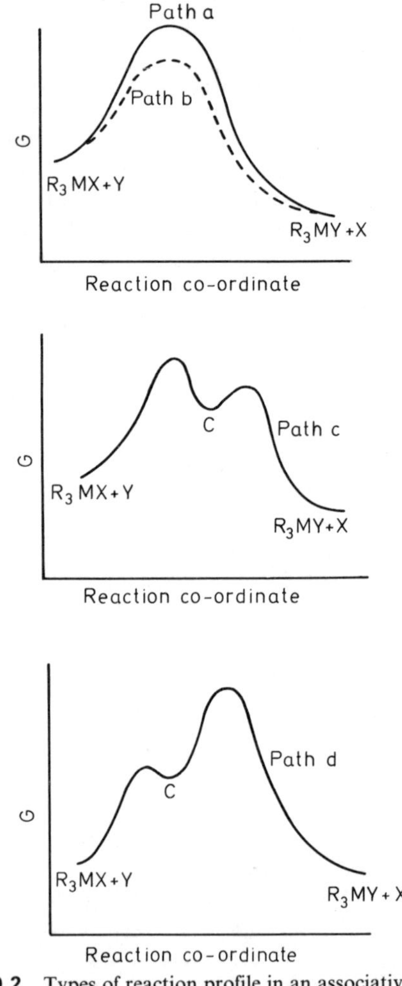

Figure 9.2 Types of reaction profile in an associative process

reaction could be observed in the initial stages, before the stationary state is reached, then an intermediate might be detected kinetically (e.g., it would be revealed as an induction period in the hydrolysis of R_3MCl). Experiments with fast-flow apparatus have failed to reveal an intermediate: if an intermediate is formed it reaches a (low) stationary concentration within a period of the order of milliseconds[72]. The effect of electron-withdrawing groups can also, in principle, distinguish between the possibilities if we assume, as is customary, that changing a group affects only $\Delta G^{\ddagger}_{A}$ and not the other quantities in Figure 9.1. Electron-withdrawing substituents on the central atom will increase the rate for scheme c (where the central atom is more negative than in the ground state) but decrease that of d. The effect on a and b is unpredictable because it depends on how far bond-making and -breaking have proceeded in the transition state. In fact, most reactions of organo-silicon compounds are accelerated by electron withdrawal. The stereochemical consequences of d-orbital participation are discussed below.

(b) d_π-*acceptor bonding*—The type of π-bonding involving p orbitals, so important in carbon chemistry, has never been observed in stable compounds of the heavier elements of Group IVb. The heavier elements all have assessible d orbitals which can have π-symmetry, and accept electrons from lone-pair orbitals of another atom or group. Whilst there is a considerable amount of evidence consistent with this type of bonding its postulation must be made with care. For instance, structural information as a criterion of d_π-p_π bonding has been strongly criticised[73], and facile conclusions from physical measurements such as n.m.r. splitting constants have also come in for criticism[74,75]. Much effort has been made in this field, and it has been recently reviewed[61,76]. Briefly, there is disagreement among different authors as to the importance of π-bonding in the different elements. Recent u.v.[77], e.s.r.[78] and hydrogen-bonding studies[79], for example, indicate decreasing interaction in the series Si, Ge, Sn, although earlier studies indicated little difference between silicon and germanium in this respect. Photo-electron spectroscopy of Si—X and Ge—X bonds indicates that p_π-d_π bonding occurs between chlorine and both elements but not with X = Br or I[215]. π-Bonding is generally assumed to be of little importance for tin and lead. The effect of π-bonding on reactivity is difficult to predict; it will depend on the relative π-bonding abilities of the leaving group and the attacking nucleophile, and on the difference between π-bonding in the ground and transition states (see for instance, Reference 72).

(c) d_π-*donor bonding*—The elements following silicon have filled d orbitals and electrons from these can conceivably be donated to vacant π-orbitals of another atom. This type of interaction may occur, for example, in the breakdown of polymeric *N*-tributylstannylimidazole by various nucleophiles, P, S, and Cl being more effective than N, O, and F[80].

9.4.3 The siliconium ion

The importance of carbonium ion intermediates in carbon chemistry is well known, and some stable carbonium ions such as $Ph_3C^+ClO_4^-$ can be isolated. All attempts to prepare stable siliconium ions of this kind have been unsuccessful, and this is clearly related to the inability of silicon to form pπ-pπ bonds which contribute extensively to the stability of species such as Ph_3C^+.

Appearance potential data indicate that in the gas phase siliconium ion formation is rather more energetically favourable than formation of the analogous carbonium ion[81,82], but this, of course, is only one term in the dissociative pathway in Figure 9.1, i.e. one term among the many influencing the choice between $(\Delta G^\ddagger_{exp.})_D$ and $(\Delta G^\ddagger_{exp.})_A$.

A number of kinetic studies have been directed at demonstrating the existence of a siliconium ion intermediate in suitable systems. Its existence was proposed in the solvolysis[83] of $Me_3SiCH_2CH_2Cl$ (Equation 9.1) which apparently involved a large charge-separation in the transition state.

$$Me_3SiCH_2CH_2Cl \xrightarrow{ROH} MeSiOR + CH_2{=}CH_2 + HCl \qquad (9.1)$$

The conclusion that the reaction proceeded by a 'limiting siliconium ion' mechanism was later criticised[85]. Baughman[84] also suggested, however, that the rate-determining step might be the ionisation of the C—Cl bond, possibly with assistance from the *β*-silicon atom. Jarvie, Holt and Thompson[86] then made the interesting finding that the solvolytic elimination reaction of (*erythro*-1,2-dibromopropyl) trimethylsilane in aqueous ethanol is highly *trans*-stereospecific,

(Newman projection: front carbon SiMe₃, H, Me; rear carbon H, Br, Br) —slow→ H, SiMe₃, H / C—+—C / Br, Me (2) → H, H / C=C / Br, Me (9.2)

and suggested that the Me_3Si group gives anchimeric assistance to the ionisation of the C—Br bond, a non-classical silicon-bridged silacyclopropenium ion (species (2) above) being formed in the rate-determining step (Equation (9.2)). To explain the fact that *cis* elimination increased with ionising power of the medium, (15% in moist formic acid) it was postulated that (2) is in equilibrium with a classical carbonium ion (3), the proportion of which increases with increasing ionising power of the medium (Equation (9.3)).

$$\text{(2)} \rightleftharpoons Me_3Si{-}CHBr{-}\overset{+}{C}HMe \ \text{(3)} \longrightarrow \text{H(Br)C=C(H)Me} + \text{H(Br)C=C(Me)H} \qquad (9.3)$$

They also pointed out that reaction by the route depicted in Equation (9.2) should provide an opportunity for rearrangement with migration of the Me_3Si group; although this was not observed, it was found that the alcohol $Me_3SiCH_2CD_2OH$ gave mixtures of $Me_3SiCH_2CD_2X$ and $Me_3SiCD_2CH_2X$ (X = Cl or Br) on treatment with $SOCl_2$ and PBr_3, a result consistent with rearrangement of a species such as (4).

$$CH_2 \overset{SiMe_3\,(+)}{\text{—}} CD_2$$

(4)

Shortly afterwards Cook, Eaborn and Walton[87] showed that when solvolysis of the bromide $Me_3SiCH_2CD_2Br$ is allowed to proceed to about 50% completion, the recovered bromide contains $Me_3SiCD_2CH_2Br$ and Me_3

$SiCH_2CD_2Br$ in 1:2 ratio. This migration of the Me_3Si group is completely consistent with the mechanism involving anchimerically-assisted ionisation of the C—Br bond to give a silacyclopropenium ion. It is interesting to note that it is estimated that electron release by the Me_3SiCH_2 group should make the halide, $Me_3SiCH_2CH_2Cl$, some $10^{9.5}$-times more reactive than methyl chloride in dissociative solvolysis, and stabilisation of the ions $Me_3CH_2CH_2^+$, should increase in the order Si < Ge ≪ Sn. It is quite clear, therefore, that in this case a $(CH_3)_3Si^+$ ion is not formed as a kinetically-distinguishable species.

An early assertion[88] that the salt-promoted racemisation of methyl-α-naphthylphenylchlorosilane proceeded through a siliconium ion-pair is unlikely to be correct[89, 90].*

It has been suggested that a ferrocenyl group is able to stabilise a siliconium as well as a carbonium ion[91]. Thus, the Si—Si bond of compound (5) cleaves very readily on treatment with 10^{-3}M ethanolic hydrogen chloride and the scheme depicted in Equation (9.4.4) was suggested,

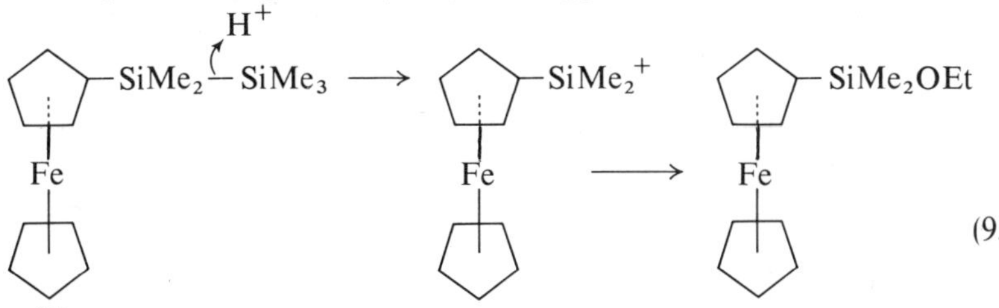

H^+

$SiMe_2—SiMe_3$ → $SiMe_2^+$ → $SiMe_2OEt$

Fe Fe Fe

(9.4)

(5)

but the role of the ethanol and of chloride ion remain as uncertainties in this interpretation. Stabilisation of $\equiv Si^+$ and $\equiv Ge^+$ species in Friedel–Crafts reactions of ferrocene by π-interaction with the ferrocenyl fragment has been suggested as a possibility[92].

It is clear that, at the moment, there is no substantial evidence for a dissociative process via a siliconium ion starting from a 4-covalent silicon species in solution. If silicon is 5-coordinated, however, the situation may well be different. Recently a number of 5-coordinate silicon complexes have been suggested as models for a 5-coordinate intermediate[99], and there is then a possibility that solvolysis reactions of such compounds could proceed by a dissociative mechanism, possibly via a 4-coordinate siliconium ion, rather than via an associative 6-coordinate transition state. The hydrolysis of nitrilophenoxysilanes (6) has been studied in water–dioxan mixtures in the presence of hydrochloric acid. The reaction product is the ligand nitrilo-

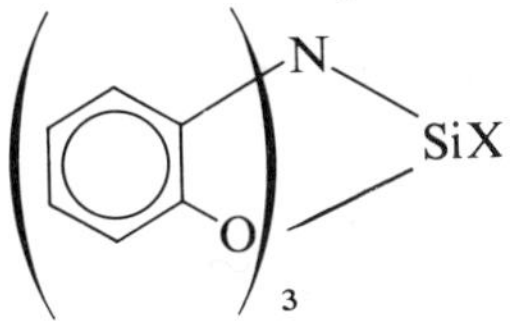

N SiX O 3

(6)

* In a recent note (*J. Amer. Chem. Soc.*, (1971), **93** 2093) Professor Sommer and his co-workers present further evidence against the operation of a siliconium ion-pair mechanism.

triphenol(Ntp) in all cases, but the mechanism depends on the substituent X. When X is Me, Ph or *p*-tolyl, the rate-determining step is the breakage of a protonated Si—O bond. When X is Cl, reaction via the ion-pair, $NtpSi^+Cl^-$, is proposed. The ion-pair may be stabilised by $(p \rightarrow d)_\pi$ bonding in the siliconium ion: the analogous germanium compound NtpGeCl does not react by a similar ion-pair mechanism[100].

Even in a FSO_3H–SbF_5–SO_2 solvent mixture it was not possible to produce silyl cations: alkoxysilanes were protonated at oxygen and underwent a slow cleavage, probably through attack by F^- [101].

9.4.4 Some aspects of solvolytic reactions

These reactions have been widely studied for a number of systems for almost two decades, and we shall consider some of the studies which reveal the main features of solvolytic reactions of silicon and Group IVb derivatives.

9.4.4.1 Base-catalysed solvolysis of silanes

A number of earlier studies[102,103] could not distinguish between a concerted mechanism and one involving formation of an intermediate: much earlier work was summarised by Steward and Pierce[105]. Weyenberg[104] has examined the deuterium isotope effect in the solvolysis of compound (7) which can undergo cleavage of the Si—H or Si—C bond to give (8) or (9) respectively:

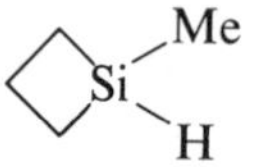

Me, Si, H (7) | (Si Me)$_2$O (8) | $(PrMeSiH)_2O$ (9)

The kinetic isotope effect k_H/k_D was 1.1, but the product isotope ratio $[(8)/(9)]^H/[(8)/(9)]^D$ was 1.9. This suggests that the rate-determining step does not involve cleavage of the Si—H or Si—D bond, and hence it is reasonable to suppose that it must be the formation of a penta-covalent intermediate from base and silane. This seems to be the strongest available evidence for the type c mechanism of Figure 9.2.

Sommer and his co-workers some time ago studied the solvolysis of a series of cyclic silanes (11)–(15). Relative rates, taking Et_3SiH as 1, are as shown[106,107] in Table 9.1. The interesting feature of this study is the enhanced reactivity of these compounds: the transition states must be different from the S_N2 transition state of carbon, and involve attack of the nucleophile from the front. The reason for the greater reactivity, especially of compound (14) was suggested to be due to the fact that the geometry of (14) around the silicon atom, assuming that the carbon atoms remained tetrahedral, was closely similar to that required for a transition state (trigonal-bipyramid or square-pyramid) leading to retention of configuration; in addition, an orbital hybridisation effect at strained bridgehead silicon led to increased s character and reactivity of the Si—H bond[108].

Experiments with an optically-active silane have shown that the reactions in relatively non-polar media proceed with retention of configuration[109].

The base-catalysed solvolysis of M—H bonds is one of the relatively few reactions where comparison with other Group IV elements has been made[110]. Trialkylstannanes react considerably more rapidly than analogous silanes,

Table 9.1 Relative rates of base-catalysed solvolysis of sila-cyclic silanes in 95% ethanol

	Silane	*Relative rate*		*Silane*	*Relative rate*
(10)	Et_3SiH	1	(13)	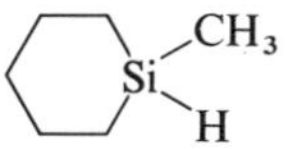 1-methylsilacyclohexane (Si–CH$_3$, Si–H)	0.1
(11)	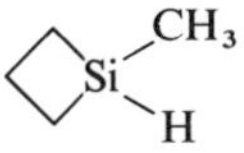 1-methylsilacyclobutane (Si–CH$_3$, Si–H)	10^4–10^5	(14)	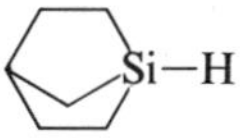 bicyclic Si—H	10^3
(12)	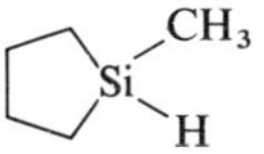 1-methylsilacyclopentane (Si–CH$_3$, Si–H)	10	(15)	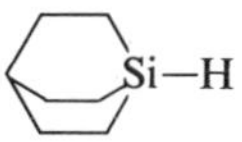 bicyclic Si—H	10

and germanes did not react with concentrated base. This unusual stability of the Ge—H bond is paralleled by the stability of the Ge—C bond towards alkali[111]. For example, the rates of alkaline cleavage of trialkylphenylallyl-stannanes,-silanes and -germanes fall in the order Sn ≫ Si > Ge, and there is apparently a substantial steric effect in the reactions of the silicon and germanium compounds[112], the Me_3Si derivatives reacting more than 10^3-times

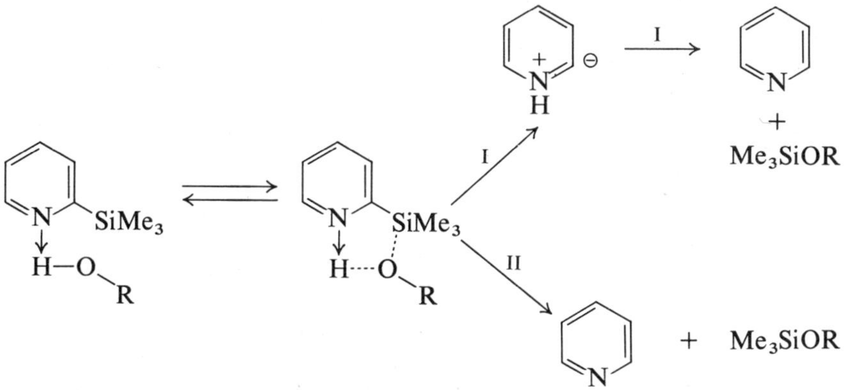

faster than those of Et_3Si. Further studies on the alkaline hydrolysis of aryl-silanes have been reported[113, 114]; the reaction of phenylsilane with nucleophiles has been studied[115], and several other nucleophilic substitutions in organosilicon hydrides have been reported[116–119]. A good deal of the background to displacement of carbon from silicon is to be found in papers on the cleavage of phenylethylnylsilanes and germanes[120], and benzyl trimethyl-silanes[121, 122]; there are also a number of later studies in this area by Eaborn and his co-workers[123]. An interesting Si—C fission reaction is observed when 2-trimethylsilylpyridine reacts with H_2O, MeOH, and EtOH. The reaction is inhibited by acids, and bases have little effect. A mechanism involving intramolecular general base catalysis (route I or II) above was proposed[124–126].

[The analogous germanium and tin compounds react similarly[127] (realtive rates Si, 1; Ge, $\sim 10^{-4}$; Sn, 22)]. This mechanism would require retention of configuration with a chiral 2-silyl group, and this is observed with optically-active 2-(methyl-α-naphthylphenylsilyl)pyridine. 2-Triphenylsilylpyridine reacts at a rate about 15-times slower than that of 2-trimethylsilylpyridine, in contrast to the approximately 15-fold greater reaction rate of Ph_3SiH with alkali, in comparison to that of Me_3SiH. Although caution is needed, as we have seen, in the interpretation of rate factors of this magnitude, this is consistent with unfavourable steric factors in transition state (16) which are not present in (17)[128].

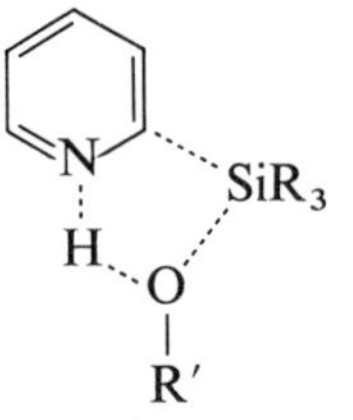

(16)

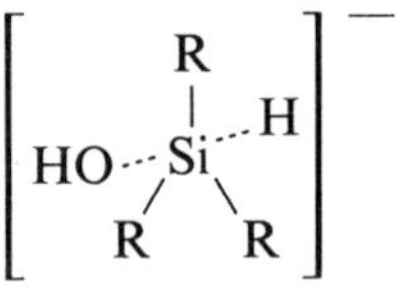

(17)

The effect of catenation produces interesting results in silane hydrolysis[129]: the relative rates of hydrolysis of Me_3SiH, $Me_3SiSiMe_2H$ and $(Me_3Si)_2SiMeH$ are $\sim$0.78:1.0:0.083, and a linear free-energy treatment shows considerable rate enhancement for the last two compounds; again caution is necessary in the assignment of the effect to any one particular factor, but d_π–d_π interaction between the two silicon atoms is favoured by the author.

9.4.4.2 Solvolysis of derivatives of the type R_3(MX X = OR, *halogen or* RCOO)

A number of studies in this area have been made, and earlier work has been summarised[65, 130, 131]. Recently-studied aspects of R_3MOR solvolysis reactions include studies of general base catalysis by acetate and phenoxide ions in the methanolysis of $PhOSiPh_3$ [132]. The solvent isotope effect for the methanolysis of *p*-chlorophenoxy-triphenysilane are $k_H/k_D = 1.25 \pm 0.06$ (acetate catalysis) and $1.35 \pm (0.06)$ (methoxide catalysis). Contrary to what is observed in the hydrolysis of carboxylates[151, 156], the plots of free energy of activation for the methanolysis of Ph_3SiOAr against the free energy of ionisation of the corresponding phenol, ArOH, are not linear, and the authors consider that there may be a change of mechanism (stepwise to concerted) or a variation in Si—O d_π–p_π bonding in the $\Delta G^\ddagger_A$ term[133]. 'Linear free energy' studies have also been made on the hydrolysis of triethylphenoxysilanes, and better log k *v.* σ correlation is obtained using the Yukawa–Tsuno equation than by the Hammett equation[134]. The alkaline hydrolysis of methylethoxysilanes and tetraethoxysilanes[135]; the hydrolysis of 1-alkyl-1-alkoxysilanthrenes[136]; and the methanolysis of methyl-2,3,4,6-tetrakis-*O*-trimethylsilyl α-D-glucopyranoside[137] have been studied.

Schott and co-workers have studied the alcoholysis of acyloxytriarylsilanes in propan-1-ol and propan-2-ol, and the hydrolysis in tert-butanol–benzene–water mixtures:

$$Ar_3SiOCOR + ROH \rightarrow Ar_3SiOR + RCO_2H$$

The reactions are catalysed by acetate ions, the rates being increased by electron-withdrawing substituents in the aryl ring, decreased by electron-releasing substituents in the aryl ring, and increased by electron-withdrawing substituents in the acyl group[138, 139]. This group has also examined the solvolysis of acetoxyaryldimethylsilanes in propan-1-ol[140].

The hydrolysis and alcoholysis of Si—N systems have been studied[147–149]. The stereochemical course of the reaction[150] is inversion, but the pyrrolidine compound,

α-NpPhMeSiN(pyrrolidine ring)

shows reduced stereospecificity. Specific oxonium ion catalysis is usually observed (see for instance Reference 149) and this is a case, rare for silicon (cf. Reference 124–126), where water is a better reagent than OH^-; but *N*-trimethylsilyl pyrrole is exceptional—it is stable to hydrolysis in neutral solution, it is relatively much less sensitive to oxonium ion catalysis, and hydroxide ion catalysis is marked[147, 148].

Solvolytic and hydrolytic reactions of this kind, involving displacement of halogen, OCOR, NRR′ from silicon and germanium, show a number of interesting features, particularly where the solvolysing agent is present in

Table 9.2 Apparent order, *n*, in water for the hydrolysis of various R_3SiX systems

Solvent	*Hydrolysis rate* = $k_{obs}[R_3SiX][H_2O]^n$			*Reference*
	R	*X*	*n*	
Acetone/water	Ph	NH_2	2.5	147–150
	Pr^n	OAc	3.5	152
	Pr^i	Cl	3.0	142, 153–155
Dioxan/water	Ph	NH_2	4.4	147–150
	Ph	OAc	4.4	152
	1-naph	Cl	4.4	153–155
Isopropanol/water	Ph	NH_2	1.2	147–150
	Pr^n	OAc	1.2	156
	Pr^i	Cl	1.2	155
$^{18}OH_2/Pr^n_3SiOH$	dioxan		4.6 ± 0.4	157, 158
	acetone		2.7 ± 0.4	157, 158

an inert solvent, and a number of studies have been directed to the comparison of the reactivity of silicon and germanium in similar structural situations in such systems. An early study[141] of germanium hydrolysis was later shown to be incorrect[142], but the early results are still widely quoted[143–145].

A characteristic feature of most hydrolyses is their great speed compared with the analogous carbon systems. Rapid-reaction techniques have to be used, such as flow/conductance (see, for example Reference 146) or flow/spectrophotometry (e.g. for SiNRR′ systems)[147], particularly where a wide range of concentrations of the solvolysing species is to be studied. Silicon and germanium have been compared in R_3MOOCR' hydrolysis reactions: ^{18}O-labelling experiments[151] show that M—O fission occurs, so that a nucleophilic attack of water at silicon is occurring (in some 6-coordinated dicarbonyl silicon derivatives attack at the adjacent carbonyl carbon, and not silicon, can occur). Germanium systems are considerably more reactive than their silicon analogues under the same conditions[152]. Steric factors are apparent, e.g. with increased branching in R, rates decrease, and do so more rapidly for silicon than for germanium. An interesting feature of these reactions, and of those of chlorides and amino derivatives, is the order of the reaction in water: Table 9.2 summarises the results.

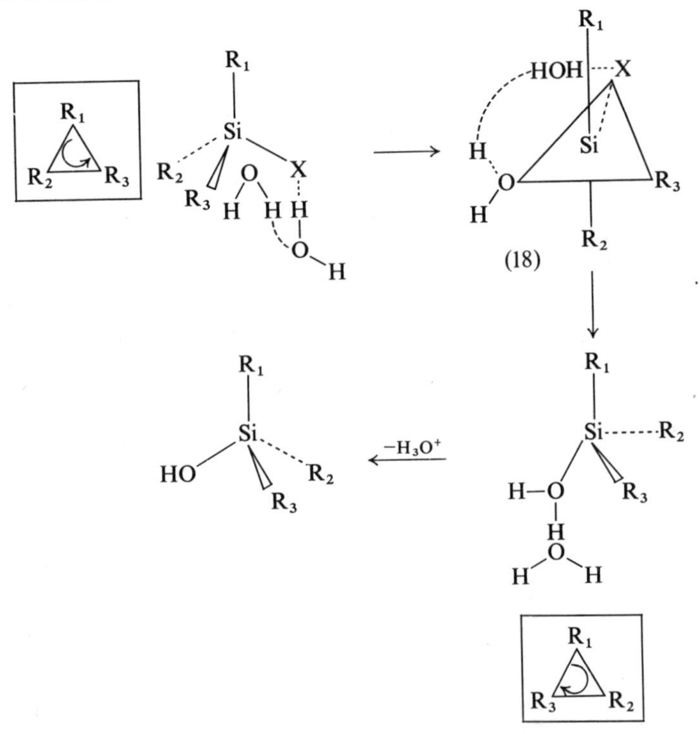

Figure 9.3 Stereochemical inversion when attack of a water polymer is between R_1 and R_2. Insets show order of R groups looking down Si—X or Si—O axes in reactant and product respectively

The order in water for the analogous germanium systems, where this has been measured, is one unit higher than those for silicon; only the Me_3Si group is exceptional[156], and with the acetate the order is 5.0. It appears that in non-hydrogen-bonding solvents more than one molecule of water is present in the transition state, whereas in propan-2-ol one water molecule is effective. The activation parameters are strongly dependent on the water concentration: at low values $\Delta H^{\ddagger}_{exp.}$ is small (and is zero for chloride hydrolyses in acetone). As the water concentration is increased so the activation enthalpy increases[156], the entropy becoming less negative. Hydrolysis of acetates is faster in propan-2-ol than in dioxan at the same water concentration, mainly because the activation entropy is less negative in the alcohol.

The structure of water in these solvents could have an important bearing on these effects but there appears to be little information to help answer such questions as: (1) are pre-formed water polymers key reactants?; (2) Does a water polymer unit grow on the polar part of the molecule before reaction?; (3) Is a polymeric unit in fact irrelevant? (see for instance Reference 71).

Table 9.3 Deuterium isotope effect for hydrolysis of R_3SiCl in acetone at 25 °C

Chloride	k_{H_2O}/k_{D_2O}
Pr^i_3SiCl	1.2
$Pr^i_2PhSiCl$	2.1
Pr^iPh_2SiCl	3.7
Ph_3SiCl	5.0
$(p\text{-tol})_3SiCl$	3.9
Ph_5Si_2Cl	5.8
Et_3SiCl	3.3
Pr^n_3SiCl	3.2

(The last four entries are subject to larger errors, *c.* ±0.3, than the first four)

Even the stereochemistry of the reaction, which probably occurs by inversion (see below, Section 9.4.5) does not help to distinguish between questions (1) and (2), since a water polymer unit could give inversion by either a straightforward rearward attack, thus answering question (1), or, as in Figure 9.3, by one end of a water polymer, closely associated with the polar Si—Cl group (e.g. H-bonded to X), moving between groups R_1R_2, R_2R_3, or R_1R_3 as shown. This assumes that no psuedorotation has occurred in (18), such as would place the more electronegative groups $(OH_2)_n$, X in axial positions (we shall return to this question later), and that M remains 5-coordinated (6-coordination is another possibility for reaction with a water polymer and may well occur, for example, in the Me_3Si derivatives referred to above). Although the order of the reaction in water is independent of the R groups, in chlorosilane hydrolysis, solvent deuterium-isotope effects are very sensitive to the nature of R (Table 9.3). These reactions also show base catalysis[152–155], and with substituted pyridines as bases steric effects on the proton-transfer stage are observed[153,154]. High orders, which cannot be accounted for by medium effects alone, are a feature not only of hydrolysis but of alcoholysis as well, and to this extent there is a resemblance to acyl alcoholysis of carbon chemistry[159,160]. The propanolysis of Ph_3SiCl in CCl_4 shows similar features[161].

For the purposes of comparison of silicon with the later members of the Group we see that hydrolysis, alcoholysis and solvolysis are quite complicated processes, and there is the added complication that, as the Group is descended, solvolytic and hydrolytic equilibria lie progressively further in favour of the reactants[142]. To the extent that orders in water for, say, hydrolysis of silicon and germanium derivatives are different we may almost regard the two elements as reacting with a different nucleophile. The most general view would be that, of a range of different structured water units existing either in the (non-hydrogen-bonding) solvent or growing near the M—X dipole; silicon and germanium make a different selection so that on average there are more water molecules associated with a Ge transition state than with a Si transition state. These difficulties are avoided in chloride-exchange reactions.

9.4.5 Radiochloride exchange in R_3MCl systems

The reaction $R_3MCl + {}^*Cl^- \rightarrow R_3M^*Cl + Cl^-$ provides a reaction system which is extremely useful for studying the effect of changing M and structure of R on reactivity, and has the advantages that arguments relating to the transition state are simplified by symmetry. In addition, the reaction is entropy driven, so that no problems arise from an unfavourable position of equilibrium. Reactions of this kind played an important role in the development of understanding of nucleophilic substitution at the carbon atom[2, 162], but their study in the case of silicon and germanium was restricted by their great speed, e.g. Ph_3SiCl undergoes exchange with $Bu_4N^+Cl^-$ in benzene at room temperature about 10^6-times faster than its (usually considered to be reactive) carbon analogue[163].

Allen and Modena studied the chloride exchange of some sterically-hindered chlorosilanes in nitromethane/dioxan, using a simple flow apparatus and quenching the reaction by precipitation of the chloride salt in pentane[164]. Haysom[165] developed a more sophisticated, and versatile, programmed flow/quench fast-reaction apparatus to investigate radiochloride exchange: he was still unable to follow the chloride exchange of less sterically-hindered chlorosilanes, but found that phenyl groups caused a large decrease in $\Delta H^{\ddagger}_{exp.}$ and $\Delta S^{\ddagger}_{exp.}$ compared with alkyl groups. Tagliavini[166] observed some startling solvent effects: addition of nitromethane or acetonitrile to acetone caused large reductions in the $\Delta H^{\ddagger}$ value for the reaction.

The use of the low-polarity-solvent combination acetone/dioxan permitted a wide range of germanium and silicon chlorides to be studied[167]: some of the results are shown in Table 9.4. The greater reactivity of germanium compared with silicon is evident, as are differences in steric effects and the relative effects of phenyl-group introduction. Phenyl-group introduction has a dramatic effect on the order of the chloride exchange at silicon, and the effect is still present whether or not the phenyl group is bonded directly to silicon (Table 9.5). Whatever interaction is present the process must be symmetrical with respect to reflection through the transition state: the first-order term suggests reaction with an Li^+Cl^- ion-pair, the half-order term a reaction with Cl^- only. The phenyl group thus promotes reaction with an ion-pair, and this could occur in the transition states (which have the requisite symmetry) shown in Figure 9.4. Activation parameters are shown in Table 9.6: the main difference between silicon and germanium is seen to be a lower $\Delta H^{\ddagger}_{exp.}$ value for germanium and a more negative $\Delta S^{\ddagger}$ value for alkyl or alicyclic R groups; the large effect of the phenyl group on both parameters is noteworthy. Transition states such as those shown in Figure 9.4 imply a

Table 9.4 Rate constants, k_1, for $R_3MCl/^{36}Cl^-$ exchange reaction
$[Li^{36}Cl] = 10^{-3}M$ Solvent = Me_2CO/dioxan 5:2 v/v (D = 6.5) k_1 = R/b

M	*R*	$10^3k_1(s^{-1})$	*Relative rate*
Si	Et	38 ± 2	5.6×10^3
	Pr^n	18 ± 0.6	2.6×10^3
	Bu^n	11 ± 1	1.6×10^3
	n-hex	8.8 ± 0.4	1.2×10^3
	Ph	0.86 ± 0.06	120
	$PhCH_2$	400 ± 16	5.2×10^4
	$PhCH_2CH_2$	140 ± 4	2.0×10^4
	cyc-hex	0.0071 ± 0.0003	1
	1-naph	0.0013 ± 0.00006	0.18
Ge	Bu^n	1300 ± 130	30
	n-hex	920 ± 50	23
	Ph	150 ± 7	3.9
	cyc-hex	40 ± 1	1

	cyc-hex	*n-hex*	Bu^n	*Ph*
k_1 Ge/k_1Si	5.7×10^3	105	120	175

Table 9.5 Radiochloride exchange: kinetic order in $Li^{36}Cl$

Solvent: acetone/dioxan	
Bu^n_3SiCl	0.5
cyc-hex_3SiCl	0.5
Ph_3SiCl	0.5+1.0
$(PhCH_2)_3SiCl$	1.0
$(PhCH_2CH_2)_3SiCl$	1.0
Ph_3GeCl	0.5

Figures for Ph_3SiCl are only approximate

Table 9.6 Radiochloride exchange: activation parameters corrected for temperature variation of salt dissociation in Me^2CO/dioxan

R *in* R_3SiCl		Si	Ge
n-alkyl	$\Delta H^\ddagger$	10	—
	$\Delta S^\ddagger$	−12	—
n-hexyl	$\Delta H^\ddagger$	—	5
	$\Delta S^\ddagger$	—	−20
cyclohexyl	$\Delta H^\ddagger$	15	6
	$\Delta S^\ddagger$	−12	−25
phenyl	$\Delta H^\ddagger$	(7)	5
	$\Delta S^\ddagger$	(−30)	−25
$PhCH_2$	$\Delta H^\ddagger$	5	—
	$\Delta S^\ddagger$	−23	—
$PhCH_2CH_2$	$\Delta H^\ddagger$	5.5	—
	$\Delta S^\ddagger$	−23	—

$\Delta H^\ddagger$ in kcal mol^{-1}
$\Delta S^\ddagger$ in cal deg^{-1} mol^{-1}

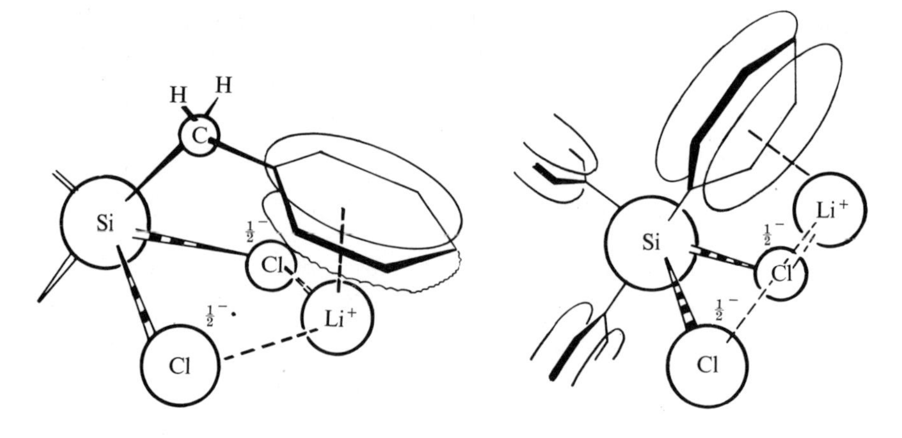

Figure 9.4 Li^+–phenyl group interaction in chloride exchange (see text)

retentive exchange pathway: we shall now examine the stereochemistry of nucleophilic displacements at Group IV elements.

9.5 THE STEREOCHEMISTRY OF DISPLACEMENTS AT SILICON, GERMANIUM AND TIN

9.5.1 General aspects

The initiating factor in the understanding of the stereochemistry of substitution was the synthesis and resolution in 1959 of the chiral silicon compound α-naphthylphenylmethylsilane (α-NpPhMeSiH, R_3Si^*H) by Sommer and Frye[168]. This marked the beginning of studies of the stereochemical course of substitution at silicon. Four years later Bott and co-workers reported a reliable resolution of α-NpPhEtGeH[169], closely followed by Brook and Peddle's resolution of α-NpPhMeGeH[170].

An n.m.r. spectroscopic study of methylneophylphenyl tin chloride (19)

$$Ph-\overset{\displaystyle CH_3}{\underset{\displaystyle CH_3}{\overset{|}{\underset{|}{C}}}}-CH_2-\overset{\displaystyle CH_3}{\underset{\displaystyle Ph}{\overset{|}{\underset{|}{Sn}}}}-Cl$$

(19)

to determine if an asymmetric organo-tin centre had sufficient stereochemical stability to allow resolution[172,173] has been carried out. The method employed was based upon the non-equivalence of the chemical shift of diastereotopic nuclei: it is clear from the Newman projection formulae of $R'R''R'''MM'A_2X$

that inversion of configuration at either M or M′ interchanges the magnetic environment of the two diastereotopic A groups:

If exchange is absent, A and A′ are anisochronous. Anisochronism is observed in dilute solution in non-polar solvents, but coalescence of the signals can be produced at a given temperature by addition of a ligand such as DMSO-d_6, or by using a polar solvent such as methylene chloride. The higher the

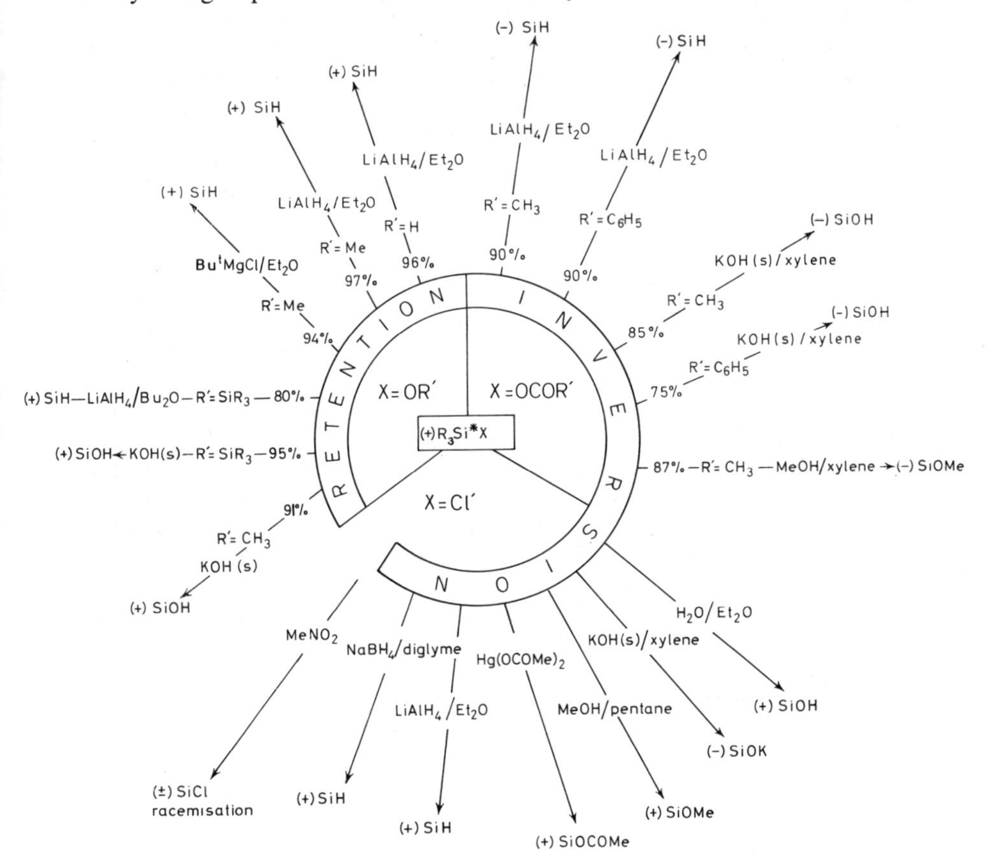

Figure 9.5 Summary of the steric course of reactions of R_3Si^*Cl, R_3Si^*OCOR', and R_3Si^*OR. Figures refer to percentage inversion or retention

concentration of (19) the lower the temperature required to produce coalescence, which results from rapid exchange of the diastereotopic nuclei to give a time-averaged signal.

The mechanism suggested as the most reasonable involves pseudorotation with inversion of configuration at tin, a point to which we shall return below.

Failure to repeat an earlier resolution of a chloride is then seen to be associated with rapid inversion of configuration about the tin atom in the conditions employed for resolution, and the original work is erroneous.

Throughout the last decade stereochemical studies on a wide range of systems, particularly those of silicon, have been made. The earlier work is the subject of a survey by Sommer[65], and a comprehensive review of more recent developments is available[171] so that only very recent developments and particular features of interest will be considered here.

Perhaps the most striking feature of the stereochemical study of a wide range of reactions of silicon derivatives is that both inversion and retention can be observed with appropriate systems (Figure 9.5), and some very simple empirical rules were soon evident[65]. Thus, for good leaving groups X (as assessed by the pK_a of the conjugate acid), for which the pK_a of $HX < 6$, reaction in a wide range of solvents gives inversion provided that the nucleophile is more basic than X. This mechanism is referred to by Sommer as the S_N2—Si mechanism. For poor leaving groups, pK_a of $HX > 10$, reaction in non-polar solvents often proceeds with retention, a mechanism referred to as S_Ni—Si, and the transition state for this is considered to be cyclic, e.g.

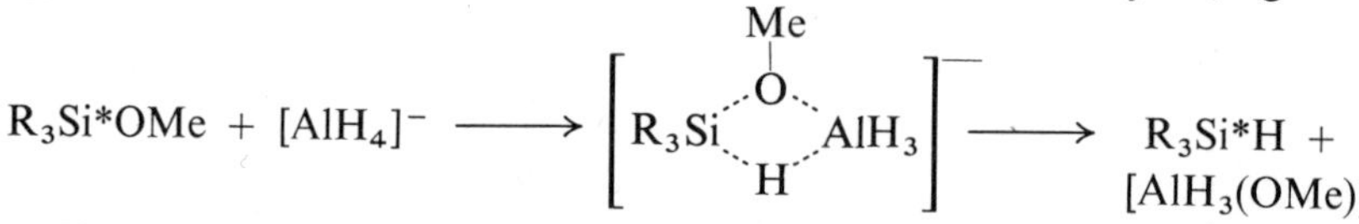

$$R_3Si^*OMe + [AlH_4]^- \longrightarrow \left[R_3Si \overset{\overset{Me}{|}}{\underset{H}{\overset{O}{\cdots}}} AlH_3 \right]^- \longrightarrow R_3Si^*H + [AlH_3(OMe)]^-$$

These rules apply not only to α-NpPhMeSiX [178] but also to (neopentyl) PhMeSiX [179], $(PhCH_2)$PhMeSiX [179], EtPhMeSiX [179] and Ph_3SiSiXPhMe [180].

The dependence of the steric course of substitution in optically-active α-NpPhMeSiX on the leaving group X, where X = Cl, F, MeO or H, has been examined. When X = Cl, reaction with alkyl lithiums gave inversion, but when X = H it is always retentive, except with benzhydryl lithium. The steric course for X = F or MeO is inversion with simple alkyls, but retention with alkyl lithium, benzyl lithium, α-methylbenzyl lithium or benzhydryl lithium. For each alkyl lithium there is a cross-over pK_a of HX, at which the steric course changes from inversion to retention: for simple alkyl lithiums this is ~4 (pK_a of HF), with $pMeOC_6H_4CH_2Li$ it is ~1.6 (pK_a of MeOH), and with benzyl and allyl lithium it is ~40 (pK_a of H_2). The mechanism of the retentive reactions is thought to be probably S_N2—Si, with either 4- or 6-membered cyclic transition states having trigonal-bipyramidal or tetragonal-pyramidal coordination about silicon[181, 182].

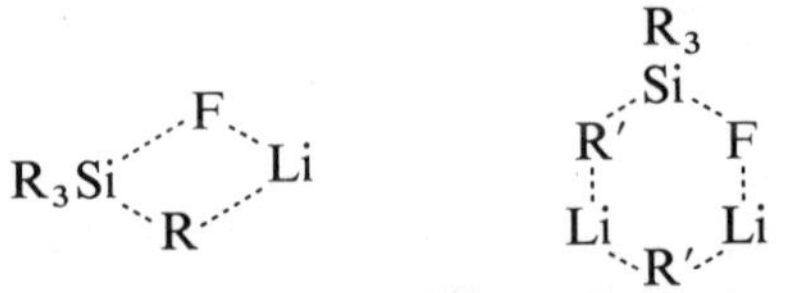

$$R_3Si \overset{F}{\underset{R}{\cdots}} Li \qquad \begin{matrix} & R_3 & \\ & Si & \\ R' & & F \\ \vdots & & \vdots \\ Li & & Li \\ & R' & \end{matrix}$$

Further studies on the steric course of Si—N reactions showed interesting substrate and reagent dependence: the reaction of α-NpPhMeSiNH_2 with PhCOOH in pentane showed 79% retention and the cyclic transition state was proposed. However when mesitoic acid was used or $R_3SiNHBu^i$, 70–80% inversion was observed and steric effects were thought to be responsible[150].

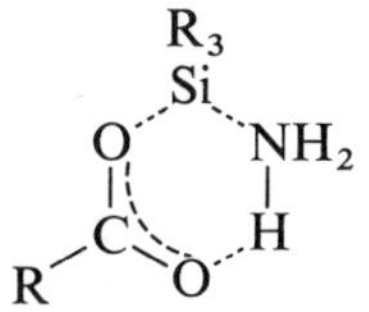

Sommer also postulated an S_N1—Si mechanism based on the fact that the rate of racemisation and the radiochloride exchange rate are the same for R_3Si^*Cl in nitromethane/chloroform solvent.* However, Corriu *et al.*[174, 175] prefer a racemisation mechanism involving an expansion of the coordination number of the silicon atom on the basis of studies of solvent-variation of racemisation rate with various NpPhRSiCl compounds (R = Me, Et, Pr^n, Pr^i. Furthermore, examination of the kinetics of racemisation by fast techniques in dioxan–acetone shows that the rate of racemisation and exchange are the same: the reaction is first order in R_3Si^*Cl and half order in LiCl. The latter is consistent with strong ion-pairing of the LiCl, only free chloride ion being kinetically active: this strongly suggests an associative, rather than a dissociative, process for the racemisation[177].

The chloride ion exchange kinetics show a number of interesting features: we saw above that when phenyl groups are present chloride exchange should proceed with retention of configuration: in benzene first- and higher-order terms are observed suggesting transition states of the type

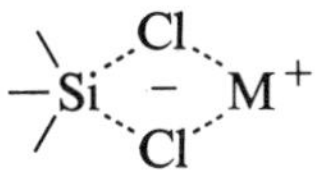

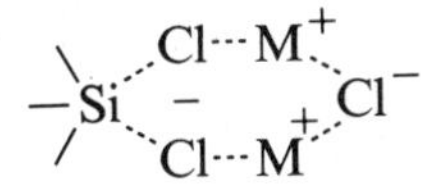

which, because of transition-state symmetry, must be retentive. This has been confirmed[163]; the rate of exchange of NpPhMeSiCl was ~100-times greater than the rate of racemisation, and the order of racemisation with respect to salt was 0.5, indicating that it involves a reaction with a free chloride ion.

As might be expected from the behaviour of analogous systems in phosphorus chemistry, when silicon forms part of a ring system the stereochemical course of the reaction may be strongly affected. It is found that the reaction of the cyclic optically-active silane(20) with lithium alkyls show a much greater tendency to react with retention of configuration than compounds of the

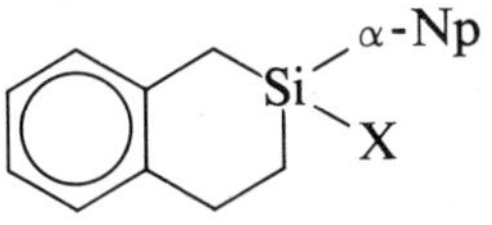

X = H, Cl, F, OMe

(20)

α-NpPhMeSiCl type. For instance, the latter normally reacts with inversion, but compound (20) frequently reacts with retention. These authors suggest that the retentive transition state (21) was favoured relative to that for inversion by electrophilic assistance, but disfavoured by non-bonding repulsions between R_1, R_2 and R_3: when two of these groups formed a ring it was

---

* In a recent note (*J. Amer. Chem. Soc.*, (1971), **93**, 2093) Professor Sommer and his co-workers present further evidence against the operation of a siliconium ion-pair mechanism.

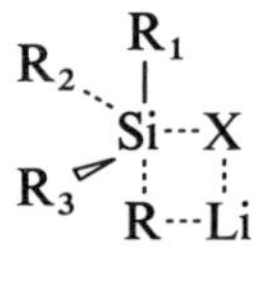

$$\begin{array}{c} R_1 \\ R_2 \cdots | \\ Si \cdots X \\ R_3 \vdots \quad \vdots \\ R \cdots Li \end{array}$$

(21)

thought that the latter would be much reduced, and hence this mechanism facilitated[183]. Further details of this work have appeared[184].

In the reaction of a number of optically-active silicon compounds with BF_3 and BCl_3 the steric course is again found to depend on the leaving group. Usually inversion is observed; with R_3Si^*OMe retention is observed, and transition states (22) and (23) have been suggested for retention and inversion respectively[185].

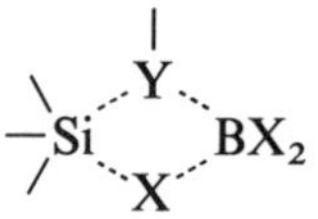

$$-Si\begin{array}{c} Y \\ \; \\ X \end{array}BX_2 \qquad X_3B \cdots X \cdots Si \cdots Y \cdots BX_3$$

(22) 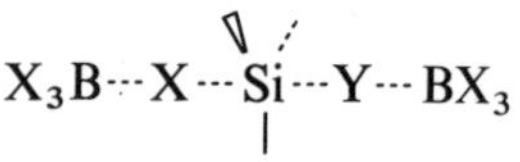 (23)

Analogous optical transformations have been found for germanium systems (though germanium systems appear to racemise more readily than their silicon analogues), and Walden cycles have been performed with optically-active α-NpPhEtGeX compounds. Inversion is found with good leaving groups, X = Cl, SPh, $SGeR_3$, NC_4H_4 and O_2CGeR_3; with poorer leaving groups X = OMe, menthoxide and $OGeR_3$, retention is observed[186]. The steric course of a reaction is clearly a function of the conditions, of the state of the nucleophile, and of the nature of the solvent so that experiments need careful interpretation. For example, earlier work on the alcoholysis and hydrolysis of optically-active chlorosilanes in homogeneous solution had shown that these reactions proceeded with complete racemisation. However, Corriu and Leard showed that the substitution is stereospecific, but the products are racemised after the act of substitution[187].

Solvent effects are marked. Peddle *et al.* found that (+)α-NpPhMeSiH reacts with lithium aluminium deuteride in Et_2O/Bu^n_2O with retention of configuration but in THF with inversion (no reaction was observed with the corresponding germanium compounds)[188]. A further illustration is provided by the base-catalysed alkoxy exchange of R_3Si^*OR [189].

$$MOR' + R_3Si^*OR'' \xrightarrow{R'OH} R_3Si^*OR' + MOR''$$

The case where R′ = R″ = Me had earlier been shown to proceed by inversion, and the rate of tritium-labelled methoxy exchange and the rate of inversion were the same (also in neutral and acid solution)[190]. By varying M, R′, R″ and solvent composition a complete spectrum of behaviour from complete retention to complete inversion was observed. The smaller the cation, the more highly branched the groups R′ and R″, and the lower the proportion of alcohol in the solvent, the more favoured was retention.

Cases exist where a rapid pre-equilibrium formation of a 5-coordinate intermediate seems to be indicated from stereochemical evidence. For example the methanol-induced racemisation of R_3Si^*F proceeds without displacement of fluoride ion[191] and a transition state (24) was suggested in which F, R, R′ and R″ are coplanar. The interesting feature of this suggestion is

(24)

that this is precisely what is envisaged to be a transition state for pseudorotation of a trigonal-bipyramidal $[RR'R''SiFOMe]^-$ complex, although the importance of pseudorotation in stereochemistry, and particularly of silicon transition states, was not widely recognised until later (see below). A similar case has been found in the racemisation of α-NpPhMeGeCl[192]: this is optically stable in hydrocarbons as well as in chloroform and carbon

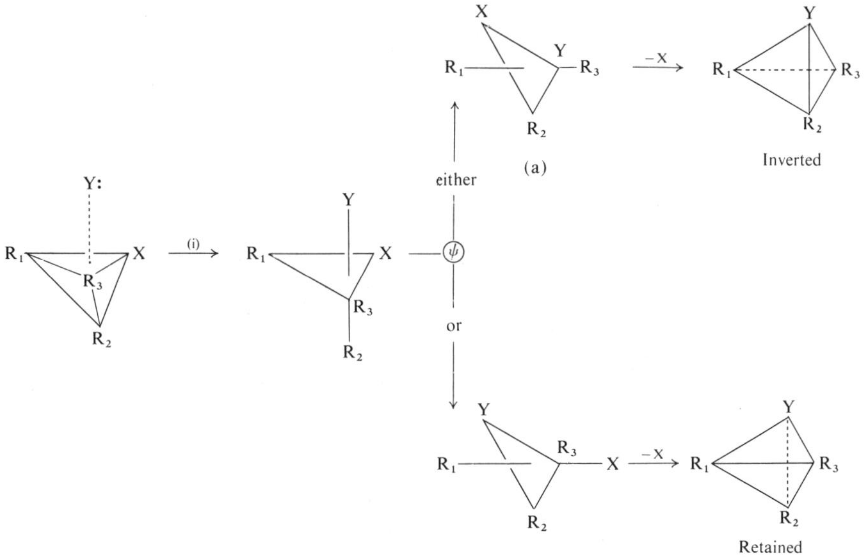

Figure 9.6 If attack of Y on $R_1R_2R_3MX$ (i) gives an intermediate of sufficiently long life then, if pseudorotations such as those shown can occur, product of either configuration can result. Note that a pseudorotation of (a) placing X and Y apical would give the identical configuration to that obtained by direct attack of Y on the $R_1R_2R_3$ (rearward) face of the initial tetrahedron

tetrachloride. Racemisation is not observed in ether, dioxan, or anisole, but is fast in THF. It is also induced by solvents with a nucleophilic atom, and alcohols and water racemise the compound without substitution (0.1% substitution even after complete racemisation). The compound is racemised much more rapidly than the silicon analogue. The authors favour a 6-

coordinated species, such as (25) in the mechanism of this process (Equation 9.5) (cf. also Reference 176).

There is also a possibility that pseudorotation of an intermediate state is involved, the necessary condition being that pseudorotation is rapid compared with the lifetime of such an intermediate. The question then arises as to what pseudorotamers are possible for a 5-coordinated species: if a pseudorotation of the type shown in Figure 9.6 occured then racemisation can occur (we note that such racemisation has nothing to do with siliconium or 'germanonium' ion formation). This figure shows two possible pseudo-

$$R_1R_2R_3Ge{-}X \xrightarrow{+2S} R_1R_2R_3Ge(S)_2{-}X \xrightarrow{-2S} R_3R_1R_2Ge{-}X \qquad (9.5)$$

(25)

rotations followed by ligand loss, and the problem crystallises into one of finding out how many pseudorotamers there are, what configurational change will occur when they lose a ligand, how they are inter-related, and what principles operate to make some more likely than others. This question has been thoroughly examined in the context of phosphorus chemistry, particularly for phosphorane-derived ground states (which clearly can be models, however crude, for an associative transition state of a Group IV element), and the transition states for substitution in PR_4^+ derivatives, which are, of course, isoelectronic with Group IV R_4M systems.

9.6 THE LABILE STEREOCHEMISTRY OF PHOSPHORUS: PSEUDOROTATION AND ITS TOPOLOGICAL REPRESENTATION

The general background to substitution and stereochemistry of phosphorus derivatives is the subject of a number of recent useful reviews[2–25, 193–198] so that only some recent aspects of labile stereochemistry will be considered which give us some insight into the role of 5-coordination in the reactivity of second row and later elements in substitution reactions.

The concept of pseudorotation is finding wide application in phosphorus chemistry, and there are several recent papers dealing with the phenomenon itself both in P[5] and As[5] systems[26–31]; its application in phosphorus acid ester hydrolysis (e.g. Reference 199–201 and references contained therein); and in the reactions of systems based on PR_4^+ [198, 202–206].

The incompleteness of Figure 9.6 raises the question of the systematic analysis of possible pseudorotations and their representation. A recent paper by Mislow[22] is particularly valuable in this respect. Pseudorotation is important because, in principle at any rate, an intermediate such as that in Figure 9.2 paths c or d may be able to exist for long enough for pseudorotation to have an important bearing on the stereochemical outcome of the reaction: such is indeed now an accepted feature of phosphorus chemistry (e.g. see References 195 and 26–31) but the notion has not yet been applied to any great extent in silicon chemistry. We note that if the phenomenon occurred readily in the transition state as it does in many P^V halides[23, 27], then stereochemically-clean reactions at Si would not be observed. Whenever such is not the case then pseudorotation of an intermediate becomes a possibility, though not, of course, the only one as we have seen (e.g., see Reference 187).

9.6.1 Representation of structural rearrangements occurring in pseudorotation

Pseudorotation is an intramolecular* ligand exchange process in an MX_5 system, resembling in some respects the vibrational motion associated with the pyramidal inversion of NH_3. A number of mechanisms is possible[205], but the most common one is thought to be the Berry mechanism in which pair-wise interchange of apical and equatorial ligands occurs through a tetragonal-pyramidal transition state (Figure 9.7), a reference or 'pivot' bond

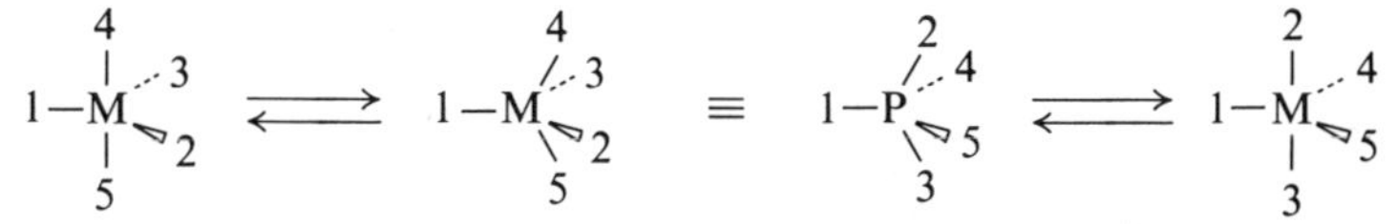

Figure 9.7 Pseudorotation occurring through a tetragonal-pyramidal intermediate or transition state. Ligands 1, 2 and 3 are equatorial and 4 and 5 apical at the left; at the right, ligands 2 and 3 are apical while 1, 4 and 5 are equatorial

which stays equatorial occupies the apex of the pyramid in the transition state. Figure 9.7 shows pivotal ligand 1 and ligands 2, 3 interchanging with 4, 5. It is clear from this (and Figure 9.6) that if the starting system is chiral this motion can have a profound effect on the stereochemical outcome of the reaction.

If we begin with an optically-active centre M(1) (2) (3) (4) (Figure 9.8) (the numbers referring to different ligands) then, formally, addition of a fifth

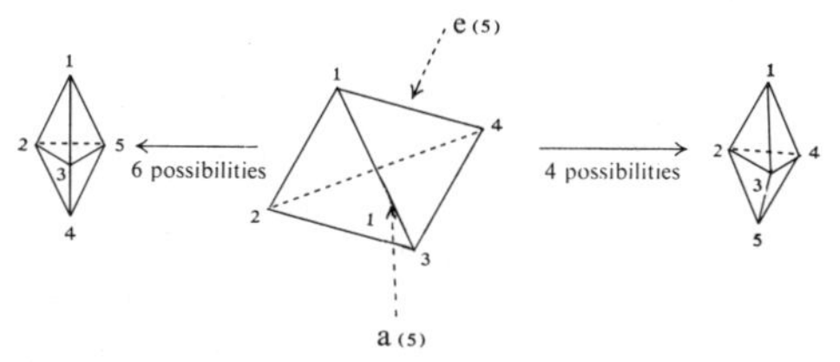

Figure 9.8 Illustration of how an edge attack on a tetrahedral structure places the new ligand (5) in the equatorial (e) position while face attack (a) places it in an apical position

different ligand (5) can occur at any four of the tetrahedral faces or six different edges: these two modes will place ligand (5) in apical or equatorial positions respectively. There are ten ways of selecting two distinct axial ligands from five, and each pair selected can have two triangular permu-

*Ligand exchange in P^V fluorides is not always intramolecular[26]; there is evidence for $(CH_3)_3PF_2$ and $(CH_3)_2PF_3$ that it is intermolecular, proceeding through a dimer as in ClF_3, BrF_3 and SF_4.

tations of the remaining three equatorial ligands, so that in the absence of special constraints (such as chelation between two ligands) there are 20 possible stereoisomers of M(1)(2)(3)(4)(5). Since each pseudorotation exchanges a pair of equatorial and apical ligands, and each isomer has three equatorial ligands which can be pivotal, there are three routes from any given isomer to three others, so that there should be 60 pseudorotation steps, but since two isomers are connected by a given pseudorotation the maximum number of inter-connecting steps is 30. A complete listing of all alternative pathways and inter-relationships is clearly cumbersome, and elegance and economy can be introduced at this point by the use of what has been called[22] a Desargues–Levi graph. This provides a topological representation of the multiple inter-conversion pathways, the set of 20

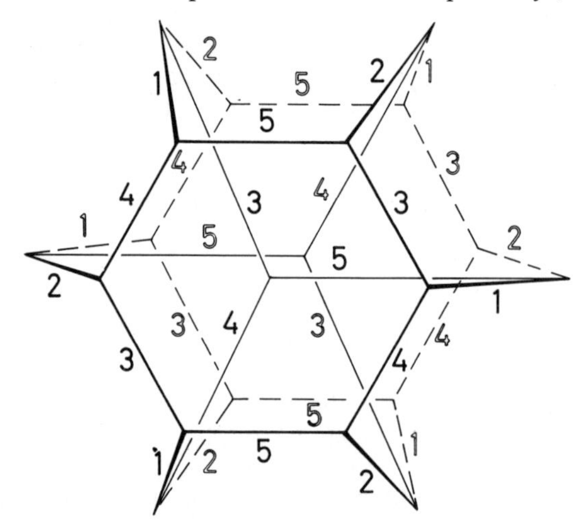

Figure 9.9 Desargues–Levi graph projected onto a plane. Isomers are represented by vertices and pseudorotations by the lines connecting vertices, the index of the pivot ligand for each pseudorotation being shown alongside the line. By permission of Professor K. Mislow

inter-converting isomers and 30 inter-connecting pseudo-rotations are represented in the graph, so that each isomer is at a vertex and each pseudorotation step is an edge. (The graph is regular (Figure 9.9) but the two-dimensional representation of the diagram inevitably produces distortions.) A pseudorotation step is represented by a numeral over an edge, related through the centre of symmetry, and each numeral refers to an equatorial pivotal ligand. The identity of an isomer is then defined except for chirality, e.g. the vertex at 1,4,5 is an isomer with equatorial ligands 1,4,5 and apical ligand 2 and 3. The convention used to represent a given isomer is simply to specify the apical ligands, 23 or its enantiomer $\overline{23}$ in this case. Chirality is denoted as in Figure 9.10; an ascending order of equatorial indices in a clockwise direction viewed from the apical ligand of lower index gives an unbarred isomer, if the order is counter-clockwise the isomer is barred. The isomers at the vertices are then as in Figure 9.11.

A particularly interesting case is when two of the ligands form a ring system, i.e. L_1 and L_2 in Figure 9.12 are joined by a 4- or 5-membered ring:

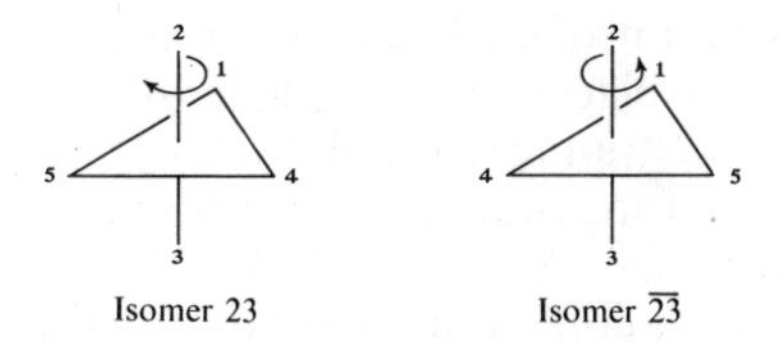

Figure 9.10

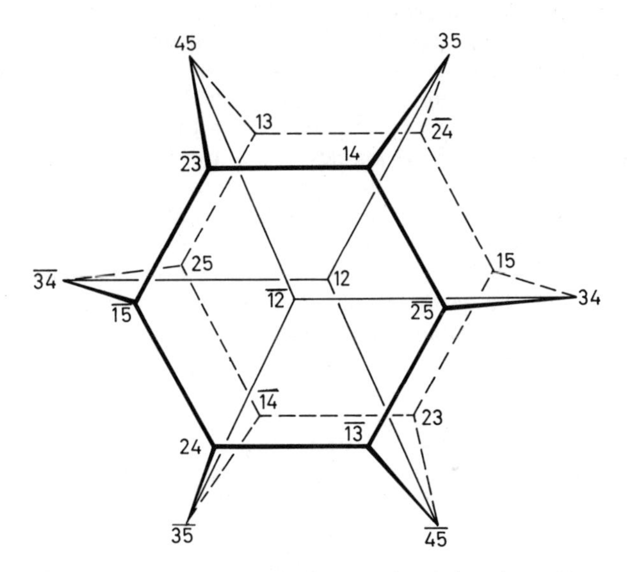

Figure 9.11 Desargues–Levi graph projection, showing designations of isomers represented by vertices. Each isomer is designated by the indices of its apical ligands; thus, 14 has ligands 1 and 4 apical, and $\overline{14}$ is its mirror image. By permission of Professor K. Mislow

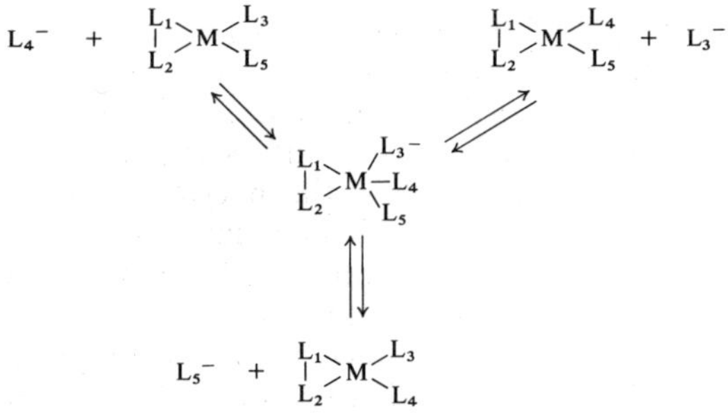

$L_4^- + (L_1, L_2)M(L_3)(L_5) \rightleftharpoons (L_1, L_2)M(L_3^-)(L_4)(L_5) \rightleftharpoons (L_1, L_2)M(L_4)(L_5) + L_3^-$

$(L_1, L_2)M(L_3^-)(L_4)(L_5) \rightleftharpoons L_5^- + (L_1, L_2)M(L_3)(L_4)$

Figure 9.12 After Mislow, Reference 22

this case has been examined in detail (Reference 22 and references cited therein) for $M = P^+$. Displacement in such compounds may be represented as in Figure 9.12 in which indices 1 and 2 refer to ring-binding sites. Since such a ring is incapable of spanning apical positions, 12 and $\overline{12}$ are eliminated from Figure 9.11 leaving the 18-vertex graph of Figure 9.13.

The last step to relate the configurations of the starting and final compounds is as follows: three surfaces are passed through the three-dimensional representation of the 18-vertex graph so that each surface divides the set of 18 vertices into two sub-sets of nine: these surfaces are shown dotted in Figure 9.13 and are referred to as σ_n, n denoting the index of the edges bisected by the plane. Also shown in Figure 9.13 are three pairs of R_4M species (e.g. R_4P^+ or R_4Si), each pair straddles a dashed line and is related to the corresponding surface in the sense that each of the six species may give rise to nine R_5M species on its side of the dashed line by the addition of a fifth ligand, e.g. *a* in Figure 9.13 can give rise to 13, $\overline{14}$, $\overline{15}$, $\overline{23}$, 24, 25, $\overline{34}$, $\overline{35}$ and 45 diastereoisomeric R_5M species which are initial products of attack by an external nucleophile and substituting group, L_5. This Figure therefore provides a stereochemical map of the displacements shown in Figure 9.12. The effect of the displacement reaction on M is to substitute L_5 for L_4: the important feature of the diagram is that retention or inversion depends upon the sector in which the ultimate R_4M species, i.e. the isomer which has lost L_4 to give product, is located with respect to σ_4. Clearly if the product falls southwest of σ_4 (giving

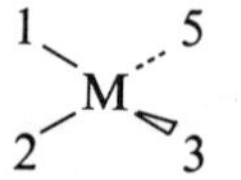

1 5
M)
2 3

the retentive substitution has occurred; if in the northeast of σ_4, (giving

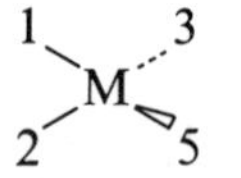

1 3
M)
2 5

an inversion has occurred.

A number of assumptions and rationalisations may be used to minimise the number of possible isomers falling in a sector. For example, if we consider the compound (20), the angle subtended at Si by the ring is 120 degrees, exactly the equatorial angle, so that if an intermediate is formed we may expect the ring to be equatorial. Not only are apical 12 and $\overline{12}$ isomers not possible for this compound but also compression of the angle to 90 degrees is improbable, so that any vertex (isomer) containing a 1 or 2 is eliminated. For a Si[5] intermediate from this compound, 13, $\overline{23}$, $\overline{15}$, 25, $\overline{14}$, 24, the hexagonal vertices, are eliminated, leaving only star-point configurations 45, $\overline{34}$, $\overline{35}$. The final product then may have the inverted configuration (45 lies in northeast sector above σ_4), or a retained configuration ($\overline{34}$ and $\overline{35}$ lie in the southwest sector below σ_4). Since retention is observed (see above) $\overline{34}$ and/or $\overline{35}$ must play a part in the reaction, and this is a case, arising because of the 6-membered sila-cyclic system which contravenes a principle often used in phosphorus chemistry, namely, that the stereochemistry (apical *v.* equatorial) of entry and departure must be the same[195, 206], although Mislow[22] has pointed out the weaknesses of the principle.

It is interesting that in 4- and 5-membered ring systems, which have been widely studied in phosphorus chemistry, the rings prefer to span one apical and one equatorial position, so that in these cases access to the star-points is restricted. This means (Figure 9.13) that pseudorotations are confined to the hexagonal tracks, which are isolated from one another by a substantial energy barrier.

We notice also that if we consider a compound $R_1R_2R_3MX$, say, and add to it a fifth ligand, Y, then at least two pseudorotations are required of an axially-arranged Y . . . M . . . X system to cross σ_4 to give retentive substitution, e.g. $45\rightarrow\overline{23}\rightarrow\overline{15}$ or $45\rightarrow 13\rightarrow 25$. Clearly, if the energy barriers along 1 and 4 and **2** and **4** (Figure 9.9) that face an intermediate at C in Figure 9.2

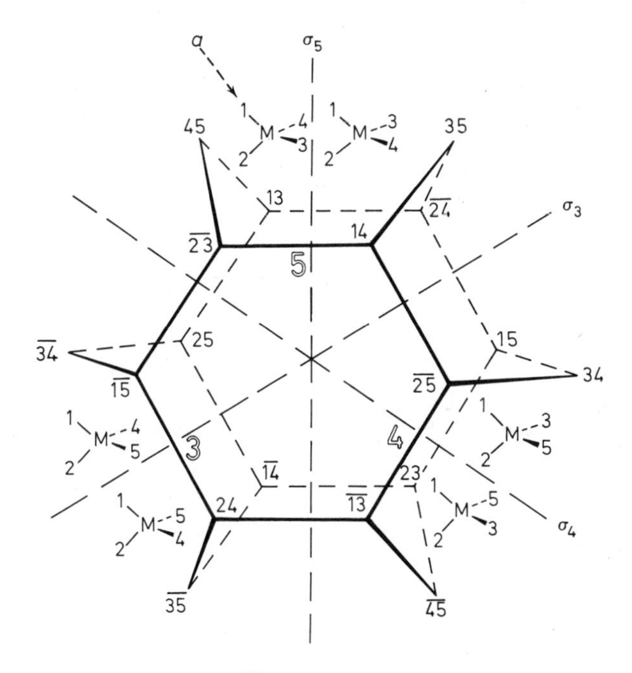

Figure 9.13

are not too large, *although rearward attack has occurred*, pseudorotation could give a retentive product. Such barriers might be of such size that the stereochemical course of the reaction could become a function of temperature, and one can envisage not just one stereochemical pathway (or traverse of the graph network) but several pathways differing in the energy barriers between pseudo-rotamers.

Examples of the approach are available[207], and Mislow (22) describes an interesting case, the stereochemistry of displacements on the system[26] demonstrating the economy of the approach in a quite complex stereochemical situation.

We shall conclude with an example on pseudorotation barriers taken from arsenic chemistry[208].

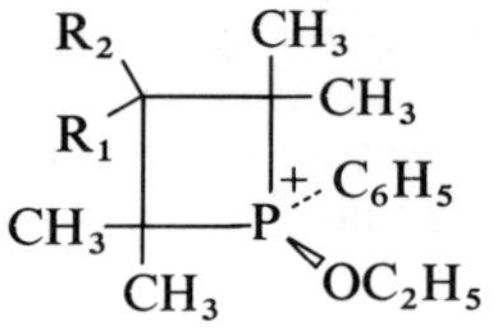

(26) *cis* $R_1 = H, R_2 = CH_3$
trans $R_1 = CH_3, R_2 = H$

For a number of compounds (27) where (a) → (e) contain respectively, X = Me,$PhCH_2$, p-$MeOC_6H_4$, Ph, and p-$NO_2C_6H_4$, and (28) (a) where X = Me, and (b) where X = Ph, two methyl proton signals of equal intensity were seen to broaden, coalesce, and finally sharpen at elevated temperatures.

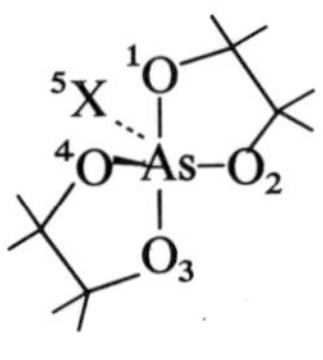

(27) 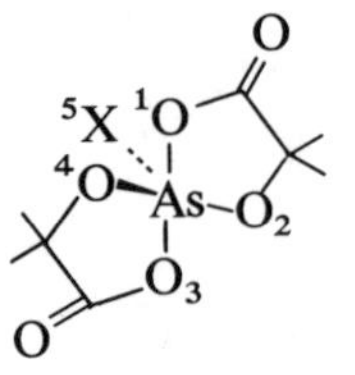 (28) 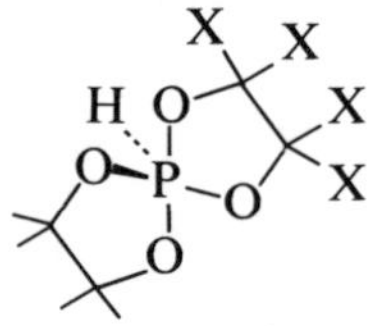 (29) (a) X = Me
(b) X = H

For (27) the Desargues–Levi graph, with substituents numbered as shown, reduces to (A) in which $13 \rightleftarrows \overline{24}$ represents interconversion between (27) and its enantiomer using X (ligand 5) as pivot. This process has been previously

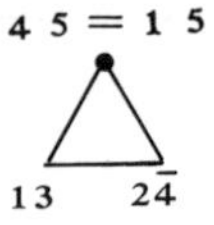

45 = 15

13 $\overline{24}$

(A)

discussed[30] for (27) and (29)[209], and if rapid on the n.m.r. time-scale reduces the four methyl signals expected for the C_2 structure to two, since the *cis* vicinal groups of each ring *syn* and *anti* to X become enantiotopic, and thus isochronous. This step, which does not require positioning the ring in the di-equatorial positions, and therefore has a low activation energy[195], has not been observed in (27) or (28), but has been found in bisphenylyl penta-coordinate systems.

A second pathway which is available for the inter-conversion between (27) and its enantiomer involves pseudorotation to an intermediate with a plane of symmetry, in which one of the rings is constrained to span the equatorial positions and t[illegible] group is apical (solid circle (A)), followed by a second pseudorotation [illegible] give the enantiomer of the starting isomer

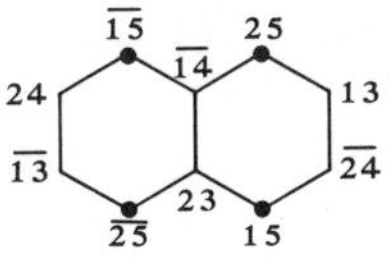

(B)

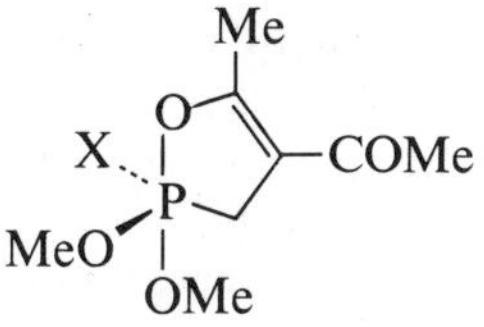

(30)

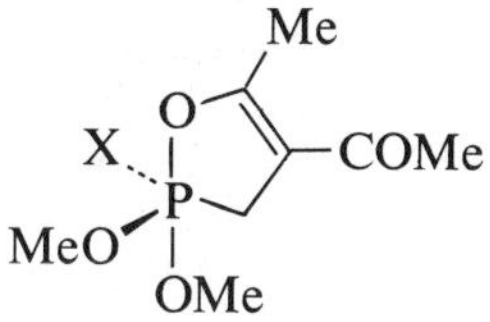

(a) X = OPh
(b) X = Ph

(31)

(a) X = OMe
(b) X = Ph

This pathway, if rapid on the n.m.r. time-scale, renders enantiotopic the *gem* methyl groups of the apical-equatorial ring and, separately, the *cis* vicinal groups of the di-equatorial ring in the trigonal-bipyramidal intermediate. Because of the necessity of placing a ring in the equatorial positions, the activation energy for this process is higher than for the one described above. Inter-conversion of 13 and $\overline{24}$ by this pathway (13 $\rightleftarrows$ 45 $\rightleftarrows$ $\overline{24}$ $\rightleftarrows$ 15 $\rightleftarrows$ 13) reduces the four methyl signals to one on the n.m.r. time-scale. The coalescence of two methyl proton signals to one is therefore associated with the high barrier.

The topological representation of pseudorotation in (28) is shown in B. N.M.R. spectra of (28a) and (28b) reveal only a single pair of methyl signals from ambient temperatures up to the coalescence point. This observation is consistent with one of the following alternatives: the predominance of one racemic pair, 13/$\overline{13}$ or 24/$\overline{24}$ to the virtual exclusion of other diastereomers; predominance and rapid inter-conversion of 13 and $\overline{24}$ (and $\overline{13}$ and 24); predominance and rapid inter-conversion of the enantiomeric pair $\overline{14}$ and 23; accidental isochrony of the methyl doublets for these stereoisomers. As in (27), pseudorotation passing through high energy (solid circle) intermediates must be rapid to render the *gem* methyl groups isochronous, i.e. the inter-conversion of 13 or 24 to their enantiomers $\overline{13}$ and $\overline{24}$, necessitates pseudorotations through the intermediates 25 and $\overline{15}$ (or their enantiomers) which possess a di-equatorial 5-membered ring.

Analogous processes in which a 5-membered ring is placed di-equatorially in phosphoranes have been observed for (29a), $\Delta G^{\ddagger}_{95}$ = 18.4 kcal mol^{-1}, and $\Delta G^{\ddagger}_{37}$ = 15.4 kcal mol^{-1} [209], and, in a case where a single ring is bound to P by one carbon and one oxygen[216], (30a) $\Delta G^{\ddagger}_{141}$ = 21.8 kcal mol^{-1}. When OPh in (30a) is replaced by Ph (30b), the high barrier increases to $>$24 kcal mol^{-1} and the lower barrier observed for these phosphoranes from *c.* 12–13 to 15.8 kcal mol^{-1}. Similarly, a change from (31a) to (31b) increases the low barrier to pseudo-rotation from 14.4 to *c.* 17 kcal mol^{-1}. The increase in the barrier height may be considered to result from positioning of the more electropositive Ph group apically in intermediate phosphoranes (30) and (31).

Evidence that the observed coalescence in the arsorane system (27) is also associated with passage through an intermediate in which the substituent X occupies an apical position (solid circle) is afforded by the observation that the barrier of (27e) is substantially less than that of (27c); in the former, the substituent X is more electronegative than in the latter. If one assumes that the effect of replacing X − Ph with X − OR is similar in the two systems, it follows that the high barrier in (27), with X = OR, should have a value of *c.* 15–20 kcal mol^{-1}. This process should be observed by n.m.r. spectroscopy using double resonance techniques above 0 °C, even with a limiting $\Delta\nu$ of only 1.0 Hz. The earlier observation of singlets in the n.m.r. spectra of (27), with X = OH and X = OMe, at −110 °C may have been due to accidental isochrony of the ring substituents, a difficulty noted

Table 9.7 Pseudorotation barriers of spiroarsoranes[208]*

Substituent X	*Limiting* $\Delta\nu$ (Hz)†	T(°C)§	$\Delta G^{\ddagger}$(kcal mol^{-1})
(27a) Me	7.6‖	155	21.8
(27b) $PhCH_2$	16.6	154	21.1
(27c) *p*-$MeO{\cdot}C_6H_4$	14.8	187	22.9
(27d) Ph	17.8	178	22.2
(27e) *p*-$NO_2{\cdot}C_6H_4$	20.6	141	20.2
(28a) Me	3.1**	140	21.7
(28b) Ph	11.4	168	22.1

**c.* 20% solute in α-bromonaphthalene; hexamethyldisiloxane used as internal reference and homogeneity standard.
†Measured in the range 60–95 °C.
§ Calculated assuming transmission coefficient of $\frac{1}{4}$, cf. Reference 29.
‖$\Delta\nu$ = 2.8 Hz in CS_2 at *c.* 40 °C.
**$\Delta\nu$ = 0.0 Hz, $\omega_{\frac{1}{2}}$ = 0.8 Hz in $CDCl_3$ at *c.* 40 °C.

in Reference 208 (Table 9.7, footnote**) and in other reports concerning the interpretation of trivalent arsenic n.m.r. spectra.

It is quite clear that the role of pseudorotation, which already occupies a prominent position in the interpretation of the stereochemistry of Group V substitutions will also assume such a role for Group IV systems as more of the stereochemistry of substitution in this Group is explored.

References

1. Capon, B. and Rees, C. W. (1969) and (1968). *Organic Reaction Mechanisms,* Ch. 9, (London: Interscience)
2. Ingold, C. K. (1953). *Structure and Mechanisms in Organic Chemistry,* (London)
3. Langford, C. H. and Gray, H. B. (1965). *Ligand Substitution Processes,* (New York: Benjamin)
4. Basolo, F. and Pearson, R. G. (1967). *Mechanisms of Inorganic Reactions,* 2nd edn. (New York: Wiley)
5. Peddle, G. J., Shafir, J. M. and McGeechin, S. G. (1968). *J. Organometal. Chem.,* **5,** 505
6. Abraham, M. H., Irving, R. J. and Johnston, F. G. (1970). *J. Chem. Soc. A,* 188, 189
7. Arnett, E. M., Bentrude, W. G. and Duggleby, P. McC. (1965). *J. Amer. Chem. Soc.,* **87,** 2048
8. Alexander, R., Ko, E. C. F., Parker, A. J. and Broxton, T. J. (1968). ibid., **90,** 5049
9. Parker, A. J. (1969). *Chem. Rev.,* **69,** 1

10. Haberfield, P., Clayman, L. and Cooper, J. S. (1969). *J. Amer. Chem. Soc.*, **91,** 787
11. Haberfield, P., Nudelman, A., Bloom, A., Romm, R., Ginsberg, H. and Steinherz (1968). *Chem. Commun.*, 194
12. Fuchs, R., Bear, J. L. and Rodewald, R. F. (1969). *J. Amer. Chem. Soc.*, **91,** 5797
13. Choux, G. and Benoit, R. L. (1969). ibid, **91,** 6221
14. Prince, R. H. and Tagliavini, G., unpublished results
15. Hague, D. N. (1969). *Comprehensive Chemical Kinetics,* Ch. 2, Ed. by Bamford, C. H. and Tipper, C. F. H. (Amsterdam: Elsevier)
16. Caldin, E. F. (1964). *Fast Reactions in Solution,* (Oxford: Blackwell)
17. Friess, S. L., Lewis, E. S. and Weissberger, A. (Eds.) (1963). *Technique of Organic Chemistry,* Vol. 7, Part II 'Investigations of Rates and Mechanisms of Reactions,' (New York: Interscience)
18. Redl, G. (1970). *J. Organometal. Chem.*, **22,** 139
19. Peddle, G. J. D. and Redl, G. (1970). *J. Amer. Chem. Soc.*, **92,** 365
20. Boer, F. P., Dorrakian, G. A., Freedman, H. H. and McKinley, S. V. (1970). ibid, **92,** 1225
21. Burgess, J. (1969). *Ann. Rep. Chem. Soc., A,* 425–427
22. Mislow, K. (1970). *Accounts Chem. Res.*, **3,** 321
23. Schmutzler, R. (1970). *Halogen Chemistry,* Vol. 2, 31, (London: Academic Press), 24 Payne, D. S. (1967). *Topics in Phosphorus Chemistry,* Vol. 4, 85 (New York: Interscience)
25. McEwen, W. E. (1965). ibid, Vol. 2, 1, (New York: Interscience)
26. Furtach, T. A., Dierdorf, D. S. and Cowley, A. H. (1970). *J. Amer. Chem. Soc.*, **92,** 5759
27. Fild, M. and Schmutzler, R. (1970). *J. Chem. Soc. A,* 2359
28. Peake, S. C. and Schmutzler, R. (1970). ibid, 1049
29. Gorenstein, D. and Westheimer, F. H. (1970). *J. Amer. Chem. Soc.*, **92,** 634, 644
30. Goldwhite, H. (1970). *Chem. Commun.*, 651
31. Casey, J. P. and Mislow, K. (1970). ibid, 1410
32. Eigen, M. (1963). *Pure and Applied Chemistry,* **6,** 97; also 3, p.96
33. Hewkin, D. J. and Prince, R. H. (1970). *Coord. Chem. Rev.*, **5,** 45
34. Baldwin, W. G. and Stranks, D. R. (1968). *Austral. J. Chem.*, **21,** 2161
35. Feeney, J., Hague, R., Reeves, L. W. and Yue, C. P. (1968). *Can. J. Chem.*, **46,** 1389
36. Gillespie, R. J., Hartman, J. S. and Parekh, M. (1968). *Can. J. Chem.*, **46,** 1601
37. Fiat, D. and Connick, R. E. (1968). *J. Amer. Chem. Soc.*, **90,** 608
38. Movius, W. G. and Matwiyoff, N. A. (1968). ibid, **90,** 5452
39. Schmidbauer, H., Weidlein, J., Klein, H. F. and Eiglmeier, K. (1968). *Chem. Ber.*, **101,** 2268
40. Tobias, R. S., Sprague, M. J. and Glass, G. E. (1968). *Inorg. Chem.*, **7,** 1714
41. Grunwald, E. and Fong, D.-W. (1969). *J. Phys. Chem.*, **73,** 650
42. Akitt, J. W., Greenwood, N. N. and Lester, G. D. (1969). *J. Chem. Soc. A,* 803
43. *Chem. Commun.*, 988
44. Van Cauwelaert, F. H. and Bosmans, H. J. (1969). *Rev. Chim. Minerale,* **6,** 611
45. Fortman, J. J. and Sievers, R. E. (1967). *Inorg. Chem.*, **6,** 2022
46. Ohnesorge, W. E. (1967). *J. Inorg. Nucl. Chem.*, **29,** 485
47. Miller, B. (1966). *J. Amer. Chem. Soc.*, **88,** 1841
48. Denney, D. B. and Adin, N. G. (1966). *Tetrahedron Lett.* 2569
49. Saville, B. (1967). *Angew. Chem. Int. Ed. Engl.*, **6,** 938
50. Pearson, R. G. (1967). *Chem. Brit.*, **3,** 103
51. Hudson, R. F. (1966). *Struct. Bond.* **1,** 221
52. Pearson, R. G. and Songstad, J. (1967). *J. Amer. Chem. Soc.*, **89,** 1827
53. Pearson, R. G. and Songstad, J. (1967). *J. Org. Chem.*, **32,** 2899
54. Ciuffarin, F. and Fava, A. (1968). *Prog. Phys. Org. Chem.*, **6,** 81
55. Tagaki, W., Kikukawa, K., Kunieda, N. and Oae, S. (1966). *Bull. Chem. Soc. Japan,* **39,** 614
56. Speziale, A. J. and Taylor, L. J. (1966). *J. Org. Chem.*, **31,** 2450
57. Chopard, P. A. and Hudson, R. F. (1966). *J. Chem. Soc. B,* 1089
58. Eigen, M. and Kustin, K. (1962). *J. Amer. Chem. Soc.*, **84,** 1355
59. Solodushenkov, S. M. and Shilov, E. A. (1945). *J. Phys. Chem., URSS,* **19,** 404
60. Lifshitz, A. and Perlmutter-Hayman, B. (1960). *J. Phys. Chem.*, **64,** 1663
61. Ebsworth, E. A. V. (1968). MacDiarmid, A. G. (Ed.), *Organometallic Compounds of the Group IV Elements,* Vol. 1, Part I

62. Shaw, C. F. and Allred, A. L. (1970). *Organometal. Chem. Rev. A,* **5,** 95
63. Cottrell, T. L. (1958). *The Strengths of Chemical Bonds,* (London: Butterworth)
64. Schott, G. (1966). *Z. Chem.,* **6,** 261
65. Sommer, L. H., (1965). *Stereochemistry, Mechanism and Silicon,* (New York; McGraw-Hill)
66. Craig, D. P. and Zauli, C. (1962). *J. Chem. Phys.,* **37,** 601
67. Craig, D. P., Maccoll, A., Nyholm, R. S., Orgel, L. E. and Sutton, L. E. (1954). *J. Chem. Soc.,* 332
68. Matwiyoff, N. A. and Drago, R. S. (1964). *Inorg. Chem.,* **3,** 337
69. Seyferth, D. and Grim, S. O. (1961). *J. Amer. Chem. Soc.,* **83,** 1610
70. Tagliavini, G. and Zanella, P. (1968). *J. Organometal. Chem.,* **12,** 355
71. Halpern, J. (1968). *J. Chem. Educ.,* **45,** 372
72. Chipperfield, J. R. and Prince, R. H. (1963). *J. Chem. Soc.,* 3567
73. Ebsworth, E. A. V. (1966). *Chem. Commun.,* 530
74. Randall, E. W. and Zuckerman, J. J. (1966). ibid, 732
75. Randall, E. W., Ellner, J. J. and Zuckerman, J. J. (1966). *J. Amer. Chem. Soc.,* **88,** 622
76. Attridge, C. J. (1970). *Organometal. Chem. Rev. A,* **5,** 323
77. Musker, W. K. and Savitsky, G. B. (1967). *J. Phys. Chem.,* **71,** 431
78. Bedford, J. A., Bolton, J. R., Carrington, A. and Prince, R. H. (1963). *Trans. Faraday Soc.,* **59,** 537
79. West, R. and Baney, R. H. (1959). *J. Amer. Chem. Soc.,* **81,** 6145
80. Janssen, M. J., Luijten, J. G. A. and Van der Kerk, G. J. M. (1964). *J. Organometal. Chem.,* **1,** 286
81. See Reference 65, Ch. 1
82. Davidson, J. M. T. and Stephenson, I. L. (1968). *J. Chem. Soc. A,* 282
83. Sommer, L. H. and Baughman, G. A. (1961). *J. Amer. Chem. Soc.,* **83,** 3346
84. Baughman, G. A. (1961). *Dissertation Abs.,* **22,** 2187
85. Bott, R. W., Eaborn, C. and Rushton, B. M. (1965). *J. Organometal. Chem.,* **3,** 455
86. Jarvie, A. W. P., Holt, A. and Thompson, J. (1969). *J. Chem. Soc. B,* 852; *idem* (1970). ibid, 746
87. Cook, M. A., Eaborn, C. and Walton, D. R. M. (1970). *J. Organometal. Chem.,* **24,** 301
88. Sommer, L. H., Stark, F. O. and Michael, K. W. (1964). *J. Amer. Chem. Soc.,* **86,** 5683
89. Corriu, R., Leard, M. and Masse, J. (1968). *Bull. Soc., Chim. France,* 2555
90. Grant, M. W. and Prince, R. H. (1968). *Chem. Commun.,* 1076
91. Kumada, M., Mimura, K., Ishikawa, M. and Shiina, K. (1965). *Tetrahedron Lett.* 83
92. Sollott, G. P. and Peterson, W. R. (1967). *J. Amer. Chem. Soc.,* **89,** 5054, 6783
93. Frye, C. L., Vogel, G. E., Hall, J. A. (1961). ibid, **83,** 996
94. Frye, C. L. (1964). ibid, **86,** 3170; (1970). ibid, **92,** 1205
95. Frye, C. L., Vincent, G. A. and Hauschildt, G. L. (1966). ibid, **88,** 2727
96. Corey, J. Y. and West, R. (1963). ibid, **85,** 4034
97. Clark, H. C., Dixon, K. R. and Nicolson, J. G. (1969). *Inorg. Chem.,* **8,** 450
98. Cook, D. I., Fields, R., Green, M., Haszeldine, R. N., Iles, B. R., Jones, A. and Newlands, M. J. (1966). *J. Chem. Soc. A,* 887
99. Boer, F. P., Flynn, J. J. and Turley, J. W. (1968). *J. Amer. Chem. Soc.,* **90,** 6873
100. Timms, R. E., personal communication
101. Olah, G. A., O'Brien, D. H. and Lui, C. Y. (1969). *J. Amer. Chem., Soc.,* **91,** 701
102. Baines, J. E. and Eaborn, C. (1965). *J. Chem. Soc.,* 4023, (1956). ibid, 1436
103. Gilman, H. and Dunn, G. E. (1951). *J. Amer. Chem. Soc.,* **73,** 3404, 4499
104. Weyenberg, D. R. (1959). *Dissertation Abs.,* **19,** 2776
105. Steward, O. W. and Pierce, O. R. (1961). *J. Amer. Chem. Soc.,* **83,** 1916
106. Sommer, L. H., Bennett, O. F., Campbell, P. G. and Weyenberg, D. R. (1957). ibid, **79,** 3295
107. Sommer, L. H. and Bennett, O. F. (1959). ibid, **81,** 251
108. See Reference 65 Ch. 9
109. See Reference 65 Ch. 6
110. Schott, G. and Harzdorf, C. (1961). *Z. Anorg. Allgem. Chem.,* **307,** 105
111. Bott, R. W., Eaborn, C. and Swaddle, T. W. (1965). *J. Chem. Soc.,* 2342
112. Roberts, R. M. G. and El Kaissi, F. (1968). *J. Organometal. Chem.,* **12,** 79
113. Schott, G., Hansen, P., Kuhla, S. and Zwierz, P. (1967). *Z. Anorg. Allgem. Chem.,* **351,** 37
114. Hetflejŝ, J., Mares, F. and Chvalovsky, V. (1965). *Intern. Symp. Soc. Commun. Prague,* 282

115. Reikhsfel'd, V. O. and Saratov, I. E. (1967). *Z. Obshch. Khim.*, **37,** 402
116. Reikhsfel'd, V. O. and Prokhovova, V. A. (1965). ibid, **35,** 182, 1826, 1830
117. Hetflejŝ, J., Mares, F. and Chvalovsky, V. (1966). *Collect. Czech. Chem. Commun.*, **31,** 586; idem (1965). ibid, **30,** 1643
118. Reikhsfel'd, V. O. (1965). *Intern. Symp. Organosilicon Chem., Sci. Commun. Suppl.*, Prague, 34; (1966). *Chem. Abs.*, 10448
119. Korol'ko, V. V. and Reikhsfel'd, V.O. (1965). *Reaksionnaya Sposobnost. Organ. Soedin. Tartusk. Gos. Univ.*, **2,** 98; (1966). *Chem. Abs.*, **64,** 1917
120. Eaborn, C. and Walton, D. R. M. (1965). *J. Organometal. Chem.*, **4,** 217
121. Bott, R. W., Eaborn, C. and Rushton, B. M. (1965). *J. Organometal. Chem.*, **13,** 448
122. Bott, R. W., Dowden, B. F. and Eaborn, C. (1965). *J. Chem. Soc.*, 4994
123. See, for instance, *Organometal. Chem. Rev. B*, Vols. 1–6, (1964–70), 'Annual Surveys on Silicon and Germanium'
124. Anderson, D. G., Bradney, M. A. M. and Webster, D. E. (1968). *J. Chem. Soc. B*, 450
125. Anderson, D. G. and Webster, D. E. (1968). *J. Organometal. Chem.*, **13,** 113
126. Anderson, D. G., Chipperfield, J. R. and Webster, D. E. (1968). ibid, **12,** 323
127. Anderson, D. G. and Webster, D. E. (1968). *J. Chem. Soc. B*, 765
128. *Idem* (1968). ibid, 878, 1008
129. Cartledge, F. K. (1968). *J. Organometal. Chem.*, **13,** 516
130. Grosjean, M., Gielen, M. and Nasielski. (1963). *Industrie Chimique Belge*, **7,** 721
131. Eaborn, C. (1960). *Organosilicon Chemistry*, (London; Butterworth)
132. Schowen, R. L. and Latham, K. S. (1966). *J. Amer. Chem. Soc.*, **88,** 3795
133. Idem (1967). ibid, **89,** 4677
134. Humffray, A. A. and Ryan, J. J. (1969). *J. Chem. Soc. B*, 1138
135. Voronkov, M. G. and Zagata, L. (1967). *Z. Obshch, Khim.*, **37,** 2551; (1968). *Chem. Abs.*, **69,** 35087
136. Voronkov, M. G. and Zelchan, G. I. (1969). *Khim. Geterotsikl. Soedin.*, 450
137. McInnes, A. G. (1965). *Can. J. Chem.*, **43,** 1998
138. Schott, G., Kelling, H. and Schild, R. (1966). *Chem. Ber.*, **99,** 291
139. Schott, G. and Deikel, ibid, 301
140. Schott, G. and Bondybey, V. (1967). *Chem. Ber.*, **100,** 1773
141. Johnson, O. H. and Schmall, (1958). *J. Amer. Chem. Soc.*, **80,** 2931
142. Chipperfield, J. R. and Prince, R. H. (1960). *Proc. Chem. Soc.*, 385; *Idem* (1963). *J. Chem. Soc.*, 3567.
143. Glockling, F. (1966). *Quart. Rev. Chem. Soc.*, **20,** 54
144. Douglas, B. E. and McDaniel, D. H. (1965). *Concepts and Models of Inorganic Chemistry*, 421, (New York: Blaisdell)
145. Benson, D. (1968). *Mechanisms of Inorganic Reactions, in Solution*, London, McGraw-Hill
146. Prince, R. H. (1960). *Z. Elektrochem.*, **64,** 13
147. Kay, P. D. and Prince, R. H., unpublished results
148. Kay, P. D. (1968). *Ph. D. Thesis*, Cambridge
149. Bassindale, A. R., Eaborn, C. and Walton, D. R. M. (1970). *J. Organometal. Chem.*, **25,** 57
150. Sommer, L. H. and Citron, J. D. (1967). *J. Amer. Chem. Soc.*, **89,** 5797
151. Prince, R. H. and Timms, R. E. (1968). *Inorg. Chim. Acta*, **2,** 258
152. *Idem*, ibid, **1,** 129
153. Prince, R. H. and Bedford, J. A., unpublished results
154. Bedford, J. A. (1964). *Ph. D. Thesis* Cambridge
155. Prince, R. H. (1959). *J. Chem. Soc.*, 1783
156. Prince, R. H. and Timms, R. E. (1968). *Inorg. Chim. Acta*, **2,** 260
157. Prince, R. H. and Grant, M. W., unpublished results
158. Grant, M. W. (1968). *Ph. D. Thesis*, Cambridge
159. Hudson, R. F. and Saville, B. (1955). *J. Chem. Soc.*, 4114, 4121, 4130
160. Hudson, R. F. and Steltzer, I. (1958). *Trans. Faraday Soc.*, **54,** 213
161. Allen, A. D. and Lavery, S. J. (1969). *Can. J. Chem.*, **47,** 1263
162. Bunton, C. A. (1963). *Nucleophilic Substitution at a Saturated Carbon Atom*, (London: Elsevier)
163. Grant, M. W. and Prince, R. H. (1969). *Nature, (London)*, **222,** 1163
164. Allen, A. D. and Modena, G. (1957). *J. Chem. Soc.*, 3671
165. Haysom, J. T. and Prince, R. H., unpublished results

166. Tagliavini, G. and Prince, R. H., unpublished results
167. Grant, M. W. and Prince, R. H. (1969). *J. Chem. Soc. A,* 1138
168. Sommer, L. H. and Frye, C. L. (1959). *J. Amer. Chem. Soc.,* **81,** 1013
169. Bott, R. W., Eaborn, C. and Varma, I. D. (1963). *Chem. Ind.,* 614
170. Brook, A. G. and Peddle, G. J. D. (1963). *J. Amer. Chem. Soc.,* **85,** 1870
171. Belloli, R. (1969). *J. Chem. Educ.,* **46,** 640
172. Peddle, G. J. D. and Redl, G. (1970). *J. Amer. Chem. Soc.,* **92,** 365; *see also idem* (1968). *Chem. Commun.,* **11,** 626
173. Redl, G. (1970). *J. Organometal. Chem.,* **22,** 140
174. Corriu, R., Leard, M. and Masse, J. (1968). *Bull. Soc. Chim. France,* 2555
175. Carre, F. H., Corriu, R. J. P. and Thomassin, R. B. (1968). *Chem. Commun.,* 560
176. Carre, F. H., Corriu, R. and Leard, M. (1970). *J. Organometal. Chem.,* **24,** 101
177. Grant, M. W. and Prince, R. H. (1968). *Chem. Commun.,* 1076
178. Sommer, L. H., Parker, G. A., Lloyd, N. C., Frye, C. L. and Michael, K. W. (1967). *J. Amer. Chem. Soc.,* **89,** 857
179. Sommer, L. H., Michael, K. W. and Korte, W. D. (1967). ibid, **89,** 868
180. Sommer, L. H. and Rosborough, K. T. (1967). ibid, **89,** 1756
181. Sommer, L. H., Korte, W. D. and Rodewald, P. G. (1967). ibid, **89,** 862
182. Sommer, L. H. and Korte, W. D. (1967). ibid, **89,** 5802
183. Corriu, R. and Masse, J. (1968). *Tetrahedron Lett.,* 5197; *idem* (1967). *Chem. Commun.,* 1287
184. *Idem* (1969). *Bull. Soc. Chim. France,* 3491
185. Sommer, L. H., Citron, J. D. and Parker, G. A. (1969). *J. Amer. Chem. Soc.,* **91,** 4729
186. Eaborn, C., Hill, R. E. E. and Simpson, P. (1968). *Chem. Commun.,* 1077
187. Corriu, R. and Leard, M. (1968). *J. Organometal. Chem.,* **15,** 25
188. Peddle, G. J., Shafir, J. M. and McGeechin, S. G. (1968). ibid, **15,** 505
189. Sommer, L. H. and Fujimoto, H. (1968). *J. Amer. Chem. Soc.,* **90,** 982
190. Baker, R., Bott, R. W., Eaborn, C. and Jones, P. W. (1963). *J. Organometal. Chem.,* **1,** 37
191. Sommer, L. H. and Rodewald, P. G. (1963). *J. Amer. Chem. Soc.,* **85,** 3898
192. Carre, F. H., Corriu, R. J. P. and Thomassin, R. B. (1968). *Chem. Commun.,* 560
193. Hudson, R. F. (1965). *Structure and Mechanism in Organo-phosphorus Chemistry,* (London; Academic Press)
194. Warren, S. G. and Kirby, A. J. (1967). *Reaction Mechanisms in Organic Chemistry,* Ed. by C. Eaborn and N. B. Chapman, 'The Organic Chemistry of Phosphorus', (Elsevier)
195. Westheimer, F. H. (1968). *Accounts Chem. Res.,* **1,** 70
196. Bunton, C. A. (1970). *Accounts Chem. Res.,* **3,** 257
197. Ramirez, F. (1968). ibid, **1,** 168
198. Capon, B. and Rees, C. W., *Organic Reaction Mechanisms* (Annual Surveys) 1965–1969, (Sections dealing with P substitution and non-carbon containing acids), (London: Interscience)
199. Khan, S. A., Kirby, A. J. and Wakselman, M. (1970). *J. Chem. Soc. B,* 1182
200. Gay, D. C. and Hamer, N. K. (1970). *Chem. Commun.,* 1564
201. *Idem* (1970). ibid, 1123
202. Driver, G. E. and Gallagher, M. J. (1970). *Chem. Commun.,* 150
203. Corfield, J. R., De'ath, N. J. and Trippett, S. (1970). ibid, 1502
204. Egan, W., Chauviere, G., Mislow, K., Clark, R. T. and Marci, K. L. (1970). ibid, 733
205. See refs. cited in Reference 22, page 321; also Ugi, I., Marquarding, D., Klusacek, H., Gokel, G. and Gillespie, P. (1970). *Angew. Chem. Int. edn.,* **9,** 725
206. De Bruin, K. E., Naumann, K.; Zon, G. and Mislow, K. (1968). *J. Amer. Chem. Soc.,* **91,** 7031
207. *See* Reference 22 and references therein and De Bruin, K. E. and Jacobs, M. J. (1971). *Chem. Commun.,* 59
208. Casey, J. P. and Mislow, K. (1970). *Chem. Commun.,* 1410
209. Houalla, D., Wolf, R., Gagnaire, D. and Robert, J. B. (1969). ibid, 443
210. Hillier, I. H. and Saunders, V. R. (1970). *Chem. Commun.,* 1183; *Idem* (1970). ibid, 1510
211. Marsmann, H., van Wazer, J. R. and Robert, J. -B. (1970). *J. Chem. Soc. A,* 1566
212. Mitchell, K. A. R. (1970). *Inorg. Chem.,* **9,** 1960
213. Bartell, L. S., Su, L. S. and Yow, H. (1970). ibid, **9,** 1903

214. Bull, W. E., Pullen, B. P., Grimm, F. A., Moddeman, W. E., Schweitzer, G. K. and Carlson, T. A. (1970). ibid, **9,** 2474
215. Ebsworth, E. A. V. and Cradock, S. (1971). *Chem. Commun.,* 57
216. Gorenstein, D. (1970). *J. Amer. Chem. Soc.,* **92,** 644